OILSEEDS —

Properties, Processing, Products and Procedures

G. NAGARAJ
Principal Investigator
Directorate of Oilseeds Research
Rajendranagar, Hyderabad - 500 030
Andhra Pradesh, India

New India Publishing Agency
Pitam Pura, New Delhi- 110 034

Published by
Sumit Pal Jain *for*
New India Publishing Agency
101, Vikas Surya Plaza, CU Block, L.S.C. Mkt.,
Pitam Pura, New Delhi- 110 088, (India)
Phone: 011-27341717, Fax: 011-27341616
E-mail: newindiapublishingagency@gmail.com
Web: www.bookfactoryindia.com

The book has been supported by the Department of Science & Technology, Govt. of India, under its USERS Scheme.

ISBN : 978-81-90723-75-6

Typeset at: Typographiya, Delhi
Printed at: Jai Bharat Printing Press, Delhi

Dr. M. V. Rao, MLC
Special Director-General, ICAR (Retd)
Special Secretary, DARE, GOI (Retd)
Vice Chancellor, ANGRAU (Retd)
President, Agri Biotech Foundation
Hyderabad

21, November, 2008

FOREWORD

India is one of the major oilseed producing countries occupying the fourth position in the world. The oilseed crops - both edible and non-edible and the cakes and other fractions from them play a crucial role in Indian economy, next only to the food crops. The edible oil crops supply essential nutrients in human and animal diets in terms of fats, proteins, carbohydrates and some essential minerals and vitamins. Vegetable oil sector is becoming more and more important in the Indian context because it has to meet the growing needs of humans, animals, industry and also the new emerging biofuel needs. The per capita consumption of edible oil is increasing with the change in the life styles and affluence of certain sectors of our population. The average per capita consumption which was only 4.5 kgs some years ago has risen to 11.5 kgs now and this is likely to rise further in future. Added to this is the ever growing population, which means India needs more and more oils and better nutritive oils in the years ahead. Oilseeds are also important constituents of animal feeds and organic manures, besides playing a vital role in industry for manufacture of paints, varnishes, lubricants and also for a whole range of consumer goods of daily use such as soaps, detergents etc. Since the country is not able to produce the needed quantities of edible oils it is resorting to heavy imports which is now touching more than 5 million tons costing Rs.10 to 12 thousand crores. Perhaps this is the highest amount spent by the Government on import of any agricultural commodity. The scenario is likely to get more complicated in the coming years because there will be less edible oil in the international markets and even if it is available it will be costier as a result of diversion of the oils for production of biofuels by the oil exporting countries. This is also leading to food shortages in the world. Hence India has to make every effort to increase the present oilseed production of 28 million tons to the projected demand of 60-70 million tons by 2020.

India is fortunate to have all the agroclimatic conditions to grow a whole range of oilseed crops which no other country is capable of. Prominent oilseed crops raised in India are groundnut, soybean, rapeseed-mustard, sunflower, safflower, niger and sesame for edible purposes and non-edible castor and linseed for industrial and other uses. It also grows a number of oil bearing tree species like coconut, oil palm, sal, mahua, kokum, melia, neem etc. To further increase domestic edible oil production rice and cotton seed are being exploited for producing rice bran oil and cotton seed oil. In recent years Jatropha, Pongamia, Simarouba glauca, Jojoba are being promoted as sources of biofuel. This biofuel sector is assuming great importance in view of the high prices of fossil fuels in which India is very deficient and also because of their biodegradable nature and less environmental problems. India occupies the second position in the world for area under groundnut, castor and rapeseed-mustard and coconut and fourth position for soybean and first position for linseed and sesame. India occupies the first position in the world for area under paddy and cotton.

In terms of human resource and infrastructure development to develop manpower and laboratory and industrial capabilities the country showed remarkable and enviable vision. A number of institutions and industries are in position and also the strategies to further augment our efforts and facilities. The oil extracting and trading sectors are one of the biggest and most vibrant sectors, employing millions of people. India also exports certain quantities of oilseeds like groundnut, sesame, niger, castor, defatted soyflour and oil cakes of groundnut, mustard etc.

In this context of oilseed scenario, any extra knowledge and information in terms of production, processing, product development and other related issues will be of great interest to scientists, technicians, policy makers, industry, farmers and students. The book by Dr G Nagaraj on **"OILSEEDS — Properties, Processing, Products and Procedures"** fills this gap and is very timely. The book covers all aspects of oilseed scenario involving major and minor oilseed crops, tree species and challenges, advances and progress in post harvest handling of the oilseeds and their products. It is very informative and comprehensive and I am sure all those people involved with oilseeds sector directly or indirectly, will be greatly benefited by this book. Dr Nagaraj has put in a lot of effort in writing this book and I congratulate him warmly for his effort. I am sure the book will be well received by all sectors of society interested in oilseeds and oil economy of India.

M.V. Rao

PREFACE

Oilseeds and their products have been playing an important role in our day to day lives since ages. They serve as sources of major and micro nutrients to humans and animals and help in their growth and health management. In this aspect they are better than cereal grains due to their better and balanced nutrient composition. As organic manures, again, they help in increased crop production. Various phytochemicals and oleochemicals are useful as medicines and cosmetics. Oils, proteins and their value added products have wide ranging industrial uses. They have recently gained importance as diesel substitutes. All these aspects with respect to the cultivated and other minor oleaginous sources have been presented and discussed in this book. Post harvest handling and processing to obtain good quality oilseed products and their utility also find a place. Antinutrients and their detoxification for better utilization of oilseed products have also been covered. Methods for evaluation of various quality parameters of oilseeds, oil and oil cakes are also included in the book. It is expected that this book will be useful to students, researchers and all those involved in oilseeds management.

I am greatly indebted to Padmashri Dr. M.V. Rao, Dr. P.S. Reddy and Dr. M.K. Chakraborty who have encouraged me in my career as an oilseed scientist in the Indian Council of Agricultural Research. I sincerely thank the Department of Science and Technology, Ministry of Science & Technology, Government of India for the financial support under their USERS scheme to write this book. I acknowledge the infrastructural facilities provided to me at the Directorate of Oilseeds Research, Hyderabad. I also thank my wife and sons for their help and support while writing this book.

This is a first major and single handed effort by me. I would greatly appreciate useful suggestions to improve the content and quality of this book in the later editions.

Hyderabad
November, 2008

G. Nagaraj
Ex. Head & P.S. Biochemistry
210, Ravathe Towers
Maruthi Nagar, Kothapet
Hyderabad - 500 060, India
Guttarla@hotmail.com
guttarla@yahoo.co.in

CONTENTS

Foreword v
Preface vii

Chapter 1 : Introduction 1

Chapter 2 : Cultivated Oilseeds 7
Groundnut 7
Rapeseed-Mustard 28
Safflower 43
Sunflower 58
Sesame 70
Soybean 85
Niger 99
Castor 109
Linseed 123
References 133

Chapter 3 : Tree Borne and Minor Oil Sources 167
Palm oil 168
Rice Bran oil 178
Corn oil 189
Cottonseed (*Gossypium*) oil 195

Coconut 202
Olive 209
Neem (*Azadirachta indica*) 213
Jatropha *223*
Karanj *(Pongamia glabra)* *230*
Mahua *(Madhuca)* *236*
Simarouba 242
Kokum *(Garcinia indica)* *248*
Melia-The Bead tree 250
Sal (*Shorea Robusta)* *253*
Rubber (*Hevea braziliensis)* *260*
Kusum *(Schleichera trijuga)* *261*
Crambe 265
Mango *(Mangifera indica)* *268*
Moringa 273
Tumba 277
Tobacco seed oil 279
Jojoba (*Simmondsia Chinensis*) 283
References 288

Chapter 4 : Processing ________ 301
Post Harvest Technology 301
Oil extraction 314
Ghani extraction 315
Oil expeller 317
Solvent extraction 323
Extrusion expeller 326
Super critical fluid extraction 328
Palm oil production 331
Refining of oil 336
References 346

Chapter 5 : Oilseed Products 349

Nutritional quality of edible oils 349
Keeping quality of oils 359
Hydrogenated fats 363
Shortenings 365
Margarines 367
Oilseed extractions 370
Oilseed based food products 377
Oleo-chemicals 398
References 410

Chapter 6 : Antinutrients 419

Antinutrients 419
Ricin-the toxic protein 429
Mycotoxins 438
Glucosinolates 450
References 456

Chapter 7 : Procedures 465

Methods of analysis 465
Fats, oils and lipids 466
Extraction and purification of lipids 467
Chromatographic purification of lipids 468
Oil analysis by soxhlet extraction 469
Oil content (cold percolation method) 471
Oil content (NMR method) 472
Oil and moisture contents by pulsed NMR 473
Specific gravity/ density of fats and oils 476
Colour of the oil 477
Determination of free fatty acids 477
Free fatty acids - colorimetric method 479
Refractive index 480
Saponification value 481

Unsaponifiable matter 482
Iodine value 484
Iodine value by dam's reagent 485
Peroxide value (PV) 486
Fatty acid methyl esters (FAME) 488
Fatty acid composition by gas chromatography 491
Lipid profile studies 495
Determination of lipase activity 497
Phospho Lipids 498
Estimation of crude proteins 499
Protein and total N by micro-kjeldahl method 501
Protein content 503
Protein by biuret reaction 505
Protein by ultraviolet absorption 506
Free amino acids 507
Protein digestibility (*in vitro*) 508
Total carbohydrates (phenol: sulphuric acid method) 509
Total water soluble sugars (T.S.S.) 511
Total reducing sugars 513
Sugars (alkaline ferricyacide – colorimetric method) 514
Determination of total and reducing sugars 515
Determination of crude fibre 517
Estimation of cellulose 518
Starch - method-I 519
Starch - method-II 521
Antinutritional factors 522
Chlorogenic acid (HPLC method) 524
Estimation of oxalic acid 525
Estimation of phytic acid 526
Estimation of polyphenols 527
Estimation of tannins 528
Aflatoxins by thin layer chromatography (TLC) 529
Aflatoxins by elisa 533

Aflatoxins by HPLC method 535
Crude ricin 538
Ricin: RID assay 540
Ricin: HPLC method 541
TLC of sugars (raffinose, stachyose and verbacose) 541
Glucosinolates-quick test for the determination of glucosinolates 543
Gluco test method for glucosinolates 544
Glucosinolates-desulphation and HPLC 546
Analysis of sesame lignans 548
Vitamin A (HPLC method) 549
Vitamin A (spectrophotometric method) 551
Beta carotene 552
Thiamine (vitamin B1) 553
Tocopherols (HPLC method) 554
Total tocopherols-colorimetric method 555
Gross calorific value (GCV) 557
Carbon and hydrogen content 559
Mineral analysis ash content 562
Atomic absorption spectrophotometry 564
Calcium (titrimetric method) 566
Calcium (colorimetric method) 567
Phosphorus (fiske and subba row method) 568
Iron (wong's method) 570
Potassium (flame photometric method) 570
References 571

Appendix ________ 575

Index ________ 591

Chapter 1

Introduction

Human being as a hunter has depended mostly on animal food and some fruits. With the progress of civilization, he settled down and started agriculture around 7000-10,000 B.C. He started depending on plant grains to take care of his energy and nutritional requirements in addition to animal foods. Initially cereal grains must have mostly accounted for his staple food requirements. Taste and health requirements must have prompted him to diversify his crop production programme. It is this need that must have made him to cultivate oilseeds, fibre and other crops.

Oilseeds, thus, have been under cultivation since antiquity. Rapeseed and sesame have been described in the Indian Sanskrit writings of 2000 BC. Historically, extraction of oil seems to have been a batch process. Seeds kept in gunny bags or boxes were pressed to remove oil. The separated oil on clarification was used for food and non food purposes. The resulting cakes were utilized as animal feed or thrown away into fields which helped in improving the soil fertility. Human beings slowly discovered the uses of oil and other oilseed products for various food and non-food purposes.

There are several species of plants in the world whose oil can be utilized for human consumption. Though around 40 oilseeds are known, only around a dozen oilseed crops have become commercially important. There are ten seed crops and three tree crops. The seed crops are groundnut, rapeseed-mustard, sesame, soybean, sunflower, cottonseed, safflower, niger, castor and linseed. These are mostly annual crops. The three major tree oil crops are palm, olive and coconut.

Among the cultivated annual oilseed crops soybean accounts for about 62% of the total world oilseed production. As a vegetable oil source, it may not be the leader due to its lower oil content of around 18%. It is followed by rapeseed-mustard and groundnut with about 13-14% of the total oilseed production. Sunflower occupies the fourth place with about 9% of world oilseed production. These four oilseeds together account for 98% of the cultivated world oilseed production. Castor, sesame and linseed account for the remaining 2%. Cottonseed is also an important edible vegetable oil source. The seed as well as oil are by products of the textile industry. Similarly niger is a valuable edible oil source cultivated in some countries. The total world cultivated oilseed production stood at around 360 million tonnes in the year 2006. This was produced from an area of 180 million ha, at a productivity level of around 2000 kg/ha.

World Area, Production and Productivity of Oilseeds During 2006

Oilseed	Area X 1000 ha	Production X 1000 tonnes	% of total production	Yield kg/ha
Groundnut	22232	47768	13.41	2149
Rapeseed & Mustard	28474	49466	13.8	1737
Soybean	92989	22150	61.9	2382
Sesame	7539	3338	0.9	443
Sunflower	23700	31332	8.7	1322
Safflower	822	583	0.1	709
Linseed	3017	2570	0.7	852
Castor	1264	1140	0.3	902
Total	**180037**	**357699**	**100**	**1987**

Among the perennial oil sources, palm oil, in recent years has gained lot of prominence. It has almost become the major vegetable oil of the world being utilized for various food and non-food purposes. This is mainly because of higher oil yield from this tree. The production of palm oil averages around 40 million tonnes annually. The yield of oil from palm trees can be as high as 8 tonnes /ha. Coconut is another multipurpose oil source grown in the coastal regions of the tropics. The dried copra is the richest source of oil containing upto 65% oil. The annual availability of coconut oil varies around 3.3 million tonnes. The third tree oil source, olive, accounts for 1.6 million tonnes of world vegetable oil production. This crop is confined to the Mediterranean region of the globe.

Past few decades saw the emergence of hither to unknown sources of oil. Large scale requirement and local shortages have prompted the identification of many new sources. They are ricebran, corn, *sal, mahua, jatropha, kokum, melia*, rubber, mango kernel, *kusum, simarouba*, neem, *karanj, crambe, moringa, tumba,* jojoba and tobacco seed etc. Most of these oilseeds and their products have wide utility in the specialty oils sector and are mostly used for non-edible purposes. Soaps and detergents along with paint and varnish industry are the major consumers of these non-traditional oilseeds. They also serve as high value and eco-friendly lubricants.

Some non-traditional oilseeds and their products have pesticidal and medicinal uses. They are also much sought after raw material by the cosmetic industry. Vegetable oils, in recent years, are gaining importance as diesel fuel due to depleting mineral oil reserves. Oilseeds will gain greater attention in future as they are renewable and biodegradable in nature. Non-traditional oilseeds like *jatropha* and *karanj* are expected to play an important role as raw material for the production of biodiesel.

Oilseeds apart from being sources of oil have a lot of importance as nutritious food items. Oilseeds like soybean, groundnut and sesame are utilized as food and in the preparation of various value added and nutritious food products. They even have neutraceutical value. Components of the oilseeds like oil, protein, carbohydrates, minerals, vitamins and trace constituents have wide and varied uses. The major constituents, oil, protein and sugars have a role in calorie nutrition of the human beings. Animal products have some negative impact on health. Animal fats increase the blood cholesterol levels and cause obesity. Hence their use is under scrutiny. Though animal protein is more balanced in amino acid composition, it is costly and is not available to a major chunck of the population. Ethical considerations and cruelty against animals also go against the use of animal products.

It is in this background that vegetable sources are gaining importance as nutritional sources. Cereal crops account for a major portion of our food. However, they are richer in one major component of food, namely, starch and hence are imbalanced with respect to nutrition. Oilseeds are next in importance and are more energetic and nutritious than cereal grains. They contain most of the food components like carbohydrates, proteins, oils, minerals, vitamins, essential fatty acids (EFA) and essential amino acids (EAA) and other nutrients. The oilseeds are generally richer in oils/fats ranging from 20-55%. The other major components are proteins and sugars ranging from

20-30% each. Minerals, vitamins and other micro- nutrients account for around 5- 10% of the seeds.

Oilseeds have a lot of economic importance as sources of oil, protein and other food and non-food constituents. Human beings need to consume 20% each of their nutritional energy through oils and proteins. Oilseeds supply both these constituents along with carbohydrates, minerals, EFA, EAA and vitamins required by the human body. They add taste to other foods. Some oilseeds like groundnut, sesame, sunflower kernels and niger are widely utilized in various food preparations. These oilseeds are directly edible and can be eaten with minimal processing. They have lot of snack food value all over the world.

Soybean, though highly nutritious, needs processing prior to it's utilization as a food item. Many products of soybean are marketed world wide. Some oilseeds like rapeseed, linseed and safflower are consumed as food items in limited quantities in specific locations. The presence of antinutrients and or non-palatability etc. hinder their utilization. Lot of progress has been made to reduce or detoxify the antinutrients present in the oilseeds.

Oilseeds like castor are unique with respect to their utility. Castor serves as the source of one of the best natural lubricant especially for the aviation industry. Most non-edible oilseeds are utilized by the industry in the manufacture of various value added products like soaps, detergents, paints, varnishes, cosmetics, lubricants, pesticides etc. With improvements in standard of living along with mechanization and automation taking place world over, the demand for vegetable oils will increase, year after year. Search for new seeds with natural and engineered composition to suit specific applications will continue. Genetic engineering will play a major role in the years to come in catering to the world food and oil supplies.

Oilseeds and their derivatives constitute a significant part of global business. Around 300 millions tones of oilseeds are produced globally every year. 90-100 million tonnes of oil get traded all over the world. Around 150-160 million tonnes of oil meals mostly as by products of oil extraction industry are also available. In monetary terms the business relating to oilseeds and their products add upto around $ 110 billion.

Oilseeds and their products vary greatly in their chemical composition and quality. The season and the location under which they are grown as well as genotypes have an effect on the composition and quality of the oilseeds. Post-harvest technology and processing of oilseeds also have great influence on the quality of oilseeds and their products. Understanding the quality and its evaluation are also important to all the stake holders, namely, the producer, processor, policy makers and the consumers. This book is written keeping in view all these aspects.

The chapters that follow deal with the cultivated oilseeds, tree borne and minor oilseeds, post harvest technology, processing, products and their utility. The last chapter deals with methods of oilseed quality evaluation. Efforts have been made to include most of the known oilseeds. Information on the seed, oil and the protein qualities have been presented. Mineral, vitamin and other micro-nutrient compositions have also been included. Food and non-food products along with antinutrients find a place. This book should cater to the needs of students, researchers, producers, technologists and planners.

Chapter 2

Cultivated Oilseeds

Groundnut

Groundnut is believed to have originated in Brazil and later spread to many countries all over the world. It is known to humanity since 950BC. Groundnut crop and its products have multiple uses. Being a leguminous crop it enriches the soil with nitrogen through symbiotic biological fixation by *rizhobium*. The economic product, groundnut in shell, contains the seeds, which are edible. The shell as well as the haulms serve as fuel and fodder. The seed oil serves as a good cooking oil while the cakes are good sources of protein for animals and even humans. Groundnut (*Arachis hypogaea* L,) belongs to *leguminosae* family and *papilionacea* sub family. It consists of two sub species each with two varieties. The first one *hypogaea* has the variety *hypogaea* (virginia group) and variety *hirsuta* kohla while the second one *fastigiata* has the variety *fastigiata* (valencia group) and the variety *vulgaris* (spanish group). *Hypogaea* varieties are spreading types with branching and lack infloresence on the main stem. They are of longer duration and have fresh seed dormancy . Varieties of the *fastigiata* are of erect type with branching and flowering on the main stem also. The crop is of shorter duration than *hypogaea* types, but with little fresh seed dermancy.

The major groundnut producing countries of the world are China, India, Nigeria, U.S.A. Indonesia, Argentina, Sudan, Senegal and Myanmar. The cultivation is spread in tropical and sub-ropical regions. Groundnut is a self-pollinated crop and pollination takes place in the morning. After fertilization an elongated stalk namely peg develops which enters the soil, bends horizontally and swells to become a pod. Within the pod, two, three or four kernels are formed. The crop can be grown under rainfall conditions ranging from 50-125cm, with good distribution during flowering and pegging stages. Sandy loams, loamy soils, and black soils with good drainage are suitable for growing the crop.

Groundnut is raised mostly as a rainfed *kharif* crop, being sown from May to July, depending upon the monsoon rains in India. In the *rabi* season, groundnut is sown in the southern states during November-December, mostly in the rice fallows. Summer groundnut is sown during the second fortnight of January upto the first fortnight of February. To control seed borne diseases, seeds should be treated with Thiram or Mancozeb (3g/kg of kernel) or Carbendazim (2g/kg). Normally about 120 kg seeds/ha for bunch and 100-110 kg seeds/ha for semi-spreading and spreading varieties are adequate. The spacings are 30x10cm or 45x10cm or 30x15cm for different groundnut types. For irrigated groundnut, the recommended spacings are 22.5x10 cm or 30x7.5cm.

Groundnut needs more phosphorus, sulphur and calcium. Seed inoculation with efficient strains of *rhizobium* can partially meet the nitrogen requirement of the crop. FYM or compost, 10-15 t/ha should be applied well before sowing and should be incorporated into the soil. NPK at 20:50:20 kg/ha is also applied before sowing.

The erect types can be harvested in 110-125 days, while the spreading types are harvested in 140-160 days. The bunch and semi-spreading varieties are usually harvested by hand-pulling when there is adequate moisture in the soil. The spreading types, on the other hand, are harvested by digging with a spade, or by ploughing or working a blade harrow. The pulled-out plants are stacked for a few days for drying and the pods are stripped afterwards. In bunch types, the pods are stripped by hand from the dry plants. In case of runner types, the plants are allowed to dry and then beaten with sticks or fed to a thresher to separate the pods from vines. The stripped pods are cleaned and dried to a safe moisture content of around 5%. It is desirable to store groundnuts in gunny bags as pods rather than as kernels. Higher humidity favours fungal growth. When *A.flavus* infects the groudnuts, it poses health problems due to prodution of aflatoxins. Hence proper storage is the most important aspect in groundnuts.

Groundnut Shell and Seed

Groundnuts as they are produced have an outer thick woody shell. Inside it are normally embedded 2 or 3 kernels. The seed (65 to 75 % of the pod) consists of two cotyledons and a germ with an outer thick brown skin called testa. The colour of the testa varies from red, brown, purple to white depending on the type and variety. Cotyledons constitute the bulk of the kernel ranging from 92 to 94%, and the testa and germ account for 3-4% each of the seed weight. The chemical composition of these different parts of the seed are presented in Table 1.

The testa is rich in carbohydrates and crude fibre with little or no oil. The germ is rich in oil (39-43%) and proteins (26-28%). The cotyledons are also rich in oil (36-54%) and proteins (21-36%). The cotyledons also contain higher levels of carbohydrates (6-24%). Both the germ and cotyledons contain very low levels of crude fibre (<2%). The germ has 8% reducing sugars and 12% sucrose.

Table 1: Chemical composition of groundnut kernel (g/100g)

Constituent	Testa	Germ	Cotyledon
Moisture	9.01	-	3.9-13.2
Protein	11-13.4	26.5-27.8	21.4-36.4
Oil	0.5-1.9	39.4-43.0	35.8-54.2
Total carbohydrates	48.3-52.2	-	6.0-24.9
Reducing sugar	1.0-1.2	7.9	0.1-0.4
Sucrose	-	12.0	1.9-6.4
Starch	-	-	0.9-5.3
Crude fibre	21.4-34.9	1.6-1.8	1.6-1.9
Ash	1.9-4.6	1.9-3.2	1.8-3.1

Source: Cobb and Johnson (1973), NDDB (1982)

There are four botanical types of groundnuts, namely, Virginia runner (VR), Virginia bunch (VB), Valencia and spanish bunch (SB) types. Many varieties are available which belong to the four botanical types. Variations in physical and chemical properties exist between varieties within the group as well as between the groups. About forty varieties have been analysed for their seed composition (Nagaraj, 1991). Ranges of chemical components of the seed of the four botanical types are presented in Table 2.

Virginia bunch followed by Spanish bunch have higher oil and energy contents. The average oil content of VB is 51.4% and SB is 48.7% as against

47.2% is the other two types. Virginia runners and valencias have higher protein and lower oil contents. The soluble sugars in Virginia runners are higher followed by virginia bunch and spanish bunch, with valencias having the lowest value. Thus it is evident that virginia runner groundnuts with higher sugars and lower oil, are most useful as table varieties.

Ten varieties have been screened for some of their physico-chemical characteristics (Nagaraj *et al, 1984).* TG 1 and TG 17 have a shelling percentage of 65 and 68% and 100 kernel weight of 72 and 68g respectively. These varieties could have better value as HPS(hand picked selections) varieties. The HPS varieties are useful for table purposes and have lot of importance for export from India. The hundred seed weight of other varieties range from 34-58g. TMV10, GAUG 10 and JL 24 have higher oil, protein and sugars.

Similar studies carried out on some of the ICRISAT Varieties (Jambunathan *et al., 1992*), namely, ICGS-1, ICGS-55, ICGS-11, ICGS-44 revealed the following information. ICGS 21 has the highest hundred seed weight, (69.39) indicating its bigger size. It also has higher oil content (about 53%) while J-11 has higher sugars. *Rabi* (post rainy-October to February) groundnuts have higher hundred seed weight and protein content than *kharif* (rainy season-June to October) groundnuts.

Table 2: Seed composition of groundnut genotypes

Character	VR	VB	Valencia	SB
Oil %	46-51 (47.2)	46-54 (51.4)	46-48 (47.2)	47-51 (48.7)
Protein %	21-28 (23.6)	21-26 (23.0)	22-25 (23.7)	20-29 (22.9)
Soluble sugars %	9-14 (11.6)	9-14 (9.8)	7-9 (7.9)	7-16 (9.8)
Energy values (J/100g)	2450-2730 (2646)	2620-2780 (2708)	2623-2658 (2645)	2650-2720 (2676)

Source: Nagaraj (1991a) Figures in parentheses are average values

Oil content is higher (48%) in seeds of size 7 to 9mm diameter (Misra and Gaur, *1982*). Varieties from U.S.A. have 44.8 to 58.3% oil (Young_and Hammons, *1978)* while Indian varieties have 40.8 to 54.4% (Nagaraj and Misra, *1984)*. Semi-spreading varieties have 49.8 $\pm$ 1.7% *(*Sekhon *et al., 1980),* while spanish bunch varieties have 47-52% oil *(*Sehkon *et al.,1972*). Wild species have a broader range of 46.5 to 63% oil content (Cherry, 1977). High oil varieties with oil content ranging from 52-56% have been identied after screening many germplasm accessions(Nagaraj *et al.*, 1986). These genotypes will have a lot of importance in developing high oil varieties.

Groundnut Oil

Groundnut oil is useful in deep fat frying. It brings out the flavour of herbs and spices without contributing it's own. Commercially the oil is extracted from groundnuts using *Ghanies,* hydraulic pressing, expeller pressing and solvent extraction. Oil content varies from 40-54% depending upon the type and variety. The Characteristics of groundnut oil are presented in Table 3.

Table 3: Characteristics of groundnut oil

Characteristic	Range
Melting point °C	0-3
Iodine value (I.V.)	82-106
Acetyl value	8.5-9.5
Saponification value (S.V.)	188-195
Free fatty acids (FFA)%	0.02-0.6
Unsaponifiables %	0.3-0.7
Refractive index (R.I.) ND 20°C	1.4697-1.4719
Density at 20°C (g/cc)	0.910
Colour visual	Light yellow
Lovibond	Yellow 15-25
	Red 1-2

Source: Williams (1966), Cobb and Johnson (1973).

It has a low melting point and can remain as a liquid at low temperatures. Thus it has good scope as a salad oil. The density is 0.9 and similar to other oils. The iodine value is medium indicating it to be a semidrying oil. FFA is low and normally not a problem in groundnut oil. It is a light coloured oil and is normally a good cooking medium. Expelled and double filtered oil is generally utilized and is easily accepted by the consumers. Refining is not necessary and is rarely resorted to by the industry.

Fatty Acid Composition of Groundnut Oil

Groundnut oil is rich in unsaturated fatty acids. Oleic and linoleic acids are the two unsaturated fatty acids accounting for 38-56% and 16-38% each respectively. Among the saturated fatty acids palmitic acid is the major fatty acid with about 10-16%. The other fatty acids present are myristic, stearic, arachidic, behenic, lignoceric and eicosenoic acids. All of them together account for about 10% of the total fatty acids. Fatty acid composition of groundnut oil has been studied extensively by many workers (Badami *et al.*, 1979, Nagaraj, 1991, Sekhon *et al.*, 1980, Ahmed and Young, 1982, Mazingo and steele, 1982). Salunkhe *et al.*, (1982) have discussed the variation available in different genotypes. The range of variation available in fatty acids of groundnut oil is presented in Table 4.

Table 4: Fatty acid composition of groundnut oil

Fatty acid	Range	Average
Saturated fatty acids		
Myristic (14:0)	0.01-0.23	0.09
Palmitic (16:0)	9-24.9	12.60
Stearic (18:0)	0-5.5	1.7
Arachidic (20:0)	2-10.2	4.2
Behenic (22:0)	0.7-3.9	2.1
Lignoceric (24:0)	0-2.8	0.3
Unsaturated fatty acids		
Palmitic (16:1)	0.6-3.3	1.4
Oleic (18:1)	38.7-56.2	47
Linoleic (18:2)	16.2-38.4	29.9
Eicosenoic (20:1)	0.74-2.27	3.0

Source: Rao and Rao (1981), Cobb and Johnson (1973).

Oils from ten varieties of groundnut (*Arachis hypogaea*) have been investigated for chemical characteristics. Oil content range is 41-51, I.V=87-93, S.V=188-193 (Toliwal *et al.*, 1993).

Forty released groundnut varieties from India have been examined for their fatty acid composition and related oil stability (Nagaraj, 1991). Considerable variation in fatty acid composition and related characteristics exist among the genotypes both within the four botanical types as well as between the types. Ten groundnut varieties analysed have the following fatty acid composition 16:0=12-20, 18:0=3-13, 18:1=37-53 and 18:2=20-37 (Toliwal *et al.*, 1993).

Fatty acid compostion of different parts of the seed like, cotyledons, embryo, testa and pericarp have been reported (Cobb and Johnson, 1973). Cotyledon oil is rich in oleic acid(53%) while pericarp oil is rich in linoleic acid (38%). Embryonic axis has equal levels of oleic and linoleic acids (36-38%) while testa has 32% linoleic and 28% oleic acids. Testa oil also has higher levels of behenic and lignoceric acids (3-4%). This information reveals that blanching or removal of testa, before eating would make it a safer food than whole groundnut seeds.

Seed size has also been reported to have correlation with lipids and oleic acid (Raheja *et al., 1986*). Badami *et al.*, (1979) who have analysed 15 varieties have reported the following results; oil content 40.8 to 46.3%; I.V., 86.5 to 104-8, protein (in cake), 35.7 to 55.1%, saturated fatty acids 19.0-36.3% unsaturated fatty acids 62.4-80.7%.

Larger variation exists in the oleic acid levels of the varieties. It ranges between 38-56%. The linoleic acid content also varies considerably (27-40.7%). Among the oil quality characteristics the average iodine values range from 95-98 with the lowest being in virginia bunch and highest in valencias. Oleic to linoleic acid ratios have been considered as indices of stability in groundnut oil (Worthington *et al.*, 1972; Young *et al.*, 1972; Brown *et al.*, 1975). A value of 2 is considerd ideal indicating very good oil stability. Oil stability index is highest in Virginia runner (1.6) followed by Virginia bunch (1.5) with lower values in the other two types (1.1 to 1.3). Nutritional quality of Spanish bunch varieties appear to be poorer. They have lower O/L ratios of 1.0 to 1.35.

Table 5 : Fatty acid composition and oil quality of groundnut genotypes

Fatty acid	VR 12*	VB 6*	Valenicia 4*	SB 18*
Palmitic (16:0)	11.1-15.5 (12.6)	10-17 (13.1)	10-14.5 (13.3)	11-16 (14.3)
Stearic (18:0)	1-5 (2.7)	0.6-5.5 (2.9)	1-3.7 (2.2)	3-6 (4.3)
Arachidic (20:0)	1.1-2.4 (1.6)	1.0-2.0 (1.4)	1-2.2 (1.5)	1-2.3 (1.6)
Behenic (22:0)	0.5-1.4 (0.9)	0.6-1.0 (0.8)	0.6-1.0 (0.9)	0.6-1.6 (0.9)
Oleic (18:1)	45-56 (49.7)	41-53 (47.6)	41-51 (46.1)	38-50 (44.7)
Linoleic (18:2)	27-37 (31.6)	28-38 (34)	32-40 (35.9)	30-36 (34.7)

Figures in parentheses are average values, * = No of varieties studied (Nagaraj, 1991a)

In general palmitic and stearic acids are higher in Spanish bunch types while Virginia runners have higher oleic and lower linoleic acids. Valencias followed by Spanish bunch and Virginia bunch types have higher linoleic acid levels. The stearic acid levels of the varieties range between 1-6% and arachidic acid between 0.6-3.6%. Oleic acid levels have been found to be higher in bigger size kernels (Raheja *et al.*, 1986) which must be due to proper maturity of these seeds. Oil stability, higher oleic acid and lower unsaturated fatty acids in Virginia runner types, make it the most ideal oil for deep fat frying purposes.

Fatty acid composition of wild species namely , *A monticola, A. cardenasii, A. stenosperma, A. duranensis, A. sp.* GK 3001, A. sp. GKP 9993, *A. Villosulicarpa, A. pusilla* and *A. hypogaea* have been reported by Nagaraj and Murthy (1979). Other workers have reported oil, protein, fatty acids, amino acids and antinutrients of some wild species (Jambunathan *et al.*, 1993). The wild species have higher oil, linoleic, behenic and lignoceric acids than cultivars. The protein and amino acid compostion, however are similar.

Groundnut oil contains very low levels of linolenic acid (0.02-0.3%). Only a few workers have reported its presence (Worthington and Holley, 1976, Worthington *et al.*, 1972, Worthington, 1977, Tairo, 1985). Due to its very low level, it rarely gets detected and hence it is not reported by many workers. At this low level it does not interfere with the oil stability, for which groundnut oil is highly popular.

Factors Influencing Composition and Quality

Among the various agricultural production factors, irrigation, micro nutrients and maturity have some influence on the composition and quality of groundnuts. The effect of irrigation on yield and seed composition have been studied (*AICORPO, 1985*). Two irrigations one each at pegging and pod development, increases the oil content and also improves the nutritional quality of oil. The keeping quality of the oil is better when irrigations are given at flowering and pod development stages. Water stress increases the I.V., acid value and S.V. (Singh and Prasad, 1978). N,P and K have a limited influence on the oil content. Phosphorus application marginally increases the oil content (Patil, 1975, Walker and Carter, 1971). Micro nutrients like Mn, Mo, Mg, S,B, Fe and Ca have an important role in increasing the oil content. Sulphur and sulpur containing compounds of some of the above metals have been reported to increase oil content ranging from 3 to 5% (Singh *et al., 1978*). No such effect has been observed on oil content on calcarious black soil under foliar as well as soil application of micro nutrients except for an increase in protein content under iron application (Nagaraj, 1987). Stress at maturity has been observed to reduce long chain fatty acids (C20 and above) and alpha tocophenol which are detrimental to oil quality (Hashim *et al.*, 1993).

Maturity of the pods considerably influences the content and quality of groundnut oil. The oil content of the seeds has been found to be 34.52 and 52.17% on 70th and 100th day respectively in spreading types and 29.38 and 51.51% in bunch types (Singh and Singh, 1974). Harvesting of Valencia bunch varieties between 125-130 days, Virginia bunch varieties between 135-140 days and Virginia runners between 145-150 days for better yield of oil, protein and oil quality has been advocated (Nagaraj *et al., 1989)*. Early harvests resulted in a reduction of oil and protein yields along with impairment of oil quality. A protein named maturin has consistently been found to be higher at maturity in cultivars, and has been suggested as an indicator of maturity (Basha, 1990).

Shelf Life and Cooking Quality of Oil

No deterioration in the quality of raw oil has been observed when stored at ambient temperatures in amber bottles for 18 months (Nasirullah *et al.* 1982) or for 660 days (Yousuf Ali khan *et al.*, 1979) and after deep fat frying for 10hrs (Nataraja Murthy and Lakshminarayana, 1979) . There has been a slight decrease in iodine value (95 to 92) and a 2% decrease in linoleic acid after 18 months of groundnut oil strage. Fresh oil contains monomeric material and little dimers and polymers. The latter group of compounds increase with temperatures from 180 to 260^{0}C and duration of frying from 3 to 6 hrs (Vidyasagar *et al.*, 1974). Runner types which contain lesser linoleic acid also contain higher levels of tocopherols, which have a role in auto-oxidative stability of the oil (Fore *et al.*, 1953). In general, groundnut is a highly stable oil, suitable for long time storage as well as longer frying periods.

Groundnut is a non-drying edible oil. It is highly useful for cooking, especially, for deep fat frying proposes. It can also be converted into margarine, shortenings, mayonnaise and salad dressings. It is used by the pharmaceutical industry as a carrier of drugs and vitamins. Low grade oil is used in the manufacture of soap, cold creams, lubricants, emulsions and for insect control. Peanuts having 80% oleic acid in its oil have been identified. High oleic groundnut oil has three times more shelf life than normal groundnut oil.

Biosynthesis and Changes During Pod Maturation

In the developing groundnut seeds, very little lipid is formed till 15 days after fertilisation (DAF). During 15-35 DAF, lipids, mainly triacylglycerol (TG), accumulate rapidly with low rate of formation of proteins. From 40 to

60 DAF, lipid synthesis continues at a much reduced rate while protein synthesis increases markedly (Ghosh and Sastry, 1986).

Oil synthesis in groundnut mutants has been studied during kernel development. Dry weight, oil and protein content increases progressively with kernel development. Fatty acid composition of the oil differs between mutants and also from the parents during development. Mutant has higher oleic and lower linoleic acid than the parents. During development, palmitate decreases slightly in parents while in the mutants the decrease at maturity is lesser than that in the parents (Kumar, *et al.,* 1988).

The lipid content increases from 1% to 38.1% of pod from 70th day to 120 days of sowing. Larger accumulation has taken place during 85-95days after sowing in spanish varieties and 90-110 days in virginia types. In general during various developmental phases, the triacylgylcerol content increases. The polar lipids which constitute a major fraction of the total lipids at early pod developmental stages, gradually decrease as the non-polar lipids (triacyl glycerols) increase in composition. Triacyl glycerols increased from about 70-80% of total lipids between 60-80 days to 96-98% at maturity in spanish bunch and virginia bunch varieties (Nagaraj, *et al.,* 1989).

Mature seeds contain more stearic and oleic acids and less arachidic acid. In general both oleic and linoleic acids increase with maturity. The increase in oleic acid has been from 41 to 52% with a concomitant decrease in palmitic acid from 32 to 29%. The mature groundnut pod has the following composition: oleic acid 40-52%, linoleic acid 25-37%, palmitic acid, 8.5 to 10.5%, stearic acid 1.5 to 3.0% and arachidic acid 1.5 to 2.5%.

Groundnut Meal and Protein

Groundnut cake is a by-product obtained after extraction of oil. Expeller cake generally contains about 6% oil. The residual oil is recovered through solvent extraction to obtain the defatted cake. It is ground and allowed to pass through a 100 mesh screen. This product if free from extraneous material and aflatoxins is fit for human consumption. Because of lower utilizable carbohydrates, it is useful as a diet for diabetic patients. The cake contains 45-60% protein, 22-30% carbohydrates and 4-6% minerals (Nagaraj, 1990a*)*. The defatted meal has higher protein efficiency ratio, net protein utilization and digestibility (Kabirullah *et al.,* 1977).

The groundnut cake is used either as an animal feed or as an organic manure. It serves as a good nutritious food for milk giving animals. It should not form more than 25% of feed for lactating animals. At higher levels it produces soft body in animals and also may induce laxative action. When free from aflatoxins, the cakes are highly useful for poultry, calves and ducklings. Groundnut fluor is widely used for culturing molds in the manufacture of antibiotics The cake contains about 1-4% phosphoric acid and 1-2% potash which makes it a good organic manure. The cakes are to some extent utilized as human food mostly as fluor in combination with other fluors like wheat.

The protein content of groundnut ranges from 22 to 32% depending upon the variety, location and year (Young and Hammons, 1978). The protein can easily be extracted from the cake with dilute alkali, and later precipitated by lowering the pH. Water, 70% alcohol, 10% NaCl and 0.25% KOH have also been tried for extraction of proteins. The protein extraction is usually carried out using defatted fluor. The fluor is ground to different particle sizes to produce grits or fluor which is then extracted with different solvents to obtain protein concentrates or isolates.

Bitter principles, odoriferous compounds and antinutrients are removed prior to protein extraction through water or aqueous alcohol leaching. The isolated protein has 16.2% N and 0.74% ash and 0.06% lipid. Presoaking of split groundnut kernels in 4% NaCl overnight followed by hot water washing helps in removing the water solubles and flavour components. Protein isolate obtained from the treated kernels has high solubility, white colour and is free from nutty flavour (Subramanian *et al.*, 1981). Protein isolates have been used in the manufacture of meat and cheese analogues (Chen *et al.*, 1979, Cross and Nicholas, 1979, Mc Waters and Heaton, 1979).

Groundnut protein consists mainly of two globulins namely arachin (93% of defatted seed protein) and conarachin. When a salt extract (0.1% NaC1 buffered with 0.01 M sodium phosphate pH 7.9) is treated with saturated $(NH_4)_2$ SO_4 arachin gets precipitated first at 40% saturation followed by conarachin between 40-85% saturation. Since globulins contain 18.3% N, the conversion factor for proteins of groundnut should be 5.46 (Jones, 1931).

Protein Quality

As with any other vegetable proteins, lysine, methionine, threonine and isoleucine are the limiting amino acids. The biological value (BV) for groundnut protein is 65 with the main limiting amino acids, lysine and threonine (Hackler, 1977) and the essential amino acid index is 69.The protein efficiency ratio of edible cake protein and a protein isolate have been observed to be 1.60 and 1.53 respectively (Anantharaman *et al., 1989*). The BV increases from 60 to 69% when supplemented with 0.31% methionine, 0.37% lysine and 0.15%

threonine and further supplementation with 0.1% tryptophan results in a BV of 83%. Methionine, lysine, and threonine are the limiting amino acids as measured by the rat bioassy and protein efficiency ratio of amino acid supplemented diets (Miller and Young,1977). Groundnut protein is rich in arginine (3.5g/100g nuts). This amino acid is linked to wound healing as also reduction of the undesirable LDL (low density lipoprotein) cholesterol. Hence groundnuts have a role in lower heart disease risk.

Immature groundnuts contain higher levels of arginine, followed by proline(Young *et al.*, 1974). Keeping this aspect in view, arginine content has been proposed as an index of maturity, for harvesting the indeterminate groundnuts(Young and Mason, 1972). Similarly, a protein named, maturin, has been considered as an index of seed maturity, after studying the protein patterns of 20 peanut cultivars(Basha, 1990). Groundnut protein is a good source of most of the essential aminoacids, required by the human being. Groundnut roasted flavour is due to the interaction of glucose and fructose with amino acids like glutamic acid, glutamine, aspargine, phenylalanine, aspartic acid and histidine (Woodroof, 1983; Young *et al.*, 1974).

Table 6: Amino acids of groundnut protein.

	Compostion Percentage		
Amino acid	**(1)**	**(2)**	**(3)**
Glycine	5.90	6.24	5.86
Alanine	4.05	4.26	4.11
*Valine	4.40	4.66	4.32
*Leucine	6.55	7.09	6.32
*Isoleucine	3.60	3.73	3.45
Serine	4.95	5.38	5.32
* Threonine	2.70	3.18	2.52
Tyrosine	4.00	4.51	3.28
Proline	4.70	5.05	5.86
*Methlonine	1.40	1.30	1.29
Cysteine	1.45	1.33	1.23
*Lysine	3.45	4.05	5.86
*Histidine	2.22	2.92	2.22
Arginine	11.60	13.19	11.31
Aspartic acid	11.70	12.51	11.46
Glutamic acid	18.90	21.75	21.83

*Essential amino acids (EAA).
Source: (1) Mosse and Pernollet (1982), (2). Indian *Rabi*
(3) Indian *kharif* Jambunathan *et al.*, (1992).

Protein content, quality and products have been reviewed in detail by Mc Watters and Cherry, (1982) . Functional properties of groundnut protein, its isolation, amino acid composition of arachin, conarachin and total protein and preparation of various products including protein isolates and concentrates have been presented and discussed by them. The need for limiting the levels of aflatoxins (30-50 ppb) in the groundnut fluors have also been stressed by them.

Indian groundnut protein and its amino acid make up has been examined in detail (Jambunathan *et al.*, 1992). It is similar to other groundnut proteins. Amino acids especially the sulfur amino acids are present at levels similar to that of groundnut protein from other locations. Arginine, Aspartic acid, glutamic acid, glycine and leucine are the major amino acids of Indian groundnut.

Carbohydrates

The total carbohydrate content of kernel and meal are 10- 20% and 38% respectively. The reducing sugar content ranges from 1.2 to 1.8%, while the total sugars range from 7.4 to 14.2% in 25 varieties (Gupta *et al., 1982*). Sucrose is the most important carbohydrate. It ranges from 2.86 to 6.35% depending upon genotype (Holley and Hammons, 1968). Glucose, fructose, arabinose and galactose are the other minor sugars present. Oligosacharides, stachyose and raffinose have also been isolated from groundnut. Tharanathan *et al.*, (1975) have reported the following carbohydrate composition; starch 12.5% hemicellulose A and B, 0.5 and 3.5% respectively, cellulose 4.5%; surcrose 13.9% raffinose 0.89%; stachyose 1.56% verbacose 0.41%; glucose 0.8% and fructose 0.41%. Groundnut genotypes with lower levels of stachyose and raffinose can improve the taste properties. However, stachyose has a role in providing frost resistance to plants. Glucose decreases with seed maturity (Farfan *et al.*, 1978).

Vitamins and minerals

Groundnuts are a rich source of vitamins E, K and B complex. It is especially rich in thiamine(B_1). Niacin content has been reported to be 16.7 to 21.7 mg/100g in Indian groundnut (Chaturvedi and Geervani, 1986). The tocopherol content has been reported to be 0.93 mg/g oil. with 36% α and 64% γ-tocopherol (Rao, *et al.*, 1965). High performance liquid chromatographic (HPLC) analysis has given the following tocopherol composition: alpha=152.2, beta=4.6 and gamma=164.6 μg / g oil (Renuka devi *et al.*,2000). The carotenoid content is around 1μg/litre oil (Pattee and Purcell,1967). The total ash content ranges from 1.95 to 2.10% (Abdel Rahman, 1982). The concentration of individual minerals have been reported, Ca=55-64, P=288-348 Fe=1.3-1.5 and Zn=4.4-6.0 mg/100g in groundnut (Khan, 1989). Indian groundnuts have P= 345-550, K=512-700, Ca=58-95, Mn=189-212, Zn=2.6-5.2, Fe=3.6-6.0 and Mn=0.9-1.3 mg/100g seeds (Jambunathan *et al.*, 1992).

Table 7: Vitamin content of groundnts.

Vitamin	mg/100g oil
Fat soluble vitamins	
Vitamin A (I.U.)	26
Vitamin E	26.3-59.4 mg/100g oil.
α-tocopherol	11.9-25.3
β-tocopherol	10.4-34.2
γ-tocopherol	0.58-2.50
Water soluble vitamins	
Thiamin	0.99
Riboflavin	0.13
Pyridoxine	0.30
Niacin	12.8-16.7
Choline	165-174
Folic acid	0.28
Inositol	180
Biotin	0.03
Pantothenic acid	2.72
Vitamin C	5.80

Table 8: Mineral content of groundnuts.

Minerals	mg/100g
Phosphorus	250-660
Potassium	500-890
Calcium	20-90
Magnesium	90-340
Sulphur	190-410
Zinc	3.4-5.0
Manganese	1.3-3.2
Copper	0.6-1.9
Iron	2.1-7.0
Boron	1.2-1.8

Some Processed Foods

Groundnut is a constituent of many household preparations. They are relished by rural as well as urban populations. Some common preparations are *chutneys*, dry powders, sweets (*laddus, chikkis*) and as an ingredient of fried-rice and even *chapaties*. Lucas (1979) has reviewed food uses of groundnuts. Groundnut has a substantial advantage in that, it needs minimal preparation to produce pleasant and acceptable foods. There is no need for any energy consuming extraction and purification steps. Some popular and simple groundnut preparations (Lucas, 1979) are as follows:

Boiled nuts: Unshelled immature nuts are boiled in medium brine and eaten fresh or frozen and canned.

Roasted nuts and kernels: Mature nuts are soaked in brine followed by roasting. Whole or chopped kernels are roasted and used directly. Use of partly or fully defatted nuts is also getting popular now-a-days.

Roasted or oil fried groundnut seeds can be incorporated into many foods. Such groundnut foods are tastier as well as highly nutritious. *Chutney,* the Indian delicacy, is a spicy, preparation. The roasted, blanched seeds are ground with chillies, tamarind, salt etc. to make the *Chutney.* It is a common item served at home as well as in restaurants in India, especially, South India.

Peanut candies: More than 50 types of candies are made of groundnuts, the most popular ingredient in American candies.

Peanut butter: This is a popular edible product in the U.S.A. There are three types: a) old fashioned, b) smooth and c) chunky butter, mostly varying in particle size of the groundnuts. Manufacture of the butter involves roasting of the kernel to a controlled brownness, followed by cooling and removal of the skins (blanching). Spoiled and discoloured kernels are removed at this stage. Then the kernels are ground to a paste or butter according to the product desired. Salt, stabilizers and other optional ingredients like sweeteners are blended with butter prior to cooling and packaging. Groundnut butter is mainly used as a spread for bread or biscuits. It is also used in cookies, in sandwiches, in candies and in frostings or icings. Groundnut butter is also used in nutty bars, cream patties, chocolate bars, wafers, butter toffees, etc. Groundnut butter is also available in powder form, the product is very smooth with milk flavour and assumes the characteristics of groundnut butter when a small amount of water is added. Groundnut butter is also a good bird feed. Groundnut butter popularity is increasing all over the world. It is available on the super market shelves of many countries due to globalization.

Groundnut Cheese: Groundnut cheese is a novelty item that may compete in price and quality with animal cheese. Cheese like products have been made from groundnut protein isolate just as cheese is made from cow's milk. It has good quality protein, is easily prepared and low in cost. A processed cheese spread has been prepared from peanut protein based toned milk in India. It has a smooth consistency and milky flavour. It compares favourably with standard cheese spread in composition and organoleptic qualities.

Peanut flour: Flour of high quality (60 percent protein, 0.75 percent crude fibre, 4.5 percent ash, 22.5 percent carbohydrate) is found suitable in soups, cold drinks, hot drinks, hot cakes, breads, cakes, instant puddings, congealed salads, ice cream and confections. It is especially beneficial for treating hemophiliacs. Bleeding is controlled by incorporating 150-200g in the daily diet. It causes shrinking of the capillary blood vessels preventing bleeding. This activity lasts for two days. It is an excellent protein supplement for wheat flour compensating some of its deficiencies particularly protein. Groundnut flour is used in many parts of the world to prepare beverages and baby foods(Orr, 1972). It is now-a-days being added to cereal fluors to increase their protein content.

Pre-cooked full fat peanut flakes: Blanched cotyledons (without germs) are dried to 2 or 4% moisture and are then ground to a fine paste. Water is added while beating and the slurry is dried to get flakes. They have high keeping quality and find use in formulated foods of high fat and protein.

Laddus and chikies: These are Traditional Indian products made from groundnut and jaggery syrup. They are very popular in Indian house holds. Now-a-days they are available on the super market shelves. The negative effects of high oil in groundnuts get reduced due to the addition of jaggery or sugar.

In India some innovative food products containing groundnut have been developed. Some important processed foods (Parpia, 1991) developed are i) *Balaahar*, a vitamin fortified mixture of wheat flour, groundnuts and chickpea, ii) pre-cooked dehyrdrated food made from groundnut cake, chickpea, green gram and wheat, iii) infant food made from 75% groundnut, wheat and skimmed milk powder, iv) multipurpose food made from 75% groundnut fortified with minerals and vitamins, v) *miltone*: milk toned with groundnut protein. This is cheaper (by 30%) than traditional toned milk and vi) Yoghurt like product made from miltone.

Non-Fermented Cheese Analogues: Caseinate which controls the rheological and melting properties of cheese can be replaced at 40-50% levels with groundnut protein isolate to produce analogues resembling cheddar and processed cheese. Even frozen desserts, meat analogues, beef patties, etc. also can be made using groundnut protein.

Protein Isolates: Protein isolates containing 92% protein have been prepared. These can be incorporated into baby foods, milk powder other formulated foods (Sheik *et al.,*1983).

Low Fat Groundnuts

Low fat groundnuts have been prepared worldwide by squeezing or extraction using solvents (Nagaraj and Kailash kumar, 1986), Ramachar, *et al.*, 1974, Balakrishnan *et al.,*1984). These are of interest to those canscious of weight reduction or low fat foods. The product will have low fat(15-30%)and high protein(35-40%). Groundnts are mixed with popular foods globally to improve quality, nutrition and acceptability.

Groundnut Nutritional Characteristics

Groundnut is a rich source of energy because of its high oil and protein. It can supply about 5.6 and 5.8 calories per gram in the raw and roasted forms respectively. Apart from its high oil and protein, it contains 18% carbohydrates. It is also a source of minerals and vitamins (Table 9). Due to it's rich and concentrated levels of all nutrients, it serves as a space food. Some minerals like Ca, Mg and Fe though present in higher levels, are in an unavailable form (as salts of phytic and oxalic acids) and do not reflect their nutrient availability. Vitamins B_1, B_2 and niacin are present at a considerable level. But others like A, C and D are the limiting ones in groundnut. Major factor in favour of groundnut is that it has a good digestibility in both raw and roasted forms. Groundnuts have been reported to contain a phytochemical namely resveratrol at 200-250 μg/100g seeds. It is a phenolic compound and has a role in reducing heart disease risk. Resveratrol is present in red wine at levels of 800μg/glass. French people who consume wine regularly, are reported to have lower incidence of hear attacks. WHO has recommended consumption of 30 g/day pulses, nuts and seeds. Groundnuts are very much useful for that purpose.

Atherogenicity

It is well known that groundnut oil induces atherosclerosis. High molecular weight fatty acids and the triacyl glycerol stucture of groundnut oil seem to be the reason for the same (Kritchevsky *et al.*, 1973).

Groundnut Flavour

Groundnuts have a typical nutty, sweet and a slightly bitter flavour. The sweetish "green" flavour of the raw nuts is mainly due to sucrose (5%) and compounds like hexanal, pentanal and octanol, though more than 800 compounds have been detected in roasted groundnut (Ahmed and young , 1982, Young and Hovis, 1990). The flavour of the roasted nuts is due to sugar amino-acid complexes and carbony1 compounds (Cobb and Johnson, 1973). The flavour is generally associated with the oil and hence deoiled cake is generally bland. Maturity, harvesting time, curing and storage methods including irrigation and pesticide residues influence the flavour and aroma of nuts. Acceptance of groundnuts is mainly related to the roasted flavour.

Table 9: Nutritional characteristics of groundnut.

		Content\100g	
Characteristics	**Raw**	**Roasted**	**Defatted flour**
Calories	564.0	582.0	371.0
Proteins	26.0	26.0	45.0
Fat (g)	47.5	48.7	5.8
Carbohydrates (g)	18.6	20.6	30.0
Calcium (mg)	18.6	20.6	30.0
Phosphorus (mg)	401.0	40.10	800.0
Iron (mg)	2.1	2.2	3.5
Thiamine (B1) (mg)	1.14	0.32	0.75
Riboflavin (B2) (mg)	0.13	0.13	0.35
Niacin	17.2	17.2	2.5

Source: Burn and Huffman (1975).

Antinutrients

Groundnuts contain some anti-nutritional factors like trypsin and enterokinase inhibitors (Leiner and Kakade,1980, Lau *et al., 1980)*, hemaglutinins, goitrogen, phytates, polyphenols, allergens etc. (Leiner, 1962). Goitrogens are present in the testa while trypsin inhibitor and hemaglutinins are present in the kernel. Moist or dry heat generally reduces the antinutrients mainly trypsin inhibitors. Goitrogens and heamaglutinins are present in low levels and do not pose any health problems.

Groundnut contains goitrogens in the testa. There have been reports of the presence of a trypsin inhibitor and a haemagglutinin in the kernel. The heat treatment of the kernel does not produce any improvement in its nutritive value.

Trypsin inhibitor activity in peanuts have been reported to be about one half to one fifth of the activity found in soybeans (Lusas,1979). This activity in peanuts is significant enough to cause peancreatic hypertrophy in rats receiving 15 per cent of protein intake from peanuts. Lectins or phytohemagglutinins have been found in peanuts (Leiner, 1979) and roasting of peanuts does not destroy them (Lau *et al., 1980*). A number of other pharmacologically and toxicologically important compounds have been reported to be present in peanuts. These include hemostatic agents and goitrogenic glycoside, arachidoside in peanuts. Oilseed proteins constitute the most highly allergenic food groups. Peanuts have also been shown to be highly allergenic. Many people claim untoward effects of peanut consumption, especially new peanuts. These untoward effects range from plain nausea and headaches to total unconsciousness and may be related to hypersensitivity reactions and idiosyncracies.

Oxalic acid has been reported at 0.4 mg /g of nuts(Oke,1967). At this low level it does not have any adverse effects. Oligosaccharides like stachyose and raffinose present in groundnut are of concern due to their flatulence.

Aflatoxins

Greater concern is occasioned by the possibility of exogenous toxicity in groundnut products. It has been shown that some strains of *Aspergillus flavus* growing on groundnuts elaborate compounds which are extremely toxic to some species of monogastric animals including poultry. The general name aflatoxin has been given to these compounds. In 1960, around 100,000 turkey poults on British farms died from a disease of unknown etiology. Investigations revealed that all the stricken birds had received feeds containing groundnut meal and the other animals including pigs and calves could also be similarly affected by such rations. In practice aflatoxin B1, is encountered most frequently, the others being absent or present in negligible amounts, but exceptional B1, accompanied by significant amounts of either G1 or B2 may be found. They are extremely potent hepato-toxic agents and carcinogens as demonstrated in various animal experiments. The amount of aflatoxins involved are extremely small, contaminations of the order of a few parts per million being significant. Thus the meal implicated in the turkey debacle contained about 7 parts per million of aflatoxin. This is a very high figure, and it is probable that for many purposes contamination levels exceeding 0.05 parts per million would be unacceptable.

In humans, low levels of aflatoxin may cause various nutritional interactions resulting in sub-optimal growth. There are indications that aflatoxins may interact with nutrients at the biochemical level. Any doubts that remained about the hazardous nature of aflatoxins to human health have been cleared by the unfortunate related episode in India. In 1974, people of a corn eating ethnic group in India who reportedly consumed 0.25 to 15.6 ppm (2 to 6 mg) aflatoxin along with contaminated corn developed acute toxicosis with 106 deaths.

It is best to avoid production aflatoxins on groundnut. This is best done by controlling the growth of *A.flavus*. Proper storage of harvested nuts at less than 5% moisture takes care of this requirement. Once the presence of aflatoxins is discovered it is better to avoid use of the same for food purposes. Though some methods for detoxification are available, such products may not get acceptance as food items. They can at best be utilized as animal feeds.

Non-Edible Uses of Groundnut Oil

Groundnut oil finds a limited use in the manufactue of cosmetics, candles, leather dressings, creams and as a tallow substitute. It also serves as a source of glycerine (9.8% of oil). It is useful as olive oil substitute due to its richness in oleic acid. It can alleviate infantile paralysis. The oil is reported to contain an inflammatory compound. The oil esmulsion is an insecticide and has been found useful as tobacco suckericide.

Groundnuts as Animal Feed

Narmally groundnut cake is widely used as animal feed. It is more useful for drought and milch animals. Groundnut haulms are also highly useful as animal feed. Pelleted and dehydrated groundnut vines are better than grass. Groundnut butter serves as attractive feed for birds.

Groundnut Shells

The shell content of the pods vary from 20-40%. The spreading varieties contain higher proportion of shells than bunch varieties. The major constituent of shells is cellulose (65%) followed by soluble carbohydrates (21%), proteins (7%), minerals (5%) and lipids (1%). The shells have very little feed value. The shells are presently being utilized as fuel either directly or in combination with dung. Other uses are as a filler for fertilizer mulch for growing plants, roughage for cattle feed when mixed with molasses and mineral supplements, litter for poultry houses, abrasive for polishing steel and aluminum, insulation for farm buildings, as a base for making crowns for beverage bottles, filler for plastic wood, veneer boards and as high grade activated charcoal.

There is a great scope for utilizing the shells for more useful purposes. The important products that can be made from them are i) activated charcoal, ii) biogas, iii)alcohol, iv) extender resins, v) cork substitute, vi) chemical pulp for making hard boards and vii) compressed blocks.

Groundnuts Haulms

Groundnut haulms are widely used in India as cattle feed. Haulms are very nutritious feed for cattle. They contain proteins (8-15%) crude fibre (22-35%) carbohydrates (38-46%) and minerals, (9-17%). The digestibility of nutrients is around 53% and that of crude protien is 88% in animals. Haulms can release an energy content of 2337 cal/kg of dry matter. Groundnut roots are also useful to farmers in enriching the soil fertility. Part of the nitrogen fixed through symbiosis (by *Rhizobium)* in groundnut remains in the soil, thus reducing the cost of nitrogen fertilization for the next crop. About 150 kg nitrogen gets fixed in the soil, due to *Rhizobium* when groundnuts are grown.

Pelleted Groundnut Vines

Tests have shown that dehydrated and pelleted groundnut vines are a valuable by product. Pelleted vines are far superior to bermuda grass in digestible nutrients and possibly superior as a source of carotene.

Non-nodulating Groundnuts

The non-nodulating groundnuts do not have the capacity for symbriotic nitrogen fixation. Such groundnuts have been observed to be low in oil (46%) and protein (16%) and high in α-amino nitrogen, soluble sugars, lysine, alanine, methionine, threonine and hydrolytic enzymes with similar iodine values to that of normal cultivars (Pancholy *et al.,* 1982). Chlorophyll a and b have also been found higher during first three weeks of growth stage (Basha, 1987) in the non-modulating groundnut leaves.

Rapeseed-Mustard

Rapeseed - Mustard group of crops belong to the genus *Brassica*, which consists of 37 species. They are mainly annual or biennial crops. Important group of *Brassicas*, grown are brown mustard (*B. Juncea*), yellow Sarson (*B. campestis*), Toria (*B.campestris*), rocket salad or *taramira* (*Eruca sativa*), *gobhi sarson (B. napus)* and *Jungli Rai* or wild mustard *(Brassica tournefortii)*. Different species of Brassicas are grown in 53 countries all over the world. They are the earliest domesticated crops known since 5000 BC. Rapeseed mustard seeds are a good source of a widely used edible oil. The seeds also find varied uses in food preparation and pickles. The cakes are useful as animal feed. The tender plant serves as a green leafy vegetable.

Brassicas seem to have originated in the Middle East or North Africa. At present Canada, China, France, Germany and India are the important countries that grow rapeseed-mustard. Asia accounts for 59% world area and 49% world production. In general it is grown in subtropical and tropical countries. In India, the crop is grown in the post-rainy (*Rabi*) season, namely, September -Octorber to February –March . It grows well in areas having 25-40 cm rainfall. The soils can be loams to heavy loams with a pH around 7.

Seed rate ranging from 3-15 kg depending on the type of rapeseed-mustard and the soil or time of planting and spacing etc. Seeds are sown at a spacing of 30x10cm or even 30 x 60cm depeding on the location and soil type. Seeds are treated with Thiram or Bavistin at 2.5 g/kg seed for protecting against seed borne diseases. 40 kg N, 40 kg P_2O_5 and 30 kg K_2O are applied at the time of sowing. Irrigated crop may need more fertilizers. FYM application (5-10 tonnes/ha) and green manuring are also practiced to boost yields. Sulphur @ 20 kg/ha and boron @ 1 kg are also required by the crop. One irrigation at flowering boosts the yields. A second irrigation at siliqua stage also is useful.

Rapeseed –mustard includes annual herbs ranging in height from 0.45m to 1.75m. Roots in general are long and tapering. The inflorescence is carymbose raceme. The flowers bear a hypogynous syncapous ovary. The fruit is siliqua. The pods are two, three or four valved depending on the number of capsules, in the ovary. The flower opens during 8AM to 12noon.

The crop matures in 75-90days. The crop is harvested when 75% of the pods turn yellow with a moisture content around 30%. It is sun dried and manually threshed to separate the seeds. The yields range from 800-1500 under rainfall and 2000-2500 under irrigated conditions. Dried seeds with less than 8% moisture can safely be stored in gunny bags.

Seed Composition

The seeds contain 16-19% hull. The oil content of the *Brassica* group of oilseeds ranges from 38-44%. The average oil content of these types are presented in Table 10.

Table 10: Oil content of rapeseed-mustard types.

Rapeseed- mustard	Oil %
Raya	40
Brown *sarson*	43
Toria	35
Yellow *sarson*	41
Taramira	28

Source: Anand *et al.*, (1979).

The smaller round shaped mustard seeds are mostly used for oil extraction, which is generally done in *ghanies* and expellers. The hulls contain 16% oil while the kernel contains 44-47% oil. The whole seed oil content is 40-42%. Oil content of 20 different *crucifer* types has been reported to range between 17-48%. The cultivars of *sarson* and *toria* have higher oil content (Anand *et al.*, 1979). Oil content is controlled by less number of genes and hence it is easier to increase the oil content of the seeds. The variation in some physical and chemical properties of major types of rapeseed-mustard is presented in Table 12.

Table 11: Oil content of rapeseed-mustard seed components.

Component	*Brassica napus*	*Brassica campestris*
Seed coat %	16.5	18.7
Seed oil %	41.5	40.0
Hull oil %	16.0	16.2
Kernel oil %	47.1	44.0

Source: Applequist and Ohlson, (1972).

Several cultivars and germplasm lines have been screened for their oil content and other quality factors. The oil content varies from 38.9 to 44.6% in *B.Campestris.* Nine Indian mustard genotypes have oil content of 40-43.5%. Another set of five genotypes of rapeseed have an average oil content of 44% (Ahuja, 1990). The protein content of the mustard cake ranges from 37-44% while the ash content ranges from 5-9%.

Yellow seeded varieties have higher oil and protein contents (Downey and Robbelen, 1989) at the expense of fibre (thinner seed coat). The yellow coloured seeds have around 40% oil and 46% protein as against 34% oil and 37% protein in black seeds(Ochodzki and Piotrowska, 2002).

Table 12: Composition and characteristics of rapeseed-mustard.

Type	Oil%	Protein % in cake	Ash % in cake	I.V.	Refractive index at 40°C
Toria	31.2 to 44.4	37.3 to 44.2	5.4 to 9.0	74.1 to 98.4	1.4612 to 1.4662
Yellow Sarson	40.1 to 47.5	37.0 to 42.2	6.3 to 8.1	100.2 to 101.5	1.4769 to 1.4800
Mustard	32.3 to 40.3	38.5 to 42.2	6.0 to 7.8	100.2 to 101.5	1.4604 to 1.4627

Source: Sarkar and Bhattacharya, (1987).

Rapeseed Oil

Rapeseed oil is mainly used as salad and cooking oil. It has a good shelf life. The crude oil contains 95-98% triacylglycerols. Rapeseed oil contains upto 3-5% phospholipids. The characteristics of high erucic rapessed oil is presented in Table 13. The smoke point ranges from 220-230°C, the flash point ranges from 275-290°C. The viscosity of high erucic acid rapeseed (HEAR) oil is around 80 while it is around 70 for LEAR (low erucic acid rapeseed oil). The rapeseed oil contains up to 50 mg/kg chlorophyll, and 35 mg/kg sulphur.

The crude mustard oil is dark brown in colour and has a characteristic sharp taste and pungent odour. It is generally refined by treating with sulphuric acid which precipitates impurities and colouring matter. After settling, the oil is repeatedly washed with hot water and when thus refined, it is yellow in colour. The refined oil has a pungent, mustard-like odour which can be removed by deodorization. But the deodorized oil is inclined to revert in flavour and acquire an unpleasant taste different from that of the undeodorized oil.

Table 13: Physical characteristics of rapeseed oil.

Characteristic	HEAR oil	Canola oil
Specific gravity 25/25^{0}C	0.906-0.910	0.914-0.917
Refractive index 30^{0}C	1.469-1.485	1.465-1.467
Iodine number	87-122	110-126
Saponification Value	170-200	188-192
Unsaponification value	0.750-1.219	0.5-1.2
Colour	Yellow brown	-
Titre	11.5-15^{0}C	-
Acid value	0.26-2.53	-
Viscosity (kinematic at 20^{0}C)	80-82	67-70
Flash point	278-282	275-290
Smoke point	226-234	220-230

Source: Sonntag (1979), Shenolikar (1985).

Mustard oil is slightly more viscous and has a lower saponification value and a higher refractive index compared to other vegetable oils. The LEAR oil is however having higher S.V. of 190 indicating the absence of high molecular weight erucic acid. The HEAR oil has lower I.V. being richer in monounsaturated erucic acid while LEAR oil has higher I.V. due to higher levels of linoleic acid. Mustard oil does not readily thicken when heated or exposed to air and is placed between non-drying and semi-drying oils. The oils from different *brassicas* closely resemble each other.

Development of Oil in Rapeseed

The formation of oil in the seed begins with the seed formation, namely after the ovary is fertilized and goes on increasing till maturity. Under normal temperatures rapid oil accumulation takes place when the seed is 20 days old. In *toria,* the oil content increases from 5.7% at 20 days to 41.64% at 40 days after flowering. In the case of *brown season* these values are 4.56 and 44.88%. At low temperature the accumulation has been maximum only at 50 days of seed development. Under frost and cold, the oil accumulation is slow and the oil formed is less. When exposed to frost, the oil content at 40 days has been 16.3 as against 41.3 and 39.6% in medium and early bloom crops in *toria*. In the case of *sarson,* it is 22.5% in late bloom as against 37.4% and 41.1% in medium and early bloom types (Singh, 1958).

Biosynthesis of Triacyl Glycerols (Oil)

In cruciferous oilseeds, common long chain fatty acids are formed by *de-novo* synthesis via acyl carrier protein (ACP) derivatives. Acetyl-ACP and malonyl-ACP yield palmitoyl ACP, which is subsequently elongated to stearoyl ACP and desaturated to oleoyl-ACP. The acyl- ACP derivatives are then hydrolyzed and the fatty acids converted to their coenzyme A (CoA) derivatives prior to further metabolism. It seems likely that the oleoyl moieties from oleoyl-CoA are transferred to phosphtidyl choline and then desaturated to linoleoyl moieties. Oleoyl-CoA is also elongated by malonyl-CoA to yield very long chain acyl-CoA derivatives, such as gadoleoyl-CoA and erucoyl-CoA. Triacyl glycerols containing both long chain and very long chain acyl moieties are synthesized mainly via the glycerol-3 phosphate path way.

Erucic acid is synthesized by the elongation of oleoyl-Co A via eicosenoic acid (Pollard and Stumpf, 1980). High erucate strain of rapeseed possesses the oleoyl elongase while it is absent in low erucate varieties (Stumpf and Pollard, 1983) and the conversion of oleate to eurcate occurrs in particulate fraction of seeds (Agrawal and Stumpf, 1985). Since the undesirable linolenic and erucic acids are the end products of a biosynthetic pathway, a reduction in these fatty acids is possible if the enzyme responsible for the synthesis of these fatty acids is eliminated from oleic acid and the linolenic from linoleic acid. The presence of two different pathways for the conversion of oleic to linoleic acid, and linoleic to linolenic acid suggests the possibility of genetic manipulation of one without affecting the other two fatty acids.

The glycerides of the oil consist of mono-oleo-di erucin, 54%; di-oleo-mono-erucin, 28%; and monosaturated mono-oleo-mono-erucin, 18%;

Fatty Acid Compsosition

Rapeseed-mustard oil is generally rich in erucic acid (37.9-57%). It is a 22 carbon monounsaturated fatty acid. The double bond occurs at the 9th carbon atom from the methyl end of the chain (22:1 ω 9). It also contains linolenic acid (4.7-13%). The other important fatty acids present are linoleic acid (14%) and oleic acid (13%) along with 16:0, 18:0 and 20:1 (5-10% each). The variation in fatty acid composition of different types of *Brassica* species has been reported by Anand (1976) and Shenolikar (1985). The fatty acid composition of three main species has been presented in Table 14. The unsaturated fatty acids like linoleic and linolenic acids are concentrated in the 2-position of the glycerol moeity (Applequist, 1972).

The total saturated fatty acids in high erucic acid rapeseed-mustard oil is low at 7%. The monoenoic acids account for about 70%. This group includes 45% erucic, 25% oleic and 10% eicocosenoic acids. The essential fatty acids, namely 18:2 and 18:3 are present at around 15 and 10% respectively. The traditional mustard oil contains 30-51% erucic acid and 13-23% oleic acid while the low erucic acid lines have high oleic acid upto 70%(Kaushik and Agnihotri, 2000). The saponifiable lipids, namely, triglycerides account for 91-99% of the oil. The phospholipids are present upto a maximum of 3.5% of the oil.

Variation in oil content and fatty acids in *B. napus* (*Gobhi sarson*) genotypes has been examined by Ahuja, *et al.,* (1990). Oil content varies from 41.0 to 45.4% with a mean value of 43.4%. Fatty acid composition is as follows: palmitic, 2.1 to 4.2%; oleic, 11.5 to 20.4%; linoleic 11.9 to 16.0%; linolenic + eicosenoic 15.0 to 21.4%; and erucic acid 40.3 to 46.0%. *Gobhi sarson* contains 2% higher level of oil along with 4 to 7% higher level of oleic acid and 7 to 8% lower erucic acid than *Raya* and *Toria* cultivars. Nutritionally, *B. napus* lines are better because of lower erucic acid and higher oleic and linoleic acids.

Reduction of 18:3 from the present level of 8-10% to less than 3% while maintaining or increasing the level of 18:2 has been attempted to improve the quality of rapeseed oil. Lower 18:3 will improve the keeping quality and higher 18:2 will improve the nutritional quality of the oil. The chemical mutation produced one agronomically weak *B. napus* line with 5% linolenic acid (Rakow, 1973). Based on this the Stellar variety with less than 3% linolenic acid has been developed. Harvesting at yellowish brown and pale yellow stages has been considered good for obtaining higher oil yield and better oil quality (Gambhir *et al.,* 1983).

Table 14: Fatty acaid composition of rapeseed species.

Fatty acid	Conventional Cultivars		Zero erucic acid cultivars
	B. napus	*B. campestris*	
16:0	3.8	3.2	9.2
18:1	11.2	26.6	59.8
18:2	13.7	17.5	19.4
18:3	8.1	8.8	10.2
20:1	9.6	11.8	0.2
22:1	52.3	31.0	0.3

Source: Downey and Robbelen (1989).

Nutritional Quality of Rapeseed-Oil

Erucic acid (22:1 ω 9) present in rapeseed oil appears to be responsible for myocardial fibrosis (Abdul Latif and Vles, 1973) and hypocholesterolemia. The nutritional quality of the oil would be substantially improved if the erucic acid content could be reduced (Kramer *et al,.1983*). The identification of plants with essentially zero levels 22:1 in *B. napus* (Stefansson *et al., 1961)* and *B. campestris (*Downey, 1964) resulted in the development of low erucic acid cultivars. From the point of view of nutritional quality, LEAR oil is considered next only to olive oil, due better balanced fatty scid composition. It has been reported to reduce total plasma and LDL cholesterols and hence LEAR oils are gaining wide acceptance.

Fatty acid composition of thirty three genotypes of *Brassica campestris* var. *toria* indicate considerable genetic differences for all the six fatty acids namely palmitic, stearic, oleic, linoleic, linolenic and erucic acids. The oil content is around 41% with a range of 39-42%. The major fatty acid, erucic acid has a range of 50-56%. There is negative correlation between erucic acid and oleic, linoleic and linolenic acids. This is of great significance in breeding with very low erucic acid content for improving the nutritional quality of rapeseed- mustard oil (Paranjit Singh *et al.,* 1995).

Table 15: Oil content and fatty acid profile of Raya, Toria and Gobhi sarson cultivars.

Variety	Oil %	16:0	18:0	18:1	18:2	18:3+20:0	22:1
Raya	41.8	2.3	1.1	9.0	13.8	16.7	51.1
Toria	41.7	2.0	1.9	11.8	13.4	16.5	50.3
Gobhi sarson	44.6	3.8	1.1	17.2	13.0	18.9	42.9
CD at 0.05	0.45	0.07	0.14	0.93	0.35	0.94	1.29

Source: Ahuja *et al.*, (1990).

Rapeseed oil has good digestibility (99%). Not much difference has been observed regarding health aspects when human subjects have been fed HEAR or LEAR oils. This indicates that at normal levels HEAR oil is safe and apprehensions regarding cardiac problems may or may not be true (Kramer *et al.*, 1983). At a somewhat low level of consumption of rapeseed oil prevailing in India, it does not appear to pose any health problem (Shenolikar *et al.*, 1976). Primates have more tolerance to eurcic acid. However it is better, zero erucic oils are fed to infants and young animals.

Table 16: Fatty acids in Brassica campestris var.toria.

Component	Mean	Range
Oil content % (seed)	41.12±0.42	39.45 to 42.48
Palmitic acid %	3.56±0.02	2.17 to 6.88
Stearic acid %	0.79±0.05	0.35 to 1.69
Oleic acid %	12.90±0.03	10.17 to 18.11
Linoleic acid %	14.82±0.29	10.46 to 17.78
Linolenic acid %	12.83±0.28	8.47 to 17.44
Erucic acid %	55.11±0.50	50.13 to 59.99

High ereucic acid in mustard oil is responsible for early myocardial lipidosis and the low erucic acid rapeseed oil is devoid of these short term deleterious effects on the myocardium. LEAR oil is similar in quality to soybean, safflower and sunflower oils with respect to reduction of plasma cholesterol and LDL cholesterol. Conola oil is considerd next only to olive oil because of its higher oleic acid.

In India some low erucic acid varieties have been identified. Varieties like 24-254, 43-56 and 51-64 have been found to contain 13.1 to 13.9% erucic acid as against levels upto 45% in standard cultivars. The fatty acid composition of low and high erucic types is presented in Table 17.

Table 17: Fatty acid composition of low and high erucic Brassica speices.

Species and Selections	16:0	18:1	18:2	18:3	20:1	22:1
B. napus						
Cultivars	3-4	8-23	11-16	6-11	6-14	41.54
Low 22:1 Sel	4-5	23-40	24-31	10-15	0-19	0-11
B. campestris						
Cultivars	2-3	9-34	11-18	6-12	8-12	24-61
Low 22:1 Sel	4-7	48-55	27-31	10-14	0-1	0
B. juncea	2-4	7-22	12-24	10-15	6-14	18-49

Source: Ahuja, (1986).

Application of 2: 4D and maleic hydrazide and incresel decreases 22:1 levels. A line containing 35% 18:2 and 15% 22:1 has been developed at Ludhiana (Brar, 1980). Some of the mutants of *B. Juncea* have the following composition (Ahuja *et al.*, 1984); oil content, 37.6-43.9%; palmitic, 1.8-4.6; oleic, 8.6-17.2; linoleic, 11-8.23; linolenic, 3.3-10; eicosenoic, 5.8-12.6 and erucic, 42.8-55.6. Studies carried out in India have not revealed any adverse effects of mustard oil (Shenolikar and Tilak, 1980).

Industrial Uses of Oil

The oil from high erucic types enters the industrial oil market where 22:1 is valued as a slip agent for plastic manufacture. Erucamide is the main product derived from erucic acid. It is useful as a anti-slip, antiblock and printing ink additive. The long chain fatty acids have uses in the cosmetics and pharmacentical industry. In recent years rapeseed oil is finding use as a biodiesel fuel (Korbitz, 1995). Negative correlation of other fatty acids with 22:1 has been considered to be useful in breeding varieties for industrial (high 22:1) and edible (low 22:1) purposes (Ahuja *et al.,* 1984).

Erucic acid on ozonolysis gets converted to to brassylic acid and pelargonic acid which are useful as plasticizers and lubricants. Brassylic acid is used in manufacturing polyamide, polysters, polyurethanes, fibres, adhesives, resins, surfactants, photographic and recording materials.

Unsaponifiables

Rapeseed-mustard oil contains 1.3-1.5% unsaponifiables. The sterol content ranges from 0.53 to 0.98% of oil. Brassicasterol, a C28 sterol is characteristic of *Brassica* oils. The major sterols however are sitosterol (47%) and campesterol (36%), followed by brassicasterol (13%). The total sterols are 8.8 g/kg oil and the esterified sterols are 4.5 g/kg oil. Plant sterols have some role in lowering serum cholesterol (Subbaiah, 1973). Total sterol content of rapeseed-mustard is around 800mg/100g, with free sterols at around 325 and bound sterol at 475mg/100g oil. Sitosterol (400), campesterol(300) and δ^5 avenasterol (100mg/100g oil) are the main sterols of rapeseed-mustard oil. *Eruca sativa* contains 59% beta sitosterol, 24% campesterol, 11% brassicasterol and 6% stigmasterol (Srinibas *et al.,* 2001).

Rapeseed oils, high as well as low in erucic acid, contain higher levels of gamma tocopherols (426-431 mg/kg) than alpha tocopherols (160-292mg/kg) (Ackman, 1983). Carotenoids are present at 100 μg/g in the oil. Chlorophylls are also present at 500-800 μg/g fresh matter while, they can come down to 20 μg/g at full seed maturity (Uppstrom, 1995). The gums namely, lecithin from rapeseed are bitter in taste and hence commercially do not have much ulility. The sulphur content of the oil range from 5-35 mg/kg oil.

Rapeseed Cake and Proteins

The protein content of the seed ranges from 11-42% (King *et al., 1977*). The oilseed meal contains 36-38% protein, 12% crude fibre and 7% ash. Cold pressed rapeseed cake contains less protein at around 30% with sugar content of 28-31% and 12% crude fibre (Leming and Lember, 2005). Such cakes have higher metabolizable energy of 14.5 MJ/ kg due to higher oil content. The composition of rapeseed meal is presented in Table 18.

Table 18: Composition of rapeseed meals.

Constituent %	*B. napus*	*B. campestris*
Protein (N x 6.25%)	43.9	35.2
Oil	6.4	7.0
Ash	5.9	6.8
Crude fibre	13.7	15.5

Source: Jones, (1979).

Protein content of Indian mustard varies from 22-43% (Gupta *et al., 1982*). The *Brasisca* species contain mainly two proteins, 12S and 17S (Bhatty *et al., 1968*). The 12S globulin (cruciferin) accounts for 60% while 17S protein (napin) accounts for 20% of the total proteins. The amino acid composition of rapeseed protein has been analysed and reported (Gupta *et al., 1982,* Janicki *et al., 1972*). It is balanced with respect to most of the amino acids. It is richer with respect to methionine and cysteine. When compared to FAO/WHO reference protein, the rapeseed protein is deficient in methionine, phenylalanine and isoleucine.

Table : 19: Amino acid composition of rapeseed protein.

Amino acid	g/16g	Amino acid	g/16g
Lysine	5.43	Alanine	5.07
Histidine	3.47	Cysteine	1.13
Arginine	6.40	Valine	5.61
Aspartic acid	8.29	Methionine	0.39
Threonine	4.94	Isoleucine	4.36
Serine	4.91	Leucine	7.76
Glutamic acid	18.10	Tyrosine	2.37
Proline	6.99	Phenylalanine	3.93
Glycine	5.33		

Source: Janicki *et al.*, (1972).

The tryptophan content of seed and meal is 2.22% and 1.3% respectively (Pennacchiotti *et al., 1973*). The content of methionine ranges from 652 to 1152 mg/16g N. Cysteine and cystine in seeds range from 775 to 1116 and 2870 to 3661 mg/16 gN, respectively. Allylisothiocynate ranges from 0.405 to 0.423% (Dixit and Shukla, 1990).

The seed cake is at present used mostly as fuel or fertilizer and to a small extent as livestock feed. Prasad and Rao (1983) have reported higher energy values of rapeseed meal for pigs than for poultry. Protein concentrate from mustard meal has a higher PER than casein (Shenolikar, 1985). Rapeseed meal is a protein supplement. Reduction of hull content improves the meal quality. Dehulled rapessed meal contains 43-45% protein with a digestibility of 81-86% as against 76% for the normal meal.

Carbohydrates

Defatted dehulled meal contains about 48% carbohydrates. The fraction soluble in 80% alcohol comprises 14% with 5% low molecular weight sugars and 29% residual fraction (Siddiqui and Woods 1977). The composition of carbohydrates in *Brassica* species is presented in Table 20. The main sugars are sucrose, stachyose, raffinose, glucose and fructose. The raffinose and stachyose contents in the rapeseed meal are higher. The sucrose content is lower (1.7%) while the fructose content is around 2.1% in *B. Juncea* types. The raffinose content is higher even in *B. Campestris*. Forunately mustard seeds are not directly edible and hence it may not pose much problem relating to flatulence or taste properties.

Table 20: Carbohydrates of rapeseed-mustard.

Component	*B. Juncea*	*B. campestris*
Total carbohydrate%	36.7	39-48
Free sugars%	12.6	5-10
Polysaccharides%	2.2	4.5
Pectic substances%	6.5	14.5
Cellulose%	2.8	7.0
Composition of sugars	(as %)	
Stachyose	2.1	1.52-2.43
Raffinose	1.2	0.31-0.34
Melibiose	0.68	-
Sucrose	1.68	2.26-7.49
Galactose	1.80	-
Glucose	0.96	0.28-0.40
Fructose	2.10	0.15-0.51

Source : Sindhu Kanya and Kantharaj Urs, (1983), Siddiqui and Woods, (1977)

Minerals and Vitamins

Rapeseed cake is rich in calcium, phosphorus, potassium and sulphur. Among vitamins choline is present at 6700 mg/kg oil. Niacin is present at 160 mg/kg oil. Tocopherols (vitamin E) are present at 14 mg/kg oil. Other vitamins are lesser in quantity in comparison to seeds like groundnut. The mineral contents of rapeseed are better than that for soybean meal. The mineral and vitamin contents of rapeseed meal are presented in Table: 21.

Table: 21: Rapeseed meal-minerals and vitamins.

Mineral	%
Calcium	6.64
Phosphorus	1.03
Potassium	1.24
Sulphur	0.86
Magnesium	0.52
Iron	0.144
Manganese	0.05
Zinc	0.07
Vitamins	**mg/kg**
Choline	6700
Biotin	1.1
Folic acid	2.3
Niacin	160
Pantothenic acid	9.5
Riboflavin	5.8
Thiamine	5.2
Pyridoxine	7.2
Vitamin E	14.0

Bell and Keith (1991) and Bell, 1993.

Compared to other oilseeds, such as groundnut and safflower, mustard seed is found to be rich in calcium, magnesium, iron, zinc, manganese, lead and chromium. Rapeseed contains S at 6.5mg/kg of oil due to the presence of glucosinolates. The dehulled, defatted meal samples of rapeseed varieties are more concentrated in minerals and trace elements. Iron content of meal samples on dry weight basis varies from 17.0 to 27.8 mg per 100g meal and zinc content from 60 to 141 μg per g. Bioavailability of the rapeseed meal iron as judged by the ionisable iron content expressed as percent of total iron (49-57%) is found to be very high. The rapeseed oil contains 1190 mg/kg P, 3.52 mg/kg Fe, 6.5 mg/kg S and some minor quantities of Zn and Pb also.

Antinutrients in Rapeseed

There are two problems in utilization of rapeseed meal though it is balanced with respect to amino acids. First, the fibre content is high and second it contains toxic factors mainly glucosinolates. The glucosinolates and their breakdown products have been reported to cause enlargement of thyroid, adrenal gland, kidney and liver haemorrhage. In the presence of moisture, the enzyme myrosinase hydrolyses glucosinolates to glucose, sulphate, isothiocyanates etc. The hydrolysis products like ally1 iso-thio cyanate and 3-buty1 iso-thio cyanate are responsible for the characteristic odour and pungent taste of mustard oil. *Ghani* pressed mustard oil is generally preferred by certain population groups in India as it has higher pungency and better flavour. Ahuja *et al* (1975) have reported a non-chemical method to obtain a pungent mustard oil.

Glucosinolates are thio-esters of glucose or other sugars. About 90 different glucosinolates have been isolated from plants of the *cruciferae* and related families. Normally they are not non-nutritional. They become antinutritional after hydrolysis by thioglucosidases, which are present in the plant. Rapeseed itself contains about 16 different glucosinolates.

Seven glucosinolates have been identified in rapeseed (Anand *et al.*, 1979 and Jones, 1979). Of these, 3 buteny1, 4 penteny1 and 2-hydroxy-3 buteny1 glucosinolates are the most important. The total glucosinolate content ranges between 1-3%.

Table 22: Glucosinolate levels of Brassicas.

Glucosinalates	mg/g meal	
	B.napus	*B. campestris*
Buteny1 iso thio cyanate	2-7	3.0
Penteny1 iso thio cyanate	6.8	2.3
Viny1 oxazolidine thione	8.7	2.3

Source: Jones, (1979).

One of the major problems associated with rapeseed-mustard is the presence of allyl-isothiocynates varying from 0.6% in Bronowski to 16.2 percent in Target cultivars (Weiss 1983). Although most of the glucosinolates remain in the rapeseed meal, the oil also contains a small proportion of allyl-isothiocynate and glucosinolates which imparts characteristic pungency and odour to the oil. The high content of glucosinolates in the meal makes it unfit as animal feed also. Feeding trials conducted have shown convincingly the deleterious effects (Bijster *et al.*, 1978; Vermorel *et al.*, 1988; Buchmann and Wenk, 1989).

Although several attempts have been made to develop processes that could eliminate thioglucosinolates, as yet no economic process is available. There is a vast scope for isolating proteins from meals since it is extremely rich in protein. It has been suggested that heat treatment of meal could inactivate myrosinase and prevent production of nitriles. However, even after heat process, the meal is not suitable as animal feed. An alternate approach has been the alkali extraction of proteins which, however, is not economical. Differences in the thioglucosinolate content in different cultivars/accesions offers scope for breeding rapeseed mustard with low thioglucosinolate content.

The variety RSL 1 contains a lower level of 1.5% glucosinolates (Gupta *et al, 1988)*. Simple methods to detoxify the cake have been developed, which involve essentially thermal deactivation of the enzyme (Kantharaj urs *et al., 1979;* Ahuja and Sherry, 1979). Plant Breeders have developed varieties with low or zero levels of glucosinolates. Similarly, double zero (low glucosinolate and low erucic acid) and triple zero types (with low cellulose, glucosinolate and erucic acids) have also been developed.

Fibre Content

The content of crude fibre, mainly cellulose and hemicellulose amounts to about 15% in the defatted seed meals. This value is high compared to that, of soyabeans. Crude fibre (insoluble polysaccharide) is mainly concentrated in the hulls which account for 12-20% of the seed weight depending on variety and shape of the seed. The large amounts of crude fibre in rapeseed meal are a consequence of the small seed size with large proportion of seed coat rich in fibre (Appelqvist, 1972). High content of rape and mustard meals contribute to the overall reduction of the feed value. The cellulose of the white mustard seed hulls appears to be unique and responsible for the thickening power of the ground hulls. Sosulski (1981) has observed that defatted rapeseed meal has 40% protein with 11.5% crude fibre whereas yellow mustard has 47.1% protein and 7.7% crude fibre content.

The higher proportion of hull contributes to higher fibre in rapeseed meal and results in a metabolisable energy (ME) value for animals that is lower than that of soyabean meal (Claudinin and Robbelen, 1981). Yellow seed cultivars are lower in crude fibre content and contain less lignin than the dark hulled cultivars (Sosulski, 1979). It has been shown that protein digestibility in poultry and pigs gets reduced when dark coloured hulls are included at a high level in the diet (Mitau *et al.*, 1983). The chemical compostion of the hull fibre fraction is important since fibre containing lignin has an adverse effect on protein and amino acid digestibility in non-ruminant animals. It is suggested that the lignin content of hulls should be minimised. This would entail selection of yellow-hulled varieties which have the added advantage of a low fibre content.

Phenolic Constituents

Phenolic components present in rapeseed and mustard meals contribute to the specific, sharp, bitter flavour of rapeseed flour and also produce undesirable dark coloured complexes when oxidized (Fenton *et al.*, 1980). Principal phenolic components that have been identified are p-hydroxy benzoic, trans-cinnamic and sinapic acids. Small quantities of p-coumaric, ferulic, caffeic and chlorogenic acids are also present in dehulled seeds (Kozolowska *et al.*, 1983). Rapeseed-mustard has sinapine content ranging from 0.49 to 0.9% on fat free dry weight basis (Kerber and Buckloh, 1982). Yellow seeds have low phenolic content. In *s. alba*, p-hydroxy benzyl glucosinolate and to a lesser extent kaempferol glycosides bound to the plant matrix and probably a glucose are present. These compounds cause fishy taint to egg yolk of certain brown-egg-laying hens fed on rapeseed meal (Butler *et al.*, 1982). This is on account of the interaction of choline, a genetic factor in the hens which adversely affects the synthesis of trimethyl amino oxidase and a substrate in the rapeseed meal which drastically reduces the activity of hepatic trimethyl amino oxidase. Yellow seeds have low phenolic content.

Phytic Acid

Rapeseed and mustard meal contain phytic acid ranging from 4-6 percent. Excess phytic acid interferes in the metabolic activities of essential trace elements like zinc and copper (Nwokolo and Bragg, 1977). This is due to strong chelating action of phytic acid which can bind mono and divalent metal ions to form complex phytates (Erdman and Forbes, 1977).

In view of some of the antinutrients present, rapeseed meal is not considered a good feed material, inspite of higher levels of protein and also a balanced amino acid make-up.

Food Uses

Rapeseed oil is a widely used cooking medium all over the world. Due to its very good shelf life, mustard oil is utilized in preparation of all types of pickes in India. Whole seed mustard powder also is an essential ingredient of good quality pickles. The seeds are regularly used in seasoning curries in India. This is done mainly to add flavour to the preparations.

Safflower

Safflower (*Carthamus tinctorius* L.) is a minor oilseed crop of the world. India is the largest producer of this crop. It is famous not only for its seed and seed oil, but also for its brilliantly coloured petals as a source of orange red dye. Safflower is believed to have originated in the Mediterranean and Persian Gulf. It later spread to Egypt, Ethiopia, Europe and Asia. Safflower belongs to the *Asteraceae* family. It is cultivated on a commericial scale in India, U.S.A Mexico, Ethiopia, Australia, China, Argentina, and Russia and to a limited extent in Pakistan, Italy, Spain, Portugal, and Iran.

Dried red or orange safflower petals are a substitute for saffron. They have recently gained importance in medicine for healing many ailments. The highly unsaturated seed oil initally has been used in manufacturing paints and varnishes. Recently its high content of linoleic acid, an essential fatty acid has made it the most nutritious edible oil. Similarly, all other parts of the plant have a lot of uses.

Safflower is a perfect dryland crop with an admirable productivity under limiting moisture conditions. It is a low risk crop with less incidence of diseases and pests. Each plant can produce up to 100 seeds which are white in colour. In India it is grown mainly in the *Rabi* season, namely September/ October to March-April in the states of Maharastra, Karnataka and Andhra pradesh. It is day neutral but thermosensitive. It comes up well in drier areas in fairly deep and well drained soils with a pH of 5-8. It is generally grown as an intercrop with other rainfed crops like wheat, sorghum, coriander, chickpea, linseed and mustard or as a sole crop under receeding moisture conditions. It is mostly a self pollinated crop.

Seed rates ranging from 7.5 to 20 kg are practiced depending on spacing from 45x20cm to 30x15 cm in different soils and locations. FYM at 5-10 tonnes are applied followed by 20-50 kg N, 15-50 kg P_2O_5 and 10-30 kg K_2O. Life saving irrigation at 5-8 cm is given at critical stages. The crop comes to harvest in 120-180 days depending upon genotype and location or soil type.

The crop is harvested when the leaves turn brown. Plants are threshed by beating with sticks. Seed yields ranging from 800-1200 kg/ha are normally obtained. Higher yields of 1500-2000 kg/ha are possible under favourable moisture conditions. The seeds contain 30-40% hull and 28-32% oil. The kernel contains 50-55% oil. The cake, preferably, the decorticated seed cake is useful as animal feed. The spiny nature of the crop is coming in the way of its wider cultivation.

Seed Composition

Safflower seed has a high proportion of hulls which has an adverse influence on it's total oil and protein content. The seed characteristics are presented in Table 23. The seeds generally contain upto 45% hull and 28-32% oil, with about 15% protein. The ash content ranges from 3-7%. The crude fibre is present upto 40% which is mostly concentrated in the hull portion. The hull content is highly and positively correlated with protein, fibre and sugars but negatively correlated with seed oil. The oil is negatively correlated with fibre and protein indicating that thin hulled types can contain more oil (Suman kumar *et al.*, 2002).

Table 23: Characteristics of safflower seed.

Component	% constituent	
	(1)	(2)
Hull	33-45	-
Kernel	55-65	-
Moisture	7-8	5.3
Oil	31-32	27.5
Protein	14-15	14.9
Ash	5-7	2.3
Crude Fibre	32-34	40.6

Source: (1) Weiss (1983); (2) Lata and Prakash (1984).

Traditional safflower varieties generally contain about 50% each of hull and kernel only. Oil, protein and ash are higher in kernel while crude fibre is higher in the hull. Thin hulled varieties contain upto 55-70% kernel with an oil content of 35-45% and 17% protein (Kulkarni *et al, 1985*). The hull or pericarp contains 70% cellulose, 21% lignin and 1% ash. A completely dehulled kernel contains 60% oil and the decorticated meal contains 60-70% protein (Kulkarni, 1978). A decorticator working on impact dehulling has been considered to be highly useful for decortication of safflower seeds (Shamantaka Sastry, 1989). The partial hull type identified by Urie (1986) may result in the development of cultivars with oil content upto 50%. Knowles (1969) has reviewed the variation in chemical composition of different safflower genotypes having different combinations of genes namely Th and Stp.

Table 24: Seed composition of safflower genotypes.

Genotype	Hull colour	Hull %	Oil %	Protein %	Fibre %
Th Th Stp Stp	Smooth White	40	39	20	41
th th stp stp	Very thin Grey	20	47	40	17
Th Th stp stp	Brown Striped	25	46	34	25

Source: Knowles (1969).

Seven safflower varieties namely A1, A300, Bhima, HUS 305, JSF1, Manjira and SI44 grown during *Rabi* 1990-91 and 1991-92 seasons have been examined for their variation in oil content, when grown at seven locations in India namely Annigeri, Dhandhuka, Hyderabad, Indore, Phaltan, Raichur and Sholapur (Nagaraj, 1994a). Among the genotypes the thin hulled type HUS 305 has higher oil content of 34.1% followed by S 144 (31.6%) and Bhima (30.5%). The most popular and India's national variety A1 has the lowest oil content of 28.0%. The other varieties have oil in the range of 28-30%. Among locations, Hyderabad and Raichur have given higher oil content of about 32% while Annigeri and Indore have shown lower levels of around 29% with the other locations having oil content of around 30%.

Thirteen released Indian safflower varieties have been analysed for oil, protein, soluble sugar and fatty acid profile (Nagaraj, 1994b). The ranges of the constituents are oil (25.3-36.2%), protein (9.7-18.6%) and soluble sugars (10.6-14.8%). The major fatty acid is linoleic acid (72-78%) with oleic acid levels of 14-18%. Among the varieties studied S144, JSF1, BLY 652 have shown higher oil yield while A1 and Manjira have given higher protein yields. A1, JSLF 113 and JSLF 118 have higher oleic acid with better oil stability while JSF1, A300, HUS 305 and S 144 have higher linoleic acid and better nutritional quality.

Another set of twenty four safflower genotypes have been analysed for oil, protein and fatty acid composition. The oil and protein contents range from 25.1 to 34.1%, and 8.6 to 16.2% respectively. The major fatty acid is linoleic acid (72-80%) followed by oleic acid (11-19%). Palmitic (6%) and stearic(2%) acids are the other two fatty acids of safflower oil. In general safflower genotypes have lower oil stability values (0.14-0.25) with higher nutritional quality indices (7.7-10.3), indicating the superiorty of safflower oil. The iodine values of all the genotypes are also higher (126-138) again because of the higher linoleic acid levels in the safflower oil (Nagaraj, 1994b).

The oil content of safflower based on analysis of 50 lines has been found to be 30± 2% with a range between 21.2 to 40.1% (Kumar *et al, 1982).* Varieties from Pakistan have an oil content of 29.7 to 32.0% oil (Raie *et al, 1985).* High oil percentage in safflower depends on 1) thin hulls, (2) high seed number/ plant and (3) high oil content of seed (Sangale *et al., 1982*). Seeds of primary branch have higher oil content than those from secondary branches (Metha *et al., 1979*). However, early flowering heads have lower oil content than later flowering heads. Oil content reaches a maximum at 35 days after flowering (Abel-Rehman and Yousuf, 1987*)* while oil synthesis has been higher at 32^{0}C than at 10^{0}C (Ichihara and Noda, 1980*).*

Nitrogen application decreases the oil content in safflower (Sharma and Verma, 1982) while K application could increase the oil content from 24.0 to 26.0% (Sayed and Al-Sayeed, 1978*).* Oil content is dependent on number of achenes followed by 1000 achene weight and number of heads (Alba and Greco, 1977) Gene action has been considered to be predominantly additive for oil content, seed weight, seed number, days to flowering and plant height. Breeding methods, such as biparental mating followed by reciprocal recurrent selection, have been suggested for the simultaneous improvement of seed yield and oil content (Ramachandram and Goud 1981). The oil is obtained by crushing the seed in bullock or power operated *ghanies* in India. Partially refined and degummed oil is used for edible purposes.

Oil Extraction

Safflower oil is normally expressed using rotary expellers, the resulting cake contains around 6% oil. Solvent extraction has been considered unsuitable, due to the floating of hull material (Knowles, 1955). *Nutan (new) ghani,* is better and more efficient than the traditional *ghani* for oil extraction (Datar, 1951). Processing of safflower seed prior to oil extraction like decortication, storage of seed and good quality oil and protein cake have been examined by Sastry (1991). It is sometimes extracted by dry distillation. The seeds are put into an earthen pot, inverted over the mouth of another pot covered with a perforated pot and buried in the ground. The upper pot is heated by buring a heap of fuel. The seeds get roasted and the oil trickles down. Such oil may not be of good quality and is only useful for non-edible purposes.

Oil Quality

Safflower oil is a drying oil with high levels of linoleic acid (75-80%). It forms a fast drying, non-yellowing film with low wrinkling characteristics. However, in view of its reported role in reducing blood cholesterol levels, it is now-a-days being utilized as an edible oil. The characteristics of safflower oil are presented in Tables 25 and 26.

Table 25: Safflower oil quality.

Constituent	Range
Free fatty acids	0.15-0.01%
Iodine value	136-146
Saponification value	186-195
Refractive index at 25^{0}c	1.473-1.476
Unsaponifiables	0.3-1.3%

Source: Vibhakar *et al.,* (1981).

Among edible oils, safflower oil has higher a iodine value, indicating its drying characteristics. It is next to linseed oil with respect to it's I.V. Freshly extracted oil has very low levels of free fatty acids. It's FFA increases an storage. The unsaponifiables are also low copaired to other oils.

Table 26: Fatty acid composition of safflower oil.

Fatty acid	% composition
Palmitic (16:0)	5.0 - 7.5
Stearic (18:0)	1.4 - 1.8
Arachidic(20:0)	1.2 - 3.6
Oleic (18:1)	6.1 - 7.6
Linoleic (18:2)	74.9 - 78.4

Source: (1) Daulatabad *et al.,* (1981) (2) Nagaraj (1994).

Safflower oil is the richest source of linoleic acid (75-78%), an essential fatty acid. Its oleic acid content is around 6-8%. The total unsaturated fatty acids thus account for about 85%. Palmitic acid is present from 5- 8% along with stearic and arachidic acids at 2% each. The total saturated fatty acid content is normally around 10% and less. The low sturated fatty acids especially the palmitic acid and the high linoleic acid (EFA) make safflower oil the best nutritions edible oil.

Safflower varieties have a fatty acid composition of myristic 0.9-3.1, palmitic 9.4-12.0, stearic 2.3-5.5, oleic 14-15 and linoleic acid 65.9-73.4% (Raie *et al., 1985*). The effect of five locations in India on the composition and quality of two varieties has been examined (Muralidharudu *et al.,* 1989). Cooler locations have higher 18:2 (77%) than other locations (72%). Oleic acid content ranges between 13.7 and 19%.

Standard varieties of safflower are rich in linoleic acid, which makes safflower oil highly unstable and unsuitable for long term storage and repeated fat fryings. A variant type rich in oleic acid has been discovered and reported (Horowitz and Winter, 1965, Urie *et al.*, 1979). The oleic acid content is 67% and linoleic acid content is 25%(Corelto *et al.*, 1980). Among six cultivated varieties Fenix has highest oleic acid (20.8%) and lowest linoleic acid (67%) while Briggs has 83.5% linoleic and 8.1% oleic acids (Masson *et al.*, 1981). Oxidative stability of the oil is inversely related to linoleic acid. High oleic types have better oil stability (Purdy, 1985).

During seed development linoleic acid predominates in every lipid class while 18:3 decreases with increasing maturation and is absent in fully mature safflower seeds. The saturated fatty acids occupy 1 and 3 positions on glycerol molecule while 99% of 2 position is occupied by 18:2 (Ichihara and Noda, 1980 and Mazuelosvela, 1981). The occurrence of 18:2 at the second corbon of glycerol moiety, imparts some minimal oxidative stability to the oil and also makes the 18:2 easily absorbable by the human digestive system. Based on the compositional variations among oleic and lionleic types, Knowles (1989) has suggested breeding of thin hulled safflower varieties along with high oleic types for better oil stability.

Wild safflower (*Carthamus oxycantha*) seeds have been analyzed for oil (28%), protein (17%) and fibre (36.2%) contents (Banerji *et al.*, 1999). The oil is rich in linoleic (70%) and devoid of linolenic acid, which is similar to that of cultivated safflower. Among the wild species *C. Oxycantha* has 57.2% linoleic acid while *C. palaestinus* and *C. lanata* have 68 and 61% linoleic acid (Nagaraj, 1997).

Unsaponifiables

Fatty acid distribution and composition of non-saponifiables including tocopherols has been reviewed by Sonntag (1979). The tocopherol levels are as follows: alpha=223, beta-7.0; gamma=33 and delta tocopherol=3mg/kg. Indian safflower oil samples contain tocopherols: alpha=450-486, beta=44-56, gamma=46-64 mg/kg oil (Nagaraj *et al.*, 2001). The major constituents of unsaponifiables of safflower oil are sterols (62.5%), 4-methyl sterols (7.5%), triterpene alcohols (13.9%) and hydrocarbons (12.6%). The sterols are composed of cholesterol, campesterol, stigmasterol, alpha-sitosterol, 5-

avenasterol, alpha-spinasterol, 7-stigmasterol and 7-avenasterol. The triterpene alcohols consist of cycloartanol, cycloartenol, 24-methylenecycloartanol and 24-cyclobranol. 4-methylsterols consist of obtusifoliol, gramisterol and citrostadienol. (Tao-Hai Rong *et al.*, 2000).

Oil Cakes and Protein

The safflower meal is high in fibre and protein. Its protein is low in lysine. It is mostly used in ruminant rations. Decorticated safflower seed cake can be considered as a high protein diet for humans. However, much work needs to be carried out on this aspect. Different grades of safflower cake have been examined for their composition and as feed material for chicken (Raj *et al., 1983*).

The protein content which is around 18% in the seed, gets concentrated to 24.5% and 28% after expeller and solvent extraction of the seed. Expeller cake contains upto 17% oil while it becomes negligible in the solvent extracted meal. Sugar and crude fibre increase in the meal while the energy content decreases due to reduction in oil content.

Table 27: Composition of safflower seed and oil cakes.

Constituent	Seed	Rotary pressed meal	Solvent extracted meal
Protein %	18.5	24.5	28.0
Oil %	33.9	16.7	0.7
Crude fibre %	29.8	21.3	34.3
N-free extract%	15.5	34.2	31.9
Calcium %	0.3	0.3	0.5
Phosphorus %	0.2	0.3	0.4
Metabolizable energy (kcal/kg)	3804	2506	1470

Source: Raj *et al.*, 1983.

Safflower meal could be included upto 5% of the diet of chicken. With lysine and methionine supplementation, it could be increased to 20% without significantly affecting weight gain or efficiency of feed conversion (Raj. 1980).

Table 28: Composition of safflower cake.

Constituent	Composition
Moisture %	7-8
Digestible energy (Kcal/g)	1.59
Fibre %	31.0
Fat %	6.0
C.P.%	19.7
Ca %	0.2
Available P %	0.6
Lysine %	0.5
Methionine %	0.1
Cysteine%	0.3
Available amino acids%	65

Source: Weiss (1983).

Protein isolates with protein content of 87-96% have been extracted from the meal with NaOH and by precipitating with HCl at pH 5 or 6. Lysine is a limiting amino acid and N-digestibilities are 75-85 in the meal and 90-96% in isolates. Processing upto 110^{0}C does not affect the PER of the meal (Betschart and Saunders, 1978). The protein precipitated at pH 6.0 is more soluble in acidic pH and is better suited in the preparation of wheat flour breads. Incorporation of 5 and 10% safflower protein isolate into breads has resulted in increases of approximately 25 and 30% in protein respectively (Betchart *et al, 1979*). Protein isolates prepared through micellization and isoelectric precipitation techniques have been examined for their properties by Parades and Odorica (1986). Isolates prepared by micellization are more suitable.

Solvent extracted meal has 28% protein, 0.7% oil, 34% crude fibre and 32% sugar while the rotary pressed meal has 25% protein, 17% oil, 21% crude fibre and 34% sugar. The metalolizable energy is higher in the latter (2506 K cals/kg) than in the fat free meal (1470 K cals/kg). Lysine and methionine supplementation, is necessary to improve the safflower feed quality (Raj, 1980).

The excessive hull content of safflower meal reduces its nutritive value. The meal is graded into two fractions namely low fibre and high fibre meals. The former is used for feeding monogastric animals while the latter is used for feeding ruminants. The low fibre meal contains 42% crude protein and 16% crude fibre while the high fibre meal contains 20% curde protein and

30% crude fibre (Vohra, 1989). As a manure the safflower cake is very useful, however, it is slower to decompose than groundnut (Daji, 1943).

Nutritional Quality

Safflower protein like any other vegetable protein is deficient in lysine. The amino acid composition is presented in Table 29. Arginine is the major amino acid followed by leucine, valine phenyl alanine, isoleucine, threonine and lysine. It is better than groundnut with respect to its arginine which has a role in wound healing and reduction of LDL cholesterol. In conjunction with it's high EFA the seeds should be considered highly nutritious. However it is not edible due to other reasons.

Table 29: Amino acid composition of safflower meal.

Amino acid	% in air dried meal
Arginine	3.8
Cysteine	0.5
Histidine	1.0
Isoleucine	1.7
Leucine	2.7
Lysine	1.3
Methionine	0.7
Phenylalanine	1.9
Threonine	1.4
Tryptophan	0.6
Valine	2.4

Source: Weiss (1983)

Safflower meal is mildly cathartic and is bitter to taste. The cathartic properties are due to the presence of 2-hydroxy arctin beta glucoside and the bitterness caused by meta-i-resinol beta glucoside. Both can be extracted with water to obtain a desirable product (Lyon *et al., 1979)*. The following new compound namely N-Feruloyl and N-(P-coumaroyl) serotinine and (P-coumaroyl) serotinin mono-β glucosides have been isolated from the meal (Sukumara *et al., 1980*).

The total phenolics in fresh samples are less than 100 mg/100g while it is around 2% in stored samples. Studies have been carried out for removal/ reduction of phenols using many plysico-chemical treatments like drying, autoclaving, alkali solutions, organic solvens etc. None of them are very effective and reductions upto 20% only could be achieved (Nagaraj, 1999 and Praveena *et al.*, 1998). Other anti-nutrients like oxalic and phytic acids have been found at 10 and 0.3 mg/100g of seed respectively (Nagaraj, 1999).

Carbohydrates

Safflower seeds, hulls and kernels have available carbohydrates at a level of 4.5%, 3.0% and 2.8% respectively while the unavailable carbohydrate is 34.8%, 88.3% and 1.2%. The carbohydrate composition of safflower meal is presented in Table 30.

Table 30: Safflower seed carbohydrates.

Constituent	Deoiled kernels	De oiled hulls
Crude fibre	5.7	49.1
Lignin	3.2	24.0
Pentosans	4.9	21.5
Cellulose	2.3	35.0
Soluble sugars and organic acids	19.9	4.6
Raffinose	1.1	0.1
Sucrose	1.4	0.2

Source: Kohler *et al.*, (1966) Saunders (1970).

The hulls are richer in insoluble polysaccharides like cellulose, lignin and pentosans. The deoiled kernel is richer in soluble sugars. Sucrose and raffinose are present at around 1% each in the kernel meal.

Safflower Leaves

Tea made from safflower foliage is used as abortifacient and other remedies in Afghanistan, Pakistan and India (Li Daje, 1996, and Weiss 1983). Safflower seedlings have been recorded to be blended freshly and applied to swellings and fractures in China (Zhang, 1997). Young leaves and thinnings are eaten boiled as a vegetable side-dish with curry or rice in India, Pakistan and Myanmar. Thinnings are high in vitamin A, iron, phosphorus and calcium. The green leaves of safflower contain 3540 μg carotene, 0.04mg thiamine, 0.10mg riboflavin and 15 μg vitamin C per g (Deosthale,1981). Safflower leaves have 28% proteins, 44% sugars, 9% fibre and 12% ash, which is very similar to coriander leaves. (Sahasrabuddhe, 1950, Weiss, 1983).

Safflower is grazed or stored as hay or silage (Zhang, 1997). The forage is palatable and it's feed value and yields are similar to or better than oats or alfalfa. Safflower straw is used similar to cereal straws (Smith, 1996, Wichman, 1996). Two or three rows of safflower around a cereal field can help keep free grazing cattle out of the grain (Chavan, 1961). It is estimated that the safflower crop generates annually around one tonne of stalks per hectare with the total ligno cellulosic biomass of about 7 lakh tonnes per annum. This biomass could help in meeting the acute shortage of ligno cellulosic raw material by the paper and hard boards industry. The proximate analysis of safflower stalks is presented in Table 31.

Table 31: Proximate analysis of safflower stalk.

Contents	%
Mosisture	13.3
Ash	2.9
Hot water benzene solubles	10.7
Lignin content	4.0
Alpha cellulose	14.4
Beta cellulose	14.1
Gamma cellulose	17.7

Safflower Petals and Flowers

Safflower florets have been widely used to colour and flavour soups and rice as well as cloth. Safflower dyes are particularly important to the carpet weaving industries of Eastern Europe, the Middle East and the Indian subcontinent. Carthamin dye has been used extensively to colour cloth until 19th century till cheaper aniline dyes became available. Hebrew writings since 2nd century A. D. have described the use of tablets of carthamin dye for food colouring and medicine. Soot from charred safflower plants has been used to make *kohl*, the Egyptian cosmetic (Weiss. 1983). In Egypt, a dye from safflower petals has been used to colour cotton and silk and as an ointment used in religious ceremonies and to anoint mummies prior to binding.

Additions of safflower florets are well known for their use as food colourant which has been a widespread ancient tradition. Safflower floret pigments are good replacements for synthetic food colour (Kulkarni *et al.*, 2001, Sarojini *et al.*, 1995). Rice, soup, sauces, bread and pickles take on a yellow to bright orange colour from the florets. Cosmetic rouge can be made from carthamin dye mixed with French chalk, and the Japanese cometic (Weiss,

1983) and lipsticks include safflower colouring (Smith, 1996, Zhang, 1997). Addition of 3 per cent yellow pigment extracted from petals of safflower in sweetned yoghurt has more acceptability as compared to a commercial sample (Kulkarni *et al.*, 1997).

Safflower petals contain carthamin and yellow pigments in different proportions depending on the colour of flower (Rudometova *et al.*, 2001). The water soluble yellow dye, carthamidin, and a water insoluble red dye, carthamin, which is readily soluble in alkali, can be obtained from safflower florets (Weiss, 1983). The safflower florets contain carthamidin to the extent of 28-36% and carthamin to the extent of 0.3-.0.6% (Zhang, 1997). Yellow florets contain little or no red dye (Smith, 1996). Mid season pickings from the Dacca area of Bangladesh are considered to be of high quality and yields of 70-140 kg/ha are normal. Florets can be collected after the crop matures, so that dye and oilseed can be obtained from the same crop (Chavan, 1961). Florets contain 0.3 to 0.6% carthamin.

Colouring of 1 kg of cotton yarn crimson requires 1 kg of dye. Rose pink requires 500 g and light pink 250 g (Weiss, 1983). Safflower yellow is present in the dried florets to the extent of 26-36 per cent (Nandini Nimbkar, 2002). The petal powder not only imparts attractive yellow colour to food items but also enhances the nutritive value of the products (Anonymous, 2004). Petals after dye extraction are subjected to bioenrichment employing a microbial consortium through anaerobic fermentaion technique. The bio-enriched product shows a 35% increase in protein content as compared to control. The enriched material can very well be employed in poultry feed ration (Anonymous, 2004).

The safflower corolla contains an yellow pigment upto 25-30% and a red pigment upto 0.5-0.6% (Han Yunzhou, 1997, Jianguo *et al.*, 1993). In China, herbal tea is prepared from safflower petals. The safflower petals contain yellow B and red pigment, safflower factor, safflower glycoside and safflower benzyl glycoside. The petals and dyes taste bitter, warm and non poisonous (Ramanjaneyulu, 2003). Spineless varieties have been used as cut flowers in Western Europe, Japan and Latin America (Li Dajue, 1993). Safflower florets are also nutritious containing about 11% amino acids and many vitamins and minerals. Productivity of petals by different varieties is presented in Table 32.

Table 32: Average petal productivity of differet varieties of safflower.

Variety/hybrid	Petal yield kg/ha
NARI - 6	202
Sharda	84
Bhima	75
JSI 7	169
JSI 97	256
JSI 103	219

Safflower petals, which are presently waste products, are available at around 100kg/ha. Safflower petals have a lot of medicinal and therapeutic values. This crop has been cultivated for a long time as the source of red and yellow pigments which are constituents of safflower petals. The pigment has been utilized as food and fabric colourant until the synthetic dyes pushed them to the background. Health problems particularly, cancer, associated with synthetic azo dyes are likely to further stimulate the utilization of natural chemicals such as the safflower dyes. In the light of this background Srinivas *et al.,* (1999) have examined the safflower petals of Indian origin for their carthamin, oil, protein and fatty acids. The alkali soluble pigment carthamin varied between 0.7% and 1.0%.

Proximate analysis of petals and elemental profile in petals of different varieties and fatty acid composition of petals are presented in Tables 33 and 34. The organic and inorganic constituents together may provide nutritional and medicinal values to petals (Anonymous. 2004, Nagaraj *et al.,* 2001, Srinivas *et al.,* 1999). The concentration of elements in the safflower petals is 5 to 40 ppm Cu; 7-37 ppm Zn which have beneficial effects for various ailments in many of the Indian *ayurvedic* preparations. The total absence of As and Hg with safe levels of Cd, Cr, Ni and Pb enable safflower petals for use in various food preparations (Anonymous, 2004).

Table 33: Proximate composition of safflower petals.

Constituent	Composition (%)	Constituent	Compostion (%)
Moisture	4.9	Decanoic (10:0)	3.9
Protein	1.9	Lauric (12:0)	1.8
Oil	5.0	Myristic (14:0)	3.4
Ash	10.4	Palmitic (16:0)	15.3
Crude fibres	12.2	Stearic (18:0)	3.1
Carthamin	0.83	Oleic (18:1)	7.8
		Linoleic (18:2)	46.7
		Gamma linolenic (18:3)n-6	2.7
		Alpha Linolenic (18:3) n-3	16.8

The fatty acid composition has revealed the presence of low molecular weight fatty acids like 10:0 (decanoic), 12:0 (dodecanoic) and others like 14:0 (myristic) fatty acids in addition to the normal fatty acids namely 16:0 (palmitic), 18:0 (stearic), 18:1 (oleic), 18:2 (linoleic) and 18:3 (linolenic) fatty acids. Safflower petal oil has been found for the first time to contain medium chain fatty acids to an extent of 5.0% & gamma linolenic acid (n-6, 18:3) to a level of 2 to 3%. The main fatty acid is linoleic acid (18:2) with about 40-52% followed by 18:3 with 15-19% and 16:0 with about 14-16%. The unsaturated fatty acids account for 70-75% of the total fatty acids(Srinivas *et al.*, 1999).

The medium chain fatty acids have a useful biological activity. They have minimal contribution to the plasma or adipose tissue lipids. They get oxidized rapidly in the liver and also are antimicrobial. The most important finding of this study relates to the identification of the gamma linolenic acid in the safflower petal oil (Srinivas *et al.*, 1999). This is an essential constituent for the human beings. It plays an importnt role in various biochemical pathways in our body, prevents cracking of skin, skin aging, improves the skin texture etc. It is antifungal, antimicrobial and useful in the treatment of skin infections like eczema (Nagaraj, 2002). Low molecular weight fatty acids are useful in parenteral and infant nutrition while gamma linolenic acid is a health promoting nutrient.

Table 34: Mineral composition of safflower petals.

Element	mg/100g
Ca	530
Mg	287
Fe	7.3
Cu	0.8
Zn	1.0
Mn	0.8

Safflower has a bitter herbal taste but the Institute of Botany of the Chinese Academy of Sciences in Beijing has developed a non-bitter, sweet smelling tea, which contains aminoacids, minerals and vitamins (Weiss, 1971). The main active ingredient in safflower medicines is safflower yellow (carmine), which is water soluble, but alcohol extracts are used in some preparations. Recent pharmacological studies has shown its use in menstrual problems and cardiovascular diseases (Li Dajue, 1996). Knowles (1965) have reported that flowers soaked overnight are applied wet to reduce allergy rashes in Egypt.

Safflower Pollen

The pollen is esteemed in China because it is easily collected and contains many nutrients. A large quantity of the safflower pollen contained in the corolla of safflower, whose extracting rate can be about 4%, has a high nutritional rate (amino acid 19.1%) and microelements (Han Yunzhou, 1997).

Vitamins and Minerals

The green leaves of safflower contain 3540 μg/g carotene, 0.04 mg thiamine, 0.10 mg riboflavin and 15μg/g vitamin C. The ash content of the seed is 2.1 to 3.5% (Deosthale, 1981). Safflower seeds and oil have also been analysed for micro nutrients and vitamins (Nagaraj, 1999). The Zn content ranges from 63-92, Cu content 24-27, Mn content 8-37 and Fe content 103-380 μg/100g. The thiamine content is 500 and β-carotene 750mg/100g. The total tocopherols in the oil is 550-594 mg/kg., with the alpha torocopherol dominating at around 470mg/kg. Gamma trocopherol which imparts high temperature stability is low at around 50 mg/kg.

Table 35: Elemental profile of different varieties of safflower (ppm).

Variety	Cd	Cr	Cu	Ni	Pb	Zn
JSI	0-1	5.6-10.4	20.6-23.2	1.6	12.6-21.6	30-33.6
PBNS 12	0.-0.4	0-22.6	16.4-23.4	3-4.6	0.8-23.4	33-34.6
Sharada	0-2.4	-	14.4-42.2	1.6-3.0	-	28.6-32.4
JSI 97	0-1.6	2.6-8.4	22.4-24.8	2.6-8.4	-	32.4-37.2
JSI 103	-	1.6-3.0	17.8-18.0	1.6-3.0	-	25.2-26.4
Annigeri	-	-	28.6-6.46	-	-	26-26.2
NARI SH	-	-	12.6-19.8	-	-	16-20.4
JSI 7	-	-	17.4-22.4	-	-	14.4-21.2
NARI 6	-	-	4.8-13.4	-	-	7-25.4
Bhima	-	-	8.4-10.4	-	-	7.2-8.8

The alpha tocopherol is higher in the oil of 180^{0}C roasted seed at 520 mg/kg as against 441 mg/kg oil from the seeds roasted at 140^0 C (Lee Young Chul *et al.*, 2004). The stability of the oil increases due to reoasting. Apart from the seeds and oil cake which have food and feed value, the safflower plant as a whole is a palatable fodder relished by the cattle. The tender shoot is used as pot herb and green salad.

Utilization of Products

Safflower seed packets and garlands of florets have been found with 4000 year old mummies. Safflower has been used in the Middle East, India and Africa for purgative and alexipharmic(antidote) effects, as well as in a medicated oil, to promote sweating and cure fevers. The major product namely oil is highly nutritious and edible. It has gained much importance because of its high level (80%) of linoleic acid responsible for reduction of blood cholesterol levels. It is particularly valuable in the case of people suffering from obesity and cardiac ailments. From the chemists point of view it is a good source of pure linoleic acid (Parker *et al, 1955).*

The oil also finds use in protective coatings, alkyd resins and paints. The oil forms into a fast drying non-yellowing film with low wrinkling characteristics. The florets serve as a source of dyes like carthamin which is used as saffron substitute *(*Chavan, *1961).* The flowers are also used in medicine as they have some influence on the circulatory system to suppress the inflammations and to relieve muscular fatigue. Safflower petals have been utilized to make some Indian recipes (Sarojini *et al., 1995).* The tender shoot and leaves are used as pot herb and salad.

Though astringent and bitter in taste, the fried seeds are used as snack items in some parts of India. Oil cakes serve as animal feed. The hulls have been experimentally utilized in the preparation of fibre boards. Thus it can be seen that the uses of safflower are multifold, (1) as a vegetable and as a source of edible oil for home (2) drying oil and carthamin for trade, (3) as a manure to soil, (4) forage crop for animals and (5) natural fence against trespass of cattle.

Sunflower

Sunflower is a widespread oilseed crop of the world. It performs well in the temperate regions. But it is grown in all the six continents of the world. It is widely adapatable to all types of agro-climatic conditions and soil types. Sunflower originated from Southern United States and Mexico. Soviet Union has been the pioneer in its cultivation as an oilseed crop. Sunflower has gained world wide importance because of its seed oil with good cooking and nutritious qualities. The oil cake and the hulls also have varied uses while the stalks and flower heads are useful as fuel and source of chemicals. Sunflower has been introduced into India in 1969. It is a short duration crop with wide adaptability, photoperiod insensitivity, high yield and cheaper cultivation. Because of these advantages it is cultivated on a large scale, all over the world.

Sunflower belongs to the family *compositae* and the genus *Helianthus*. It is a cross polinated crop and hence a lot of sunflower hybrids are available. The sunflower has got its name because its heads are heliotropic (tendency to bend towards light-Sun). It is related to the auxin level in the actively growing plant parts.

The crop requires cool climate during germination and seedling growth and warm weather later on. A 300-500 mm rainfall is sufficient for good crop. Sunflower grows well on neutral to moderately alkaline soils. A fertilizer dose of 35:50:35 NPK kg/ha and 60:75:60 NPK kg/ha repectively for rainfed and irrigated conditions is recommended. Irrigation at critical stages namely flower bud initiation, lower opening and seed filling stages is useful for obtaining good yields. Depending on the genotype used the crop comes to maturity between 85-110 days. Seed yields ranging for 800-1000 kg/ha are obtained under rainfed conditions while 2000-2500kg/ha are realised under irrigated conditions.

The sunflower heads are harvested when the leaves start drying and withering and the heads turn from green to lemon yellow . The heads are sun-dried for 2-3days. Then they are threshed to separate the seeds. The seeds with 9-10% moisture are stored in gunny bags. Sunflower which has been an ornamental plant, has become an important oilseed plant because it contains high quality and easily digestible oil. The seed is rounded at the base, and pointed at the top. Beneath the outer hull (pericarp) there is a papery white testa. The kernel consists of two cotyledons attached to a protruding radicle and embryo. The kernel constitutes 60-70% of the total seed. The rest namely 30-40% is accounted for by the hull.

Seed Composition

The oil content of the whole seed is 28-35%. The kernel oil content ranges from 48-53%. The seed protein content ranges between 14-19%. The crude fibre and ash contents vary between 16-27% and 2.4 to 3.0% respectively (Nagaraj, 1990, Muralidharudu, 1995). The 'Peredovic' variety has been reported to contain 48.1% oil (Dorrel, 1978). Different locations also have shown variations in oil content ranging from 38-48% (Muralidharudu and Haripriya, 2000). Indian varieties have been reported to contain oil upto 48.5% (Nagaraj, 1991).

Eighteen high oleic F1 hybrids, five germplasm accessions including wild species and twelve cultivars belonging to *H. annus* have been analyzed for oil and protein contents and fatty acid composition (Muralidharudu and Virupakshappa, 1998). The oil content of high oleic lines range from 23.6 to

36.3% while that of cultivars range from 24.1 to 35.6%. The protein content of eight cultivars range from 21.5 to 29.4% and that of high oleic types range from 17.2 to 22.0%.

Japanese sunflower seed has been observed to have an average oil content of 38.8% (Nagao and Yamazaki, 1983). Oil content has shown an increase from 12% on 10th day after flowering to 43% on 40th day (Lakshminarayana *et al.,*1980). Seeds from the post rainy crop *(Rabi* season) has higher oil content (Verma *et al.,* 1977). Among the seed components, protein accumulation reachs maximum earlier than oil, which has touched the maximum at 33 days after anthesis (Singh *et al.,* 1988).

A study carried out on six varieties at six locations in India has revealed the oil content range to be 46.6 to 26.4%, protein 34.9 to 26% and carbohydrates 13.0 to 7.0% in the whole seeds. Some locations have shown higher oil (40-44%) and better oil quality (Nagaraj *et al.,* 1990). This must be due to better soil fertility, better crop and influence of season. Studies carried out over four seasons (two *kharif* and two *rabi*) with three varieties and two fertiliser levels have revealed that the national check hybrid namely KBSH1, in general has higher oil content of around 40% as against around 30 % in MSFH17 hybrid and Morden, a short duration variety (Nagaraj and Reddy, 2006).

Table 36: Sunflower seed characteristics.

Character	Composition
Hull %	21-27
Oil (kernel) %	48-53
Protein %	14-19
Souluble sugars %	7.5-9.4
Crude fibre %	16-27
Ash %	2.4-3.0

Source: Dorrel, 1978.

Oil and protein content in whole seed varies from 29.33 to 44.19% and 20.36 to 27.68% respectively and in the kernel, these constituents vary from 46.19 to 57.25% and 23.66 to 33.77%. Kernel percentage in the seed is 71 with the rest (29%) being the hull. Oil and protein in the kernel are correlated negatively. A sharp increase in oil level could probably be achieved through selections for thin hull types (Mandal and Ranbir Singh, 1993).

Sunflower seeds grown in India have been observed to contain 36% oil, 22% protein, 18% carbohydrates, and 6.5% ash (Nagaraj and Srinivas, 2002). Oil content of sunflower seeds grown in India has shown a decline to around 30-32%. Shorter photoperiod, lower temperature and lesser number of sunshine hours during post anthesis period of the crop have been considered to be the reasons for the lower oil content (Chandru, *et al.*, 2000).

Sunflower Oil

Sunflower oil is a premium oil because of its light colour, bland flavour, high smoke point and good nutritional quality. Sunflower oil is expressed in crushers. Oil from whole seeds is dark in colour and hence needs to be refined prior to its marketing. Crude sunflower oil is light amber in colour, while the refined oil is pale yellow. The oil is mostly refined to an appealing golden yellow colour. Crude oil has some antioxidants (45mg/100g) which get reduced during refining. Continuous heating for more than one hour of oils like sunflower oil is not advisable (Khatoon,2001). The FFA content is normally low. It has higher I.V. due to higher unsaturation. The unsaponfiable content is lower which contains some phospatides and and mucilaginous materials. Sunflower oil is an excellent salad oil rich in unsaturated fatty acids. Some characteristics of the oil are presented in Table 37.

Table 37: Some characteristics of sunflower oil.

Charateristic	Value
Specific gravity	0.9909 - 0.9187
Refractive index 40^{0}c	1.4636 - 1.4681
Colour (Y+5 R)	1.0 - 12.7
Acid value	0.7 - 25.7
Iodine value	101.4 - 135.0
Saponification value	193- 200
Unsaponifiables	0.20 -1.93

Source: Kamath *et al.*, 1979.

Last week of April to third week of May sown crop yields oleic rich oil while June sown crop has lower oil content (Bhattacharya *et al., 1982*). Oil extracted from rainfed crop has a better keeping quality than oil from irrigated crop (Balasaraswathi, 1983). Both these crops are exposed to higher temperatures and hence rich in oleic acid with better keeping quality. Nagaraj and Reddy, (2006) have shown higher levels of linoleic acid in the seeds obtained from *Rabi* (October-March, mostly, winter) season The quality of the oil does not show any deterioration when stored in amber bottles even upto 700 days (Yousuf Ali Khan *et al.*, 1980). Similarly the oil from dehulled seeds has longer shelf life (Thayumanavan, *1981)*. Influence of temperature and irrigation during seed maturation on quality of oil also has been noticed (Afzalpurkar and Laxminarayana, 1980a and 1980b). Oil content increases from 12% on 21st day after flowering (DAF) to 45% on 43rd DAF with a marginal decrease on 56th DAF. Iodine value and total unsaturated fatty acids show a similar trend (Afzalpukar and Lakshminarayana, 1979).

The oil content increases from around 25% under immature stage to 34% at maturity followed by a significant decrease to 30% at over maturity. Oleic acid decreases from 43% to around 37% during various maturation stages. The major fatty acid, linoleic acid increases from 50 to 55% during the same period (Nagaraj, 2003).

Table 38: Effect of maturity on oil content and composition in sunflower.

Maturity stages	Oil (%)	Oleic (%)	Linoleic (%)
15 days before maturity	24.6	42.6	49.5
10 days before maturity	32.2	41.4	52.1
5 days before maturity	32.4	39.8	52.4
maturity	34.3	37.7	54.7
5 days after maturity	29.7	36.8	54.8

Source: Nagaraj, (2002).

Fatty Acid Composition

Sunflower oil has high 18:2 and moderate levels of 18:1 and less than 15% saturated (16:0 and 18:0) fatty acids. The oleic acid content has been reported to range from 42-57% and that of linoleic acid content between 33-48% (Chakrabarty *et al, 1979)*. Seeds produced in cool climates normally have 70% of 18:2 in their oil while those from hotter climates have about 30% 18:2. Oleic acid has shown an opposite behaviour (Kinman and Earle, 1964). Linoleic acid ranges from 64-70% in oil produced world wide (Campbell, 1983). Linoleic acid increases from around 40% on 10th day of flowering to 68% on 40th day, while with an initial increase to 30% on 30th day oleic acid decreases to 27% on 40th day (Lakshminarayana *et al.*,1980). Indian sunflower oil, in general, has 52% linoleic, 41% oleic, 4.6% palmitic and 2.3% stearic acids (Nagaraj and Srinivas, 2002).

Table 39: Fatty acid composition of sunflower oil.

Fatty acid	Composition
Myristic	5.0-7.0%
Palmitic	3.0-5.0%
Stearic	0.3-0.8%
Arachidic	0.6-0.8%
Oleic	22-50%
Linoleic	40-67%

Source: Bernardini, (1985).

Eighteen high oleic F_1 hybrids, five germplasm accessions including wild species and twelve cultivars belonging to *H. annus* have been analyzed for oil and protein contents and fatty acid composition (Muralidharudu and Virupakshappa, 1998). Among the wild species, EC-331346 has higher oleic acid (68.8%) followed by *Helianthus peteolaris* with 47.9%. High oleic types have oleic acid ranging from 83-90%.The linoleic acid varies from 20.7 to 66.7% in wild species. Sunflower inbreds have been observed to contain 65-70% linoleic acid while populations and hybrids have 50-55% linoleic acid. In general, high variability has been observed for fatty acids among the inbreds, populations and hybrids (Ranganatha, *et al.*, 2006).

Effect of Season on Fatty Acid Profile

Nagaraj and Reddy(2006) have studied the fatty acid composition of three sunflower genotypes under two levels of fertilizes over two seasons, namely *Kharif* and *Rabi* in India for two years. The average fatty acid composition of *Kharif* sunflower oil is palmitic=5.6, stearic=1.9, oleic=45.9 and linoleic=46.6% while for *Rabi* it is palmitic=7.6, stearic= 3.2, oleic=20.9 and linoleic=68.2%. *Kharif* sunflower seed oil has equal levels of oleic and linoleic acids(around 46%). *Rabi* seed oil has very low level of oleic(21%) and very high level of linoleic(68%) acids. The lower temperatures that prevail during *rabi* season favour higher synthesis of linoleic acid, the polyunsaturated fatty aicd (Mathur and Sharma, 1989). The desaturase enzyme activity becomes higher during low temperatures (*Rabi)* and lower during higher temperatures (*Kharif)*. Similar observations on the effect of temperature on the effect of oleic and linoleic acids have been made by Gupta and Wagle (1989). The higher summer temperatures favoured the accumulation of oleic acid while the cooler winter temperatures favoured linoleic acid. The ratio of oleic to linoleic acid at maturity in winter is 1:2.5 while in summer it is 3:1.

Similar observations on the effect of temperature on sunflower oil quality have been made by Nagaraj *et al.,* (1990). Due to higher level of oleic acid, Indian sunflower oil is considered more balanced (Gosh Choudhuri, 2003) and similar to sesame oil in its fatty acid composition. For obtaining more nutritious sunflower oil, richer in linoleic acid (the EFA) it is better to grow the crop under *rabi* conditions in India.

Higher proportion of linoleic acid, the essential fatty acid plays an important role in reducing the blood cholesterol and hence sunflower has special significance as a health promoting oil. People suffering from multiple sclerosis lack linoleic acid, which sunflower oil supplies in good abundance and hence is useful for it's treatment. Sunflower oil has higher rating of cholesterol reducing value next only to safflower oil.

As linolenic acid is either absent or very low (<1%), sunflower oil is resistant to oxidation. Hence off flavours present in soybean oil are absent in sunflower oil. Sunflower oil contains around 1% gums (lecithins). It is better to remove them prior to refining. These gums are present mostly in the hulls and hence oil extracted from dehulled seed is of better quality.

High Oleic Sunflower

Using chemical mutagenesis, high oleic sunflower with 50-89.5% oleic acid has been developed. High oleic sunflower oil with 80% oleic acid is available. It has greater oxidative stability and is useful as a frying oil in the preparation of snack foods. High oleic oil also serves as a source of oleic acid, useful in preparation of cosmetics and pharmaceuticals (Fick, 1989). In India also, a high oleic sunflower genotype has been developed with oleic acid levels ranging from 62-82% (Giriraj and Nagaraj, 2003). These lines have 2-5% each of palmitic and stearic acids and 11-34% linoleic acid.

Sunbutter-A Peanut Butter Alternative

The American Academy of Allergy, Asthma and Immunology estimates that 3 million Americans are allergic to peanuts and/ or tree nuts. As children are more allergic, some schools seem to be disallowing peanut butter in their cafeterias. Sunflower seeds are a good source of protein, fibre, vitamin E, zinc and iron, and hence, they are a popular peanut alternative. Sunbutter is prepared similar to that of peanut butter. Sunbutter made from the kernels is really good and palatable. It has a mild, but distinctive sunflower seed flavour.

Salad Oil

Sunflower oil is an excellent salad oil rich in unsaturated fatty acids. Sunflower, safflower and corn oils need dewaxing before they meet the criteria of salad oil. High oleic safflower and sunflower oils with about 80% oleic acid, serve as stable cooking and salad oils. There is a substantial market for cold pressed sunflower oil in many countries. The use of unprocessed oils is gaining popularity due to nutritional considerations. Sunflower oil is widely used in manufacturing margarines as well as in formulating many oil blends.

Lecithin

Higher phophatidyl choline in sunflower lecithin helps it to be utilized in food stuffs. It's use in the manufacture of foods and cosmetics can be increased by refining and fractionation and/or modifications. The phopholipid content in sunflower oil ranges from 0.02 to 1.5%, with an average of 0.75%. The composition of phospholipids varies greatly with phosphatidylcholine ranging from 12.7 to 64.2%. Sunflower lecithin has a tendency to undergo oxidative deterioration and hence its usage is getting reduced.

Tocopherols

Various forms of tocopherols namely alpha, beta, gamma and delta forms together constitute vitamin E. They have antioxidant properties. The alpha form is predominant in sunflower (608 mg/kg) followed by beta (17) and gamma (11) (Mullor-Mulot, 1976). Sunflower seed oil of Indian varieties have been analysed for tocopherols, α = 406-485, β= 35-56 and γ= 4-15 mg/kg oil (Nagaraj *et al.*, 2001). Among all oils, sunflower oil has the highest level of α-tocopherols (Campbell, 1983). At normal temperatures the antioxidant properties of different tocopherols remain similar. But at elevated temperatures, the order of activities is gamma, beta and then alpha tocopherol. Hence, the anti-oxidant activity during frying in sunflower oil is lower. For human nutrition, 6mg of alpha tocopherol per g of linoleic acid is considered as a normal requirement (Mullor, 1968).

Sunflower Meal and Proteins

The composition of sunflower meal is similar to that of any other vegetable oilseed meal except that it has higher fibre and ash contents which tend to reduce the metabolizable energy. The defatted meal is used mostly in feedstuffs. The nutritional value of the meal varies according to the method of its extraction or the hull content. The composition and characteristics of different meals is presented in Table 40.

Colour of the meal varies from grey to black depending upon the content of hulls and heat treatment during oil extraction. Non-dehulled or partially dehulled meal is useful in feeding ruminant animals while partially or completely dehulled meal is useful for feeding swine or poultry (Fick, 1989).

Table 40: Composition of sunflower meals.

Constituent	Partially Dehulled	Undehulled	Decorticated
Protein %	37	29	42
Fat %	9.1	1.0	10
Carbohydrate%	22.4	32	28
Fibre %	15.8	29	15
Ash %	6.5	6.56	8

Source: Bernardini, (1985).

Whole seed meal is low in protein (29%). It increases to 37% in partially dehulled meal and to 42% in the decorticated seed meal. Fibre content and carbohydrates are higher in undehulled meal due to the presence of hulls. Fully dehulled meal has higher oil (10%). This is because whole kernel oil extraction process is inefficient. Protein content of the sunflower seed ranges from 20-23% and cake ranges from 33-37%. This protein composition does not show much variation among different genotypes (Krishnaveni and Sadasivam, 1994). Sunflower protein contains 55-60% globulins, 17-23% albumins, 11-17% glutelins and 1-4% prolamines (Mosse and Pernollet 1972). Growing sunflower under dryland conditions increases the salt soluble globulins (reserve proteins) and lowers the protein quality (Borodulina and Supra Nova, 1976). The amino acid composition of sunflower protein is presented in Table 41.

Table 41: Amino acid composition.

Amino acid Essential	g/100g	Amino acid Non-essential	g/100
Arginine	56.2	Alanine	24.5
Histidine	14.4	Aspartic acid	54.9
Isoleucine	29.2	Glutanmic acid	143.0
Leucine	38.4	Glycine	33.9
Lysine	19.5	Proline	31.1
Methionine	11.4	Serine	26.3
Cysteine + cystine	13.4	Tyrosine	14.3
Tryptophan	10.0		
Phenyalanine	29.4		
Valine	34.9		
Threonine	22.3		
Total	**279.1**	**Total**	**327.9**

Source: Tkachuk and Irvine (1969).

Sunflower protein is balanced with respect to all amino acids except lysine. The essential amino acid index is 68 as against 79 for soybean and 100 for egg while the protein score is 63 (Sosulski and Sarwar, 1973).

The amino acid composition of Indian sunflower protein has been analysed and reported (Nagaraj, 1999a). The major amino acids are glutamic acid (23%), aspartic acid (10%) and leucine (7%). The other amino acids are threonine, serine, proline, glycine, alanine, valine, cysteine, isoleucine, tyrosine, phenylalanine, lysine and histidine. Their concentrations range from 1-6%. In general, like in any other vegetable protein, lysine and sulfur containing amino acids are lower in sunflower.

Table 42: Amino acid levels in Indian sunflower.

Amino acid	g/100g protein	Amino acid	g/100g protein
Aspartic acid	10.37	Cysteine	1.58
Threonine	4.67	Methionine	–
Serine	5.68	Isoleucine	3.62
Glutamic acid	22.63	Leucine	7.31
Proline	4.74	Tyrosine	3.02
Glycine	5.65	Phenyl alanine	5.30
Alanine	5.37	Lysine	3.57
Valanine	5.00	Histidine	2.75

Source: Nagaraj, (1999a).

Carbohydrates and other Nutrients

The sugars of dehulled, defatted meal are 0.6% glucose, 2.3% sucrose, 3.2% raffinose and 0.8% trehalose (Ceigla and Bell, 1977). Pure hull contains upto 26% reducing sugars (14% xylose, 7% arabinose and 2% galactose) which are less available than those of kernels (Cancalon, 1971). Raffinose, the flatulence factor, is present at a higher level in sunflower.

Sunflower is a good source of water soluble B complex vitamins namely nicotinic acid (318.7mg/kg), thiamine (37.8), pantothenic acid (44.8) and riboflavin (3.6). Sunflower is a richer source of the first two B complex vitamins than any other oilseed meal (Cladinin, 1958). HPLC analysis has revealed alpha tocopherol as 464.1 and beta tocopherol as 15.2 µg/g oil (Renuka devi *et al.,* 2000)

The mineral composition of Indian sunflower seed is N=3%, P= 0.1%, K=2.0%, Ca=0.6%, Mg=0.4%. The micronutrients present are (µg/g) Zn=78,

Cu=30, Mn=47 and Fe=360 (Nagaraj, 2002a). Morrison, (1957) has reported higher levels of the minerals, Ca = 0.26%, P = 1.22%, K = 1.08% and Cl = 0.19%. Sunflower seed thus is a good source of meneral nutrients required for human nutrition.

Antinutritional Constituents

The cholrogenic acid content of sunflower seeds range from 1.4 to 4.0% (Dorrel, 1976). It is responsible for the production of dark colour in the meal due to oxidation. Changes in chlorogenic, caffeic and quinic acids during maturation has been reported by Afzalpurkar and Lakshminarayana (1981). Very low level of 0.33% chlorogenic acid has been reported by Tyagi and Bhowmick(2001). Aqueous extraction has been found effective in removing this phenol from the seed.

Phenol content of Indian sample varies between 2.6 to 3.8%. Hybrids contain lower level of phenols. Removal of these phenolic compounds by washing with acidified acetone (0.5 M HCl : Acetone, 1:9) for improving the meal and protein quality has been carried out (Nagaraj, 1990b). Some more methods for removal of phenols have been proposed by Rahma and Narsinga Rao (1981). The total phenol content of fresh whole seed cake ranges from 80-87 mg /100g while that of hulls is around 25mg/100g and the aged samples have 2.6%. Among detoxification methods attempted, 0.5% $Ca(OH)_2$ and 0.5% $NaHCO_3$ soaking overnight reduces the phenols by 60-90% (Nagaraj, 1999a). Acidfied solvents reduce phenols from around 0.85-1.9% to 0.2 to 0.4% (Rahma and Narasinga Rao, 1981, Krishnaveni and Sadasivam, 1994).

Utilisation

Sunflower cakes of various grades and the protein isolates have been utilized in the preparation of some Indian snack food items like *pakodi*, biscuits and *chapatti*. Sunflower cake flour at 10% level enhances protein content by 3-4% and protein isolate at 5 % level increases it by 15-25% and at 10% it is 18-29% higher. Taste panel acceptability has been 50-90% for the snack foods as against 100% for the control food items. There is scope for utilization of sunflower products in food items (Praveena *et al.*, 1999, Nagaraj *et al.*, 2000). Sunflower oil could be modified for industrial utilization as oleochemicals or as biodiesel. Sunflower seeds as well as different parts of the plant have various uses. These have been compiled and listed by Sudhakar babu (2003).

Confectionary Sunflower

Sunflower used for table purposes other than oil extraction should be of a bigger size (ex: 8mm length and above). Such bigger size seeds can be used similar to that of groundnut kernel. Sunflower seeds in shell are salted and roasted and used for eating as snack food. Hundred kernels weighing 8.0g or more are preferred for snack food use and also for use by the bakeries. Spain is an important producer of confectionary sunflower. Sunflower with hulls after roasting are cracked between teeth and the kernels eaten. As in the case of groundnuts, confectionary sunflower also should have lower oil and higher sugar and protein contents for better paltability and acceptability. Low oil (4%) and high protein sunflower has been prepared using solvent extraction (Narasimhachar *et al.*,1984).

Sunflower Heads and Pith

Sunflower thalamus, left after removal of seeds has some use as animal feed. It can be incorporated upto 40% in the feed for milch animals. The heads contain 12.5% crude protein, 2% lipids, 11% ash and 12% lignin. They are richer in inorganic micro nutrients (183 mg/kg). An infusion made from sunflower heads helps relieve chest pain. Pectin content of the heads ranges from 150 to 250g/kg and can be extracted for use by the food industry. Sunflower pith has more buoyancy than cork or raindeer's hair. The pith is used to treat pains. It is placed in the area of pain and burnt to reduce pains/ infections.

Sunflower Stalks and Hulls

Sunflower stalks are normally left in the field and ploughed in as crop residue. Sometimes, micro-organisms like *Trichoderma viridae* are added *in situ* to the sunflower stalks for fast decomposition (Sudhakara babu, 2003). Addition of farm yard manure supplements the nutrient requirement of the subsequent crop. The dried stems also find use as fuel in rural areas. The stems contain high amounts of P and K and hence the ash serves as a nutrient to the soil. The stems also are a source of commercial fibre for rough paper, hard board and sound proof tiles.

Sunflower hulls constitute of 20-30% of the seeds. They are rich in fibre and low in fat and ash. Sometimes hulls are removed prior to oil extraction. Hulls also surve as fuel either directly are as pellets. They also have some use in manufacturing ethyl alcohol and furfurol and as substrate for growing yeast. The hulls also are useful in lining plywood. They also surve as organic manure as they are rich P and K.

Sunflowers

The flower of sunflower has since long been used for ornamental purposes. Both the artist and the common men have drawn a lot of inspiration from the attractive sunflower. The word sunflower itself has become a brand name and is commercially used to sell products and organisations. The flowers serve as sources of pigments, particularly, the yellow and purple dyes. The sunflower has been a source of honey. The bees also serve in pollination, seed setting etc. They also suck the nectar and supply the same to human beings. The flower extracts appear to have some pesticidal activity.

Bird Feed

Sunflower seeds, particularly the black oil-type, are the favourite of most seed-eating birds. Sunflower seed mixed with millet and other grains is used as bird feed. The black oilseed varieties are also sold separately, and usually are favoured by birds over the striped confectionery seeds. The high oil content of sunflower seeds provides an excellent source of energy for birds.

Animal Feed

Sunflower seed proteins are deficient in lysine (5-6.5g/kg). The biological value of proteins of sunflower seeds is lower than that of soybean or of proteins from animal sources. But sunflower seed proteins contain high levels of sulphur amino acids. Sunflower seeds contain 160 to 310g/kg of crude fibre, which may help avoid milk fat depression in high-producing cows fed with high-energy rations.

Sesame

Sesame, *sesamum indicum* L. is an ancient oilseed crop (Joshi, 1961.,Weiss 1971). It is an annual crop cultivated in tropics to temperate zones. In general it is a crop of the developing countries grown by poor farmers which is the reason why, it has not received much research attention. India is a major producer of sesame, followed by China, Myanmar , and Sudan. These four countries account for 75% of the area of the crop. The remaining 25% is dispersed in countries like Ethiopia, Nigeria, Uganda, Mexico, Venezuela and Turkey. India alone accounts for about 40% of production from 27% of the total world area under sesame. India produces around 0.5m tonnes of sesame annually.

Sesame seed has a high food value because of its higher contents of good quality edible oil and nutritious protein. The sesame oil is a rich source of poly- unsaturated fatty acids with a long shelf life. Its protein is rich in sulphur containing amino acids. The seed also contains useful carbohydrates. It is also a rich source of vitamins and mineral elements like Ca, and P. The seed can be eaten raw. Simple processes like roasting improve its palatability. It is used to make many food items and to garnish foods like bread, biscuit, snacks and sweet meats. Because of its greater and varied utility it is considered the "Queen of Oilseeds".

Sesamum seems to have originated in India or the Iran-Afghanistan area. Some people suggested that Ethiopia could also be its centre of origin. Sesame belongs to the family *Pedaliaceae.* It is a self pollinated crop. The crop is mostly confined to well drained light soils, with limited cultivation in the black soils. It is a short duration crop grown under rainfed conditions, in seasons like *kharif, Rabi* and summer in India. A seed rate of 4-7 kg/ha is sufficient, for sowing at a spacing of 30x15 cm. FYM @ 10 tonnes and NPK @40-40-20 kg/ha are normally applied.

For maximum yields, sesame requires fairly high temperatures and evenly distributed rainfall during its growth period. It cannot withstand frost, prolonged drought, wet weather or water logged conditions especially at flowering and pod developmental stages. Irrigations may be given at intervals of 12-15 days depending on the soil type, weather conditions and season. For good capsule filling and yields, irrigation at flower initiation and capsule formation are most critical. Low temperatures at flowering may lead to premature flower drop or production of sterile pollen.

The crop is harvested when the leaves turn yellow and the capsules greenish-yellow. The capsules contain numerous small ovate seeds. The crop is usually threshed by gentle beating of well dried plants with sticks. Depending on the type of cultivation, seed yields ranging from 300-3000kg/ha can be obtained. Seeds have good storable value and do not present any problem.

Sesame Seeds

Sesame seeds are very small in size and are 4 mm long, 2mm wide and 1 mm thick. They are pearl shaped, ovate, small, slightly flattened, and somewhat thinner at the hilum. The seed contains testa (hull), endosperm, cotyledons and embryo. The hull content ranges between 10 and 20%. It is responsible for the colour, bitterness, higher fibre and high oxalate contents of the seed. The seed colour ranges from white to brown. The white varieties generally contain higher proportions of hull.The sesame kernels are generally white in colour. Sesame seeds can be stored at room temperature for about five years and can be kept viable for ten years under low temperature and humidity.

Seed Composition

Sesame seed has a high food value because of its higher content of oil and protein. In addition, it has sufficient quantities of carbohydrates, fibre and minerals. The proximate composition of seed is presented in Table 43.

Table 43: Seed composition of sesame.

Constituent	Composition
Moisture	6-7%
Proteins	20-21%
Oil	48-55%
Sugars	14-16%
Fibre	6-8%
Minerals	5-7%

Source: Bernardini, (1985)

Among edible oilseeds sesame seeds have the highest oil content up to 60% along with about 20% proteins. The soluble sugars are also present up to 15%. Its fibre content is less at around 8%. The mineral content is higher at 7%. The whole seeds are edible and utilized as such in most preparations due to its balanced composition of the major nutrients.

Analysis of 35 varieties in India for oil and proteins has revealed their ranges to be 32-58% and 18-26% respectively (Chavoramony and Padmaja, 1982). Other workers have reported somewhat lower range of oil content (38-41%) with an I.V. of 96-112 and refractive index of 1.4598 to 1.4667 (Sarkar and Bhattacharya, 1987). Black seeded Pakisthani varieties have 20-24% protein and 52-54% oil(Khan and Hussain, 1988). Fifty forms representing ecotypes divided between 2 groups have been analysed for seed oil content. Forms from narrow leaved group contain highest oil (59.4-59.9%) with 3% differences between ecotypes and 6% differences within ecotypes (Yarosh and Ivanenko, 1984). Analysis of 20 genotypes grown in India has revealed the oil content to be 41-55%. The refracive index of the oil is 1.4710 while iodine value is between 104-108 indicating medium unsaturation (Nagaraj, 1991).

Additive gene action controls the inheritance of oil content in sesame (Mosjidis, 1982). Oil formation begins within 5 days after fertilization and reaches 52% after 15 days(Saha and Bhargava, 1984). Processing of whole seeds have given 43.6% oil and the oil from decorticated seed has low FFA (Surendranath *et al., 1985)*. The refining losses in sesame oil ranges from 6.3 to 8.4%. The oil retains its quality without appreciable increase in FFA. during storage over a period of 300 days (Sunrendranath *et al.,* 1982).

The variation in sesame oil content is both genotypic and environmental. Seeds from later maturing cultivars have higher oil content than those of early maturing ones (Yermanos *et al.,* 1972, Lee *et al.,* 1991). Black seeds have lower oil than white or brown sesame seeds. It is important to note that white seeds, despite higher hull has higher oil. In India black seeds are prefered in auspicious functions as also for preparations during that period. This might be because of lesser oil content. The lower oil content is due to thicker seed coat (Tashiro *et al.,* 1990 and Nasriki, 1995). Seeds from basal capsules of main stem have higher oil than apical capsules or those from side branches. (Mosjidis and Yermanos, 1985). Seeds from wild species have lesser oil content of 27-36% (Were *et al.,* 2001).

Table 44: Chemical composition of whole and dehulled sesame seed.

Constituent	Whole seed	Dehulled seed
Moisture %	5.4	5.5
Protein %	18.6	18.3
Fat %	49.3	53.4
Carbohydrate %	21.6	17.6
Crude fibre %	6.3	2.4
Ash %	5.3	5.3
Energy (cal/100g)	563	582
Calcium (mg/100g)	1160	110
Phosphorus (mg/100g)	616	592
Vitamin A (I.U)	30	-
Thiamine (mg/100g)	0.98	0.18
Riboflavin (mg/100g)	0.24	0.13
Niacin (mg/100g)	5.4	5.4

Source : Weiss (1983).

Colour and size of the seed also influence the content of oil in the sesame seed. White or light coloured and small sized seeds contain more oil than dark coloured and large sized seeds (Seegler, 1983). White seeded types have 55% oil as against 47.8% oil in black sesame seeds. The oil content is negatively correlated with hull content(Tashiro *et al.*, 1990). Similarly smooth seeds have higher oil(Yermanos *et al.*,1972). Higher doses of nitrogen fertilizer in the field reduces the oil content (Singh *et al.*, 1960). This may be because of increase in protein in the seed due to N fertilization and the consequent dilution that reduces oil. Among wild species, *Sesamum augustifolium* has higher oil content than *S. Indicum* (Uzo *et al.*, 1985). The composition of whole and dehulled seeds is presented in Table 44. The dehulled seed (kernel) is richer in oil while the whole seed is richer in carbohydrates, cellulose and calcium. The energy content of kernel is marginally higher, due to higher oil content.

Oil Extraction

Generally 80% of the seed produced is utilized for oil extraction. In India 23% out of the same is crushed in *Ghanies* at the village level (Achaya, 1986). The rest (about 75%) of the sesame seeds are crushed in a continuous screw press or hydraulic press and solvent extraction (Godin and Spensley 1971). The oil is extracted in three stages. The first pressing is made in cold to obtain a good quality oil. It has a light colour and agreeable taste and odour. The second pressing is done after heating under high pressure. The oil obtained is coloured and needs refining. The third pressed oil is of inferior quality and is inedible. Kernel oil is costly and premium oil much preferred for use in preparation of pickes and home cooking. It also serves as *ghee* (butter oil) substitute.

Sesame hull is somewhat difficult to remove. The hull contains undigestible fibre. Hence it is better, the hull is removed, especially, if the oil meal is to be utilized as food. Normally it is soaked in water overnight and rubbed between rough surfaces to remove the husk. Both the oil and cake quality, gets improved, when the dehusked seed (kernel) is utilized for oil extraction. Protein concentrate prepared from dehulled sesame meal, serves as a good source of sulphur containing amino acids.

Oil biogenesis is fast. It begins within 5 days after fertilization and maximum accumulation (52%) is achieved within 15 days (Saha and Bhargava, 1984). Oil yield, I.V. and S.V. decrease due to leaf curl disease (Prasad *et al.*, 1985). Sesame seed is found to be associated with mycotoxin producing fungi (Reddy and Reddy, 1983).

Sesame Oil Quality

Sesame oil in the crude form has amber to yellow in colour while refined oil is pale yellow with a nutty flavour and is exclusively used for edible purposes. Sesame oil has high unsaponifiable matter which mainly consists of tocopherols and antioxidants. Because of this the oil is highly stable and does not turn rancid or develop objectionable taste either during storage or during deep frying. Freshly extracted oil has low acid value. It's iodine value is higher indicating semidrying chasracteristics. It contains unsaponifiables ranging from 1-3%.

Table 45: Quality characteristics of sesame oil.

Constituent	Composition
Oil %	48.8-49.1
Specific gravity	0.992-0.924
Refractive index 25^{0}C	1.4760-1.4770
Acid value	1.1- 2.2.
I.V.	103.6-112.8
S.V.	188.6-191.2
F.F.A. %	1-3
Unsaponifiables %	1.4-2.5

Source : Tyagi and Vasistha (1983).

Sesame Oil Composition

Sesame oil contains about 80% unsaturated fatty acids composed mainly of oleic (18:1) and linoleic (18:2) acids. The rest 20% are saturated fatty acids, consisting of palmitic (16:0) and stearic (18:0) acids. Sesame oil contains mainly 18:1 and 18:2 in almost equal proportions(about 40-45% each). Fatty acid composition of 165 varieties from various countries have been analysed by Lee and Kang (1980). Varieties from Korea and U.S.A. are low in 16:0 and 18:0 and high in 18:2. Unsaturated fatty acid content is higher in varieties from temperate regions than in those from tropical regions and are also higher in late than early maturing varieties. Similarly 18:2 is more in yellow seeded varieties than white seeded varieties and in varieties setting 3 capsules than in varieties setting single capsules at each node.

Analysis of samples from India have revealed the following fatty acid composition (Sarkar and Bhattacharya, 1987*):* 16:0=7-13%; 18:0=7-11%; 18:1=41-46% and 18:2=35-45%. Glyceride composition of sesame oil has been studied by Tyagi and Vasistha (1983). The mono unsaturated-diunsaturated glycerides range between 33.3 and 40.6% while triunsaturated glycerides range from 50.2 to 61.5%. Smith (1971) has reported 44% oleic and 42% linoleic acids with 13% saturated fatty acids. Arachidic (20:0) and linoleic (18:3) acids are minor components (Rao and Rao, 1981).

Seventy genotypes in India have been analysed (Raheja *et al.*, 1989) and the following fatty acid ranges have been reported. 16:0 = 9.35 – 15%; 18:0 = 3.5 -7%; 18:1 = 40-48%; 18:2 = 30 – 45%. The iodine value ranges between 96.7-117.9. The oleic/ linoleic ratio which is a measure of oil stability for the genotypes is 0.9 –1.60. Oleic acid has negative correlation with linoleic acid and iodine value. Fourteen cultivars have shown the ranges of 18:1 = 39–49, 18:2= 35-45%; 16:0=9-16% 18:0 = 4-5.0% (Amin and Kothari, 1989).

Nagaraj (1991b) has reported the following fatty acid composition for14 Indian genotypes: 18:1 = 45.3-41.9; 18:2 =41.7-37.6; 16:0=8.2-9.8: 18:0 =5.5-7.7 with around 0.5% of 20:0. The nutritional quality index namely the ratio of 18:2 to total saturated fatty acids range from 2.31 to 2.65. Similar such information on fatty acids have also been reported by Sarkar and Bhattacharya (1987), with an oleic acid level of 41-46%, linoleic, 35-45% and palmitic, 7-13%. Tyagi and Vasistha(1983) have reported the following ranges after analysing 15 genotypes 16:0=8.6 – 11.9%; 18:0 = 3.5-8.9%; 18:1=34.6 to 51.3% and 18:2 = 27.9 to 47.2%.

Table 46: Fatty acid composition of sesame oil.

Fatty acid %	(1)	(2)	(3)
16:0	8.2-9.8	9.3-14.6	8.4-11.9
18:0	5.5-7.7	3.5- 6.8	3.5-7.9
18:1	41.9-45.3	39.9-48.3	34.6-47.3
18:2	37.6-41.7	29.9-45%	27.9-47.2
20:0	0.3-0.7	–	0-1.4
18:2/SFA	2.31-2.65	–	–

Source: (1) Nagaraj (1991) (2) Raheja *et al.*, 1989.(3) Tyagi and Vasistha (1983).

Mutants (in M_2 generation) obtained through gamma ray irradiation have shown 2.7% lesser oleic acid and 2.4% higher linoleic acid. Through mutation two promising lines with 52% linoleic acid and 92% unsaturated fatty acids have been selected (Lee *et al.*, 1984). Effect of capsule position on sesame seed lipids has been examined. In the initial stages of development palmitic and stearic acids decrease from basal towards apical portion. Brar (1977) has reported a progressive increase in 16:0, 18:0 and 18:1 with a concomitant decrease in 18:2 of the oil from basal to top nodes of the plants. These effects are ascribed to the changes in environmental conditions during capsule development. Increasing day/night temperatures from 21/ 15.5^{0}C to 26.5/ 21^{0}C significantly increases the degree of unsaturation of the high linoleic acid genotypes (Brar, 1990). A negative correlation between 18:1 and I.V. and positive correlation between 18:2 and I.V. has also been observed (Brar 1982).

Basal and intermediate capusles have heavier seeds, more total lipids, neutral lipids and less phospho-and glycolipids than the apical ones in sesame. Palmitic and stearic acids decrerase from basal towards apical portion. Seeds of basal capusules have less oleic acid and more linoeic acid at all the stags of development than apical seeds. In the intial stages linoleic acid is considerably higher in apical than basal and intermediate capsules (Gupta, 1990).

The proportion of the four main fatty acids do not vary among the cultivated sesame and the wild species. Only *S. Alatum* appears to have slightly higher 16:0 and 18:1 with lesser levels of 18:2. *S. Agustifolium* and *S. Radiatum* have slightly higher level of 18:2 and 18:0 with less than 10% palmitic acid.

Table 47: Fatty acid composition of sesame species.

Sesame	16:0	18:0	18:1	18:2	Reference
Cultivated	9-14	3-7	33-54	30-59	Kamal - eldin, 1993
S. Alatum	11-13	5-6	44-48	33-37	Raleja, *et al.* 1993 Yermanos *et al.*, 1972
S.Aagustifolium	8-9	6-8	34-36	45-48	-do-
S. Radiatum	8-10	8-10	36-38	40-47	-do-

About 75-80% of the seed produced in India is utilized for oil extraction with 20-25% of it being utilized for edible use within the country or outside. About 2% of the production goes back to the soil for sowing purposes.

The aroma components have been identified as straight chain aldehyde or ketone derivatives(Lyon, 1972). Sesame flavour constituents are octanol, 2,4 undecadienol, 3 methyl butanol, 2,3 dimethyl pyrazine, 2,5 dimethyl pyrazine, 2 ethyl pyrazine and 2,5 diethyl pyrazine. The sweet roasted flavour has been attributed mainly to 2-furfuryl alcohol (Elsawy *et al.*, 1988).

Sesame Meal and Proteins

Whole seed expeller cake contains around 35% protein while dehulled expeller cake contains 41% protein. These cakes contain upto 13% oil due to inefficient oil expulsion. Solvent extracted cakes contain lower levels of oil ranging from 1-3% and higher levels of protein up to 46% and carbohydrates upto 35%. The mineral contents in all types of cakes range between 5-10%.

Whole seed cakes contain upto 4% oxalic acid while dehulled cakes have low levels of 0.5% only. This indicates that the anti-nutrient is concentrated in the hull. The composition of sesame cakes is presented in Table 48.

Sesame seeds contain about 19.0% protein (Joshi, 1961). A range of 17-32% protein with an average of 26% has been reported for U.S varieties (Lyon *et al.*, 1972). Similar range of 21-30% protein in sesame seed has also been reported (Yen *et al.*, 1986). Sesame meal contains 40-45% protein and 6-8% crude fibre with a lysine content of 2.7% (Lennerts, 1989). Solvent extracted sesame fluor has the highest protein content of 56% (Surendranath *et al.*, 1982).

The protien isolates have been tested on weanling rats. Lysine supplementation improved their protein efficiency ratio(Guerra *et al., 1984*). Low fat, high protein sesame seed has been prepared which has 26-33% oil and 31-36% protein (Surendranath *et al., 1984).* Dehulling of sesame seed in a saline solution reduces total free oxalates and calcium while roasting and grinding has no such effect (Toma *et al., 1979).*

Table 48: Composition of sesame cakes.

Constituent	Expeller	Press Cake	Alcohol Extracted		Hexane Extracted	
	1	2	1	2	1	2
Moisture %	8.1	8.3	8.6	8.9	8.6	8.8
Oil %	13.5	12.7	3.1	3.4	0.8	1.1
Protein %	35.1	41.3	38.2	45.8	39.6	46.7
Ash %	8.9	4.8	9.4	5.0	9.7	5.2
Crude fibre %	5.3	3.1	5.9	3.1	6.1	3.2
Oxalic acid %	3.0	0.4	3.5	0.5	3.6	0.5
Carbohydrates%	26.1	29.4	31.3	33.4	31.6	34.5
Calcium %	1.8	0.4	2.0	0.4	2.1	0.5
Phosphorus %	0.8	1.0	0.9	1.1	1.0	1.2

Source: Weiss, (1983), 1= undehulled cake; 2= dehulled cake.

The various fractions of sesame protein are as follows: albumins, 8.6%; globulins, 67.3%; prolamins, 1.3% and glutelins, 6.7%(Rivas, 1981). The molecular weights of the protein fractions ranges from 17,000 to 51,000. The sesame globulin consists of two fractions (Nath and Giri, 1957) namely α-glubulin (60-70%) and β-globulin (25%). Hasegwa *et al.*, (1978) have found mostly 13S gloubulin in protein isolates of sesame meal. Consesamin glolulin from sesame has been isolated from sesame. It is rich in acidic amino acids especially glutamine and hydrophobic amino acids (Rajendran and Prakash, 1989).

Among oilseed proteins, sesame protein is the most nutritious in that it is a rich source of sulphur containing amino acids like methionine (Smith, 1971, Brito, 1981, Narsinga Rao, 1985) and tryptophan (Johnson *et al.*, 1979). However, like any other vegetable protein, it is deficient in lysine. It is also deficient in threonine, isoleucine and valine. The essential amino acid composition of sesame protein is presented in Table 49.

White seeded and black seeded varieties do not differ in their amino acid composition. However, the NPU, true digestility and biological values for the white cultivars are 54, 87 and 62, while they are 57,83 and 61 for the black seeded cultivars (Khan and Hussain, 1988). In studies with Wistar rats, sesame meal protein at 10% of the diet has shown a protein efficiency ratio of 1.4, NPU, 30.5%; digestibility, 83.8% and BV, 36.4% while for untreated meal the values are 1.25, 35.7%, 84.7% and 42.2% in that order (Salgado and Goncalves, 1988).

Table 49: Essential amino acid composition of sesame protein.

Amino acid	mg/g
Arginine	547
Histidine	121
Isoleucine	261
Leucine	461
Methionine	160
Phenylalanine	461
Threonine	194
Tryptophan	91
Valine	244

Source : Weiss, (1983)

Sesame cake is a valuable feedstock. It also serves as a nourishing human food. Low quality cake is utilized as organic manure. Indian *ghani* cake reduces the availability of minerals like Ca, Mg, Zn, and Fe, apart from rendering proteins insoluble. Phytates and oxalates present in the cake reduce the availability of minerals like Ca, and also impart bitter taste to the cake due to formation of calcium oxalates and phytates. The sesame cake also contains 19.3 mg/g iron, 1.0 mg/g of thiamine, 5.3 mg/g of niacin and 0.4 mg/g of riboflavin.

Sesame is widely processed into several high protein products such as flakes, fluors, protein concentrates and isolates (Sastry *et al.*, 1969). Some aspects of sesame protein have been reviewed by Johnson *et al.*, (1979). Different solvents namely phosphate buffer (pH 7.66), 2% and 10% NaCl have been examined for extraction of proteins (Hasegawa *et al.*, 1975, Obuko *et al.*, 1979). NaOH (0.04M) and NaCl (0.5 to 1.0M) are observed to be good extractants of protein (Rivas *et al.*, 1981). Protein isolates have better PER than the meal (Taha *et al.*, 1989). Protein isolates and concentrates do not have undesirable pigments and off-flavours (Johnson *et al.*, 1979).

Heat treatments like autoclaving do not have any effect on protein and lysine contents (Sastry *et al.*, 1974). However, it increases the methionine levels (Villegas *et al.*, 1968). Protein isolate with 6.65 mg phytate has been shown to have higher PER than that with 14.5 mg phytate (Taha *et al.*, 1987). Though sesame cake is nutritious and edible, it is generally utilized as animal feed. As animal feed it is better than or equivalent to groundnut and soybean cakes.

Carbohydrates

Sesame seed contains 21-25% carbohydrates. The soluble sugar content is around 5% and most of it is of the the reducing type. The crude fibre content is low compared to the other oilseeds (3-6%) which is mostly present in the outer seed coat (Narasinga rao, 1985). The carbohydrate composition has been reported by Wankhede and Taranathan, (1976).Glucose and fructose are the major sugars and sucrose is present at a very low level of 0.2 to 0.3%.

Table 50: Carbohydrate compostion of sesame.

Sugar (%)	Seed (1)	Defatted fluor (2)
D. Glucose	1.55	3.63
D. Galactose	0.65	0.40
D. Fructose	0.24	3.43
Sucrose	0.34	0.17
Raffinose	-	0.59
Stachyose	-	0.38
Planteose	0.06	0.23
Sesamose	-	0.14
Other sugars	-	0.16
Total sugars	-	11.26

Source: (1).Aquilar and Torres, (1969), (2).Wankhede and Taranarthan, (1976).

Stachyose and raffinose are undesirable in that they impart bad taste properties and cause flatulence. In this respect sesame is better than peanut and soybean which have higher levels of these two sugars.

Minerals and Vitamins

Sesame is a good of source of minerals. The total ash (mineral) content of sesame seeds range between 5-7%. Deosthale(1981) has reported 1% Ca, and 0.7% P. Calcium is present in the seed coat. The bio-availability of Ca from sesame is 65% as against 100% for $CaCO_3$ (Poneros - Schneider and Erdman, 1989) which is because of the binding effect of the oxalic acid present in the seed. P and K in the seeds of different sesame varieties vary from 4968 to 5991 mg/kg and 3375 to 5875 mg/kg, respectively. The micronutrients Zn, Fe, Mn and Cu content range from 28.3 to 72.8, 58.6 to 375.3, 14.4 to 24.4 and 47.9 to 61.9 mg/kg, respectively (Murthy *et al.*, 1997).

Sesame is a good source of niacin, folic acid and tocopherols (Gopalan *et al.*, 1982, Weiss, 1983). Sesame oil is rich in tocopherols. A range of 29.4 to 52.8mg total tocopherols per 100g oil has been reported (Muller-Mulot, 1976, Speek *et al.*, 1985). Crude sesame oil has 190 µg/g gamma tocopherol (Fukuda *et al.*, 1981), which has a role in high temperature stability of the oil. HPLC analysis has revealed very high level of 600 µg/g of gamma tocopherol (Renuka devi *et al.*, 2000).Other antioxidants in conjunction with gamma tocopherols impart very high stability to sesame oil.

Antinutritional Constitutents

Sesame, especially it's kernel is edible and is devoid of anti-nutritional factors. It is eaten raw or roasted. The seed and kernel are used in confectionary and garnishing of bakery products. The seed coat/hull contains, the antinutritional constituents namely oxalate and phytic acid (Prakash and Nandi, 1978). Phytates range from 2.0-5.2 mg/100g and oxalates range from 4-46mg/100g seeds (Toma *et al.*, 1979). Oxalic acid can be removed by treatment with H_2O_2 at pH 5.0. The high proportion of oxalic acid renders mineral elements like Ca unavailable. Phosphorus is made unavailable by phytic acid. Dehulling process of sesame in a saline solution reduces total free oxalates and calcium. Roasting and grinding do not have any effect (Toma *et al.*, 1979). Kernels have a high nutritional value than the whole seeds. Hence decortication is a widely accepted practice prior to the consumption of the sesame.

Table 51: Antinutrional constituents in sesame.

Sesame	Phytic acid mg/100g	Oxalates mg/100g
White seed	4.7	45.9
Dehulled seed	5.2	27.9
Roasted seed	4.7	43.7
Dehulled, roasted seed	5.1	24.3
Red sesame butter	4.8	4.3
White sesame butter	4.9	22.7
Sweet red sesame mix	2.2	22.2
Sweet white seed	2.0	13.4

Source: Toma *et al.*, 1979.

Antioxidants

Sesame oil has two minor constituents, sesamin and sesamol. They constitute the bulk of the unsaponifiable fraction of sesame oil. These constituents impart high stability to the oil. These compounds also give characteristic colour reactions, because of which the presence of sesame oil in oil blends or as an adulterant can easily be detected. The TLC and HPLC methods for estimation of these antinutritional constituents have been reported (Yoshida and Kashimoto 1979, Huang *et al.*, 1984).

Sudanese oils contain 0.14-0.48 % sesamolin and 0.32- 0.59% sesamin (Eloeid and Rosie, 1983). Sesamolin and lignan antioxidants in 14 Japanese strains have been reported. Black strains have higher levels (Fukuda *et al.*, 1988). The sesamin in oil is 0.07 to 0.6% while sesamolin is 0.02 –0.48%. (Tashiro *et al.*, 1990). Sesamol has been detected in roasted sesame oil. Sesamol and tocopherol are higher in oils extracted from seeds roasted to temperatures of 200-220^{0}C (Yen, 1990). The content of free sesamol is 0.0021 - 0.0058% of total lipids which increses during storage. Sesamolin is 0.11 to 0.19% of total lipids and it tended to decrease during storage (Choi and Cho, 1983). Sesamoline is inactive and sesamol is very acive as antioxidant (Jikugawa *et al.*, 1983).

Sesame oil is unique in containing several lignan antioxidants as well as some tocopherols. (Kamal-Eldin (1993) and Namiki (1995). The lignans, e.g., sesamin and sesamolin and their derivatives, such as sesamol and sesaminol, as well as other related compounds, prevent the oxidation of sesame oil and give it the characteristic very long shelf life and stability. The average lignan contents of sesame germplasm are 2.09 ± 1.02 mg/g of sesamin, and 1.65± 0.61 mg/g sesamolin(Kim- KwanSu *et al.*, 2006).

The content of lignans and tocopherols in cultivated sesame and 3 wild species is presented in Table 52. Namiki (1995) reports similar and lower values for sesamin and sesamolin in connection with seed color. Sesamin and sesamolin are present in the cultivated species and in *S. alatum, S. radiatum, and S. augustfolium*, but both are rare in *S. alatum*, while sesamolin is rare in *S. radiatum* (which is rich in sesamin).

The genetics of lignan biosynthesis and levels have not been investigated thoroughly. Black seeded accessions have a significantly lower amount of sesamin and there is no significant difference in the sesamolin content between accessions with white, brown, and black seeds (Tashiro *et al.*, 1990 and Namiki 1995). Wide variations in lignan content among the germplasm accesions needs to be thoroughly evaluated for further utilization in breeding programmes (Kamal-Eldin, 1993).

Table 52: Lignans and total tocopherols in sesamum species.

Species	Lignan (% in oil) Sesamin	Sesamolin	Tocopherols, Total (% in oil)
S.indicum	0.55 + 0.17	0.50+0.08	0.054
S. alatum	0.01	0.01	0.032
S.augustifolium	0.32+0.04	0.16+0.01	0.073
S. radiatum	2.40+0.10	0.02	0.080

Source: Kamal- Eldin, (1993).

Utilization

Sesame seeds can be eaten raw, roasted or parched. They are incorporated into a wide variety of foods and sweet meats, *halva* etc. Sesame is also used to decorate bread, pastry etc. The oil is used mainly for culinary purposes. Sesame oil is well balanced with respect to all the three groups of fatty acids namely, saturated, monounsaturated and polyunsaturated fatty acids. Moreover, the tri unsaturated linolenic acid is absent. Hence this oil is most suitable for deep fat frying purposes. Snacks fried in sesame oil can keep longer without development of rancidity. The sesamin and sesamol present in the oil act as synergists for pyrethrin and rotenone, which are potent insectides. The oil is also used for body anointing and conditioning of the skin. The cake also can serve as human food due to its better amino acid make-up and other nutrients. Cooked leaves are sometimes eaten and also utilized to prepare folk medicine. The oil also is used as a carrier of medicines.

Quality and utility-wise sesame seeds are better than groundnut, soybean or for that matter any other edible oilseed. Dehulled seeds are directly used in different types of foods like *halva, laddu* and *chikki.* They are eaten whole after roasting. Ready to eat instant foods have been prepared using sesame flour as an ingredient of other cereal, pulse and oilseed flours. Sesame seeds are directly used to garnish bakery products. Sesame flour is used as a methionine supplement. A nutritious beverage can be prepared using 70% soy protein and 30% sesame protein. Like groundnut, sesame can be eaten raw or roasted. Many household items like sweets, chutneys, powders etc., are prepared from sesame. The seed and cake can be utilised in making bakery and confectionary products. Sesame seeds are used in the treatment of burns, gout, piles and poultice. Seed extracts reduce joint pains. Sesame seeds are digestive, rejuvenative and anti-ageing. The seeds provide 640 calories/100g seeds. They are rich in minerals, vitamins, E.F.A. and E.A.A.

Sesame seeds are also rich in vitamins E, A, B complex and minerals like calcium, phosphous, iron, copper, magnesium, zinc and potassium. The seeds are also rich in the sulphur amino acid methionine and tryptophan which have a role in liver and kidney health as well as utilization of B complex vitamins. This unique chemistry, along with abundant levels of all the essential fatty acids makes sesame the most balanced food item. In India sesame seeds are made into various snack foods, along with other nuts and spices and given to children, pregnant women and lactading mothers, to improve their nutrition and health. Sesame seeds unlike groundnut do not have any problems related to allergy.

Sesame oil has many medicinal uses. It is a laxative, anthelminthic, thermogenic, emollient and demulcent. It has antibacterial, antifungal and antiviral properties. It nourishes blood, calms down nervous spasms, alleviates headaches, dizziness and numbness caused by deficient blood supply. It has been used in traditional medicine to treat ailments such as burning sensation, backache, blurry vision, migraine, vertigo, constipation, hemorrhoids, dysentery, amenorrhea, dysmenorrhea, leucoderma, tooth decay, hair loss, weak bones, osteoporosis, emaciation , dry cough, blood in urine, stiff joints. (David and Vasantlad, 1986, Vogel, 1989, John, 1994, Michael, 1999 and Anon, 2000).

Sesame oil is useful in maintaining youthful skin. It protects the baby skin against rashes. It is the best body anointing oil. Body massaging with sesame oil, detoxifies the body and retains body heat at desirable levels. Sesame oil penetrates deep into the body benefiting all tissues to balance *vata, pitta* and *kapha.* The oil is a U.V. protector when applied externally. It also acts against free radicals (mostly oxygen) reducing peroxidation and thus prevent tumor initiation and promotion. Sesame oil with equal quantity of molasses improves the reproductive ability in women and also cures problems relating to menopause. Sesame and linseed decoction serves as an aphrodisiac. Sesame plant also has some medicinal uses. Sesame roots, stems, leaves, flowers and fruits and the ash soaked in water are used to treat dyspepsia.

It is thus evident that sesame and its products are more versatile in their utility. Research needs to be intensified to prepare more food products, as in the case of soybean and groundnut.

Soybean

Soybean (*glycine max* L. Merrill *)* is the major oilseed crop of the world accounting for nearly 50% of the area and production of oilseeds. It provides 60% of the vegetable protein and 30% of the oil in spite of the lower oil content in the seed. Soybean has originated in China during 2800 B.C. It has later spread to Korea, Japan, Europe and U.S.A. At present U.S.A. tops in acreage and production of soybean followed by Brazil, China, Argentina, Indonesia, Canada, Paraguay and Italy.Soybean seeds are sources of nutritious vegetable protein and edible oil. The protein and oil have multiple uses as food and non-food items. A variety of products can be made at home as well as by the industry.

Soybean is known to Indians since ages. Black-seeded soybean or *Kalitur*, is being grown in North-Indian Kumaon and Garhwal hills of Uttar pradesh and scattered pockets of the central India. Soybean is utilised as a source of food in the Kumaon hills and also in the Naga hills. Soybean has recently been reintroduced into India. It has become an important oilseed crop of the country. India, with 6.50 million hectares of soybean growing area and 7.86 million tonnes production ranks fifth in the world. The average yield in India is about 1210 kg/ha. It contributes 25 to 30% to India's oilseeds production.

Soybean belongs to the legume family. Soybean plant is an erect annual type with sparse branching. It is a self pollinated crop. Soybeans can be grown in fairly wide range of climates and soils. It grows well in sandy loam to clay soil. It is somewhat photosensitive and can be grown in rainy , spring, and summer seasons. As a rainy season crop it can be sown in June-July and harvested in October. As a spring - summer crop, it can be sown during February-March. As a *Rabi* crop it can be sown during October-December period.

Seed rate of 75-100kg/ha at spacings of 30x5 or 45x5 cm is practiced. NPK at 20-80-20 kg/ha is normally applied. FYM at 10t/ha and S at 20kg/ha are also applied. The crop duration ranges from 90-140 days depending on season and variety. The pods are straight or curved, with one to three seeds. The pods are harvested when they become golden yellow, black or gray. Threshing can be done mechanically or manually. Yields ranging from 1600-2000 (rainfed) or 2000-3000 (irrigated) can be obtained. Seeds, sun dried to less than 10% moisture can be stored in steel bins and polythene or gunny bags.

Seed Composition

Soybean seeds have two major parts, the seed coat and the cotyledons. The hundred seed weight ranges from 13-20g (Celik et al., 2001). The soybean seed consists of 8% hull, 90% cotyledons and 2% hypocotyls. The protein and oil account for 60% of the seed with about 30% carbohydrates. Minerals are present to an extent of 5%. The oil content ranges from 14-23% and the protein 32-42%. The protein content of some soybean varieties range from 30- 36% with the corresponding nitrogen values being 16.5-17.9g/100g protein (Zarkados, et al., 2007) The variation is mainly due to location, season and genotype effects. The protein fraction exceeds that of oil fraction in soybean. The proximate composition of soybeans is presented in Table. 53.

Table 53: Proximate composition of soybean.

Constituent	Whole seed	Cotyledons	Hulls
Crude protein %	32-42	42-43	9.0
Crude fat (oil) %	14-23	20-24	1.0
Carbohydrates %	31-37	27-33	86.0
Ash %	4.9-5.0	4.2-4.5	4.5

Source: Carter and Hopper, 1942.

Oil Content and Quality

The average oil content of soybean seeds is generally well below 20%. The oil content is influenced by temperature. Higher temperatures increase the oil content from 19% at 21^{0}C to 23% at 29^{0}C. The oil contains about 88% neutral lipids, 10% phospholipids and 2% glycolipids.

Table 54: Characteristics of soybean oil.

Characteristic	Range
Free Fatty Acid, % oleic	0.57-1.06
Peroxide value, meq O_2/kg	5.91-9.41
Iodine value	118-132
Saponification value	188-196
Unsaponifiables	0.72-1.46%
Specific gravity 25/25^{0}C	0.919-0.922
Refractive index 25^0 C	1.436-1.471

Source: Celik et al., 2001.

Crude soybean oil is obtained by pressing or by solvent extraction methodology, with the latter predominating in view of the low oil content in the seeds. Good quality crude oil has a light amber colour which upon alkali refining, is reduced to light yellow colour. Freshly extracted oil has low level of free fatty acids. It is liquid at ambient temperatures and serves as a salad oil. Due to it's higher I.V., it is categorized as a drying oil. It is a light weight oil with a specific gravity of 0.92. The unsaponifiable content of soybean oil is around 1.45%, with phosholipids being the major constituents. It's S.V. is similar to that of other vegetable oils. The oil is mainly utilized in the manufacture of mayonnaise and shortenings. It is also to some extent utilized in the manufacture of drying oil products. A by-product of soybean oil processing, lecithin is used in many products as an emulsifying agent.

Fatty Acid Composition

The saturated fatty acid content ranges from 11-16%. Soybean oil predominantly contains linoleic acid (48-58%) followed by oleic acid (19-30%). Palmitic acid is present upto 12% while stearic acid is present up to 6%. Soybean oils contain 5-9% linolenic acid. The fatty acid composition of soybean oil is presented in Table 55. The soybean oil is somewhat unstable because it contains both the highly unsaturated fatty acids, namely, the linolenic and linoleic acids. The content of these fatty acids greatly change with the environment, especially the temperature (Hou-Guang Yun et al., 2006). Lower temperatures increase the unsaturated fatty acids. Flavour reversion and rancidity are important problems in soybean oil which are mainly due to the higher levels of 18:2 and 18:3. The ratio between MUFA and PUFA ranges from 0.30- 1.39 again indicating instability of the oil (Celik *et al.*, 2001). Industrial processes have been developed to partially hydrogenate the soybean oil.

Partial hydrogenation and winterization of Indian soybean oil has been attempted by Taneja *et al.,* (1986) and Laxminarayana *et al.,* (1987). This oil has I.V. of 110^0 C and remains clear at 20^0 C with winterization at 10^0 C. The reduction in linolenic acid has been made from around 7to 1%. Hydrogenation results in the production of undesirable trans-fatty acids. Hence to overcome the same, genetic research has been directed towards producing low linolenic soybean (less than 4%). Similar efforts to reduce the undesirable palmitic acid are also going on (Cardinal *et al.,* 2007). However, it needs to be mentioned that linolenic acid is an important essential fatty acid and has its own importance in human metabolism. Lower level of saturated fatty acids, moderate levesl of MUFA and higher EFA make soybean oil, a highly nutritious oil.

Table 55: Fatty acid composition of soybean oil.

Fatty acid	% Composition
Palmitic	7-12
Stearic	2-5.5
Oleic	19-30
Linoleic	48-58
Linolenic	5-8.8

Oil Cakes and Protein

Soybean meal is the most widely utilized oil meal. It serves as a protein supplement for all classes of animals. The meal contains 44 to 50% crude protein. Since soybean is mostly processed through solvent extraction, the resulting meal generally contains very little oil. The proximate composition of cakes is presented in Table 56.

Soybean meal is an excellent protein supplement for lactating cattle and calves. Undehulled press cake contains 41% protein while the solvent extracted meal contains 44% protein. The dehulled cake contains around 50% protein. Soybean proteins are often added to improve the functional properties of other food products such as emulsification, water absorption and adhesion. Well-processed soy protein products can be used as meat and fish protein substitutes. In conjunction with cereal based diets, soy protein flour can improve the nutrition of children and adults.

Table 56: Composition of soybean cake.

Component	Composition
Moisture	10.9%
Crude protein	48.5%
Crude fibre	3.0%
Fat	1.0%
Ash	6.0%

Soybean contains relatively higher levels of lysine and hence can be used to supplement cereal based diets. Methionine and cysteine are the limiting amino acids. It is richer in glutamic acid, leucine arginine, proline, serine, phenylalanine, valine and isoleucine. Arginine is linked to wound healing properties and reduction of the bad LDL cholesterol. The amino acid composition of soy protein is presented in Table 57.

Table 57: Essential amino acids of soybean.

Amino acid	Content (g/16g)
Lysine	6.4
Methionine	1.1
Cysteine	1.4
Tryptophan	1.4
Threonine	3.9
Isoleucine	4.6
Leucine	7.8
Phenylalanine	5.0
Valine	4.6
Arginine	7.3
Histidine	2.6
Tyrosine	3.8
Serine	5.5
Glutamic acid	11.8
Glycine	4.3
Proline	5.5

Soybean contains on an average 40% protein, which is higher than that of any other legume (chickpea = 15-30%, peas = 21-33%, cowpea = 20-39%, pigeonpea = 19-28%, greengram = 21-23% and lentil =20-30%). Soy protein is on par with egg or milk protein with respect to its quality, especially its amino acid make up. It has 50-60% BV, 60-75% digestibility coefficient and 2.0 PER. It has higher lysine and like any other vegetable protein, it is deficient in sulphur containing amino acids. Blending of soybean flour with sesame meal improves its protein quality.

Nutritive Value of Oil

Compared to any other pulse or legume, the oil content in soybean is higher and is around 20%. Other pulses have it around 1 to 5% only. The oil is more unsaturated (78%). The PUFA (linoleic and linolenic) content is around 58% with 20% oleic acid. Because of this, it is considered a very good nutritious oil. The oil contains around 2% lecithins, a food emulsifier. During processing, it is separated out, and serves as a useful by-product. Because of high unsaturation and around 7% linolenic acid, soybean oil has flavour and odour reversion. To avoid the same, the oil is sometimes, partially hydrogenated to confer better storability and oxidative stability. The oil has sufficient quantities of tocopherols (Vitamin E).

Carbohydrates

The total carbohydrate content of soybean seeds ranges from about 18 to 30%. It contains 13.2% available and 17.6% nonavailable sugars. The defatted flour mainly contains 7-8% sucrose, 5-6% stachyose and about 1% raffinose. Flatulence arises due to the presence of the latter two sugars. The hulls contain 40% fibre (cellulose). The starch content is very low (less than 5%).

Minerals and Vitamins

The soyflour contains about 6% minerals. The seeds have Ca, Mg, P, Fe, Cu and Na. The contents are Ca=226 mg, P = 546mg, Mg=236mg, Cu=2.4mg and Na=27.9mg/100g. Soybean is a rich source of vitamin A and B group (thiamine, riboflavin and niacin). It is rich in water soluble vitamins and folic acid.

Soybean oil contains maximum vitamin E constituents among vegetable oils. α- tocopherol=144, β-tocopherols =16, γ- tocopherol=870 & δ-tocopherols =342, mg/kg oil with the total being 1370 mg/kg. γ and δ tocopherols are especially higher. γ-tocopherols impart high temperature stability to oils. Soybean oil, deodouriser distillate contains tocopherols upto 15%. The oil also contains phytosterols at 0.35%. It can be concentrated in the deodouriser distillate to 18%. Recent HPLC analysis has vevealed the tocopherol composition as follows.: alpha=77.7, beta=1.7, gamma=797 μg/g oil, indicating again the very high level of gamma tocopherol (Rukminidevi, *et al.*,2000).

Table 58: Vitamin content of soybean seeds.

Vitamins	Content
Betacarotene (mg/g)	0.2-2.4
Thiamin (mg/g)	11.0-17.5
Riboflavin (mg/g)	2.3
Niacin (mg/g)	20-26
Pantothenic acid (mg/g)	12.0
Pyridoxine (mg/g)	6.4
Folic acid (mg/g)	2.3
Choline (mg/g)	3.4
Ascorbiric acid (mg/g)	0.2
Tocopherols mg/kg oil	900-1370

Food Uses of Soybean

Soybean has a number of nutritional advantages over other food legumes. Soybean is an excellent health food. It contains about 40 per cent protein, 23 per cent carbohydrates and 20 per cent cholesterol-free, EFA rich oil along with sufficient levels of minerals and vitamins. Soybean has twice as much protein as pulses. Whole soybean as well as partially/fully defatted cake/meal can be used for making various soybased food items. Partially or fully defatted beans can be used in making baked products (chapatis, bread, biscuits, bun, rusk and cake), texturised soy protein, protein isolates and concentrates, extruded snack foods etc. Soybean products can solve the malnutrition problem in children most effectively and economically in the developing countries. The oil apart from being rich in EFA is low in saturated fatty acids.

Soybeans can be processed into soyflours, protein concentrates, protein isolates and fermented food products. Soyflour is a finely ground product processed from full fat cotyledons or defatted flakes. Soyflour is similar to non-fat milk power. Soy protein concentrate contains 70% protein while the protein isolate contains about 90% protein. Other products made from soybean are (i) beverages like whole milk or blends and milk extender, (ii) coagulated products like *tofu* (*paneer*), (iii) fermented products like *tempeh*, soy sauce, *miso*, yoghurt and cheese analogues, (iv) sprouted cooked beans (v) fried and roasted nuts (vi) fortified foods like bread, biscuits, cakes etc., (vii) substitute products like *besan*, frozen dessert, meat analogues etc. (ix) textured products like nuggets, chunks etc. and (x) high protein foods like biscuits and granules.

Soybeans can be processed in different ways for production of a variety of soyflours viz, full fat soyflour, medium fat soyflour, high fat (refatted) soyflour, lecithinated soyflours, etc. with varied nutritive values.

Table 59: Proximate analysis of soyflours (per cent)

Constitutent %	Soy-flour type			
	Full fat	Medium fat	Defatted	Lecithinated
Moisture	5.0	5.5	5.0	5.5
Protein (Nx6.25)	41.5	46.0	53.0	45.2
Fat	21.0	6.5	0.9	16.4
Curde fiber	2.1	3.0	2.9	2.4
Ash	5.2	5.5	6.0	5.3
Carbohydrates	25.2	32.50	32.20	25.20

Full Fat and Medium Fat Soyflours

Full fat soyflour is essentially dehulled, blanched and ground soybeans. It is produced by dry heat treatment given by extrusion equipment. Where the expelling unit is coupled with the extruder, edible quality medium fat soyfluor along with good quality oil is produced.

Full-Fat Soyflour

Soybeans are processed to retain all of the fat present in the soybean. The bean is subjected to a mild heat treatment to minimise enzyme activity and retain maximum nitrogen solubility and functional properties for use in specified applications like supplementation of baking products. Alernatively, if functional property of soyflour is not mportant, different modes of heat treatment namely, wet heat, moist heat and dry heat can be adopted. For wet method, varied approaches of blanching have been suggested either for higher nitrogen solubility index (NSI) or lesser beany flavour. Process gives good quality soyflour with NSI of about 18-19 per cent. Some of the approaches of blanching suggested are:

1. Ten min blanching of water soaked (4 hr) sound soybeans in boiling water or boiling of unsoaked beans for 20 min.,
2. Two hour heat treatment of pre-soaked (4^{0}C, 24h) soybean seeds at 70^{0}C,
3. Blanching of cotyledons at 99-100^{0}C in water for 30 min.
4. Immersion cooking of whole soybeans at 90^0 and 100^{0}C for 25 and 5 min respectively.

In moist heat treatment method, the hot water blanching step is replaced by steaming or steam blanching. Approaches suggested for obtaining good quality soyflour include:

1. Open steaming of cracked beans to 100°C completed in 5 minutes,
2. Steaming of cotyledons (17-19% dry bean moisture content) at 99°C for 25 seconds,
3. Adjusting cotyledon moisture to 16.3 per cent with pH 9.8 buffer and then heating in steam at 91°C for 10 seconds
4. 15 min steaming (100°C) of overnight soaked whole soybeans in 0.5 per cent sodium bicarbonate solution at 8.5 pH etc.

The dry heat treatment may be accomplished either by direct air heating, roasting or extrusion cooking. Steps adopted depend on desired end product quality. For good quality full fat soyflour following approaches are suggested:

1. Dry heat treatment of cracked soybeans at 99°C for 62 min.,
2. Dry heat conditioning at 100°C followed by extrusion cooking,
3. Dry heating of soy-splits for 15 min. at 140°C at 10 per cent moisture content or 10 min heating at 15-20 per cent moisture content,
4. Roasting of whole soybeans for 210 seconds at 25-57 per cent moisture content,
5. Roasting of whole soybeans in salt bed at 272°C for 20 seconds,
6. Roasting of whole soybeans at 12 per cent moisture content in a hot sand bath (217°C) for 90 seconds
7. Extrusion cooking of coarsely ground soybean at 125-130°C without pre-conditioning at 10 per cent moisture content.
8. Extrusion cooking of coarsely ground material at 139°C after conditioning to 16 per cent moisture content.

Textured Vegetable Protein

Texturised protein, a unique product made by extrusion, from soy flour, is one of the most successful applications of Soy Flour. A 100% vegetarian product of high protein value used in a variety of dishes such as *Thakkaali,* Stuffed Parathas, *Kababs,* South India *Upma, Sambar, Pulao,* Peas Curry, *Dahi Vada,* Soup, *Kheer,* Sombrero Spread, Patties, *Shahi Pulao,* Crispy *Samosas,* Burgers, etc., thereby enhancing their nutritional value.

Exturders also are able to produce a meat analogue that has a remarkable similarity in appearance, texture and mouthfeel to meats. The utilization of extrusion cooking throughtout the food industry has shown that a variety of products can be made on extrusion equipment. Some of these products include breakfast cereals, breading, snacks, instant rice, instant pasta, starch modifications and animal and aquatic feeds.

Protein Isolate

Protein isolate is another value-added product extracted from defatted soy flour. Soy protein isolate is generally manufactured by the aqueous alcohol extraction system or by alkali extraction. The new soy protein concentrate manufacturing technologies have created the ability to render soy protein concentrates more functional, such as increased solubility, water and fat holding capabilities, high emulsification capacities. etc.

Soy protein isolate is highly digestive and it contains all the essential amino acids required in human nutrition. It is a high quality isolated protein of high molecular weight. In the process of manufacture of soy isolate, the natural soybean oligosaccharides, sucrose, stachyose and raffinose are removed. Therefore, soy isolates are known to reesult in very negligible intestinal gas formation when consumed internally.

Besides being a concentrated protein of high quality, it has several desirable functional characteristics and /or properties such as moisture binding, emulsifying, stabilising, dough-forming, adhesive and thickening, foaming and foam stabilising, etc. that may be used to advantage in various food and pharmaceutical applications.

Soy Fortified Biscuits

Process for making soy fortified biscuit has been standardized for adoption at home and industry levels. The soybean is used in the form of fulfat soyflour, 30% (Baker's percent) fortification of wheat maida with soyflour gives highly acceptable good quality biscuits containing about 12% protein and 24% fat. About 1.8 kg biscuits can be made from 1000g maida. The process for the production of soy fortified biscuits at home/cottage/small scale level consists of creaming of sugar and shortening followed by mixing of other ingredients, sheeting of mixed dough, cutting of sheeted dough using a die mould and baking at 200^{0}C for 15 minutes. These processes are similar the normal biscuits manufacture. The soy fortified biscuits have a great potential for combating protein-calorie malnutrition at low cost. The biscuits have been accepted very well by the consumers.

Soy Paneer

Soy *paneer*, popularly knows as *TOFU* in the Orient, is a coagulated and pressed soy protein, used as a vegetarian food ingredient. Milk *paneer* is very popular among high income group urban people in India for making vegetarian *curry* and *paneer pakoda.* However, the cost of milk-paneer is very high, and hence, is out of economic reach of the majority of Indian people. Soy *paneer* is an appropriate alternative to milk-*paneer* as the cost of the former is very low, and gives all the nutritional advantages. Soy paneer, therefore, has a great potential to replace costly milk-paneer and provide an opportunity to low and middle income population to enjoy the delicacy of paneer.

The home/cottage level process advocated for the production of soy paneer consists of cleaning and grading, dehulling, soaking of soysplits in water (containing 1% $NaHCO_3$ w/v basis) with bean to water ratio on 1:3 at room temperature for 4 hours or for one hour at 60^0C, grinding of soaked and thoroughly washed splits using hot water (bean to water ratio= 1:8). Boiling of soy-puree for 15-20 min, and then removing the okara through filtering, cooling the milk thus obtained to about 70^0C, addition of 0.02M $CaSO_4/MgCl_2$ and stirring the mixture 2 or 3 times, filtering of proteinate complex through muslin to cloth lined paneer forming boxes, pressing the soy paneer at 10g/cm^2 for about an hour, removing the soy paneer blocks to cold water and keeping it there, after washing, for about 30min, and then storing the paneer in water till use. Using the above process, 1.5-2kg soy paneer could be obtained from one kg cleaned and dry soybean. However, yield of soy paneer depends on the moisture content. Soy paneer at about 72 % moisture, contains approximately 14% protein, 8% oil, 3% carbohydrates and 2% ash.

Soybean Sprouts

Soy sprouts have been used as food in the Orient since ancient times. They may be used as salad (uncooked), boiled in water, or may be fried in fat and may be used alone or mixed with other edible vegetables. Sprouting improves the nutritive value.

Soy Fortified Baked, Extruded and Infant Foods

5-25% soy fluor can be used in bread cakes and cookies, doughnuts, confections, snack foods, macaroni products and infant foods. The quality of the foods, especially, the protein content increases, thus improving the nutritional quality.

Soy Milk and Its Products

Soy milk is a popular drink in China. It is a low cost, high protein drink suitable to lactose intolerant children. Soy milk is also the base for preparation of *Tofu (paneer)*, soy-yoghurt, soy ice cream etc. These items are cheaper than milk items.

Fermented Soy Foods

Fermented soyfoods are used as flavouring agents to generally blend, low protein rice diets. Fermentation involves a substrate consisting of soybean and cereal food and an inoculum consisting of bacteria, yeasts and moulds. Fermented soyfoods are soy-yoghurt, *tempeh*, soy sauce, *miso, natto, tofu*, etc. Of these soy sauce and *tempeh* are very popular.

Soy Fortified Indian Foods

Indian foods like *paratha, halwa, laddu* and cookies have been prepared with 50% soy fluor in place of *besan* and other fluors. Full fat soy fluor has also been used in preparations like *chapathi, puri, pakoda, sev,* soy-*sattu, idli* and *dosa*. Green beans also serve as vegetable. All the foods are acceptable and have great potential for wider use.

Soybean oil Derivatives

Soybean oil is nearly an ideal oil for processing to meet the requirements of the food processor, food service operators and in the household. It's advantages far out-weigh the disadvantages relating to shelf life and flavour reversion. It has low refining losses, it is a natural winter oil, contains heat sensitive pigments and forms large easily filterable crystals. It's high iodine value permits hydrogenation to obtain products of varying utilities, rich in PUFA and EFA. It is available in large quantities at reasonable rates. It's disadvantages are it contains a triunsaturated fatty acid namely linolenic acid (6-8%) which decreases its cooking qualities and it is a beta crystal former.

Soybean oil is a natural winter oil which means that it will remain clear at cool temperatures for long periods before clouding. Therefore, only minimal processing, is reqired to produce a cooking or salad oil.Partial hydrogenation followed by winterization and deodorization of soybean oil is employed to improve flavour stability while maintaing a clear liquid oil at cool temperatures.

The main derivatives of soybean oil are lecithin, margarine, shortening (*vanaspati*), fatty acids, glycerol, etc. The use of soybean oil as cooking and salad oils and margarine is increasing as the people are becoming more health conscious and wish to replace animal fats and butter with cholesterol free vegetable oil. Fast food companies world over are switching from animal fats to vegetable oil for frying foods. Hydrogenated soybean oil is utilised in shortenings for baking and frying in food industry, hotels, restaurants, and homes. On global basis, soybean oil for cooking and salad uses accounts for 45% of soybean oil consumption. The use of soybean oil for baking and frying accounts for another 35% of its consumption, while margarine products accounts for 16%.

Table 60: Constituents of crude and refined soybean oil.

Constituents	Crude soybean oil	Refined soybean oil
Triglycerides, %	96.0	99.0 or more
Phosphatides, %	2.0	Trace
Unsaponifiable matters, %	0.6	0.3
Free fatty Acids, %	0.5	Trace
Other substances, %	0.9	0.7
Total	**100.00**	**100.00**

Hyderogenated Soybean Oil/Fat

Hydogenation increases the melting point of oil/fat and resistance to oxidation and flavour deterioration. Dual purpose cooking/salad oil is made by a light hydrogenation so as to remain liquid but have improved flavour stability. Winterisation follows the hydrogenation so that unsaturation could be removed more without losing the liquid character of the finished oil, especially at refrigerator temperature. Shortenings, margarines, confectonery fats and other specialised food fats usually involve a greater degree of hydrogenation and/ or a mixture of hydrogenated products.

Margarine is a water-in-fat emulsion in which the water droplets are kept separated by the fat crystals. It contain a minimum of 80% fat. The other constituents are water, emulsifier, salt, preservative, antioxidant, colour, flavour and vitamins. It is used as table spread in place of butter or as an addition to butter.

Shortenings are made from one or more partially hydrogenated vegetable oil base stock or mixture of such stocks with animal fat base stock. They are used in baking to tenderise baked products by preventing the cohesion of wheat gluten strand during mixing.

Soylecithin is obtained from crude soybean oil by means of hydration, centrifugation and drying. Lecithin content in the oil is about 2% and the commercial lecithin has less than 1% moisture. Soylecithin is a multi-functional ingredient for food and industrial uses. The major applications are in food industry, cometics, pharmauceuticals, dietary supplement, paints/ coatings, plastic and rubber, glass and ceramics, paper and printing, etc.

Antinutritional Factors

Soybean contains many antinutritional factors like protease inhibitors, lectins, phytates, polyphenols etc. Some of these inhibitors present no problems.

A few of them like trypsin inhibitor and hemagglutinins can be deactived under moist or dry heat treatments. Phytate which makes minerals non-available is present from 1-2%. Stachyose and raffinose present in the seeds cause flalulence. Most of the antinutrients are absent in products like protein concentrates and isolates.

Tailoring Varieties for Feed and Food Uses

Soy beans have great potential for use as human food because of their high level of good quality proteins and their unique functional and nutritional properties. Soybean seed has an ideal chemical composition (Wilson, 1999). The health benefits of soy-protein relate to the reduction of cholesterol levels and menopause symptoms and the reduction in risk of several chronic diseases, i.e. cancer, heart disease and osteoporosis. The cholesterol-lowering effect of soybean has been attributed to isoflavones (Potter, 1998). The isoflavones namely daidzein and genistien present in soybean may reduce risk of a number of cancers (Broihier, 1997) and also may help prevent and treat osteoporosis (Anderson *et al.*, 1995).

It is evident from recent studies that soybean alleviates pains of women physiologically. Soybean isoflavones, resembling sex hormone estrogens but being much weak, can greatly help alleviate menopause symptoms viz, hot flashes and night sweats that often interfere with vitality. Soy-hulls contain approximately 65% dietary fiber and offer a source of fiber when used in various food applications (Shogren *et al.*, 1981). Soybean also contains lecithin, a mixture of various phospholipids, which is a natural emulsifier combining many characteristics that improve baked foods (Aust, 1993). Lecithinated soy-flour is an important ingredient in reduced-fat bakery foods.

Despite its virtues, there are constraints associanted with food uses of soybean (Liu, 1999). Soybean has a beany flavour which is not liked in some countries. This is due to the presence of lipoxygenases which catalyse the hydroperoxidation of some unsaturated fatty acids and lipids. The oxidation products further decompose to middle chain aldehydes and alcohols which are responsible for the undesirbale grassy-beany flavour and bitter taste in soybean products. Varieties which have low proportion of unsaturated fatty acids may also be desirable for reducing beany flavor. On an average, the current soybean varieties contain about 60% polyunsaturated fatty acids viz, linoleic (50%) and linolenic (9-10%) acids in seed oil.

Trypsin inhibitor proteins, especially the Kuntiz trysin inhibitor (SBTI-A), are considered the major anti-nutritional factors in raw mature soybeans. Mutants lacking in trypsin inhibitor are available (Hymowitz and

Hadley, 1972; Orf and Hymowitz, 1977 and 1979) and can be employed for heat-inactivation of the inhibitor.

Soybean Products

The processing and value addition in case of soybean is of paramount sinificance. The technology towards this end is evolving at a very fast pace. Some of these new processing techniques include modified seed preparation methods, use of new solvent extraction systems to deactivate the enzymes prior to processing, and use of various physical and chemical methods to remove phospholipids, colouring pigments, free fatty acids, undesirable flavour and aroma from the crude oil.

Utilization of membrane technology in edible oil refining, industrial membrane processing methods for production of soya milk, protein isolate, new types of textured soy-protein products and supercritical fluid extraction for removing chlorophyll and lecithin and for selectively producing functional and natural products have been developed. Although all this frontline technology cannot be commercially and readily adopted, there are several feasible new technology areas. One such area is the adoption of the expander processing technique prior to solvent extraction to reduce energy costs and increase production capacities of oil mill. This process also results in oil and defatted meal products with improved quality.

"SoyCow" machine for soymilk/soypaneer productions has been developed jointly by Child Haven International (CHI), Canada and Prosya Inc., Canada. This machine produces 15 litres of nonbeany soy milk without chalkiness and with excellent soluble protein yield in one batch by using 2 kg of soybean grain.

A Fertility Builder Crop

Soybean cultivation does not deplete soil fertility. On the other hand, it enriches the nitrogen status of the soil as it is a leguminous crop due to it's nitogen fixing capacity. It is szimilar to groundnut in this matter. Amount of nitrogen fixed by the soybean ranges from 70 to 120 kg/ha.

Niger

Niger (*Guizotia abyssinica L.*) is a minor oilseed crop mainly grown in Ethiopia and India. The genus *guizotia* belongs to the family *compositae (Asteracea).* Niger is an annual dicotyledonous herb. Niger appears to have originated in Ethiopia. It spread to East African highlands and India. It is a major oilseed crop of Ethiopia accounting for 50% of its total oilseed production and a minor oilseed of India accounting for 5% of its total oilseed production. Being a minor oilseed crop of the world, niger received limited research efforts. It is cultivated as a sole or mixed crop by poor farmers and also by tribal farmers on soils with poor fertility.

Niger seed contains 35-40% oil, which is edible and highly nutritious. The whole seed also is edible. The cake serves as an animal feed. The seed, oil and the cake are utilized for various purposes at home as well as by the industry. The seed is edible with no known anti-nutrients. It's oil is rich in linoleic acid, the essential fatty acid, (up to 70%) and its cake is balanced in most of the essential amino acids. The plant and it's parts have some medicinal uses and also serve as fuel and organic manure. Niger has an added advantage of yielding oil in marginal and submarginal lands in and around forests.

Niger is a temperate region crop that has adapted to a semi- tropical environment. It is a short day plant. A rain fall of 100-130 cm is sufficient for a good niger crop. In India it is grown in *kharif* (June-November) under rainfed conditions. Niger can grow on wide range of soil types, but appears to thrive best in clay loams and sandy clays. The crop prefers light textured soil and common heavy clays and black cotton soils are not suitable for niger. The optimum pH range is 5.2 - 7.3. Generally, 5-8kg/ha seed is required when sown in pure stands. Row widths vary from 20 to 30 cm depending on soil moisture and amount of fertilizer applied.

The most usual method of sowing by farmers in India is by broadcasting. FYM at 5t/ha is applied. NPK at 20:20:10 can be applied. The crop matures in about 80-145 days. The crop is harvested when the leaves dry up and the head turns blackish. After drying in the sun for about one week by stacking, the crop is threshed by beating with sticks. The seed yields range from 400-500 kg/ha, with a potential upto 1000kg/ha. The dried seeds are stored in gunny bags in houses with good ventilation. India produces around 115,000 tonnes and Ethiopia 7000 tonnes of niger annually. India earns around Rs.30 crores through export of niger seed (Damodaram and Hegde, 2003).

Seed and Oil Quality

Niger seed is an achene 3-5 mm in length and 1.5 mm in width and lanceolate in shape. The testa is hard, glossy and normally black. One thousand seed weight ranges from 1.6 to 6.0 g with a mean value of 3.8 g. The information on niger seed and oil quality characteristics have been compiled and reviewed by Hilditch *et al.* (1994), Maiti *et al.*, (1988) and Nagaraj, (1990b). In general the composition is as follows: the seed oil 30-43% (mean 40%) protein 10-30% (20%), and soluble sugars, 7-18% (12%). The seed also contains 10% crude fibre and 4% ash. On the higher side, the seeds may contain up to 60% oil.

Sahasrabuddhe (1925) has observed the oil content of 10 niger samples to range from 40.9 to 45.65 with a moisture content of 2.2 to 3.9%. Sahasrabuddhe and Kale (1933b) have analysed and reported oil quality values such as specific gravity (0.92), melting point (8°C), acidity (4) and saponification value (176). Sahasrabuddhe and Kale (1933a) studied oil synthesis in niger and have reported that sugars got converted to low molecular weight fatty acids, higher molecular weight saturated fatty acids, unsaturated fatty acids and later glycerides are formed. Enzymes in the seed are responsible for the transformations. Higher activity of the enzymes result in higher oil accumulation in the seeds.

The triglyceride composition of the niger oil has been determined by Vidyarthi and Mallya (1940) which is as follows: trilinolein 2%, oleo-dilinolein 40%, dioleolinolein 30%, myristo dilinolein 2%, myristo oleo-linolein 3%, palmito-dilinolein 6%, palmito oleo-linolein 11%, stearo-dilinolein 2% and stearo-oleo-linolein 4%. Niger oil contains 93% neutral lipids 4.9% glycolipids and 0.6% phospholipids(Ramadan and Morsel, 2003). The major pospholipids are phosphatidylcholine (ca. 8%), phosphatidylethanolamine (22%), phosphatidylinositol (14%), and phosphatidylserine (ca. 8%) (Ramadan and Morsel, 2003a)

Dunn and Hilditch (1950) have furnished information on niger seed oil and its quality. It is a pale yellow oil. A yield of 35% is obtained, when clean niger seed from Southern Rhodesia has been extracted with light petroleum ether. The other characteristics of the oil are saponification value =193, iodine value =135, free fatty acids = 1%, unsaponifiables = less than 1%, refractive index at 25°C = 1.4723 and at 40°C = 1.44679. They have used salt - alcohol fractionation, iodine number and also spectrophotometric methods to study the fatty acid composition. The results are as follows: palmitic = 7.9-8.7, stearic = 10.6-10.4, arachidic = 1.0-0.9, oleic = 7.0-6.9, linoleic = 72.6 - 72.8 and linolenic acid - 0.9-0.8%.

Nasirullah, *et al.*, (1982) after analysing eight niger samples from Maharashtra and Gujarat states of India have reported the following composition: oil content =30-33.4%, protein = 26-30.6% and moisture = 1.7 to 3.0%. The fatty acid composition of the oil is as follows: oleic acid = 13.4-39.3%, linoleic acid = 45.4 - 65.8%, palmitic acid - 5.8-13%, stearic 5.0-7.55 and arachidic acid = 0.21-1%.

Table 61: Niger seed composition and oil quality.

Constituent	Composition (1)	(2)
1000 seed weight (g)	1.6-6.0	-
Oil %	30-43.2	32-32.4
Protein %	10-30	26-30
Soluble sugars %	6.9-18	-
I.V.	138.2	112.8-129.0
Refractive index 25°C	1.4715	1.4695-1.4722
Unsaponifiables	-	0.2-1.7
Colour of the oil	2.97 + 0.5 R	2.9 Y + 0.6R

Source: (1) Nasirullah *et al.*, (1987); (2) Nagaraj (1995)

Mechanical extraction and solvent extraction of niger seed have given oils, differing in fatty acid composition (Ahmed, 1989). The former has 83% linoleic and 3.5% oleic acid, while the latter has 71% linoleic and 12% oleic acids. The differences might be due to higher temperatures (steam) used during solvent extraction which might have oxidized and reduced the linoleic acid.

Oil content in niger has been shown to be affected by locational altitude and/or temperature which would make selections for higher oil content difficult (Westphal and Kelber, 1973). Oil content is also influcenced by hull thickness. The hull content of 25 accessions varied from 13.5 to 36.6%. Low hull varieties have higher oil, protein and low crude fibre (Getinet and Belayneh, 1989)

Riley and Belayneh (1989) have furnished information on oil quality. The niger oil is pale yellow in colour with a sweet odour and pleasant "nutty" sweet taste. However, on storage at moderate room temperatures for more than six months, the oil has thickened and become rancid. The fatty acid composition of niger oil is similar to that of safflower and sunflower, with high content of up to 75% linoleic acid, except that it may contain a small amount of (2%) lignoceric acid. Indian and Ethiopian oils are reported to differ markedly in linoleic acid content, with the Ethiopian oil averaging 70% and extending upto 85% under cool conditions. Indian oil, has 52 to 74% linoleic acid, with the difference being made up by oleic acid. It is believed that the temperature under which the seed matures is the major determinant of this variation.

Nagaraj (1990b) has studied the fatty acid and amino acid composition of five niger varieties. Highest oil content of 39.4% has been observed with a range of 30.9 to 39.4%. The refractive index of the oil is around 1.4715 with a bright yellow colour (mean value 2.9Y+5R). Oleic acid and linoleic acids are the main fatty acids of niger varieties which together accounted for about 85% of the total fatty acids. The oleic acid ranges from 25-29% while linoleic acid ranges from 52-58%. Palmitic and steatric acids are present at around 7to 8%. The stearic acid, a healthier fatty acid than palmitic acid, is higher in niger oil than that of other oils. Arachidic acid at 0.5% level and an unidentified fatty acid at 1.5% have also been reported.The linoleic acid content of these Indian varieties is lower than that observed in varieties from other countries which is around 75%. The differences are mainly because of the higher temperatures prevailing in India during seed formation and maturation stages.

Fourteen niger genotypes have been analysed for seed oil, protein, iodine value and fatty acid composition (Dasthagiraiah and Nagaraj, 1993). The oil content is 37-38%, protein 21-30%, iodine value 137-141, and the linoleic acid ranges from 75-78%. The nutritional quality index (linoleic acid/saturated fatty acid ratio) of the oil is 4.5-5.3. The niger has been grown at Chintapally, Andhra Pradesh a hilly tract in the *Rabi* season with cooler climate. Hence, higher linoleic acid has been observed in those samples

Dutta *et al.* (1994) collected niger seed samples from different regions in Ethiopia for determination of oil content, fatty acids, tocopherols and sterol composition in the seed oil and analysed them by gas-liquid chromatography and high performance liquid chromatography methods. Large variation in oil content, ranging from 29-39% has been observed. More than 70% of the fatty acids is linoleic acid (18:2) in all the samples analysed. The other predominant fatty acids are palmitic (16:0), stearic (18:0) and oleic (18:1) acids at a range of 6 to 11% each. Total polar lipids recovered after preparative thin layer chromatography comprises a small fraction of the total lipids. They have higher 16:0 and lower 18:2 contents than the triacylglycerols.

Table 62: Fatty acid composition of niger oil.

Fatty acid	**Composition %**	
	(1)	**(2)**
Palmitic (16:0)	9.2	6-9.4
Stearic (18:0)	6.6	5-7.5
Oleic (18:1)	10.9	13.4-39.3
Linoleic (18:2)	72.9	45.4-65.8
Arachic (20:0)	0.5	0.2-1.0

Source: (1) Nasirullah *et al.*, (2) Nagaraj (1999b)

Nagaraj (1994) has analysed niger samples from eight locations in India for their fatty acid composition. Oleic to linoleic acid ratio has shown wide variation with respect to location. Linoleic acid is higher (60-70%) when the mean maximum temperature is less than 30°C and mean minimum temperature (less then 20°C) is lower. At higher temperatures, oleic acid levels are higher (around 40%). These as well as other studies thus point out that niger fatty acid profile is thoroughly temperature dependent.

Nine niger genotypes developed at Semiliguda (Orissa state of India) have been evaluated (Venkat Rao *et al.,* 1998) for plant and seed characteristics. Genotype ONS-8 and ONS-123 have given highest yields while *Pittaguda* local has given the lowest yield. Genotype, ONS-1 and *Pittaguda* local have higher oil content of 39%. The seed protein content of the genotypes varies from 20-23%. The main fatty acid linoleic acid range from 74-76%. Among the genotypes evaluated GA 11 and ONS-124 are considered better from the point of view of yield and oil quality.

Studies on Indian and Ethiopian niger samples reveal that, the Ethiopian samples contain higher oil and linoleic acids. However, some Indian genotypes grown under cool climate also have higher linoleic acid (up to 76%). The higher oil content in Ethiopean samples might be due to the longer crop duration while higher linleic acid must be due to the low temperatures that prevail during crop growth. The quality of such oil becomes equal to safflower oil. In addition, the less harmful stearic acid is present at higher level in niger oil making it healthier than safflower oil.

The Unsaponifiables

Álpha tocopherol is the predominant tocopherol in all samples with 94-96% of the total amounting to 630-800 μg/g oil. More than 40% of the total sterols has been observed to be β sitosterol, at 2000 μg/g oil. The other major sterols are campesterol and stigmasterol ranging from 11 to 14%. The 5 and 7 avenasterols are in the range of 4 to 7% each.

Niger Germplasam

The importance of niger germplasm as a source for breeding better quality niger varieties has been stressed by Belayneh (1991). Nagaraj *et al.,*(1994) and Sharma et al., (1994) have evaluated 417 niger genetic resources for their yield, morphological and biochemical characteristics. The days to 50% flowering range from 40-70 days (mean 62.5) while the days to maturity range from 90-111 days (mean 106). The plant height has varied from 100-197 cm with a mean of 142 cm. The number of branches are from 3-17 with an average of 11. The capsules range from 30-110 / plant (average 63/ plant). The thousand seed weight has ranged from 1.6 to 6.0 g with an average of 3.8 g. The per plant yield has shown variation from 0.7-7.3 g with an average of 2.6 g. The mean oil content is 40.2% ranging from 30-43.2% while for protein the values are 20.1 and 10-30% and for sugars, they are 11.9 and 6.9-18.1%. The information furnished provides wide variation in the characters examined. It reveals the scope for breeding varieties with high yield as well as better seed quality.

Getinet and Teklewold (1995) studied the genetic, agronomic and seed quality characteristics of 241 niger germplasm collections from different parts of Ethiopia. The maturity of the niger accessions range from 132 to 168 days. Seed oil content range from 39.8 to 46.9% with linoleic acid being major fatty acid (76.6% of total fatty acids). The results indicate that genetic differences for maturity exist among the *noug* niger accessions. Oil content variation is continuous, without clear separation of accessions into oil content groups. It has been concluded that the genetic variation observed among these accessions could be utilized in breeding programmes to develop high yielding, well adapted, high oil content *noug* cultivars for production in Ethiopia.

Nagaraj (1999) analysed 78 specialty germplasm accessions of niger for their fatty acid profile along with some plant and seed quality characters. Nineteen lines mature earlier than 105 days, while 4 have more than 4.8 g thousand seed weight. Seed yield per plant is more than 4 g in 21 accessions. 70 have more than 40% oil, 43 have more than 20% protein and 10 have more than 14% sugars. The average fatty acid composition of the 78 germplasm lines is 16:0=9.2, 18:0=6.5, 18:1=10.9 and 18:2=72.9. It is concluded that there is scope for breeding niger varieties with higher linoleic (more nutritious) acid content. Since niger oil contains higher stearic acid which is less harmful than palmitic and myristic acids, niger oil could be considered as more nutritious and healthier oil.

Seed weight and days to maturity are negatively correlated indicating that late maturing varieties produced seed with less weight (Nagaraj, 1999). Sugars are negatively related to seed weight. Oil percent is highly significantly and negatively correlated to stearic acid. This is understandable as stearic acid is an intermediate and will get transformed to oleic and linoleic acids as soon as it is formed which are the major constituents of the oil. Sugar is positively correlated to linoleic acid. As a precursor involved in oil synthesis sugar can be expected to be positively correlated to linoleic acid which is the major constituent (73.4%) of the oil. Stearic and oleic acids are highly negatively correlated to linoleic acid. The former two being precursors of the major fatty acid linoleic acid the correlation is on the expected lines. The levels of the former two acids get depleted or transformed quickly to 18:2 in the oil. Hence, higher levels of 18:0 or 18:1 would mean lower levels of 18:2 and *vice versa.* Seed yield does not have any correlation with the characters studied. These studies thus reveal that there is scope for selection as well as breeding improved varieties with higher seed weight, lesser duration and better seed oil quality with higher or lower levels of linoleic or oleic acids.

Table 63: Correlations among plant, seed and oil quality characters.

	Yield	Days to Maturity	Seed weight	Oil percent	Protein percent	Sugar percent	Palmitic	Stearic	Oleic	linoceic
Yield	1,0000									
Days to maturity	0.0910	1.0000								
Seed weight	-0.1551	-0.3155	1.0000							
Oil percent	-0.0980	-0.1360	-0.0523	1.0000						
Protein percent	0.0689	0.0630	0.1646	0.0760	1.0000					
Sugar percent	0.0520	0.0978	-0.2220	-0.1430	-0.1046	1.0000				
Palmitic	0.0423	0.1084	0.0209	-0.0363	-0.761	-0.2036	1.0000			
Stearic	-0.1291	0.0385	-0.0266	-0.2812	-0.1608	-0.0464	-0.0443	1.0000		
Oleic	-0.1175	-0.1976	0.0770	-0.1386	-0.0946	-0.1775	-0.0971	0.2262	1.0000	
Linoleic	0.1387	0.1382	-0.0666	0.1960	0.1217	0.2360	-0.1354	-0.4725	-0.9307	1.0000

Niger Cake

Chavan (1961) has reported the information on niger cake and its importance as cattle feed. Some of its characteristics are as follows: digestible fats, 5.2%, digestible fibres 3.9%, nutritive ratio 1.3% and starch equivalent 51. Niger cake appears to have a fairly high percentage of soluble carbohydrates in addition to its high protein content. The starch equivalent provides a measure of relative values of feeding stuffs for the production of growth, fat, milk, or work, and is also useful for comparing the relative prices of various food stuffs; the price per ton divided by the starch equivalent gives a fair measure of the relative cost of the cakes. Niger cake, as a feed, has a fairly high nutritive value. It is very much relished by the cattle in the Deccan and Konkan regions of India.

Nagaraj (1995) has furnished information on niger cake. Niger cake is dark in colour. It contains 24 to 34% proteins, 4 to 14 percent oil, 6 to 24% crude fiber, 20 to 28% sugars and 8 to 12% ash. Niger cake is mostly utilized as animal feed. The cake contains 5% N, 2% P_2O_5 and 1.5% K_2O and the low grade cake can be used as manure. Protein concentrates can be prepared from niger meal. The protein seems to be more or less balanced in its amino acid composition. It is deficient in leucine, lysine and threonine. Blending of niger cake with other cakes with more imbalance of amino acids is suggested for improving the nutritional quality of those cakes.

Ethiopian niger protein quality has been evaluated using chemical score and essential amino acid score (Haile, 1972). It has been observed that niger protein is somewhat deficient in methionine, lysine, cysteine, isoleucine and leucine. But over all, only lysine is observed to be really deficient. Nagaraj (1990b) analysed the deoiled cakes of five niger varieties for protein content and amino acid composition. The crude protein content of the cake ranges from 37- 43%. The sulfur containing amino acid, methionine is present upto 2.3% while cystine is present upto 2.4%. The total lysine content of the varieties range between 3 and 3.7%. Glutamine is the major amino acid (19%) followed by arginine (8%), asparatic acid (8%) and leucine (6%). Niger protein appears to be deficient in leucine and lysine (Table 64).

Table 64: Amino acid composition of niger protein.

Amino Acid	Range (g/100 g protein)	Amino acid	Range (g/100g protein)
lanine	3.3-3.4	Arginine	7.8-7.0
Aspartic acid	7.6-7.9	Cysteine	2.0-2.4
Glutamic acid	18.5-18.8	Glycine	4.3-4.4
Histidine	2.4-2.9	Isoleucine	3.6-4.0
Leucine	5.2-5.7	Lysine (total)	3.1-3.7
Methionine	1.9-2.3	Phenylalanine	4.1-4.7
Proline	3.5-4.0	Serine	4.2-4.4
Threonine	2.6-2.7	Tyrosine	2.6-3.1
Valine	4.0-4.5		

Source: Nagaraj (1990b).

Daji (1943) who evaluated eleven types of cakes for their chemical composition has furnished information on niger cake composition, moisture 7.6%, fat 6.4%, protein 37.0%, phosphoric acid 0.88%. The higher percentage of inorganic constituents in the niger cake provides it an advantage. A better idea of the nutritive value of oilcakes can be obtained by computing their actual digestible nutients.

A lipoprotein concentrate has been prepared by extracting niger seed with hot water/ethanol and NaCl solution (Eklund, 1971b). The lipoprotein has 4% moisture, 12% ash, 46% protein, 20% fat, 7% crude fibre and 11% soluble sugars. The energy content is 400 Kcal/100g. They also have given a conversion ratio of 5.9 to calculate protein content of niger from its nitrogen content.

Niger expeller cake is normally used as a fuel, and in India it is sometimes used as a fertilizer. In general, ethiopian cake is found to be higher in fiber (24%), lower in protein (24%) and total digestible nutrients than other oilseed cakes. Indian niger cake is found to be lower in fiber (14%) and higher in protein (30%) (Seegler, 1983).

Utilisation

Niger seeds are edible. Their seeds fried in *ghee* (milk fat) can accompany other foods to improve their palatability. They are parched and mixed with pulses as snack food. A spicy item called chutney, prepared by mixing red chilli powder with roasted and pounded niger seeds is a common recipe in some parts of rural India (Nagaraj, 2002). Niger seeds are parched and ground with water to make a drink. The seeds can be sprinkled on bread (Riley and Belayneh, 1989). The seeds have good market value in the export market. It is utilized as bird feed in some countries (Subramanian, 2003).

Niger cake is a very useful cattle feed. In performance it is similar to groundnut and linseed cakes (Sinha *et al.*, 1983; Roy choudhury and Mandal, 1984). Niger cake in addition to being rich in proteins, is also rich in soluble sugars. Niger cake and its protein are balanced with respect to essential amino acids. Its quality is similar to that of sesame cake. Hence, attempts are needed for wider utilization of niger cake as a food ingredient (Nagaraj, 2002). Low grade cake can be utilized as manure or fuel (Vaughan, 1970).

Better grade niger seed oil is used for edible purposes. It is also used as adulterant, particularly for sesame and mustard oils. Niger oil is used for cooking, lighting, anointing, painting and cleaning of machinery (Chavan, 1961, Patil and Joshi, 1978., Patil and Patil, 1981). Low quality oil is used as an illuminant, lubricant and for preparation of soaps and detergents (Vaughan, 1970). It is also used as a body oil for smooth skin and supple joints (Nagaraj, 1995). Niger oil is useful in birth control and treatment of syphilis (Belayneh, 1991). Higher unsaturation of niger oil makes it a useful raw material for paint and varnish manufacture.

Riley and Belayneh (1989) have given information on utilization of niger seed and oil. Ethiopian niger is extracted in small mechanical village expeller mills. Such oil is highly prized for edible purposes which usually commands a premium over the other available oils. Some oil is extracted in the home by parching the seed, grinding it into a fine powder, adding hot water and stirring. This process allows the oil to float to the surface and be skimmed off. Water may also be added to the parched, ground seed to make a drink. Other minor uses include mixing the parched seed with pulses as a snack food, and making a stew or "wot" with ground seed and spices (Seegler, 1983). Niger flour is also baked into a bread or sprinkled on bread prior to consumption. In India the oil is extracted by bullock-powered *"ghanies"* or in mechanized mills. Usually the oil is sold for edible purposes in the local market in its pure form, but may also be mixed with other oils.

Vles and Gottenbos (1989) have reported that dietary fats, rich in linoleic acid prevent cardiovascular disorders such as coronary heart disease, atherosclerosis and high blood pressure. Also linoleic acid derivatives serve as structural components of the plasma membrane and as precursors of some metabolic regulatory compounds. The high linoleic acid content makes the niger oil nutritionally valuable and safe for human consumption. A niger based agar medium has been used to distinguish algae or *fungi* (Paliwal and Randhawa, 1978).

Castor

Castor (*Ricinus communis*) is the most important non-edible oilseed crop. It belongs to *Euphorbiacea* family. The members of this family contain a poisonous milky sap or latex that exudes copiously from cut stems or leaves. In-fact the most important world source of natural rubber comes from several members of the *Euphorbiacea* family. Castor plant is considered native to Ethiopian region of tropical East Africa and India. It has become naturalized in tropical and warm temperate regions throughout the world and is being cultivated widely. It also grows wildly (as a weed) as a perennial plant. Though the seeds are considered poisonous, they are the source of a large number of economically important products and are probably the most ancient commercial products. Seeds and various plant products like stems, roots, leaves and flowers have wide ranging uses in industry as well as at home.

Knowledge of the medicinal uses of castor plant is found in the ancient Indian Sanskrit literature. The use of castor oil has been mantioned in *Susrutha Ayurveda*. Seeds of castor have been found during excavations in Egypt, Sudan, India, Iran and North West Asia. In Egypt seeds have been found in the pyramids dating back to 4000 years. Written evidence is available about its cultivation in Mediterranean region and China dating back to around 1000 years.The interest of the ancient people in castor has been mainly for the medicinal uses of seeds, roots and leaves and the use of oil in perfume and lighting.

Castor is grown as a perennial crop in Africa, Eastern and North-Eastern India, as a biennial crop in South America and as an annual crop in Western and peninsular India. Castor is normally cross-pollinated, wind being the major agent in the tropics. India leads the world in production of castor. It is also an important crop of China, Brazil, Paraguay, Thailand, Ethiopia and Pakistan. It's Latin name *Ricinus* means 'dog tick' because of the resemblance of the seed to the canine pest.

Castor crop is a drought tolerant crop. It can grow under low rainfall conditions ranging from 50-70cm. It needs temperatures above 20^0C with low humidity for good crop yields. It is mainly grown in *kharif* (June to February-March of the next year). It can grow in all types of soils ranging from red to sandy loams and alluvial soils. Before sowing seeds the land is prepared by ploughing 2 or 3 times during May-June. Immediately after rains, the seeds are sown in the field at a spacing of 45x30cm and a seed rate of 15kg/ha. The seed is preferably treated with Thiram or Bavistin @3g/kg to protect the plant from seed borne diseases. Fertilisers at the rate of 20kg N, 40kg P_2O_5 are applied as basal dose prior to sowing. After seedling emergence around 37-45 days and 65-70 days 20kg N is applied depending on necessity. Wherever necessary depending upon non-occurrence of rains protective irrigation is given at intervals of 20 days. Other practises like weeding, interculture are done to obtain a good crop.

Castor, by nature a multiple branching type plant, produces 4 to 5 spikes (1 each) on the branches. The crop requires 180-240 days for total harvest. The first spike can be harvested during 90-120 days and the next three or four pickings can be had at 30day intervals when the capsules turn yellow and start drying. The spikes are cut and dried in the sun for a few days. Threshing is usually done by beating with sticks. The harvested seeds are collected either pick wise or as bulk of all the picks. The seed yields under rainfed conditions range from 800-1000kg/ha while irrigated crop can yield 1500-2000 kg/ha seeds. Storage is normally not a problem in castor.

Castor fruits are spiny globular capsules. They are hard and brittle when ripe. There are three seeds in each castor fruit. The seeds are flattened oval in shape with brittle testa, inside which the oleaginous kernel is present. The seed colour varies from white, brown, grey, faint chacolate, deep chacolate, purple or buff to black or red. Usually the different colours exist as molting on the testa. The seed size varies from 4 to 25mm in length to 5-16mm in breadth. The 100 seed weight ranges from 10 to 100g with an average of 30g. The kernel accounts for 70-83% of the seed.

Seed Composition

The oil content of the seeds range from 40-55%. The oil content of Brazilian varieties range between 39.6 and 59.65 and the Indian varieties range between 46.0 and 51.8% (Patterson, 1989). Hill and Diley, (1951) have observed that castor seeds contain 45% oil with a range of 35-51% with the large seeds containing more oil. Laxminarayana *et al.,* (1984) have reported the chemical composition of castor seeds. The oil content of different genotypes have been reported to range from 46-54%.The oil content of some newly released Indian cultivars including hybrids range from 45-55% (Nagaraj, 1990). The seed also contains protein (18-20%), carbohydrates (2%), ash (2-3%) and phenols (2.5%).

Table 65: Castor seed composition.

Constituent	% Composition
Hull	25-30
Kernel	70-75
Seed oil	40-55
Kernel oil	64-71
Protein	18-23
Soluble sugars	5
Crude fibre	25
Ash	2

Earlier formed castor seeds contain more oil than the later ones, which means that the seeds from the main stem have higher oil (Sethi, 1931). The oil content increases from around 2% (immature) to 53% (mature) over a period of 81 days during seed maturation phases (Kulkarni and Ramanamurthy, 1977). The oil in the seeds starts getting synthsized 20 days after flowering, more than 2/3rds getting synthesized in the next 20 days. Synthesis of protein occurs over a longer period. It has also been established that oil content varies from bloom to bloom in some varieties. The oil content as well as free fatty acids have not shown any variation over a three year period of storage at Coimbatore (Kulkarni and Ramnamurthy 1977).

The nitrogen free extract or the sugar content of the seeds is around 3 to 7%. The crude fibre or cellulose content is 23-27%. Castor proteins contain around 60% globulins. The phosphorus of the seeds is mostly present in the form of phytin. The phospholipids, mainly, phosphatidylethanol amine is present at around 9-12%. Uric acid is present at 6 mg/100g, while HCN is present at 7ppm in the seeds. Many enzymes namely lipase, amylase, invertase, maltase, endotripsin, glycolic acid oxidase, ribonuclease, fat soluble zymogen, catalase, peroxidase and reductase are also present in dry castor seeds and or germinating seeds.

Castor oil is generally obtained by crushing the whole or decorticated seed in expellers, hydraulic presses, rotary mills, screw presses, village *ghanies* and domestic method of boiling. The latter method consists in boiling the seed meal in water and separating the supernatant oil. This oil is highly valued as medicinal oil. The yield of oil through the above methods is as follows: expellers and hydraulic presses, 40-41%; rotary mills and screw presses, 39%; village *ghanies*, 35% and domestic method, 30%. Generally oil for medicinal purposes is obtained by hyraulic pressing or by expeller pressing at low temperatures. The seed is first pressed at between 32^0 to 38^0C to give a pale yellow straw-coloured oil which is bleached by ultra violet light and filtered to give medicinal oil. The cake is re-pressed at higher temperature to extract more oil. The oil is termed 'seconds' oil. The third pressing produces still more coloured oil, both of which go for industrial use.

The crude castor oil is refined using hot water/steam and filtered through activated earth. The refined castor oil is non toxic and highly stable. The acidity of the oil is very low (less than 3). The castor oil is nearly colourles, having mild taste and odour. Castor seeds stored for two years yield darker coloured oil. Similarly higher temperature extraction of seeds at 150^0C have given darker coloured oil. Oil from rainfed or irrigated seeds do not have any differences in colour (Nagaraj and Srinivas, 2002). Caster oil has high viscosity, specific gravity, and acetyl value. It is soluble in alcohol and acetic acid and insoluble in non polar solvents like hexane and petroleum ether, kerosene etc. It is dextrorotatory (α_D= + 7.6 to 9.7^0).

Castor Oil

Since castor oil is inedible and for a long time, it used to be cheap, there has been a tendency to adulterate the same with other premium edible oils, like groundnut and sesame oils. A need has been felt to identity castor oil when adulterated. A TLC (thin layer chromatography) test for screening castor oil detection has been proposed (Nasirulla, *et al.*, 1982). Sajid Hussain (1992), apart from reviewing various detection methods, has developed a new NMR method to detect oil in the seed sample down to 3% level.

Oil Quality

Castor oil is one of the few naturally occurring oils which is like a pure compound. It is unique in that it contains 85-90% ricinoleic acid. This fatty acid has an 18 carbon backbone with a hydroxyl group on the 12th carbon (from the carboxyl end) and a cis-double bond between carbons 9 and 10. The physical and chemical properties of castor oil are unique because of the presence of ricinoleic acid, an organic compound, with a hydroxy group, a double bond, a caboxylic group and a long aliphatic chain. Some of the characteristics of the oil are I.V.=81-91, S.V =176-187, density=0.945-0.965, free fatty acids=2% and refractive index=1.478-1.4810 (Nagaraj, 1990).

Castor oil consists entirely of triglycerides with only 0.3 to 0.7 percent unsaponfiables. The fatty acid composition of castor oil is: palmitic 0.8-1.2%, stearic 0.7-1.2%, oleic 2.0-3.3%, linoleic 3.4-4.7%, ricinoleic 87.7 to 90.4% and dihydroxy steric acid 0.6-1.4%. Fatty acid composition of various castor genotypes has been analysed (Laxminarayana *et al.*, 1984, Ravi Kishan *et al.*, 1995). Ricinoleic acid is present to an extent of 87% followed by linoleic acid (6.1%) and oleic acid (4%). Three other fatty acids namely arachidic, palmitic and stearic acids are present at around 1% each. Muralidharudu and Nagaraj, (1997) and Nagaraj, (1999) have reported the fatty acid constituents with respect to the ricinoleic acid in different genotypes (varieties and hybrids). Their studies have shown that ricinoleic acid is present to an extent of 85-90%, followed by linoleic (4-7%) and oleic 3-7% along with 1-3% each of palmitic, stearic, linoleic and dihydroxy stearic acids.

Variation in ricinoleic acid has been studied (Nagaraj and Srinivas,2001). Twenty genotypes have ricinoleic acid in the range of 80-90%. Improved varities and hybrids have higher levels ranging from 86-90%. Oil from fully mature seeds has 90% ricinoleic acid while oil from immature and over mature seeds has 87% . Similarly oil from seeds of primary branch has 88% ricinoleic acid. Other factors like irrigation and fertilizers do not have much influence on the ricinoleic acid levels. These factors need to be kept in mind while utilizing castor for manufacture of value added products.

Castor oil is inedible and has cathartic action because of the presence of the hydroxy fatty acid. The cathartic action is exerted on the intestinal mucosa or the Auerbach plexus and is independent of the absorption of its constituents from the intestine. In some parts of Tamil Nadu and Orissa states of India, castor oil in small quantities is being utilized for edible pruposes. (Ravi Kishan, *et al.*, 1995). Oil for such purposes is separated by boiling the seed powder in water. The floating oil is scooped out, filtered and water removed from it by heating. This oil is called "cooked" oil and is a premium oil used at household level as a purgative. Apart from direct consumption, in limited quantities, it is a common adulterant in all-edible oils, whenever it is cheaper.

Table 66: Castor oil characteristics and fatty acid composition.

Physicial characteristics	Range	Fatty acid	Per cent (Range)
Iodine value	81-91	Palmitic	0.9-1.2
Saponification value	176-187	Stearic	1.0-1.2
Density	0.945-0.965	Oleic	3.2-3.3
Free fatty acids	2%	Linoleic	3.4-3.7
Refractive index	1.4748-1.4810	Ricinoleic	89-89.4
Hydroxyl value	155-169	Dihydroxy Stearic	1.3-1.4

Uses of Castor Oil

Castor oil is mainly used as a lubricant because of its property to remain liquid at very low temperatures (-32^{0}C), high density and viscosity (18 times that of other vegetable oils). The hydroxyl group binds to metal surfaces making castor oil a tough lubricating oil. The oil in India is used as a body anointing agent. It occupies a pride of place with coconut and sesame oils as a hair oil. The oil also finds use in the treatment of guinea worm abcess and flatulence. It has industrial uses in the manufacture of sebacic acid, surface coatings, disinfectants, cosmetic and pharmaceutical preparations apart from its main use as a lubricating fluid. The major outlet for castor oil in India is in soap making and to a smaller extent in paints and varnishes (Bringi *et al.*, 1985).

Castor oil is a versatile industrial raw material because of the presence of hydroxyl group, a double bond, carboxyl group and a long aliphatic chain. Many chemical reactions can be performed with this ricinoleic acid to obtain wide ranging chemical derivatives (Achaya, 1972; Laxminarayana, 1983). The major derivative of castor oil used in the industry is hydrogenated castor oil (HCO), also called castor wax or opal wax. It is a hard, brittle, high melting waxy solid obtained through controlled hydrogenation. Dehydrated castor oil (DCO), sebacic acid, ethoxylated castor oil, polyols, 2- octanol, undecylinic acid, heptaldeyde, turkey red oil etc. are the other most important products of castor oil.

These derivatives and or the castor oil have wide ranging uses in the manufacture of lubricants, plastics, adhesives, waxes, polishes, coating applications, inks, paints, varnishes, cosmetics, perfumes, flavours, fragrances, textile dyes etc. The development of chemical processes and varied chemical reactions/ derivatives that can be obtained from castor oil are detailed by Achaya (1972), Laxminarayana (1983), Bringi *et al.*, (1985) and Kulkarni and Ramanamurthy (1959). Hydrogentaed castor oil, its importance and preparation has been mentioned by Sovani and Vasistha (1995). Castor methyl esters also have been prepared (Azhar *et al.*, 1999). Fatty acids which are industrial raw materials have been prepared from castor oil through hydrolysis by microbial lipases (Ghosh and Bhattacharya, 1997).

Castor Oil and its Chemistry

Castor oil has relatively high purity (high for naturally occurring material). Hence it can be subjected efficiently to a number of chemical processes to yield high purity chemical derivatives. The chemical bases for such reactions are the three points of functionality existing in ricinoleic acid. a) the carboxyl or ester groups, b) the single point of unsaturation (double bond) and c) the hydroxyl group. These functions are utilized in a wide range of esterifications. Hydroxyl group can be acetylated or alkoxylated. The unsaturation can be altered by hydroxylation or expoxidation (Forbes, *et al.*, 1940). The hydroxyl group can be removed by dehydration to increase the unsaturation of the compound, which can further be made to yield drying compounds and polymers. The hydroxyl position is so reactive that the molecule can be split at that point by high temperature pyrolyis and by caustic fusion to yield useful products of short chain length (Achaya, *et al.*, 1964). Castor oil is used in the manufacture of cleavage products like sebacic acid, undecylenic acid and heptaldehyde and nylon through dimerization, polymerization etc., (Naughton 1974).

There are a large number of industrial applications for castor oil and its derivaties, and new ones are continually being discovered (Naughton, 1974). When dehyrated, castor oil is converted into quick-drying oil used extensively in paints and varnishes (Achaya, 1972). Its water resistant qualities make it ideal for coating fabrics and for protecting coverings, insulation, food containers and gums (Fru, 1969 and Kirk, 1923). Castor oil is used in Nylon, as a primary raw material for the production of sebacic acid, which is the basic ingredient in the production of nylon and other synthetic resins and fibres (Belanger, *et al.*, 1966). The superior oiliness of castor oil and its ability to cling to very hot moving parts make it an outstanding racing oil lubricant for high performance engines. Castrol-R, racing motor oil, for high-speed automobile and motor cycle engines is made from castor oil.

Castor oil is the most unusual product, among all the vegetable oils. It is, by far more versatile due to presence of ricinoleic acid, which constitutes about 90% of the oil. Mostly oils have one or two applications, such as for edible purposes (soybean, peanut, rapeseed, sunflower, olive, corn, cotton seed etc.). Some have other uses as in coatings, inks, lubricants, detergents &

soaps (Naughton 1974). In contrast, castor has many more uses, with a wide diversity of commericial applications, all directly related to the unique hydroxy fatty acid namely ricinoleic acid. Castor oil is non-toxic, renewable and biodegradable. It's oil is used in medicines as purgative and also as hair oil (Irwin, 1990).

Castor Oil in Medicines

Castor oil is one of the old-fashioned remedies for everything from constipation to heart burn. It is indeed a very effective cathartic or purgative (laxative) for which purpose it is still used to this day. Castor oil is also used as a personal lubricant as soothing emollient for dry skin, dermatitis, other skin diseases, sunburn, open sores, etc. (Hawkins, 1989). Divesified uses of castor have been compiled by Nagaraj (2003a).

Castor oil improves digestion, cures women's reproductive disorders and increases liver activity. Small quantities of pure and warm castor oil (half to one tea spoon for children and two teaspoons for adults) with a glass of warm water acts as a mild laxative within two hours. Lactating mothers can apply castor oil on their breasts to increase milk secretion, reduce sagging and prevent nipple cracking and take castor oil to relieve constipation of their babies. Application of castor oil on the navel and lower abdomen relieves premenstural pain. Intake of larger doses of castor oil may induce abortion. External application of castor oil provides relief and healing to anal fissures. Castor oil along with *Aloe vera* juice reduces piles. It prevents fungal infections of the foot and also reduces itching and sweating and helps in healing wounds, bedsores, burns and cuts. Castor oil also helps in healing corns when it is regularly applied.

Application of warm castor oil also reduces joint pains, arthritis and varicose veins. Castor oil along with a decoction of dried ginger, reduces rheumatic pain and sciatica. Joint pains get reduced when castor oil coated castor leaves are heated and the hot oily leaves are tied round the aching joint twice daily. Castor oil prevents hair loss and also helps in retaining blackness and luster of hair. Massaging castor oil on the head reduces headache. One teaspoonful of castor oil with a pinch of salt in a glass of water when gargled twice a day, for four days reduces tooth pain. Undecylinic acid is useful in the preparation of antileprosy drug, similar in action to choulmoogric acid. Undecylinic acid also has bactericidal and fungicidal properties. The zinc and sodium salts of this acid are more potent. Potassium ricinoleate inhibits blue-green algae and is toxic to catfish. Sodium ricinoleate is useful in preparation of tooth paste due to its bactericidal properties.

Cosmetics

Castor oil is non-comedogenic i.e. it does not contribute to formation of acne and hence is the most preferred constituent in cosmetics. Non-comedogenic esters popular as emollients are cetyl ricinoleate, cetyl acetyl ricinoleate, propylene glycol ricinoleate, glyceryl triacetyl hydroxy stearate, octyl hydroxy stearate etc. Due to its uniform triglyceride composition, castor oil is the preferred medium for prepartion of many cosmetics. It is a mild moisturizer with excellent film forming and pigment wetting properties. It is used as an occlusive agent in skin care products. As an emollient, castor oil works for softening the skin. Castor oil is a well known important constituent (20-40%) in lipsticks. Castor oil is a highly popular hair oil and a skin care agent. Some of the above castor oil derivatives along with wheat germ oil are good emulsifiers.

Castor oil and its products are widely utilised in making toilet soaps, shaving creams, nail polishes, shampoos, sun creams, hair conditioners, antiperspirants and talcum powders. In India a major portion of castor oil is utilized for soap manufacture(Laxminarayana,1983). Cetyl ricinoleate gives a silky, but non-oily feel on the skin and moisturizes dry skin. Propylene glycol ricinoleate imparts gloss and is an effective wetting agent. "Quats' prepared from castor oil are incorpated into shampoos as hair conditioners. Castor oil soap also finds use in disinfectant formulations such as *'Dettol'*. Sodium ricinoleate is useful in the manufacture of transparent soaps. Castor soaps (sodium and potassium ricinoleate) are non-irritating and skin friendly. Heptaldehyde and undecylinic acid are two important pyrolytic products of castor oil widely utilized by the perfumery industry.

Undecylinic Acid

Undecylinic acid, a derivative of castor oil is a very useful perfume chemical. This undecylinic acid is a component of human perspiration. Due to its role in disinfection, undecylinic acid is widely utilised in skin and hair care products. Salts and esters of undecylinic acid are used in the treatment of athlets' foot and dandruff. Undecylinic acid has fungicidal and bactericidal properties. It is also a very useful perfumery compound. Undecylinic acid and sebacic acid (another pyrolysis product) are useful in the manufacture of nylon-11 and nylon-10 which have good molding properties. Nylon-11 is the highest value added product. It is mostly manufactured in France from imported castor oil. Nylon-11 is highly useful as synthetic fibre, engineering plastic and powder coating. Nylon-11 is a better plastic than petrol based plastics.

Table 67: Perfumery chemicals from castor oil.

Undecylinic acid derived	Heptaldehyde derived
1. Methyl, ethyl and allyl esters	1. Amyl cinnamic aldehyde
2. Aldehydes	2. Heptanol
3. Alcohols	3. Gamma-Nara lactone
4. Undecalactone	4. Methyl nonenoate
5. 12- Oxa hexadecanolide	5. Heptyl heptanoate
6. Muskone	6. Methyl hexylketone
	7. Dihydrojasmone

Castor Cake

The cake or pomace is a rich source of nitrogen and minerals. Because of the toxic constituents like ricinine, ricin and an allergen in the cake it is unfit for edible purposes. Except poultry which accepts castor cake without any harmful side effects, it cannot be fed to any other animal. A ration containing 10% castor bean meal has been found to be harmless for the sheep (Purushotham *et al*, 1986). Castor cake is mainly used as an organic fertilizer (organic matter 85%). When applied to the soil the cake releases the nutrients (NPK) slowly and over a long period.

Table 68: Composition of castor cake.

Constituent	Decorticated	Undecorticated
Oil	9	6.5
Protein	46	20.5
Carbohydrates	26	49.0
Minerals	10.5	15.0

Source: Weiss, (1983)

The cake contains oil (6-9%), protein (25-40%), sugars (25%) and minerals (10%). Castor meal generally contains about 6% N, 2.5% P_2O_5 and 1.5% K_2O. Polysaccharides of castor have been analysed and reported (Anjaneyulu *et al.*,1983). Utilization of castor cake has been discussed by Mulky (1985). The chemical compostion along with minerals of the seed and cake has been studied (Nagaraj, 1999). The ash contains 25% P_2O_5 17% CaO, 14%K_2O, and 10% MgO and 6% Fe_2O_3(Laxminrayana, 1993).

Table 69: Essential amino acids of castor.

Amino acid	CBH	Ricin Protein	DCH 177	DCS 9	FAO/WHO levels
Lysine	2.1	3.9	1.8	2.3	5.5
Threonine	3.0	4.8	5.9	4.1	4.0
Methionine + cystine	2.0	1.3	0.5	1.9	3.5
Valine	3.0	4.8	0.7	5.0	5.0
Leucine	5.1	7.7	7.5	6.9	7.0
Isoleucine	3.3	5.0	4.9	4.9	4.0
Phenylalanine	4.7	7.2	5.8	6.4	6.0
Histidine	1.0	1.6	-	-	-
Arginine	7.2	5.2	9.2	7.8	-

CBH= bean heamagglutinin

Like any other vegetablre protein, castor is deficient in lysine. It is also deficient in sulphur containing aminoacids. It is better with respect to all other essential amino acids. It is rich in the wound healing arginine. However, it is well known that castor is neither useful as food nor as feed due to the presence of toxins.

Protein based surfactants can be made from castor cake. Protein isolates have been examined for their surfactant properties Castor cake is useful as an organic manure with insecticidal properties. In its properties as an organic manure, it is intermediate between synthetic(chemical) fertilizers and animal manure (compost). It has good qualities of both fertilizers and manures. Castor meal (fully defatted) is superior to castor cake and neem cake with respect to availability of nitrogen to crops. Castor cake is most suited as organic manure to irrigated crops.

The protein isolates can be used in the manufacture of fibres, plastics and wall distempers. Surfactants can be prepared by acylation of amino acids or peptides and by amidification with alkyl amines. The toxic constituents in castor like ricin, ricinine and allergens may have namaticidal and termiticidal properties and hence in recent years, the cakes have gained lot of importance as organic manure for many crops in different types of soils. It is evident that castor cake has triple benefits (1) as slow release fertilizer (2) as soil amendment to improve the physical properties and (3) as insecticide/ nematicide. Among various grades of castor cake, deoiled castor cake is better for denitrification.

Phenolic Compounds

The phenolic constituents impart objectionable properties like sourness, bitterness and astringent taste to oilseeds (Ariale, *et al.*, 1966). Phenols also form complexes with minerals and proteins, lowering their nutritional value. They bind with proteins reversibly by hydrogen bonding mechanism or irreversibly by oxidation to quinones. They also combine with reactive groups of protein molecules and make them unavailable (Loomis and Batallie, 1966). Polyphenols may also form complexes with carbohydrates and their affinity for polysaccharides is strongly dependent on molecular size, conformational mobility, shape as well as water solubility of polyphenols (Cai Ya, *et al.*, 1980). They also possibly inhibit the absorption of iron (Gilloly, *et al.*, 1983).

Castor cake has been analysed through HPLC to contain some phenolic compounds, namely, p-coumaric acid, feruli acid, o-coumaric acid, syringic acid and cinnamic aicds. These compounds are important for normal plant growth, development and defence against infection and injury (Srinivas and Nagaraj, 2007).

Ricinine

Ricinine, the less toxic alkaloid, has been conclusively shown to be N-methyl-3-cyano 4-methoxy-2-pyridone by extensive degradation studies. Ricinine content in the castor seed is around 0.3% and it is one of the few alkaloids having a cyano group. It is optically inactive and soluble in alcohol. Its melting point is 201.5^{0}C. It is useful as an insecticide. It is highly bitter in taste and irritates the throat. It is mildly toxic to human beings. In killing insects its activity is similar to lead arsenate. Under the trade name *Sprakast*, it is being marketed (Hanumantha Rao, 1970).

Allergens

Substances causing allergy in castor seed have been recognised since 1914 (Aliare, 1914). The allergenic principle has been found to be different from ricin. It is heat stable, soluble in basic lead acetate solution, insoluble in 75% ethanol, and non-toxic to unsensitized animals (Spies, *et al.*, 1943 and Coulson, *et al.*, 1950). This allergenic material has been designated as CB-1A (castor bean allergen). It is present up to 12.5% of weight of the castor seed cake. Inhalation of this allergen causes allergic coryza and bronchial asthma, and severe reactions if injected into the skin of sensitive persons. The principal allergens of castor bean have been resolved into 4 bands by disc gel electrophoresis (Moris, *et al.*, 1965).

Ricin

Ricin is one of the major toxic constituents present in castor seed cake. Its presence has limited the usage of castor cake. Ricin is a highly stable protein and keeps for months without deterioration (Osborne, 1962). Before separation ricin is present in combined form with RCA (Ricinus Communis Agglutinin). The RCA is also a protein with a molecular weitht of 120, 000. Ricin in crude form agglutinates not only blood but also cholesterol suspensions of negatively charged particles (Gunn, 1922 and Pascucci, 1972). Ricin is a water soluble protein. The molecular weight is around 65,000 consisting of a toxic A chain covalently coupled to a galactose-binding B chain (Olsnes and Pihl, 1983). Hence ricin is classified as lectin (sugar binding protein).

Ricin loses its agglutinating property once it is purified as single protein but retains its toxicity (Funatsu *et al.,* 1962). RCA agglutinates blood cells of humans and animals but is not toxic. Pure ricin does not agglutinate blood cells but it is toxic to animals and human beings. Ricin is toxic because it inhibits protein synthesis by interacting with r-RNA (ribosomal nucleic acid). The toxicity of ricin has been used in various medicinal applications like in curing tumor cells and also in treating HIV virus (Ghetie and Vitetta, 1994 and Jones *et al.,* 1998).

Ricin and its polypeptides have been purified and their amino acid composition has been examined. The molecular weight of ricin sub-units are 29,000 and 28,000. The total helical content of ricin is around 53.6% (Srinivas and Nagaraj, 2008a). The biochemical properties of ricin at different stages of seed maturity has also been examined. Hemagglutination, SDS-PAGE and U.V.spectrophotometric studies have revealed total absence of RCA protein in the immature seed. Interestingly, ricin extract on SDS-PAGE has shown only one protein band with molecular weight of 29,000 corresponding to A chain of ricin. The toxic component is synthesized first followed by the B chain synthesis (Srinivas and Nagaraj, 2008b).

Ricin agglutinates red blood cells at 100-2000 mg/100 ml and can be lethal to animals and humans. It is present in the range of 15-201 mg/100g in the cake. The ricin content varies among genotypes, maturity levels and raceme positions (Nagaraj, 1999). Treatments like alkalies (0.5% $Ca(OH)_2$, 0.5% $NaHCO_3$ and autoclaving, reduce the ricin to non-detectable limits (Srinivas and Nagaraj, 2000). Castor toxins have been found to be neutralized when treated with sal tannins (Gandhi and Mulky, 1994).

Utility of Ricin (Therauptic)

Ricin has been used to target specific cells such as cancer cells by conjugating the ricin toxin sub unit A to antibodies or growth factors that preferentially bind the unwanted cells. Ricin has been used as immuno-toxin to destroy T lymphocytes *in vitro*. Ricin is likely to find application in auto-immune therapy. DNA recombinant technology has been utilized and ricin has given beneficial effects in the treatment of AIDS(Ghetie and Vitetta, 1994 and Jones *et al.*, 1998).

Toxic Constituents and their Detoxification

The castor seed contains an alkaloid ricinine, which is mildly toxic and a protein, ricin, which is extremely poisonous. It also contains a heat stable allergen, which is irritating to humans. All these toxins do not get into the oil. In India long exposure must have built up resistance both in animals and human beings. Repeated intermittent steaming is a simple method of removal of the toxins. Castor cake can be detoxified through autoclaving, boiling or cooking with water, or lime water steaming, treating with salt solutions (ex. ammonium sulphate, sodium sulfate, bisulfate, formaldelyde, dilute HCl etc.). Ricin as well as the allergens get destroyed almost completely. The detoxified castor meal has been found suitable as feed for many animals. Ammoniation is another process which apart from detoxifying increases the nitrogen content of the cake. Attempts to make the cake suitable for human consumption have also been made.

Castor Hulls

A dehuller to separate kernel and hulls has been examined (Puttaraj and Narendra Singh, 1992). Oil from dehulled seed is of good quality. Castor hulls constitute 20 to 33% of the seed and contain 1.6% N, 0.8% P, 3.8% K and 10% ash. They also contain 7-7.6% dark viscous oil. The fatty acid composition (Subbaran *et al.*, 1964) is similar to that of kernel oil (castor oil). Castor hulls have use in the preparation of quality activated carbons.

Castor Stalks and Leaves

Castor stem pulp can be utilized in the manufacture of paper. The castor pulp mixed with bamboo pulp has been used for preparation of newsprint (Bhat and Shenolikar, 1946), writing, printing and wrapping papers (Bhat and Jaspal, 1951). Castor leaves are used for feeding silkworms and cattle (Reddy and Alfred, 1979). Branches and stems are used for the production of lower grade paper as well as fuel.

Linseed

Linseed is an important dual purpose crop grown for its seed oil and fibre, from the stem. India ranks first in terms of area and third in terms of production of linseed in the world. It is also an important crop of Canada, China, U.S.A, U.K. Germany, Ukraine and Russia. Linseed is believed to have originated in an area East of the Mediterranean sea. From there it spread to Northern and Western parts of the world. The seed contains oil ranging from 35-45% which is utilized in the mannufacture of paints, varnishes, inks, soaps etc. Linseed ocupies an important place in the world market due to its technical grade oil.The cake as well as other parts of linseed plant are very useful. The fibre from the stem is lustrous and blends well with wool, silk, cotton etc.

The cultivated linseed *(Linum Usitassimum L.)* belongs to the family *Linaceae.* Linseed plant is an annual herb, with a thin, erect and wiry stem, about 60-100cm high. The fibre type cultivars are slender, tall, non-tillering, and slightly branched. Seed purpose linseed cultivars, are shorter, much branched and profusely tillering. Linseed is a self pollinated crop. The fruit is a round capsule with five chambers, each containing upto two yellow or brown seeds.

Linseed is a cool season crop. In India it is grown in *Rabi* (September-March) season. It is the second most important oilseed crop grown in the *Rabi,* the first being rapeseed -mustard in India. It grows well under a rainfall ranging from 45-75cm. It can be grown in well drained, fertile, medium/ heavy soils like sandy loam, clay loam or silty clays with a pH 5-7. Linseed is grown as mono- or mixed crop. Seeds are sown in furrows at a row spacing of 20-30 cm at the rate of 20-30 kg/ha. Nitrogen at 20kg and P_2O_5 at 20-30 kg/ha are applied at the time of sowing. Linseed matures in 120-170 days after sowing. Seed yields ranging from 800-1000 kg/ha and 1600-2000 kg/ ha under rainfed and irrigated conditions respectively are obtained. The crop is harvested when the capsules turn brown and seed becomes shiny. The crop is dried and threshed by beating with sticks. The seeds dried to less than 8% moisture are stored in gunny bags.

Seed Composition

The seed is small, flat, oval, brown or yellow coloured, glossy in appearance with mucilagenous taste and oily nature. The seed contains 30-45% of fixed oil. It also contains about 6% mucilage which resides in the seed coat and about 25% of protein together with wax, resin, carbohydrates, phosphates, sterols, colour pigments like carotenoids and xanthophylls, vitamins(thiamine, riboflavin and niacin), minerals and a small quantity of a cyanogenic glucoside-linamarin. The chemistry of linseed has been reviewed by Gill (1989). Brown seeds have lower oil content (40%) than fawn (40-42%) and yellow coloured varieties. The compositional characteristics of linseed and its oil are presented in Table 70.

Table 70: Chemical composition of linseed.

Constituents	Range
Moisture	6.5%
Protein	20.3%
Fat	37.1%
Minerals	2.4%
Carbohydrates	28.9%
Energy	530K calories/100g.
Calcium	170mg/100g
Iron	370mg/100g
Carotene	2.7mg/100g
Thiamine	0.23mg/100g
Riboflavin	0.07mg/100g
Niacin	1.0mg/100g

Linseed due to sufficient levels of fat, protein and sugars can supply 530 calories/100g. It is similar to any other edible oilseed like groundnut or sesame in its energy content. The seed has sufficient levels of minerals and vitamins. It is rich in the minerals, iron and calcium and the vitamins, carotene and niacin. In view of this, and it's higher levels of alpha linolenic acid, an essential fatty acid, it is eaten in some parts of India. Fried and powdered seed is utilized in preparation of various day to day foods.

Oil Content

Linseed oil is a drying oil. It is unique among vegetable oils in that it contains more than 50% linolenic acid (18:3) an eighteen carbon fatty acid with three double bonds. About 80% of the oil is used for industrial purposes. The oil used for edible purposes should have lower iodine value (less linolenic acid). Linseed seeds on expression yield 30-35% oil. Varieties with yellow and fawn coloured seeds have higher oil content than the brown coloured seeds. Eighty six genotypes have been analysed by Bhatta *et al.*, (1985) who have reported the following ranges: oil 37.8–47%. Eighteen linseed cultivars have been analysed for their seed and oil quality (Muralidharudu and Nagaraj,

1995). The oil and protein contents range from 35.2 to 42.9, and 14.8 to 28% respectively.

Characteristics of Linseed Oil

Freshly prepared oil is a viscid and yellow liquid having a faint, but distinct odour and bland taste. The yellow seeded varieties have higher iodine value (170-195) followed by fawn (180-185) and brown (170-180) coloured varieties respectively. It has a saponification value of 189-196 and thiocyarogen value of 114-124. It's acetyl value is low at 4-10. Similarly, unsaponifiables are low ranging from 0.5 to 0.6% (Jameson, 1943) and so it's triglyceride (TG) content in high (93-98%). Linseed oil has higher iodine value and thus is a drying oil. This is because of higher levels of 18:3 in the oil. Its saponification value is similar to other vegetable oils. Freshly extracted oil has low FFA content, which increases on storage.

Table 71: Characteristics of linseed oil

Characteristic	Range	Crude oil	RBD Oil
Density 20^{0}C	0.925-0.935	0.921	0.920
Refractive index n_D20^{0}C	1.475-1.475	1.4657	1.4665
M.P. ^{0}C	-20-24^{0}C		
Flash point ^{0}C	120-135		
Viscosity	-	46.8	46.4
Iodine value	182-203	142	144
Unsaponifiable %	0.1-1.7	1.2	0.6
Saponification value	187-195	185	185
P(ppm)	1.0-3.0	300	1.0
Cholorophyll (ppm)	00-1.5	0.4	0.1
FFA (%18:1)	0.1-2.0	0.3	<0.02
TG %	94-98	93-98	96-98

The chief constistuents of the oil are a mixture of triglycerides of oleic, linoleic, linolenic, palmitic, stearic and myristic (traces) acids. Linseed or flax-seed oil is a very rich source of α-linolenic acid (about 40-60%), an omega-3 fatty acid. In addition, the oil contains varying amounts of lignan, secoisolariciresinol diglycoside (SDG). The lignan, belongs to a group of plant substances known as phytoestrogens-natural anti-cancer agents (Morton *et al.*,1994). The importance of these minor constituents in linseed needs further examination and utilization.

Fatty acid Composition

Linseed oil is a rich source of linolenic acid, an eighteen carbon fatty acid with three double bonds. The content ranges from 35-70%. Oleic acid ranges from 13-38% while linoleic acid ranges from 7-18%. It is low in palmitic acid at 5-6% of total fatty acids (Jameison, 1943). Eighty six genotypes have been analysed by Bhatta *et al.,* (1985) who have reported the following ranges: oleic 15.7 - 35.5%, linoleic 7.6 - 15.3% and linolenic acid, 40.6 -62.6%. Bajpai *et a.l,* (1985) have reported the variation in oil quality and glyceride composition of 10 varieties. Eighteen linseed cultivars have been analysed for their seed and oil quality (Muralidharudu and Nagaraj, 1995). Linolenic acid has shown variation from 41 to 57% and oleic from 19 to 30%.

Considerable variability with regard to linolenic acid (45.75 to 56.58%), oleic acid (16.53 to 27.10%), linoleic acid (12.96 to 19.16), stearic acid (4.6 to 7.06%) and palmitic acid (05.4 to 6.91%) has been observed(Rai *et al.,* 1989), with maximum fluctuations in oleic and stearic acids. The mean values reported by them after analyzing 35 varieties are linolenic=52.6%, oleic=20.2%, linoleic=15.4%, stearic=5.3% and palmitic=5.9%. The possibility for genetic advancement in 18:3 is 6.6% while for oil content it is 15% (Rai *et al.,* 1985). Development of biparental progenies at successive hybridizations has been suggested for improving oil content in linseed (Jeswani, 1985).

Forty two genotypes of linseed which included 28 seed, 12 double purpose and two flax types have been examined for their plant and seed quality characteristics (Bateria and Nagaraj, 1999). The double purpose types have given higher seed yield of 1489 kg/ha than seed types with 1355 kg/ha. The former matures in a shorter period of 206 days(against a mean maturity period of 219 days)and have higher thousand seed weight of 9.47 g. The seed types have 43% oil while the double purpose types have 42% oil. Both have around 55% linolenic acid.

Linseed oil is unique among vegetable oils in the sense that it contains more than 50% linolenic acid, a 18 carbon fatty acid with three double bonds, and the total unsaturated fatty acids constitute more than 80% of the fatty acids. The linseed oil is not edible because of the laxative properties of the mucilage associated with the seed coat and to the higher level of linolenic acid that can on oxidation cause rancidity and off-flavors (Green and Marshall, 1981). However, about 20-30 per cent of the linseed oil produced is reportedly consumed for edible purposes either solely or in combination with mustard, groundnut and coconut oils in India. Although there is scope for hydrogenation of the oil, the commercial production is low for viable scales of hydrogenation. As the oil has the least storage stability, the edible uses of linseed are limited under the present conditions.

Apart from the genetic variations, the oil content and the fatty acid compostion in linseed is influenced considereably by biotic and abiotic factors. While the oil content slightly increases with an increase in nitrogen level from 0 to 30 kg/ha, the linolenic acid content of the oil is reduced with the increasing nitrogen fertilization. Similarly, the linolenic acid content is highest under rainfed conditions and reduced with irrigation and high nutrient conditions. The linolenic acid content in irrigated and low nutrient level conditions is intermediate. In contrast, the oleic acid content is lowest under rainfed conditions and highest under irrigated and high nutrient levels (Ravikishan *et al.*,1995). This signifies the negative association between the contents of linolenic and oleic acid. Thus, it is apparent that linolenic acid is synthesized more under stress (low moisture and fertilzer) conditions.

Although, it is possible to alter the fatty acid composition through cultural practices, to a limited extent, it is necessary to develop genotypes with high linolenic acid for industrial purposes and low linolenic acid types to improve the nutritional quality of the oil. Wide variation in linolenic acid content, ranging from 31 to 61 percent, has been observed in the germplasm collections in India. Attempts have been made to develop low linolenic acid types through mutations (Green and Marshall, 1984 and Green 1986). The oil from these lines is high in desirable unsaturated linoleic acid (about 63%), but is virtually free of linolenic acid (about 1%). The polyunsaturated quality of this low linolenic acid line (*Linola*) is better than that of rapeseed and comparable to sunflower oil (Hocking *et al,* 1987). It's prospects as an edible oil may not be bright due competition from oils like soybean, safflower and sunflower.

Table 72: Fatty acid composition linseed oils.

Fatty acid	Flax (linseed) %	Low linolenic (linola) %
16:0	5.3	6.1
18:0	3.3	3.8
18:1	17.9	15.5
18:2	14.7	71.3
18:3	58.7	2.0
SFA	9.0	10.0
MUFA	18.1	17.1
PUFA	72.9	72.9

It is better to use this oil for industrial uses as is being done presently. However, this oil in small quantities can be used as a prescription oil by people suffering from obesity and candiac problems. An alternative method is to use this oil as a blend with other oils to overcome imbalances in the fatty acid composition of other edible oils (Rajani chopra *et al.*, 2004).

Tocopherols and Phytosterols

Linseed oil has higher levels of beta tocopherol than alpha and gamma. It has 15-20 ppm alpha, 200 ppm beta and 5-7 ppm gamma tocopherols. Higher levels of unsaturation and lower levels of alpha and gamma tocopherols indicate that linseed oil is highly unstable with limited shelf life, in addition to bad frying quality. This oil has 23-27 ppm campesterol, along with about 70 ppm of other phytosterols.

Medicinal Uses of Seed and Oil

Research in the recent past, all over the world is indicating the health benefits of linolenic acid (an essential fatty acid), lignans and other soluble fibres present in flax seed/linseed. Important medicinal uses of seed and oil are as follows: 1) Freshly extracted oil can be used as a laxative in doses of 30 ml. 2) A rectal injection of 60 ml of oil, given at night and in the morning has been recommended for releif against piles. 3) It can be applied externally as a soothing application for burns in the form of *carron oil* (a liniment of lime water with linseed oil. 4) Linseed oil is used as carrier of vitamines and irritant drugs. 5) Linseed tea is prepared by boiling one part of linseed with 20 parts of water until the grains become soft. The tea can be used as a demulcent in cough especially those which cause irritation of pharynx and upper part of respiratory passage. The tea is also used as a demulcent drink in intestinal or urinary catarrhs. 6) For the preparation of cough syrup mucilage of linseed (1 in 8 parts of water) is used. 7) Crushed linseed is used in the form of poultice (Cataplasmalini) to apply warmth and moisture locally, for the relief of superficial or deep rooted inflammation. It is prepared by boiling 28 g linseed with 72g (ml) of water. The poultice may be sprinkled with boric acid prior to application. The poultice mass is enclosed in muslin, the surface of poultice may be smeared with oil to keep it from adhering to the skin. 8) Linseed and its oil have anti-inflammatory action in the treatment of arthritis. 9) Feeding linseed and its oil is recommended in the treatment of hyperlipidemia and to reduce platelet aggregation, atherosclerosis, thrombosis and myocardial infarction by reducing cholesterol and low density lipids. It is also useful in reducing blood pressure. 10). It's usefulness in the treatment of anxiety, benign prostate hypreplasia, vaginitis and weight loss, is unsubstantiated by the available research data. However, veterinarians are reported to use these products to treat some animals.

Linseed in Health Foods and / or Functional Foods

Linseed becaue of it's richness in linolenic acid, directly serves as a neutraceutical and also indirectly through other foods. Linseed oil apart from being rich in 18:3 fatty acid is low in saturated fatty aicds, especially, palmitic acid (about 5%). Thus, it certainly can serve as a good salad oil, if not as a frying oil. Seeing the possible role of flax-seed in the treatment of some cancers, particularly mammary cancer in rats, at the recommendation of Canadian government, some poultry farmers are feeding chicken, feeds rich in flaxseed for boosting the omega-3 fatty acid content of eggs (ω–3 rich) of the chickens, These functional foods have captured 4% of Canadian market.

Similarly attempts are being made to develop omega- 3 rich beef by providing compound feeds rich in linseed to ruminant animals. Various branded products with either linseed oil or the seed or its cake as constituents in different proportons are already available in the market. More products are likely, in view of its various health benefits. It is anyway, already a part of the folk Indian medicine and is widely being used by rural people in India.

Industrial Utility

Linseed oil has its greatest use in house paints, exterior stains and industrial coatings since the double bonds react rapidly with oxygen in the air to polymerize into a relatively soft and flexible film. The oil sets into a hard vitreous film when applied as a thin film and exposed to air. This film has a glossy surface and is resistant to acids and alkalies. It is also used in printing inks, brake linings, concrete protection and curing restructured wood products (Formo, 1984). It is also used in the manufacture of linoleum and oil cloth and soaps. A relatively smaller amount is also employed to produce core oils, linings, packings, oil modified alkyd resins, caulking compounds, putties, leather finishing compounds, lubricants, greases, polishes, plasticizers, adhesives and pyrotechnic compositions. However synthetic products of the petrochemical industry have made inroads into the traditional markets for linseed oil and its derivatives in the paint and varnish industry.

Linseed oil is primarily an industrial oil. Farmers apply the linseed oil on the surface of their iron made implements for protection against rust as well as on wooden materials for longevity. Its oil finds use either in its raw cold pressed form or in the modified form. Linseed oil is used as a solvent for industrial stains. It is also used for seasoning bobbins (in textile industry), cricket bats and other sports goods.Linseed oil fatty acids are used as protective coatings and emulsifying agents.

Modified Oils

Raw linseed oil is slow in drying rate, has a tendency for yellowing, with poor colour retention. Hence, some processes have been developed to improve the drying properties of linseed oil by heat modification, blending or by combining.

Boiled linseed oil: It is obtained by heating raw oil at 90-150^{0}C with driers such as linoleates or resinates of lead, manganese, cobalt or zinc. The oil becomes viscous, dries at a faster rate, film formed is smoother and more lustrous and colour retention increases. Boiled oils are used in the manufacture of paints, water proof materials, patent leather and special finishes for cotton and silk fabrics. Boiled oil is sold as single, double, pale, extra pale or bleached, the names indicating difference in colour consistency and drying properties.

Blown linseed oil: It is produced by blowing air through linseed oil at 130^{0}C for several hours till the desired viscosity is reached. It is golden yellow or reddish yellow in colour with high specific gravity and has excellent drying and pigment wetting properties. It is used in the manufacture of linoleum and oil-cloth, paints, varnishes and enamels.

Stand linseed oils: It is obtained by heating alkali refined linseed oil in an inert gas atmosphere at 260-285^{0}C for obtaining the desired consistency. Stand oil is pale yellow with honey like consistency and high acid value. Stand oil gives an elastic film with improved durability and superior gloss. Thinner oils are used in enamel and varnishes while thicker oil (Litho oils) are used in the manufacture of printing ink.

Enamel oil: It is obtained when stand oi is prepared with a mixture of linseed and tung oil. It thickens faster than linseed oil and yields lighter coloured surface coating.

Chemically treated oils: Linseed oil treated with various chemicals is in demand for special applications.

Sulphurized linseed Oil : It is prepared by stirring the oil with sulphur chloride or treating the oil with flowers of sulphur at 100^{0}C. It is a thick, brownish red liquid and sets with extreme rapidity. It is used in the preparation of pigmented oil, required for wet-on-wet process of painting in which several coats of paint are applied at short intervals. The composite film is elastic and fairly durable.

Styrenated linseed oil (Ketrol L): It is prepared by reacting linseed oil with styrene. It dries to a hard, tough and alkali resistant film.

Esskol and Solinox: These are hydrogen treated linseed oils used as substitutes for tung and castor oil.

Polymerized linseed products: Treatment of a solution of linseed in chlorinated hydrocarbon (like Ethylene dichloride) with aluminum trichloride, boron trifluoride, silcon tetrachloride or titanic chloride produces polymerized products. The treatment of raw, blown or polymerized oil with a solution of phosphorus tri- or pentachloride in spirit gives products with improved drying properties.

Linseed Cake

Linseed meal contains 25-30% protein and is a good cattle feed. The cake is rich in minerals like P, Ca, and K and is also utilized as an organic manure. The cake contains a glucoside linamarin. On hydrolysis it liberates HCN which is poisonous to animals. The essential amino acid composition is presented in the Table 73.The amino acid conposition is similar to that of any other oilseed. It is rich in lysine and poor in methionine. However, linseed protein quality is better than that of rapeseed proteins.

Table 73: Essential amino acids of linseed.

Amino acid	g/16g
Isoleucine	4.96
Leucine	5.76
Lysine	3.68
Methionine	1.60
Phenylalanine	4.32
Threonine	1.60
Valine	5.12

Source: Weiss (1983).

If the toxic factor linamarin is removed, the linseed meal serves as a source of very good quality protein. It has a lot of uses in the industry as a source of mucilage and protein. Linseed cake is useful 1) as a filter aid in the processing of uranium, iron and copper sulphide and in the setting of cement slurry and kaolin clay, 2) for the preparation of plywood glues by blending with phenolic resins (4:6), 3) as a filler in thermosetting resins because it imparts improved water resistance, 4) as a corrosion inhibitor, 5) as a binding material in water paints, and 6) for stabilizing mud plaster.

Linseed Mucilage

Linseed mucilage is prepared from aqueous extract of seed (soaking in water for 24 hrs) by precipitation. It is a white fibrous mass and is friable when totally dry and gets dissolved in water giving high viscosity solution. It is used a) in cosmetic and pharmaceutical industries as a demulcent, b) in food products as a water soluble emulsifying agent as well as thickener and binder, c) as a substitute for acacia gum in stabilization of emulsion and d) as a useful base for eye ointment.

Uses of Stalk or Straw

There are two types of linseed grown namely the flax type linseed which yields seed of poor oil content, but its straw gives fibre of good length and quality and the seed type linseed produces good seed but its stalk gives fibre of poor quality. Now-a-days double purpose varieties have been developed which produce higher seed yield as well as fibre (Bateria and Nagaraj, 1999). Harvested deseeded stalks (up to first branching) are completely dried and subjected to wet retting for 2-3 weeks. Fibres are then separated from the retted stock by scutching and drying. Dried cortical tissue, separated from fibre, contains 10% wax.

Flax also serves as a source of fibre which could be spun and woven into cloth. The fibre which is composed mainly of cellulose is stronger, more durable and more resistant to moisture than cotton or wool. Therefore, it is used in gloves, foot wear, netting and sports gear. It is also used in paper manufacture and textile industry. Work is presently being intensified to breed varieties which serve as sources of oil rich seeds and stems with good quality fibre.

Flax is the oldest textile fibre. It is one of the most natural and most eco-friendly of all the textile fibres. Flax fibre has more strengh, fineness and durability than cotton fibre. It is soft, lustrous and flexible and has high water absorbency. It can be blended with cotton, hemp and jute. Blending of flax with cotton will enhance not only its textile qualities but also save cotton for other uses (Mukherjee, 1988; Mahapatra,1998). It also mixes well with wool and tussar silk. It is used for the manufacture of linen stitching including handkerchiefs, table and bed linens, dress inter lining, men's suitings, towels, tents, drapperies, upholstery, covers, surgical dressings, fire fighting hoses and water holding bags.

The dry scutched fibre on spinning produces rough and thick yarn which can be utilized for making strong ropes, which are damp resistant. The tensile strength is comparable to other fibre producing materials. The linseed fibre twines can be used for making fishing nets, wrapping purposes, house decoration purposes, etc. The yarn of linseed fiber can be utilized for the manufacture of rough textiles such as carpets, blankets, mats, mattresses etc. This yarn can be a substitute for cotton in making tents.

Linseed-jute (Linju) and flax-cotton (Linco) blended fabrics show better quality as furnishing materials than cent-per-cent jute or cotton fabrics (Pandey and Dayal, 2003). With the introduction of linseed fibre for fabrics (pre treatment of linseed fibre), substitute of cotton for the manufacture of fine textiles like suiting and shirting, bed sheets, bed cover, curtains, etc. is possible. The linco-fabrics can supplement the existing cloth production as well as generate employment especially in linseed growing areas.

Manufacture of pulp: After extraction of fibre from stalk the remaining material, about 80% of stalk, can be converted into pulp which can be utilized for the manufacture of a) high grade paper. b) cigarette paper, air mail, bond, bible and other high priced books and writing papers, c) safety paper used for currency notes. (Gopalan *et al.*,1971).

Wax: Wax is commercially extracted from the cortical tissues by organic solvents and used in shoe polishes (Gill 1987).

Conclusion

Widespread use and adaptability of linseed oil has been restricted due to the availability of cheaper petroleum products. However, linseed oil and its cake, being renewable, biodegradable, free of allergens, enriched with phyto-chemicals, would be the base stock not only for pharmaceutical, medicinal and chemical industries but also for cosmetic industry. The recent developments in healthcare foods has led to a resurgence of the uses of linseed/ fax-seed and its oil.

References

Abdel Rahman, A.H.Y. 1982. Changes in Chemical Compostion of Peanut During Development and Ripening. *Revista Italian Dulls Sostanze Grasse* 59(6): 285-286.

Abdel Rehman, A.H.Y. and Yosuf, S.A.H. 1987. Changes in Chemical Composition of Safflower Seeds During Development and Ripening. *Rivista Italiana, Dulls sostanze Grasse* 58(8): 399-400.

Abdul Latif, A.M.M. and Vles, R.O. 1973. Short and Long Term Pathological Effects of Glyceryl trierucate and of Increasing Levels of Dietary Rapeseed Oil in Rats. Nutrition Metabolism 15:219-231.

Achaya, K.T. 1971. Chemical Derivatives of Castor Oil. Journal of American Oil Chemists' Society 48: 758-763.

Achaya, K.T., 1986. Oilcakes in India- Production Systems and Individual Oil. Journal of Oil Technologists' Association of India, 18(3):59-69.

Achaya, K.T. Craig, T. and Youngs, S. 1964. The Component Fatty Acids and Glycerides of Castor Oil. Journal of the American Oil Chemists' Society 41:54.

Ackman, R.G. 1983. Chemical Composition of Rapeseed Oil. In. High and Low Erucic Acid Rapeseed. Eds. Kramer, J.K.G. and Gaur, F.D. AcademicPress. pp. 85-129.

Afzalpurkar, A.B. and Lakshminarayana, G. 1979. Changes in the Seed Characteristics and Oil Composition During Sunflower Seed Maturation. Journal of the Oil Technologists' Association of India, 11(4):83-84.

Afzalpurkar, A.B. and Lakshminarayana, G. 1980. Influence of Water Avilability and Temperature on Composition of Sunflower Seed and Oil. Journal of the Oil Technologists' Association of India, 12: 1-3.

Afzalpurkar, A.B. and Lakshminarayana, G. 1981. Changes in Chlorogenic, Caffeic and Quinic Acid Contents During Sunflower Seed Maturation. Journal Agricultural and Food Chemistry. 29:203-204.

Agrawal, V.P. and Stumpf, P.K. 1985. Elongation System Involved in the Biosynthesis of Erucic Acid from Oleic Acid in Developing *B. Juncea* Seeds. Lipids. 20: 361-6.

Ahmed, I.O. 1989. Studies on Niger Seed Oil. Oil Crops News Letter No.6.

Ahmed. E.M. and Young, C.T. 1982. Composition, Quality and Flavour of Peanuts. In *Peanut Science and Technology* Eds. Pattee, H.E. and Young, C.T. American Peanut Research and Education Society Inc, Yoakum, TX U.S.A. pp. 655-688.

Ahuja, K.L.1986. Quality of Oil and Cake of Rapeseed-Mustard. Paper Presented in the Subject Matter Workshop-Cum-Seminar on *Rabi* Oilseeds. DOR, Hyderabad. August 28th Sept., 4th, 1986.

Ahuja, M.M. and Sherry, S.B. 1979. Processing of Rapeseed and Mustard Seed Cake for Production of Proteinaceous Meal/Flour. Indian Chemists' Journal 14(4): 25-28.

Ahuja, M.M., Sattar, A. and Rajgopal, N.S. 1975. Production of High Pungent Mustard Oil by Non-Chemical Methods and Characterization of some Mustard Varietities Journal of the Oils Technologists' Association of India. 7: 107.

Ahuja, K.L., Labana, K.S., Raheja, R.K.,Badwal, S.S. 1984. Oil Content and Fatty Acids Variation in Mutants of *Brassica Juncea* L. Czern-Cross. Journal of Oilseeds Research. 1:71-75.

Ahuja, K.L., Raheja, R.K., Labana, K.S. and Gambir, T.S., 1990. Oil Content and Fatty Acid Composition of Some Promising Genotypes of *Gobhi Sarson* (*B. napus*). Journal of Oilseeds Research. 7:114-116.

AICORPO. 1985. Annual Progress Report of *Kharif* Oilseeds Workshop. D.O.R. Hyderabad.

Alba, E. and Greco, I. 1977. Analysis of Some Characters which Influence Seed Oil Content in Safflower (*Carthamus Tinctorius* L.) *Annali Della Facolta di Agraria, Universita di Bard*, 29: 391-396.

Aliare, E. 1914. Allergens and Their Effects. Annals of the Institute of Pasteur, 28: 605.

Amin, D. Kothari, I.L. 1989. Seed Composition of some New Varieties of Sesame. Journal of the Oil Technologists' Association of India. 21: 15-16.

Anand, I.J. 1976. Oil Quality *Vis-à-Vis* Aphid Resistance in *Brassicae*. Workshop Meeting of Oilseeds Research Workers. Nagpur, India.

Anand, I.J., Malik, R.S. and Rawat, D.S. 1979. Breeding Rapeseed-Mustard Oil Crops for Productivity and Improved Quality. Proeedings of the Symposium on Rapeseed and Mustard, CFTRI, Mysore, India. Pp. 58-70.

Anantharaman, K., Subramanyam, N., Bhatia, D.B. and Subramanyam, V. 1959. Processing of Groundnut Cake for Edible Protein Isolate. Indian Oilseeds Journal. 3:85.

Anderson, J.W., Johnstone, B.W. and Cook-Newell, M.E. 1995. Meta-analysis of the Effects of Soy Protein Intake on Serum Lipids. New England Journal of Medecine. 333:276.

Anjaneyulu, Y.V., Kulkarni, K. and Gouda, D.C. 1983. Polysaccharides of *Ricinus Communis* Seed Endosperm. Current Science. 52: 874-875.

Anon, 1988. Low Fat Peanuts. Journal of the American Oil Chemists' Society. 65:1406.

Anon, 2000. The Robison York State Herb Garden: Herb Gardening, Pp.147-149.

Anonymous, 2004.Safflower-Nature's Gift to Humanity. Progress Report of the NATP Project on "Value addition of Safflower Petals for Natural Dyes and Herbal Healtheare Products". Central Institute for Research on Cotton Technology, ICAR, Matunga, Mumbai.

Applequist, L.A. and Ohlson, R. 1972. In. Rapeseed. Eds L.A. Applequist, and Ohlson, Elsevion, Amsterdam. pp 1-8.

Aquilar, V.G. and Torres, E.G. 1969. Determination of Sugars in Sesame Seeds by Chromatography. *Zeit Lebensm – Unter. Forsch.* 140 : 332-35.

Ariale, S., Suzuki, H., Fujimaki, M. and Sakurai, Y. 1966. Chemistry of Phenols and Their Properties. Agricultural and Biological Chemistry, 30: 364.

Aust , K. 1993. Application of Lecithin in Bakery Foods. AIB Technologists' Bulletin. 15.

Azhar, M., Saravanan, K., Ram Mohan Rao, S. and Vijay Kale, 1999. Studies on Optimization of Reaction Conditions for the Preparation of Castor Methyl Esters. Journal of the Oil Technologists' Association of India. 31(4): 171-172.

Badami, R.C., Shivarmurthy, S.C., Joshi, M.S., Patil, K.B., Subba Rao, Y.V.,Sastry, G.S.R.and Viswanatha Rao, G.K. 1979. Characterization of 15 Varieties of Peanuts for Yield, Oil Content and Fatty Acid Composition. Journal of the Oil Technologists' Association of India. 17 : 15-16.

Bajpai, M., Sharma, A. and Vasistha, A.K. 1985. Component Acids and Glycrides in the Oils of Different Genetic Varieties of Linseed. The Journal of the Oil Technologists' Association of India. 17:15-16.

Balakrishnan, M., Taneja, V.K., Wagle, R.D. and Patel, S.M., 1984. Partially Defatted High Protein Groundnuts. Journal of the Oil Technologists' Association of India. 16(2):72-74.

Balasaraswathi, R. and Raj, W. 1983. Studies on Sunflower Oil with reference to its Keeping Qualities. Journal of Food Science and Technologists' of India. 29: 21-24.

Banerji, B. Pandey, V. and Dixit, B.S. 1999. Wild Safflower- an Alternative Source of Safflower Oil. Journal of The Oil Technologists Association of India. 31 (2): 59-60.

Basha, S.M. M. 1987. Compositional Changes in the Nodulating and Non-Nodulating Peanut Plant During Development. Journal of Plant Physiology. 128: 21-28.

Basha, S.M. M. 1990. Protein as an Indicator of Seed Maturity. Journal of Agricultural and Food Chemistry. 38: 373-376.

Bateria, S. and Nagaraj, G. 1999. Plant and Seed Characteristics of Some Seed, Double Purpose and Flax Genotypes of Linseed. Journal of the Oil Technologists Association of India. 31, No.1 pp.6-9.

Batty, R.S., Mekenzie, S.L. and Finlayson, A.J. 1968. The Proteins of Rapeseed (*Brassica napus* L) Soluble in Salt Solutions. Canadian Journal of Biochemistry, 46:1191-1197.

Belanger, R., Devoe, D. and Reynolds, T. 1966. Coating Compositions Comprising Epoxy Resin. US Patent., 3294, p.721.

Belayneh, H.1991. Oil Crop Germplasm, a Vital Resource for the Plant Breeder. In Plant Genetic Resources of Ethiopia Eds. J.M.M.Angeles, J.G. Hawkes and M.Woredo. Cambridge University Press, Cambridge. pp 344-354

Bell, J.M. 1993. Composition of Conola Meal. Feed Industry Guide. Canola Council of Canada, Winninpeg, Canada. pp 7-11.

Bell, J.M. and Keith, M.O. 1991. A Survey of Variation in the Chemical Composition of Commercial Canola Meal Produced in Western Canadian Crushing Plants. Canadian Journal of Animal Science. 71: 469-480.

Bernardini, E. 1985. Oilseeds, Oils and Fats Vol. I. Interstampa, Roma.

Betchart, A.A. 1979. Development of Safflower Protein. Journal American Oil Chemists' Society. 56(3):454-457.

Betchart, A.A. and Saunders, R.M. 1978. Safflower Protein Isolates-Influence of Recovery Conditions upon Comparative Yield. Food Science. 43: 964-986.

Bhat, S.R. and Shenolikar, G.S. 1946. Hyderabad Science and Industrial Research, Education Department, Commerce, India. H.E.H. The Nizam's Government. Hyderabad. Bulletin, 11.

Bhat, R.V. and Jaspal, N.S. 1951. Indian Forest leaflet No. 124. 1954.

Bhatta, S.K., Acharya, K.L., Raheja, R.K. and Labana, K.S. 1985. Variability in Oil Content and Fatty Acid Composition in Linseed (*Linum Usitassimum*) Annals of Biology. 1:80-85.

Bhattacharya, P., Samui, R.C., Dasgupta, S.K., Chandra Sekhara Rao, T. and Lakshminarayana, G. 1982. Effect of Date of Sowing and Harvesting on Characteristics and Fatty Acid Composition of Sunflower Seeds. Journal of Food Science and Technologists of India. 19:246-248.

Bijster, G.M., Hudalla, B., Kaiser, H., Mangold, H.K. and Vles, B.O. 1978. Nutritional Effects of Various Rapeseed Consitituents in Male Wistar Rats. Proceedings of 5th International Rapeseed Conference. Sweden, June 12-16. 2:141-143.

Borodulina, A.A. and Supranova, L.V. 1976. Change of Sunflower Seed Protein Complex-Qualitative Composition Depending upon Genotype and Ecological Factors. Proceedings of 7th International Sunflower Conference. Krasnodar, U.S.S.R. pp. 173-175.

Brar, G.S. 1977. Variability in the Fatty Acid Composition of Sesame Seed (*Sesamum indicum* L) due to the Capsule Position on the Plant and the Seed Position in the Capsule. Crop Improvement, 4:1-10.

Brar, G.S. 1980. Effects of Temperature on Fatty Acid Composition of Sesame (*Sesamum Indicum* L) Seed. Plant Biochemistry Journal. 7:133-137.

Brar, G.S. 1982. Variation and Correlation in Oil Content and Fatty Acid Composition of Sesame. Indian Jouranl Agricultural Sciences, 52:434-439.

Bringi, N.V., Mani, V.V.S. and Kasbekar, M.G. 1985. Utilization of Non-Traditional Oils. In Oilseed Production-Constraints and Opportunities. Oxford and IBH Publishing Co. pp. 589-597.

Brito, O.J. 1981. Usage of Sesame as a Source of Protein for Human Consumption. Dissertation Abstracts International. B 41:105.

Brochers, R. and Ackerson, C.W. 1950. The Nutritive Value of Legume Seeds: Effect of Autoclaving and Trypsin Inhibitor Test for 17 Species. Journal of Nutrition, 41:339.

Broihier, K. 1997. Fighting Cancer With Phytochemicals. Food Processing, 58:41.

Brown, D.F., Catter, C.M., Mattil, K.F. and Darroch, J.G. 1975. Effect of Variety, Growing Location and their Interaction on the Fatty Acid Composition of Peanuts. Journal of Food Science, 40:1055-1060.

Buchmann, M. and Wenk, C. 1989. Effects of Feeding Low and High Glucosinolate Rapeseed Meal on Metabolism and Thyroid Function of Growing Pigs. Energy Metabolism of Farm Animls, Wegeningen, Netherlands. pp. 135-138.

Burn, E.E. and Huffmann, V. 1975. Food Quality and Peanut Products. In. Peanut Production in Texas. Ed. Muller, J.E. Bulletin of Experimental Station, Texas. pp. 127.

Butler, E.J., Pearson, A.W. and Fenwick, G.R. 1982. Problems which Limit the Use of Rapeseed Meal as a Protein Source in Poultry Diets. Journal of the Science of Food and Agriculture. 33:866-875.

Cai Ya, S.H., Lilley, T.H. and Haslam, E. 1980. Chemistry and Significance of Condensed Tannins. Plenum Press, New York, 307.

Campbell, E.J.1983. Sunflower Oil. Journal of the American Oil Chemists' Society, 60(2):387-392.

Cancalon, P. 1971. Chemical Composition of Sunflower Seed Hulls. Journal of the American Oil Chemists' Society, 48: 629-632.

Cardinal A.J., Burton, J.W., Camacho Roger, A.M., Yang J.H., Wilson, R.F. and Dewey, R.E. 2007. Molecular Analysis of Soybean Lines with Low Palmitic Acid Content in the Seed Oil. Crop Science. 47(1): 304-310.

Carter, J.L. and Hopper, T.H. 1942. Influence of Variety, Environment and Fertility Level on the Chemical Composition of Soybean Seed. USDA Technical Bulletin 7: 87-89.

Ceigla, G.F. and Bell, K.R. 1977. High Pressure Liquid Chromatography for Analysis of Soluble Carbohydrates in Defatted Oilseed Fluors. Journal of the American Oil Chemists' Society. 54 : 150-152.

Celik, S., Akgul, A. and Boydak, E. 2001. Seed and Oil Composition of some Soybean Cultivars Experimented in Harran Plain. *Zaraat Fakultesi Dergisi, Ataturk Universities*. 32(4):469-475.

Chakraborty, M.M., Saha, S. and Datta, R. 1979. Cultivation of Sunflower (*Helianthus Annus L.*) in West Bengal. Studies on Changes in Lipid Composition according to Date of Sowing in *Rabi* Season. Journal of the Oil Technologists' Association of India. 11: 80-82.

Chandru, R., Nanjareddy, Y.A., Chikkadeviah, C., Uday Kumar, M. and Umashankar, R. 2000. Decline in Seed Oil Content of Sunflower Exoticcs on Introduction to India- Possible Reasons. Abstracts-National Seminar on Oilseeds and Oils- Research and Development Needs in the Millennium, Feb. 2-4, 2000, DOR, Hyderabad-30, India. pp.325.

Chaturvedi, A. and Geerani P. 1986. Bio Availability of Niacin from Processed Groundnuts. Journal of Nutrition Science and Vitaminology. 32: 327-334.

Chavan, V.M. 1961. Niger and Safflower. Indian Central Oilseeds Committe, Gandhi Bhavan, Hyderabad-1, India. pp.30-122.

Chavoramony, D. and Padmaja, A.P. 1982. Yield Potential, Oil and Protein Content of Some *Sesamum* Varieties. Oilseeds Journal. 12:67-68.

Chen, S. Wan, P.J., Lusas, E.W. and Rhem, K.C. 1979. Utilisation of Peanut Protein and Oil in Cheese Analogues. Food Technology. 33:88.

Cherry, J.P. 1977. Potential Sources of of Peanut Seed Proteins and Oil in the Genus *Arachis*. Journal of Agricultural and Food Chemistry. 25:186-193.

Cherry, J.P. 1982. Peanut Protein, Properties, Processes and Products In. Peanut Worldwide. Avi Publishing Company. Washington, U.S.A. CT 337-359.

Choi, S.D. and Cho, M.J. 1983. Changes in Fractionation Pattern of Sesame Seed Lipids and Minor Components During Storage. Journal of Korean Agricultural Chemists' Society. 26(4) : 255-268.

Clandinin, D.R., Robblen, A.R. 1981. Rapeseed Meal in Animal Nutrition-II. Non-Ruminant Animals. Journal of the American Oil Chemists' Society. 58:682-685.

Clandinin, D.R. 1958. Sunflower Seed Oil Meal. In Processed Plant Protein Foodstuffs. Ed. A.M.Attschull. Acadamic Press, New York U.S.A. pp.557-575.

Cobb, W.Y. and Johnson, B.R. 1973. Physico-Chemical Properties of Peanuts. In.Peanuts-Culture and Uses. American Peanut Research Education Society. Stone Printing Co. Raonoke, U.S.A. pp. 209-261.

Corelto, A., Pinto, F., Caro, A.D.E., Alba, E., Franchi, S.De, Magini, L., Malik, A., As-Saqui, Franco, F. and Cicanipi, A.1980. Safflower, A New Oil Plant for the South. *Informatore Agrario*, 36 (6): 9045-9057.

Coulson, S. and Stevens, L. 1950. The Allergen Content of Castor Bean and Castor Pomace. Journal of the American Oil Chemists' Society. 48: p.616.

Cross, H.R. and Nicholas, J.E. 1979. Palatability of Groundnut Beef Patties Containing Peanut Meal, Textured Soy Protein and Mechanically Processed Beef. Peanut Science, 6:115.

Daji, J.A. 1943. Oil Cake and their Uses. Indian Farming, 4(11): 553-558.

Damodaram, T. and Hedge, D.M. 2003. Oilseed Situation-A Statistical Compendium, Directorate of Oilseeds Research, Hyderabad-30, India. pp-471.

Dasthagiriah, P. and Nagaraj, G. 1993. Seed and Oil Quality Characteristics of Some Niger Genotypes. Journal of the Oil Technologists' Association of India, 25(2): 42-44.

Datar, G.V. 1951. *Nutan Ghani*. Proceedings of Symposium on Indian Oils and Fats and their Utilization. pp 95-97.

David, F. and Vasantlad. 1986. The Yoga of Herbs. pp.29-227.

Deostale, Y.G. 1981 Trace Element Composition of Common Oilseeds. Journal of the American Oil Chemists' Society, 58: 988-90.

Dixit, M.L, and Shukla, U.C, 1990. Effect of Moisture and Applied Nutrients (P, S, B and Mn,) on Quality of Indian Mustard (*Brassica Juncea L.)*. Journal of Oilseeds Research. 7:111-113.

Dorrel, D.G. 1976. Chlorogenic Acid Content of Meal from Cultivated and Wild Sunflowers Crop Science. 16:422-424.

Dorrel, D.G. 1978. Processing and Utilization of Oilseed Sunflower. In. Sunflower Science and Technology. Agronomy – a Series of Monographs Ed. Carter J.C. American Society of Agronomy. Madison, Wisconsin, pp, 407-440.

Doulatabad, C.D., Ankalgi, R.F., Vijay kumar, S. and Habib, A.F. 1981. Fatty Acid Composition of Important Safflower Varieties of Karnataka. Oilseeds Journal, 12: 29-32.

Downey, R.K. 1964. A Selection of *Brassica Campestris L.* Containing no Erucic Acid in its Seed. Canadian Journal of Plant Science. 44: 499-504.

Downey, R.K. and Robbelen, G. 1989. *Brassica Species.* In. Oilseed Crops of the World. Eds. Downey. R.K., Robbelen, G. and Ashri, A. Mc Graw Hill Publishing Co., New York. U.S.A. pp 339-362

Duarte, C.D., De, H.M.P. and Martins, R.M.D.S. 1979. Level and Characteristics of Seed Oil of Safflower in Portugal. *Grasas Y. Aceites.* 30(5): 315-321.

Dunn, H.C. and Hilditch, T.P. 1950. African Drying Oils. II. Components of Linoleic Rich Oils: Niger Seed Oil. Journal of the Society of Chemistry and Industry (London), 69:13-15.

Dutta, P.C., Helmersson, S., Kebedu, E., Getinet, A. and Appelqvist, L.A. 1994. Variation in Lipid Composition of Niger Seed (*Guizotia Abyssinica* Cass) Samples Collected from Different Regions in Ethiopia. Journal of the American Oil Chemists' Society. 71(8): 839-843.

Eklund, A. 1971a. Preparation and Chemical Analysis of a Lipoprotein Concentrate from Niger Seed (*Guizotia Abyssinica* Cass). *Acta Chemistry Scandinavia*, 25: 2225-2231.

Eklund, A. 1971b. Biological Evaluation of Protein Quality and Safety of a Lipoprotein from Niger Seed (*Guizotia Abyssinica* Cass). *Acta Physiology Scandinavia.* 82:229-235.

Elobeid, M.O. and Rosic, D.A. 1980. High Performance Liquid Chromatographic Determination of Sesamol, Sesamolin and Sesamin in Sudanese Sesame Oil. Proceedings of 6th International Congress on Food Science and Technology, London. pp 2: 3-4.

El-Sawy, A.A. and Soliman M.M. 1988. Identification of Volatile Flavour Components of Roasted Sesame Seeds. *Grasas Y. Aceites,* 39 (3): 160-162.

Erdman, J.W. and Forbes, R.M. 1977. Mineral Bioavailability from Phytate Containing Foods. Food Products Development. 11:46-52.

Fenton, T.W., Leung, J. and Claudinin D.R. 1980. Phenolic Components of Rapeseed Meal. Journal of Food Science. 45:1702-1705.

Fick, G.N. 1989. Sunflower. In. Oilseed Crops of the World. Eds. Robbelen, G. Downey, R.K. and Ashri. A. Mcgraw Hill Publishing Co. New York.U.S.A pp, 301-318.

Forbes, R. 1940. Catayltic Methods for Increasing the Unsaturation of Long Chain Fatty Compounds. Industrial and Engineering Chemistry. 32 (4): 555.

Fore, S.P., Morris, N.J., Freeman, A.F. and Brickford, W.G. 1953. Factors Affecting the Stability of Crude Oil of 16 Varieties of Peanuts. Journal of the American Oil Chemists' Society. 30: 298-301.

Formo, M.W. 1984. The Future of Linseed Oil. Flax Institute U.S. Proc. 50:14-22.

Fru, K. 1969. Union Carbide Corporation Kansas City, M.O. Castor Oil Modified Acrylic Coating and Composition. US. Patent. 3454, 509.

Fukuda, Y., Oswa, T. and Namiki. M. 1981. Antioxidants in Sesame Seed. Journal of Japanese Society of Food Science and Technology. 28:461-464.

Funatsu, M., Funatsu, G., Ishiguro, M., Nanno, S. and Hara, K. 1962. Biochemical Studies on Castor Bean Hemagglutinin. Journal of Biochemistry. 51: 288.

Furfan, J.A., Basha, S.M.M. and Young, C.T. 1978. Variation of Total Monosaccharides in Developing Peanuts (*Arachis hypogaea*, L). *Ciencia E Cultura*. 30(1): 79-82.

Gambhir, P.N. Chanora, S. and Tiwari, P.N. 1983. Date of Harvesting for Optimum Oil Yield in Indian Varieties. India Journal of the Agricultural Sciences. 53(2): 115-117.

Gandhi, V.M. and Mulky, M.H. 1994. Neutralization of Castor Toxins by Sal Tannins. Journal of the Oil Technologists' Association of India, 26(2): 39-42.

Getinet, A. and Teklewold. 1995. An Agronomic and Seed Quality Evaluation of Niger (*Gulzotia Abyssinica Cass)* Germplasm grown in Ethiopia. Plant Breeding. 114:375-376

Getinet, A.and Belayneh. 1989. Association of Hull and Embryo Weight with Oil Content in Niger (*Guizotia Abyssinica* Cass). Ethiopian Journal of the Agricultural Science, 11:41-43

Ghetie, V. and Vitetta, E. 1994. Immunotoxins in the Therapy of Cancer from Bench to Clinic. Pharmaceutical Therapy. 63: 209.

Ghosh, S. and Sastry, P.S., 1986. Triacylglycerol Synthesis in Developing Seeds of Groundnut (*Arachis Hypogaea*): Lipid Accumulation and Incorporation of Labeled Precursors During Seed Maturation. Indian Jouranl of Biochemistry & Biophysics. 23:100-104.

Ghosh, M. and Bhattacharyya, D.K. 1997. Hydrolysis of Castor Oil and Linseed Oil by Different Microbial Lipases. Journal of the Oil Technologists, Association of India. 29(2): 30-31.

Gill, K.S. 1989. Linseed. I.C.A.R. New Delhi, India. Pp. 300-316.

Gilloly, D., Bothwell, T.H., Torranc, A.P. and Macphill, W.R. 1983. Tannins in Foods. British Journal of Nutrition. 49: 331.

Giriraj, K. and Nagaraj, G. 2003. A High Oleic Acid Sunflower Genotype. Journal of Oilseeds Research. 20 (1): 172-73.

Godin, V.J. and Spensley, P.C. 1971. Oils and Oilseeds. Tropical Products Institute, London.

Gopalan C., Rama Sastri, and B.V. Balasubramanian, S.C. 1971. Nutritive Value of Indian Foods. National Institute of Nutrition. Hyderabad, India. pp.83.

Gosh Chaudhuri, P. 2003. Fatty acid Profile of Indian Sunflower Oil. 58th Annual Convention and International Conference on Technological Developments in the Field of Oils, Fats and Allied Products. Novermber 21-23, Kolkata.

Green, A.G. and Marshall, D.R. 1981. Australian Journal of Agricultural Research 32: 599-607.

Green, A.G. and Marshall, D.R. 1984. Euphytica, 33:321-28.

Guerra, M.J., Jaffe, W.G. and Sangroms, E. 1984. Protein Fractions from Commercial Sesame Seed Cake (*Sesamum Indicum*). *Archivos Latino Americanos de Nutrition,* 34: 477-487.

Gupta S.K. 1990. Moisture, Dry Weight and Lipid Composition as Influenced by Capsule Position in Developing Seeds of Sesame (*Sesamum Indicum L.*). Journal of Oilseeds Research. 7:10-15.

Gupta, S.K. and Wagle, D.S. 1989. Influence of Season and Seed Position on Fatty Acid Composition of Polar and Non-Polar Lipid Fractions of Developing Seeds of Sunflower (*Helianthus Annuus* L.). Journal of Oilseeds Research 6:181-186.

Gupta, M.L. Ahuja, K.L. Raheja, R.K. and Labana, K.S. 1988. Variation for Biochemical Quality Traits in Promising Genotypes of Indian Mustard. Journal of Research-Punjab Agricultural University, Ludhiana. 25:1-5.

Gupta, S.K., Sekhon, K.S. and Ahuja, K.L. 1982. Chemical Composition of *Brassica* Seeds. Journal of Food Science and Technology (India). 9:84-86.

Hackler, 1977. *In Vitro* Indices/ Relationships to Estimating Protein Value for Humans In. Evaluation of Protein for Humans. Ed. Bodwell C.E. Avi. Publishing Co. Inc. West Port, Cannecticut. U.S.A. pp, 55-57

Haile, M. 1972. Protein Evaluation of Niger Meal. M.Sc. Thesis Submitted to the University of Arizona, U.S.A.

Hanumatha Rao,K. 1970. Toxic Factors and their Detoxification in Castor. Journal Food Science and Technology. 7: 77-82.

Han Yunzhou, 1997. Appraisal of Corolla. In IV International Safflower Conference, Bari, Italy June 2-7, pp, 369-371.

Hasegawa, K., Maeda, K., Fugino. Y., Wakinaga, T. and Furyino, S. 1978. Gelation Characteristics of Sesame Seed Protein, Journal Agricultural Chemists, Society, Japan. 52: 341-346.

Hashim I.B., Koehler P.E., Eiten Miller R.R and Kvien C.K. 1993. Fatty Acid Composition and Tocopherol Content of Drought Stressed Florunner Peanuts. Peanut Science. 20: 21-24.

Hawkins, R. 1989. Bring Back the Castor Crop. Journal of the American Chemists' Society. 66: 1567

Heaton. T.C., Knowles, P.F., Mikkelsen, D.S. and Ruckman, J.E. Production of Free Fatty Acids in Safflower Seeds by Fungi. Journal of the American Oil Chemists' Society, 55(5): 465-468.

Hilditch, T.P., Sime, I.E., Aky, Y.A.H. and Meara, M.L. 1994. The Component Acids of Various Vegetable Fats. Journal of the Society of Chemistry and Industry, (London). 63(4): 112-114.

Hill, A.P. and Diely, P. 1951. Economic Botany. Mc Graw-Hill, New York. 20

Hocking, P.J., Randall, P.J. and Pinkerton, A. 1987. In Advances in Agronomy Ed. Brandy, N.C. Academic Press: 41:221-296.

Holley, K.T. and Hammons, R.O. 1968. Strain and Seasonal Effects on Peanut Characteristics. University of Georgia Agricultural Exp. Station Bulletin.

Horowitz, B. and Winter, G. 1965. A new Safflower Oil of Low Iodine Value. Nature, 1979(4459): 582-583

Hou-Guang Yun, Ablett, G.R., Pauls, K.P. and Rajcan, I . 2006. Environmental Effects on Fatty Acid Levels in Soybean Seed Oil. Journal of the American Oil Chemists' Society. 83(9): 759-763.

Huang, K.S., Hawar, W.S., Nam Y.J. and Min, B.Y. 1984. Quality Evaluation of Sesame by Liquid Chromatography. Korean Journal of the Food Science and Technology. 16(3): 348-352.

Hymowitz, T. and Hadley, H.H. 1972. Inheritance of a Trypsin Inhibitor Variant in Seed Protein of Soybean. Crop Science. 12: 107-108.

Ichihara, O.K. and Noda, M. 1980. Fatty Acid Composition and Lipid Synthesis in Developing Safflower Seeds. Phytochemistry. 19(1): 49-54.

Irwin, T. 1990. Technical Report on the Medicinal and Toxic Studies of Castor Oil. F344/ Rats and B6C3F1 Mice(Feed Studies). National Toxicology Program, NC.

Jambunathan, R., Gurtu, S., Raghunath, K., Sreedhar, R., Dwivedi S.L. and Nigam, S.N. 1992. Chemical Composition and Protein Quality of Newly Released Groundnut *(A.hypogaea L)* Cultivars. Journal of the Science of Food and Agriculture. 59: 161-167.

Jambunathan, R., Singh, A.K., Gurtu, S. and Raghunath, K., 1993. Amino acid Composition, Fatty Acid Composition and Levels of Protease Inhibitors in Seeds of Wild *Arachis Species, Oleagineau*. 48 (10): 415-419

Jamieson, G. 1943. Vegetable Fats and Oils. Reinhold Publishing Corporation, New York, USA. pp, 262-280.

Janicki, J., Warchalonski, J., Sobkowska, E. and NoWaskowska, B. 1972. Amino Acid Composition of Proteins of some Oil Bearing Plants (in Polish). *Rocz. Technology Chemists Zwyn.* 22:251.

Jeswani, L.M. 1985. Linseed (*Linum Usitassimum* L.). In, Oilseed Production -Constraints and Opportunities. Oxford & IBH Publishing Co. pp, 133-149.

Jianguo, Y., Delu, Z. and Jiapei L. 1993. Utilization of Safflower. Proceedings of the III International Safflower Conference. Beijing, China. 893

Jikugawa, K., Arai, M. and Kurechi, T. 1983. Participation of Sesamol in Stability of Sesame Oil. Journal of the American Oil Chemists' Society. 60:1528-1533.

Johnlust. 1994. The Herb Book, pp. 495-568.

Johnson, L.A., Suleiman, T.M. and Lusas, E.W. 1979. Sesame Protein: A Review and Prospectus. Journal of the American Oil Chemists' Society, 56:463-468.

Jones, D.B. 1931. Factors for Converting Percentage Nitrogen in Foods and Feeds into Percentage of Proteins. U.S.D.A. Circular No. 193. Washinton. D.C.

Jones, J.D. 1979. Preparation and Tests on Rapeseed Proteins. Proceedings of the Symposium on Rapeseed and Mustard. November, Mysore, India. pp, 169.

Jones, D. and Marascow, H. 1998. Antibodies for Gene Targetting and Intracellular Expression. Advances in Drug Research. 31(6):153.

Joshi, A.B. 1961. *Sesamum*. Indian Central Oilseeds Committee. Hyderabad.

Kabirullah, M., Khan, M.A. and Faruque, O. 1977. Studies on the Nutritive Value and Properties of Oilseed Meals. Bangladesh Journal of Science and Industrial Research. 12:197.

Kamal-Eldin, A. 1993. Seed Oils of *Sesamum Indicum* and Some Wild Relatives. Ph.D. Dissertation. Swedish University of Agricultural Science. Uppsala.

Kamath, K.S., Govindham, P.E., Paul, T.M., Chatterjee, K.L. and Mathew, J.V. 1979. Sunflower Seed Oil – Physico Chemical Characteristics and Industrial Use. Research and Industry, 24: 168-170.

Kantharaj Urs. M. and Kousalya, S. Murthy, 1979. Intergrated Processing of Mustard Seed for Food and Feed Purposes- Proceedings of the Symposium on Rapeseed-Mustard, Mysore, India. pp, 162-166.

Kaushik, N. and Agnihotri, A. 2000. GLC Analysis of Indian Rapeseed Mustard to Study the Variability of Fatty Acid Composition. Biochemical Society Transactions. 28(6): 581-583.

Kerber, E. and Buckloh, G. 1982. Sinapine in the Tribe *Brassicae. Angew Bot.* 56: 85-91

Khan, K.A. 1989. Utilization of Chickpea and Groundnut in Pakistan, Proceedings of a Consultants Meeting. ICRISAT, Hyderabad, India. pp, 102.

Khan, S. and Hussain, T. 1988. Amino Acid Compostion and Nutritive Value of Sesame (*Sesamum Indicum* L.) Seeds. Pakistan Journal of Biochemistry. 21: 25-28.

Khatoona Sakina. 2001.Effect of Heating on Expeller Pressed and Refined Sunflower Oil. Journal of the Oil Technologists' Association of India. 33(1):12-15

Kim-KwanSu, Lee-JungRo and Lee-JoonSeol, 2006. Determination of Sesamin and Sesamolin in Sesame (*Sesamum indicum* L.) Seeds Using UV Spectrophotometer and HPLC. Korean Journal of Crop Science. 51(1): 95-100.

King, J., Mc Neilly, T. and Thuman, D. 1977. Variation in Single Seeds of Four Varieties of Oilseed Rape. Journal of the Science of Food and Agriculture, 28:1065.

Kinman, M.L. and Earle, F.R. 1964. Agronomic Performance and Chemical Composition of the Seeds of Sunflower Hybrids and Introduced Varieties. Crop Science. 4:417-20.

Kirk, P. 1923. Chemical Engineering in the Production of Coated Fabrics. Chemical Engineering. 28(23): 1107.

Knowles, P.F. 1955. Safflower Production, Processing and Utilization. Economic Botany. 9(3): 273.

Knowles, P.F. 1969. Centres of Plant Diversity and Conservation of Crop Germplasm: Safflower, Economic Botany. 21: 156-162.

Knowles, 1989. Safflower. In Oil Crops of the World. Eds. G. Robbelen, R.K. Downey and A. Ashri. Mc Graw Hill Publishing Co. pp. 365-374.

Knowles, P.F. and Mutwakil, 1965. Inheritance of Low Iodine Value of Safflower Selections from India, Economic Botany. 17:139-145.

Kohler, G.P., Kuzmicky, D.D., Patter, R., Guggolz, J. and Herring, 1966. Safflower Meal. Journal of the American Oil Chemists' Society. 43:413-475.

Korbitz,W. 1995. Utilization of Oil as a Biodiesel Fuel. In *Brassica* Oilseeds-Production and Utilization Eds.Kimber and D.I.McGregor. CAB International, Wallingford, U.K. pp. 353-371.

Kozolowska, H., Rotkiewicz, D.A., Zodernowski, R. and Solulski, F.W. 1983. Phenolic Acids in Rapeseed and Mustard. Journal of the American Oil Chemists' Society. 60: 19-23.

Kramer, K.G., Sauer, F.D. and Pidgen, W.J. 1983. High and Low Erucic Acid Rapeseed Oils. Academic Press. NewYork.

Krishnaveni, S. and Sadasivam 1994. Fractionation and Electropheretic Separation of Sunflower Proteins. In. Sustainability in Oil Seeds. Eds Prasad, M.V.R., *et al.* I.S.O.R. Hyderabad, India. 510-513.

Kritchev, D.S., Tepper, S.A., Vesselinovitch, D. and Wissler, R.W. 1973. Cholesterol Vehicle in Experimental Atherosclerosis Part 13: (Randomized Peanut Oil). Atherosclerosis, 17:225.

Kulkarni, D.N., 1978. Proceedings of First Indian Convention of Food Science and Technologists, June 5th- 7th, CFTRI, Mysore.

Kulkarni, D.N., Joglekar, N.V. and Ingle, Y.M. 1985. Processing of Safflower and Production of Edible Grade Safflower Meal. In. Oilseed Production: Constraints and Opportunities. Oxford & IBH Publishing Co. pp. 652-656.

Kulkarni, D.N., Kulkarni, K.D., Tathe, S.B. 2001. Studies on the Extraction of Safflower Yellow B and Carthamin Red Pigments from Safflower Florets as Food Colorant. In Proceedings of V International Safflower Conference, Montana, USA, June. 23-27, p. 321.

Kulkarni, D.N., Revanwar, S.M., Kulkarni, K.D., Deshpande, H.W., 1997. Extraction and Uses of Nautral Pigments from Safflower Florets. In IV International Safflower Conference, Bari, Italy. June 2-7, pp 365-368.

Kulkarni, L.G. and Ramanamurthy, G.V. 1977. Castor, ICAR, New Delhi. Pp.46

Kumar, H. Agarwal, R.K. Singh, R.B. and Singh, R.M. 1982. Correlation and Path Analysis of Oil in Safflower (*Carthamus Tinctorius*). Malaysian Applied Biology 11 (1): 19-28.

Kumar,N. Srivastava, K.N. Sharma, N.D. and Mehta S.L. 1988. Oil Synthesis and Composition in Developing Cotyledons of Groundnut. Journal of the Nuclear Agriculture and Biology. 17:11-17.

Lakshminarayana, M. and Ramesh Babu, T. 1992. Evaluation of Five Insect Growth Regulators on the Egg Parasitoid, *Trichogramma Chilonis* (Ishii) (*Hymo. Corcyra Cephalonica (Lep. Galleridae)*. Journal of Applied Entomology, 113: 56-60.

Lata, T.S. and Prakash, V.1984. Studies on the Proteins from Safflower Seed*Carthamus Tinctorius* L.). Journal of the Agricultural and Food Chemistry, 32(6): 412-1416.

Lau, A., Ako, H. and Washburne, W. 1980. Survey of Plants for Enterokinase Activity. Biochem. Biophys. Research Committee, 92:1243.

Laxminarayana, G. 1983. Castor Seed Products. S.S. Ramaswamy Memorial Lecture. 38th Annual Convention of the Oil Technologists' Association of India. 19-20, Feb., Calcutta, India. pp. 1-34.

Laxminarayana, G., Paulose, M.M. and Kumari, B.N. 1984. Characterisics and Composition of Indian Castor Seed and Oil. Journal of the American Oil Chemists' Society, 61:1871-1872.

Laxminarayana, G., Paulose, M.M. and Rao K.V.S.A., 1987. Selective Reduction of Linolenic Acid Content in Indian Soybean Oil by Partial Hydrogenation. Journal of the Oil Technologists' Association of India, 19(1):17-19

Laxminarayana, M.R., Seetharam, A., Ramanathan, K. and Khetrapal, 1980. NMR. Current Science, 49(8): 308-309.

Lee, J.I. and Kang, C.W. 1980. Breeding of Sesame for Oil Quality Improvement- Evaluation of Oil Quality and Differences in Fatty Acid Composition between Cultivars in Sesame, Journal of the Korean Society of Crop Science, 25:54-65.

Lee, J.I., Kang, C.W. and Lee, S.T. 1984. Breeding Sesame for Improved Oil Quality 3. Variatiion in Fatty Acid Composition in Mutant Populations. Research Reports Crop Experiment Station Suweon, Korea Republic, 26(1): 134-143.

Lee, J.I., Lee, B.H., Seong, N.S. and Kang, C.W. 1991. Studies on Interspecific Hybridization in Sesame. I. Characteristics and Cross Affinity of Wild Sesame (in Korean). Korean Journal of Breeding. 22:356-360.

Lee Young Chul, Oh Se Wook, Chang Jaehyun, Kim-InHwan, 2004. Chemical composition and oxidative stability of safflower oil prepared from safflower seed roasted with different temperatures. Food Chemistry. 84(1): 1-6.

Leiner, I.E. 1962. Toxic Factors in Edible Legumes and their Elimination. American Journal of Clinical Nutrition. 11:281.

Leiner, I. 1979. Significance for Humans of Biologically Active Factors in Soybeans and other Legumes. Journal of the American Oil Chemists' Society, 56: 124.

Leiner, I.E. and Kakade, M.L. 1980. Protease Inhibitors. In. Toxic Constitutents of Plant Food Stuffs. Ed. Leiner, I.E. Academi Press. New York.U.S.A. pp. 7

Leming, R. and Lember, A. 2005. Chemical Composition of Expeller-Extracted and Cold-Pressed Rapeseed Cake. *Agraarteadus.* 16(2): 103-109.

Lennerts, L. 1989. Sesame Cake / Expeller and Sesame Oil Meal. *Muble Mischfutter Technik.* 126: 240-241.

Li Dajue, Han Yunzhou. 1993. Proceedings of III International Safflower Conference, Beijing, China 14-18 June 1993.

Li Dajue, Hans Henning Mundel, 1996. Promoting the Conservation and Use of Under Utilized and Neglected Crops: Safflower (*Carthamus Tinctorius* L.) IPGRI, U.K.

Liu, Keshun. 1999. Current Constraints in Soybean Food Utilization and Efforts to Overcome them. In. Proceedings of the World Soybean Research Conference VI, Chicago, USA, August 4-7, Ed. H.E. Kauffman, The University. of Illinois, USA. pp. 409-418.

Loomis, W.D. and Battallie, J. 1966. Mechanism of Phenols with Proteins. Phytochemistry, 5: 423.

Lusas, E.W. 1979. Food Uses of Peanut Protein. Journal of the Amrican Oil Chemists' Society. 56: 425-430.

Lyon, C.K. 1972. Sesame- Current Knowledge of Composition and Use. Journal of the Amrican Oil Chemists' Society, 49: 245-40.

Lyon, C.K., Gumbmann, M.R., Betchart, A.A., Robbins, D.J. and Saunders, R.M. 1979. Removal of Deleterious Glucosides from Safflower Meal, Journal of the Amrican Oil Chemists' Society. 56:560-564

Maiti, S., Hegde, M.R. and Chattopadhyay, S.B. 1988. Niger(*Guizotia Abyssinica* Cass) In. Handbook of Annual Oilseed Crops. Oxford & IBH Publishing Co. Pvt. Ltd., New Delhi,pp. 213-227.

Mahapatra N.N. 1998. "Grafting Flax". The Indian Textile Journal, CIX, (2): 50-52.

Mandal, S. and Ranbir Singh, 1993. Oil and Protein Contents in Sunflower Genotypes. Journal of Oilseeds Research. 10(1): 161-162.

Masson, S.L., Villaroel, G.O. and Ceron Diaz, W. 1981. Composition and Content of Oil Extracted from Six Safflower Seed Varieties Grown in Chile. *Grasas Y. Aceites.* 32(3):163-169.

Mathur, J.M.S. and Sharma, N.D. 1989. In. Recent Advances in Plant Biochemistry. Eds. Mehta, S.L. and Sane, P.V. ICAR, New Delhi. pp.340-396.

Mazingo, R.W. and Steele, J.L. 1982. Fatty Acid Composition of Peanut Genotypes in the Virginia Carolina Production Area. Proceedings of American Peanut Research Education Society Inc. 14(1):29-35.

Mazulosvela, F.1981. Determination of Esterified Safflower Oils by the Pancreatic Lipase Method; *Grasas Y. Aceites.* 32(3): 143-146.

Mcwatters, K.H. and Heaton, E.K. 1979. Quality Characteristics of Groundnut Beef Patties Expanded with Moist, Heated and Unheated Seed Meals. Journal of the Amrican Oil Chemists' Society. 56:86.

Metha, S.K. Trehan, K.B. and Prasad, B. 1979. Studies on the Primary and Secondary Branch Contribution to 100 Seed Weight, Oil and Protein Content of Safflower. Science and Culture. 45(1): 30-32.

Michael Tierra. 1999. Planetary Herbology. pp.326-327.

Misra, S.D. and Gaur, B.K. 1982. Effect of Seed Size on Oil Content in Groundnut Variety TG-1. Indian Journal of Plant Physiology. 25(1):106-109.

Miller, J. and Young, C.T., 1977. Protein Nutritional Quality of Florunner Peanut Meal as Measured by Rat Bioassay. Journal of Agricultural and Food Chemistry. 25(3): 653-657.

Mitaru, B.N., Blair, R. and Reichert, R.D. 1983. The Effect of Canola Hulls on Growth, Feed Efficiency and Proteins and Energy Utilization in Broiler Chickens. Canadian Journal of Animal Science. 63: 655-662.

Morrison, F.B. 1957. Feeds and Feeding, Morrison Publishing Co., New York.

Morrison, R.S., Joseph, R., Spies, J.R. and Coulson, E.J. 1965. Electrophoretic Pattern of Castor Bean Allergens. Journal of the American Oil Chemists' Society. 110(2): 300.

Morton, M.S., Wilcox, G., Wahlquvist, M.L. and Giffiths, K. 1994. Determination of Lignans and Isoflavonoids in Human Female Plasma Following Dietary Supplementation. Journal of Endocrinology 142(2):251-259.

Mosjidis, J.A. 1982. The Inheritance of Oil Content and its Fatty Acid Compostion in the Sesame Seed (*Sesamum Indicum* L) and the Study of their Correlations. Dissertation Abstracts International 42(10):3947 B.

Mosse, J. and Bandet, J. 1972. Biochemical Aspects of Quality and Content of Lysine in Sunflower Seed Proteins. Proceedings of 5th Intertnational Sunflower Conference, (Clemant-Ferrand). pp. 437-440

Mosse, J. and Pernolett, J.G. 1982. Storage Proteins of Legume Seeds. In. Chemistry and Biochemistry of Legumes. Ed. S.K. Arora. Oxford & IBH Publishing. Co., pp.120.

Mukherjee, P.K. 1988. Ramie Holds Vast Potential. The Indian Textile J. CVIII, 7: 42-45.

Mullor, J.B. 1968. Improvement of the Nutritional Value of Food Oils. *Rev. Fae. Ing. Quim., Univ. Nae. Litroal.* 37:183-210.

Mullor- Mulot, W. 1976. Rapid Method for Quantitative Determination of Individual Tocopherols in Oils and Fats. Journal of the Amrican Oil Chemists' Society, 53:732-736.

Muralidharudu, Y., Nagaraj, G. and Mev Singh, 1989. Effect of Location and Genotype on Safflower Oil and Oil Quality. Annals of Plant Physiology. 3(2):132-136.

Muralidharudu, Y. Sujatha, M. and Mev singh, 1993. Extent of Variation in Seed Oil Content and Fatty Acid Composition of Cultivated and Two Closely Related Wild Species of Safflower. In. Proceedings of III International Safflower Conference, Ed. Li Dajue and Han Yuanzhou. Beijing, China, June 9-13, 241-246.

Muralidharudu, Y and Nagraj, G. 1995. Seed Characteristics and Fatty Acid Composition of Linseed Cultivars. Journal of the Oil Technologists' Association of India. 27(2):173-174.

Muralidharudu, Y. and Hari Priya, Ch. V. 2000. Effect of Genotypes – Environment Interaction on the Seed Oil Content in Sunflower. Abstracts-National Seminar on Oilseeds and Oils – Research and Development Needs in the Millennium, Feb. 2-4, DOR, Hyderabad-30. India. pp.333.

Muralidharudu, Y. and Nagaraj, G. 1997. Oil in Some Indian Castor (*Ricinus communis*) Genotypes. Journal of Oil Technologists' Association of India. 29(3): 59-60.

Muralidharudu, Y. and Virupakshappa, K. 1998. Extent of Variation in Oil, Protein and Fatty Acid Composition of High Oleic Hybrids, Wild Species and Some Cultivars of Sunflower. Journal of the Oil Technologists' Association of India. 30(4):159-161.

Murthy I.Y.L.N., Mev Singh, Sastry J.A. 1997. Some Macro and Micro Nutrients in Sesame Seeds. Journal of Oilseeds Research. 14(1):124-125.

Nagao, A. and Yamazaki, M. 1983. Lipids of Sunflower Seeds Produced in Japan, Journal of the American Oil Chemists' Society. 60(9): 1656-1658.

Nagaraj, G. 1987. Effect of Foliar Spray of Micronutrients on Yield and Chemical Composition of Peanuts. Annals of Plant Physiology. 1:176-202.

Nagaraj, G. 1990a. Chemistry and Utilization. In Groundnut Monograph Ed. P.S. Reddy, ICAR, New Delhi, India. pp 555-565

Nagaraj, G. 1990b. Biochemical Quality of Oilseeds. Journal of Oilseeds Research, 7: 47-55.

Nagaraj, G. 1990c. Fatty Acid and Amino Acid Composition of Niger Varieties Journal of the Oil Technologists' Association of India. 22(4):88-89.

Nagaraj, G. 1991a. Kernel Quality and Fatty Acid Composition of Released Indian Groundnut Varieties. Proceedings of National Seminar on Making India Self-Reliant in Vegetable Oils. Indian Society of Oilseeds Research, Hyderabad, India. pp. 577-581.

Nagaraj, G. 1991b. Fatty Acid Profile of New Plant Types in Sesame (*Sesamum indicum* L.) Journal of Oil Technologists' Association of India. 23:28-29.

Nagaraj, 1993a. Safflower Seed Composition and Oil Quality – A Review. Procedings of III International Safflower Conference, Beijing, China, June 14-18. pp 58-71.

Nagaraj, G. 1993b. Seed Composition and Fatty Acid Profile of Some Indian Safflower Cultivars. Proceedings of III International Safflower Conference Beijing, China. June 14-18, 241-242.

Nagaraj, G. 1994a. Seed and Oil Quality of Safflower. Proceedings of Group Metting on Breeding Approaches for Improving Productivity in Safflower, DOR Hyderabad, India. Pp. 32-36.

Nagaraj G. 1994b. Seed and Oil Quality of Safflower Genotypes. Journal of Oilseeds Research. 11(2): 242-244.

Nagaraj, G. 1994c. Effect of Location on the Fatty Acid Composition of Niger Seed Oil. Journal of the Oil Technologists' Association of India. 26(3):75-76.

Nagaraj, G. 1995. Quality and Utility of Oilseeds, Directorate of Oilseeds Research, Hyderabad, India. Pp. 28-30.

Nagaraj, G. 1999a, Final Report on the Adhoc Project on Antinutritional Constituents in Oilseeds and Cakes, D.O.R. India. pp 50.

Nagaraj, G. 1999b. Quality and Production Characteristics of Some Specialty Genotype Accession of Niger: Journal of Oil Technologists' Association of India. 31(4): 187-188.

Nagaraj, G. 2002. Value Addition for Improving Utility and Profitability of Sesame and Niger. In: Integrated Crop Management of Sesame and Niger. Eds. S.S. Duhoon, A.K. Tripathi and H.K. Jharia, PC Unit (S&N), JNKVV, Jabalpur. India. pp 26-31.

Nagaraj, G. 2003. Effect of Seed Maturity on Seed and Oil Quality in some Oil Seeds. Abstracts. 58th Annual Convention and International Conference on Tecnological Developments in the field of Oils, Fats and allied Products. Oil Technologists' Association of India, Kolkata.

Nagaraj, G. 2003a. Diversified Uses of Castor. Ed. D.M. Hedge and G.V. Ramanjaneyalu. DOR Bulletin Directorate of Oilseeds Reasearchs Hyderabad, India. pp 1-15

Nagaraj, G. and Anjani, K. 1996. Seed Quality and Fatty Acid Composition of Newly Developed Safflower Hybrids and Their Parents, Journal of Oilseeds Research, 13 (1):106-108.

Nagaraj, G. and Kailash Kumar, 1986. Preparation of Fat-Lean Protein-Rich Groundnuts. The Oils and Oilseeds Journal. July-December, pp, 82-85.

Nagaraj G. and Misra, D.P. 1984. Some Factors Influencing Groundnut Oil Production. Journal of Oil Technologists' Association of India, 16:91-94.

Nagaraj, G. and Murthy, T.G.K.1988. Fatty Acid Composition of Wild Relatives of Groundnut. Journal of the Oil Technologists' Association of India. 20:14-15.

Nagaraj, G. and Reddy B.N. 2006. Effect of Cultivars, Fertilizers and Season on the Seed and Oil Quality of Sunflower Oil *Helianthus annuus L.* Journal of Oilseeds Research. 23(2):342-343.

Nagaraj, G. and Srinivas, C.V.S. 2001. Factors Influencing Ricinoleic Acid in Castor- an Importat Oleochemical Raw Material. 56th Annual Convention and International Seminar on Oleochemicals, Surfactants and Specialty Products. November 26-28, Mumbai, India.

Nagaraj, G. Kailash Kumar and Sheela Chauhan, 1984. Seed and Oil Quality Characteristics of Ten Groundnut Varieties. Journal of the Oil Technologists' Association of India. 16(4).

Nagaraj, G. Kailash Kumar, Sheela Chauhan, Taslim Ahmed and Bhagat, N.R. 1986. High Oil Groundnut Varieties. Indian Journal of Agricultural Sciences. 56:64-65.

Nagaraj, G., Muralidharudu, Y. and Mev Singh, 1990. Effect of Location and Genotype on the Sunflower Seed and Oil Quality. Annals of Plant Physiology, 4(1):48-53.

Nagaraj, G. Praveena, B. and Srinivas, C.V. S. 2001. Tocopherols in Some Indian Genotypes of Sunflower, Safflower and Castor. Journal of the Oil Technologists' Association of India. 33(1): 20:21.

Nagaraj, G., Sharma, S.M. and Balakrishnan, R. 1994. Studies on Yield and Some Physico-Chemical Characteristics of Niger Germplasm Accessions. In. Sustainability in Oilseeds Eds. M.V.R. Prasad *et al.*, Indian Society of Oilseeds Research Hyderabad, India. pp.523-524.

Nagaraj, G., Sheela Chauhan and Ravindra, V. 1989. Peanut Composition and Oil Quality as Influenced by Genotype and Harvest Stages. Journal of the Oil Technologists' Association' of India. 14:55.

Nagaraj, G., Srinivas, C.V.S. and Praveena, B. 2000. Nutritional Quality of Sunflower Genotypes and Utilization of Deoiled Cake. (Abstracts), National Seminar on Oilseeds and Oils – Research and Development Needs in the Millennium, Feb. 2-4, DOR, Hyderabad-30, India. pp. 336-337.

Nagaraj, G. and C.V.S. Srinivas, 2002(a). Chemical Composition and Quality of Indian Sunflower. Oil Technologists Association' of India – American Oil Chemists' Society-Asian Conference on Lipids, Fats and Oils, New Delhi. Dec. 6-8, 2002.

Nagaraj, G. and C.V.S. Srinivas, 2002(b). Some Factors Influencing Colour and Acid Value of Castor Oil. Oil Technologists Association' of India – American Oil Chemists' Society- Asian Conference on Lipids, Fats and Oils, New Delhi, India. Dec. 6-8, 2002.

Nagaraj, Muniyappa, V., Singh, S.J. and Virupakshappa, K. 1997. Occurrence of a Mosaic Virus Disease on Sunflower in Karnataka, Indian Phytopathology. 50(2):277-281.

Namiki, M. 1995. The Chemistry and Physiological Functions of Sesame. Food Reviews International. 11:281-329

Nandini Nimbkar, 2002. Revival of an Ancient Crop-Safflower, Times of India-Agricultural Journal. Nov.-Dec.

Nasirulla, K., Ankaiah, N., Krishna Murthy, M.V., Nagaraja Rao, K.V. and Kapur, O.P. 1982. Storage Study on Groundnut Oil. Journal of Oil Technologists' Association of India. 14:55.

Narasimhachar, B.L., Atchyuta Ramayya, D. Reddy, B.R and Thirumala rao, S.D.1984. Laboratory Prepared Low Oil and High Protein Groundnut and Sufflower Kernel by Solvent Extraction Method. Journal of the Oil Technologists Association' of India. 16(4): 1-2.

Narasinga Rao, M.S. 1985. Nutritional Aspects of Oilseeds. In. Oilseeds- Production, Constraints and Opportunities Eds. H.C. Srivastava, Oxford and IBH. Publishing Co. Ltd.

Nasirullah, K., Mallika, T., Rajalakshmi, S., Pasupathi, K.S., Ankiah, N.N., Vibhakar,S., Krishna Murthy, M.N., Nagaraja, K.V. and Kapur, O.P. 1982. Studies on Niger (*Guizotca Abyissinica*) Seed Oil. Journal of Food Science and Technologists' (India). 19:147-149.

Nataraja Murthy, T. and Lakshminarayana, G. 1979. Studies on Stability of Groundnut Oil During Continuous Deep Frying in a Plant. Abstracts International Congress on Oilseeds and Oil, pp.37.

Nath, R. and Giri, K.V. 1957. Physico Chemical Studies on Indigenous Seed Proteins II.Fractionation, Isolation and Electrophoretic Characterisation of Sesame Globulins. Journal of Scientific and Industrial Research. 16C: 51-58.

Naughton, F.C. 1974. Production, Chemistry and Commercial Applications of Various Chemicals from Castor Oil. Journal of the American Oil Chemists' Society. 51: 65.

NDDB, 1982. A Manual on Groundnut. National Diary Development Board, Anand, India. Pp.1-68.

Nwokolo, E.N. and Bragg, D.B. 1977. Influence of Phytic Acid and Crude Fibre on the Availability of Minerals from Four Protein Supplements in Growing Chicks. Canadian Journal of Animal Science. 57:475-477.

Obuko, K., Nishimura, N. and Shibasaki, K. 1979. Composition of Sesame Seed Protein Components and Purification of Main Globulin. Cereal Chemistry. 56:100-104.

Ochodzki, P. and Piotrowska, A. 2002. Physical properties and chemical composition of winter rapeseeds with different colour of seed coat. *Rosliny Oleist.e* 23(2):235-241.

Oke, O.L. 1967. Chemical Studies on Some Nigerian Pulses. West African Jornal of Biological and Applied Chemistry. 9: 52-55.

Olsnes, S. and Pihl, A. 1983. The Specificity and Action of Animal, Bacterial and Plant Toxins. Chapman and Hall., London. 1:128.

Orf, J.H. and T. Hymowitz 1977. Inheritance of a Second Trypsin Inhibitor in Seed Protein of Soybeans. Crop Science. 17: 811-813

Orf, J.H. and Hymowitz, T. 1979. Inheritance of the Absence of the Kunitz Trypsin Inhibitor in Seed Protein of Soybeans. Crops Science. 17: 107-109.

Orr, E. 1972. The Use of Protein Rich Foods for the Relief of Malnutrition in Developing Countries- an Analysis of Experience. Tropical Products Institue. Report No. G. 73, London. U.K. p.13.

Osborne, T.B. 1962. Medicinal and Poisonous Plants of Southern and Eastern Africa, Living Stone Ltd. London. pp 434.

Paliwal, D.K. and Randhawa, H.S. 1978. Evaluation of Simplified (*Guizotia Abyssinica*) Seed Medium for Differentiation of *Cryptococcus Neoformans*. Journal of Clinical Microbiology. 7:346-348.

Pancholy, S.K., Basha, S.M.M. and Gorbet, 1982 Composition of Seeds Obtained from Non-Nodulating and Nodulating Peanut Line and Cultivars. Peanut Science. 9: 50-52.

Pandey Ritu and Dayal, R. 2003. Flax-Jute and Flax-Cotton Blended Fabric. Asian Textile Journal. 12(7):51-55.

Parades, L.O. and Odorica, F.C. 1986. Production of Safflower Protein Isolates - Composition, Yield and Protein Quality. Journal of the Science of Food and Agriculture. 37(11):1097-1103.

Paramjit Singh, Gupta, V.P., Gupa M.L. and Satija D.R. 1995 Variability Studies for Oil and Fatty Acid Composition in *Brassica Campestris* var. *Toria.* Journal of Oilseeds Research. 12(1):30-32.

Parker, R.E., Koos and Swern, D. 1955. Biochemical Preparations. 4, Wiley, New York.

Parpia, H.A.B. 1991. Oilseed Proteins for Human Consumption. Proceeings of National Seminar on Strategies for Making India Self Reliant in Vegetable Oils. ISOR, Hyderabad, India. pp.550-557.

Pascuci, O. 1972. Medicinal and Poisoinous Plants of Southern and Eastern Africa. John Mitchell Watt and Maria Gerdina Breyer-Brandwijk, Edinburgh,U.K. pp.434.

Patil, V.C. 1975. Effect of Zinc, Iron and Calcium Under Varying Levels of Phosphorus on Growth, Yield and Quality of Groundnut. Thesis Submitted to University of Agricultural Sciences, Bangalore. India.

Patil, C.B. and Joshi, B.P. 1978. Niger Yields can be Doubled. Indian Farming. 27:9.

Patil, C.B. and Patil, B.P. 1981. Niger Cultivation in Maharashtra. Indian Farming. 30:13-14.

Pattee, H.E. and Purcell, A.E. 1967 Carotenoid Pigments of Peanut Oil. Journal of the American Oil Chemists' Society. 44:328-330.

Patterson, H.B.W. 1989. Handling and Storage of Oilseeds, Oils, Fats and Meal. Elsevior Applied Science, New York. pp. 87-180.

Pennachiotti, J., Romeo, M. and Sehmidt Hebbel, H. 1973. The Tryptophan Content of Some Chilean Foods. *Ann. Biomat.*, 125:267.

Pollard, M.R. and Stumpf, P.K. 1980. Biosynthesis of C_{20} and C_{22} Fatty Acids by Developing Seeds of *Limnanthes Alba*. Plant Physiology. 66: 647-655.

Potter, S.M. 1998. Soy Protein and Cardiovascular Disease: The Impact of Bioactive Components in Soy. Nutritional Reviews. 56:231.

Poneros – Schneider, A.G. and Erdman, J.W. 1989. Bio Availability of Calcium from Sesame Seeds, Almond Powder, Whole Wheat Bread, Spinach and Non-Fat dry Milk in Rats. Journal of Food Science. 54: 150-53.

Prakash, V. and Nandi, P.K.1978. Journal of Agricultural and Food Chemistry. 26:319-28.

Prasad, D. and Rao, A.R. 1983. Mustard Meal for Poultry- A Review. Indian Poultry Reviews. 14:21-28.

Prasad, B., Verma, O.P. and Daftri, L.N. 1985. Effect of Leaf Curl Disease on Seed and Oil Qualities of Sesame. Curent Science. 54(8): 388-389.

Praveena, B., Srinivas, C.V.S. and Nagaraj, G. 1998. Detoxification of Sunflower, Safflower and Castor Cakes. Journal of the Oil Technologists' Association of India. 30 (3): 107-109.

Praveena, B. Srinivas, C.V.S. and Nagaraj, G. 1999. Utilization of Sunflower Cake and it's Protein Isolate in Some Indian Snack Foods. Journal of Oilseeds Research 16(2) : 303-305

Purdy, R.H. 1985. Oxidative Stability of High Oleic Sunflower and Safflower Oils. Journal of the Amrican Oil Chemists' Society. 63(3):523-535.

Purushotham, N.P., Rao, M.S. and Raghavan, G.V. 1986. Utilization of Castor Bean Meal in the Concentrate Mixture of Sheep. Indian Journal of Animal Sciences. 56: 1090-93.

Puttaraj, S. and Narendra Singh, 1992. Dehulling of Castor Seeds. Journal of the Oil Technologists' Association of India. 24: 149-151.

Raheja, R.K., Ahuja, K.L., Batta, S.K., Labana, K.S. and Chowrasia, B.D. 1989. Evaluation of Some Promising Indian Genotypes of Sesame for Oil Content and Component Fatty Acids. Annals of Biology. 5(1): 33-38.

Raheja, R.K., Kanchan Gandhi, Ahuja, K.L. and Labana, K.S. 1986. Correlation Among Kernel Size, Lipid Constituents and Fatty Acid Composition in Groundnut (*Arachis Hypogaea* L.). Annals of Biology. 2:45-50.

Rahma, E.H. and Narsinga Rao, M.S. 1981. Removal of Polyphenols from Sunflower Meal by Various Solvents - Effect on Functional Properties. Journal of Food Science. 46(5):1521-1522.

Rai, M. Vasishtha, A.K. and Naqui, P.A. 1985. Estimates on Variability, Heritability and Genetic Advance in Seed and Oil Components of Linseed. Journal of Oilseeds Research. 2:240-245.

Rai ,M., Pandey, S., Naqvvi, P.A., Kerkhi S.A. and Vashishta A.K. 1989. Variability for Fatty Acid Profiles in Linseed (*Linum Usitatissimum* L.) Journal of Oilseeds Research. 6: 123-127.

Raie, M.Y., Muhammad, D. and Khan, S.A. 1985. Studies of Safflower Seed Oils, *Fette Seifen, Anstrichmittal*. 87(7): 282-283.

Raj, A.G. 1980. Safflower Seed Meal and Protein Supplement in Poultry Rations. Thesis submitted to TNAU, Coimbatore, India.(Indian Journal of Poultry Science, 1981. 16(9):89-90).

Raj, A.G. Kothandaram, P. and Kadirvel, K. 1983. Proximate Composition and Nutritive Value of Safflower Seed Meal. *Cheiron*. 12: 290-293.

Rajani chopra, Krishna Kumari, K. and Nagaraj, G. 2004. Fatty Acid Profile and Shelf Life of Linseed-Groundnut, Linseed-Sunflower and Linseed-Palm Oil Blends. Journal of Oil Technologists' Association of India. 36(1):21-24.

Rajendran, S. and Prakash, V. 1989. Isolation and Characterization of B- Globulin- Low Molecular Weight Protein Fraction from Sesame Seed (*Sesamum Indicum* L). Journal of Agricultural and Food Chemistry. 7: 127-31.

Rakow, G. 1973. *Selecktion Auf Linol. And Linolensau Regehalf in Rapssameu Nach Mutagener Behaudlung Z. Pflauzenzenzuchog*. 69: 62-82.

Ramachandram, M. and Goud, J.V. 1981. Genetic Analysis of Seed Yield, Oil Content and Their Components in Safflower. Theoretical and Applied Genetics. 60:191-195.

Ramachar, Thirumala Rao, S. D. and Reddy, B.R. 1974. Laboratory Prepared High Protein Low Oil Groundnuts. Indian Food Packer. 28(1): 79-82.

Ramadan M.F. and Morsel J.T. 2003. Determination of the lipid classes and fatty acid profile of niger (*Guizotia abyssinica* Cass.) seed oil. Phytochemical Analysis. 14(6):366-370.

Ramadan M.F. and Morsel J.T. 2003a. Phospholipid composition of niger (*Guizotia abyssinica* Cass.) seed oil. *Lebensmittel Wissenschaft und Technologie*. 36(2):273-276.

Ramanjaneyulu G.V., 2003. Diversified Uses of Safflower. Directorate of Oilseeds Research, Rajendranagar, Hyderabad.

Ranganatha, A.R.G., Anjana, R.M. and Nagaraj, G. 2006. Assessment of Variability Among Inbreds, Hybrids and Populations of Sunflower (*Helianthus annuus*). Indian Journal of Agricultural Sciences. 76(7): 441-2.

Rao, M.K.G., Rao S.V. and Achaya, K.T. 1965. Separation and Estimation of Tocopherols in Vegetable oils by TLC. Journal of the Science of Food and Agriculture. 16: 121-124

Rao, P.V. and Rao, P.S. 1981. Chemical Composition and Fatty Acid Profile of High Yielding Varieties of Oilseeds. Indian Journal of Agricultural Sciences. 51: 703-707.

Ravikishan, V., Mev Singh and M.V.R. Prasad, 1995. Composition and Variability of Oil Quality in Castor and Linseed. In. Oilseed Quality and Value Added Products. Eds. Nagaraj, G. *et al.*, D.O.R. Hyderabad. pp. 19-24

Reddy, M.V. and Alfred, J.R.B. 1979. Utilization of Castor (*Ricinus Communis* L.) Leaves by Last Instar Lavae of Silk moth *Pholosamia Ricini* Hutt. Indian Biology. 11: 36-40.

Reddy, M.V. and Reddy, S.M. 1983. Elaboration of Mycotoxins from Fungi Associated with *till (Sesamum Indicum* L.). Curent Science. 52: 613-614.

Renuka Devi, R., Suja, K.P., Jayalekshmy A. and Arumughan C. 2000. Tocopherol and Tocotrienol Profiles of Some Vegetable Oils by HPLC. Journal of the Oil Technologists' Association of India. 32(4):176-182.

Riley, K.W. and Belayneh, H. 1989. Niger. In. Oil Crops of the World. Eds G. Robbbelen, R.K. Downey and A. Ashri. Mc Graw Hill, New York, U.S.A. pp. 394-403.

Rivas, N.R. Dench, J.E. and Caygill, J.C. 1981. Nitrogen Extractability of Sesame (*Sesamum Indicum* L.) Seed and the Preparation of Two Protein Isolates. Journal of the Science of Food and Agriculture. 32: 565-71.

Roy Choudhury, A. and Mandal, L. 1984. Utilisation of Deoiled Niger (*Guizotia Abyssinica*) Cake in the Rations of Growing Pigs. Indian Veterinary Journal. 61: 608-611.

Rudometova, N.V., Pasovskij, A.P., Blohima, E.A., 2001. Method of Isolation and Identification of Carthamin in Safflower- Application Perspectives in Russian Food Products. Proceedings of V International Safflower Conference, Montana, USA, June 23-27. pp 309-313.

Saha, S.N. and Bhargava, 1984. An Evaluation of the Oil Concentration of Sesame Seeds in Relation to Developmental Stage, Node Position and Capsule Age. Experimental Agriculture. 20(2): 129-134.

Sahasrabudde, D.L. 1925. The Chemical Composition of the Food Grains, Vegetables and Fruits of Western India. Bulletin of the Department of Agriculture, Bombay. 124-125.

Sahasrabuddhe, D.L. 1950. The Chemical Composition of Food Grains: Vegetables an Fruits of Western India. Bombay Bulletin. 124: 5-10.

Sahasrabuddhe, D.L. and Kale, N.P. 1933a. A Biochemical Study of the Formation of the Oil in Niger Seed (*Guizotia Abyssinica* Cass). Indian Journal of Agricultural Sciences. 3(1):57-58.

Sahasrabuddhe, D.L. and Kale, N.P. 1933b. A Study of the Oil from Niger Seed (*Guizotia Abyssinica*). Indian Journal of Agricultural Sciences. 3(3):57-58

Sajid Hussian, 1992. Detection of Castor Oil Plant, *Ricinus Communis* L. U.P. Agricultural Life Society of India. 1(3): 248-62.

Salunkhe, D.K., Sathe, S.K. and Reddy N.R. 1982. In Chemistry and Biochemistry of Legumes. Ed. Arora, S.K. Oxford and IBH, New Delhi. pp 57.

Salgado, J.M. and Goncalves, C.M.M. 1988a. Study of Sesame (*Sesamum Indicum* L) Seed. Methods for Preparing Edible White Meal. *Archivos Lathinoamericans de Nutrition.* 38: 306-311.

Salgade, J.M. Goncalves, C.M.M. 1988b. Study of Sesame Seed (*Sesamum Indicum L).*II. Use of the Meal in Protein Mixtures. *Archivos Latinoamericans de Nutrition*. 38: 312-322.

Sangale, P.B., Pagar, S.D. and Daftardar, S.Y. 1982. Note on the Association of Seed Oil Content with Hull and Kernel and Seed Sizes of Safflower. Indian Journal of Agricultural Sciences. 52(9):613-614.

Sarkar, S. and Bhattacharya, D.K. 1987. Seed Composition of Some New Varieties of Sesamum, *Toria,* Yellow *Sarson*, Mustard and Groundnut. Journal of the Oil Technologists' Association of India, 19:13-15.

Sarojini, G. Nirmala, G. and Nagaraj, G. 1995. Utility and Acceptability of Safflower Petals as Food Ingredient. Journal of the Oilseeds Research. 12(27):299-300.

Sastry, M.C.S. 1991. Decorticating of Safflower Seeds. Proceedings of II International Safflower Conference. ISOR, D.O.R., Hyderabad. India. pp.341.

Sastry, M.C.S., Subramanian, V. and Parpia, H.A.B. 1974. Effects of Duhulling and Heat Processing on Nutritional Value of Sesame Proteins. Journal of the American Oil Chemists' Society. 5: 151-18.

Sastry, M.C.S., Subramanian, V. and Rajagopalam. R. 1969. Studies on the Wet Dehulling of Sesame Seeds to Obtain Superior Grade Protein Concentrates, Journal of the American Oil Chemists' Society. 46: 592-96.

Saunders, R. 1970. The Sugars of Safflower. Journal of the American Oil Chemists' Society, 47:254-255.

Sawaya, W.N., Khalil, J.K., Ayaz, A.L. and Mohammed, M.M. 1985. Chemical Composition and Nutritional Quality of *Helia*. Nutrition Reports International. 31(2): 389-397.

Sayed, H.I. and Al Sayeed, F.A. 1978. Effect of K Fertilizer on Yield and Oil Content of Safflower. Proceedings of Saudi Biological Society. 2:1-6.

Seegler, C.J.P. 1983. Oil Plants in Ethiopia, their Taxonomy and Agricultural Significance. Centre for Agricultural Publication and Documentation, PUDOC, Wageningen.

Sekhon, K.S., Ahuja, K.L., Sandhu, R.S. and Bhatia, I.S. 1972. Variability in Fatty Acid Composition in Peanut Bunch Group. Journal of the Science of Food and Agriculture. 23: 919-924.

Sekhon, K.S., Gupta, S.K., Ahuja, K.L. and Jaiswal, S.V. 1980. Variability in Fatty Acid Composition in Semi-Spreading Peanut Types. *Oleagineaux*. 35:409-412.

Sethi, R.L. 1931. Observations on Castor Oil Plant *Ricinus Communis* L. U.P. Agriculture and Live Stock, India. 1(3): 248-262.

Shamantaka Sastry, M.C. 1989. Decortication of Safflower Seeds, II International. Safflower Conference, Hyderabad India. Jan., 9-13, Abstracts. pp. 70-81.

Sharma, V.D. and Verma, 1982. Effect of Nitrogen, Phosphorus and Row Spacing on Yields, Yield Attributes and Oil Content of Safflower Under Rainfed Condition. Indian Journal of Agronomy. 27(1):28-33.

Sharma, S.M., Nagaraj, G. and Balakrishnan, R. 1994. Niger Genetic Resources Evaluation and Analysis. Bulletin, DOR, Hyderabad-30. India. pp 32.

Sheikh, I.A., Aslam, M. and Arshad, M., 1983. Preparation and Nutritional Evaluation of Groundnut Protein Isolate from Industrially Produced Groundnut Cake. Pakistan Journal of Science. 35: 1-2.

Shenolikar, I.S.1985, Nutritional Qualities of Rapeseed-Mustard. In. Oilseed Production-Constraints and Opportunities. Oxford & IBH Publishing Co. pp.635-659.

Shenolikar, I.S. and Tilak, T.B.G. 1980. Effect of Feeding Different Levels of Mustard Oil. I. Monekeys, Nutrition and Metabolism. 24:199-208.

Shenolikar, I.S., Tilak, T.B.G., Krishnamachari, K.A.V.R. and Nagarajan, V. 1976. Toxicological Studies on Mustard Oil. Proceedings of Symposium on Rapeseed and Mustard. CFTRI, Mysor, India. pp. 84-86.

Shogren, M.D., Pomeranz, Y. and Finney, K.F. 1981. Counteracting the Deleterious Effects of Fibres in Breadmaking. Cereal Chemistry. 58:142.

Siddiqui, I.R. and Wood, R.J. 1977. Carbohydrates of Rapeseed – A Review. Journal of the Science of Food and Agriculture. 28:530-538.

Sindhu Kanya, T.C. and Kantharaj Urs, M. 1983. Carbohydrate Composition of Mustard (*B. Juncea*) Seed Meal. Journal of Food Science and Technologists' of India. 20:124-126.

Singh, D.P. 1958. Rapessed and Mustard. Indian Central Oilseeds Committee, pp: 53.

Singh, D.P. and Singh, R. 1974. Effect of Growth Habits on Carbohydrates and Oil Content during Germination and Seed Formation of Peanuts (*Arachis Hypogeae* L.). Indian Journal of Agricultural Research. 8:61-63.

Singh, H. Gupta, M.L. and Rao, N.K.A. 1960. Effect of NPK on Yield and Oil Content of Sesame. Indian Journal of Agronomy. 4: 176-81.

Singh, P.N. and Prasad, R. 1978. Influence of Water Stress on Lipid Quality in Peanut. National Academy of Sciences Letters. 1:388-390.

Sing, S.P., Singh, V. and Singh, P.P. 1988. Characterisation of Oil and Protein Dynamics in Developing Sunflower (*Helianthus Annus* L.) Seeds, Journal of Oilseeds Research. 5:77-79.

Singh, N. Subbaiah, B.V. and Gupta, V.P. 1970. Effect of Sulphur Fertilization on the Chemical Composition of Groundnut and Mustard. Indian Journal of Agronomy. 15:24-28.

Sinha, T.N., Srivastava, J.P., Verma, A.K. and Gupta, B.S. 1983. Utilization of Niger Cake (*Guizotia Abyssinica*) as a Nitrogen Supplemet in Growing Calf Rations Indian Journal of Animal Sciences. 53:887-889.

Smith, K.J. 1971. Nutritional Frame Work of Oilseed Proteins. Journal of the Amrican Oil Chemists' Society, 48: 625-28.

Smith J.R. 1996. Safflower, American Oil Chemists' Society Press, Champaign, II, USA. pp. 624

Sonntag, N.O.V. 1979. Composition and Characteristics of Individual Fats and Oils. In. Baileys Industrial Oil and Fat Products Ed. D. Swern. John Wiley & Sons, New York. pp: 289-478.

Sosulki, F. 1979. Potential of Brassica Species as Protein Sources. Proceedings of the 5th International Rapeseed Conference. 1978, pp 365-368.

Sosulski, F.W. and Sarwar, G. 1973. Amino Acid Composition of Oilseed Meal and Protein Isolates. Canadian Institute of Food Technologists' Journal. 6:1-5.

Sosulki, F. 1981. Franctionation of Rapeseed Meal into Flour and Hull Components. Journal of the American Oil Chemists' Scociety. 56: 96-98.

Sovani, D.R., and Vasistha, 1995. Hydrogenated Castor Oil- A Versatile Industrial Chemical. International Seminar on Challenges Facing Fats, Oleochemicals and Surfactants. Oil Technologists' Association' of India, Bombay. Dec. 19-12.

Speek, A.J., Shrijver, J. and Schreurs, W.H.P. 1985. Vitamin Compostion of Some Seed Oils as Determined by HPLC with Fluorometric Detection.Journal of Food Science. 50: 121-24.

Spies, J.R. and Coulson, E. 1943. The Allergens of Castor Beans. Journal of the American Oil Chemists' Scociety. 65: 1720.

Srinivas, C.V.S. and Nagaraj, G. 1999. Factors Influencing Ricin, the Toxic Protein in Castor Cake and its Detoxification. Journal of the Oil Technologists' Association of India. 32(1): 21-23.

Srinivas, C.V.S. and Nagaraj, G. 2000. Extended Abstracts - Oilseeds and Oils - Research and Development Needs in the Millennium. Feb. 2-4. ISOR, DOR, Hyderabad. pp, 326-32.

Srinivas, C.V.S. and Nagaraj, G. 2007. Identification and Characterization of Phenolic Compounds in Castor Seed. Natural Products Research. 21(12):1073-1077.

Srinivas, C.V.S. and Nagaraj, G. 2008. Biochemical Properties of Ricin in Immature Castor Seed. Natural Products Research. 22(7):600-6006.

Srinivas, C.V.S. and Nagaraj, G. 2008a. Amino Acids of Ricin and It's Polypeptides. Natural Products Research. 22(3):258-263

Srinivas, C.V.S., Praveena, B. and Nagaraj, G. 1999. Safflower Petals - A Source of Gamma Linolenic Acid. Journal of Plant Foods and Human Nutrition. 54 (1):89-92.

Srinivas Das, Tyagi A.K. and Singhal K.K. 2001. Chemical Composition Including Amino Acid, Fatty Acid and Glucosinolate Profile of Taramira (*Eruca Sativa*) Oilseed. Indian Journal of Agricultural Sciences. 71(9): 613-615.

Stumpf, P.K. and Pollard, M.R. 1983. Pathways of Fatty Acid Biosynthesis in Higher Plants with Particular Reference to Developing Rapeseed. High and Low Erucic Acid Rapeseed Oils. Ed Krammer John, K.G. and Saver Frank, D. Pigden, Wallace Jones Academic Press Canada.

Subbaiah, 1973. Dietary Plant Sterols. Current Status in Human and Animal Sterols Metabolism. American Journal of Clinical Nutrition. 26: 219

Subbaran, M.R., Rajagopal, N.S., Paulose, M.M., Subba rao, R., Roomi, M.W. and Achaya, K.T. 1964. Journal of the Science of Food and Agriculture. 15: 645.

Subramanian, A.N. 2003. Export of Oilseed - Potential, Constraints and Remedial Measures. In. Souvenir, National Seminar on "Stress Management in Oilseeds for Attaining Self Reliance in Vegetable Oils" Eds Harvir Singh and D.M. Hegde, Indian Society of Oilseeds Resarch, Hyderabad. India. pp. 128-138.

Subramanian, N., Ramamani, S., Sastry, M.C.S. and Sreedhara, N. 1981. Quality Characteristics of Groundnut Protein Isolate as Influenced by Processing Conditions. 2nd Indian Convention of Food Scientists and Technologists, CFTRI, Mysore, India. Feb., 19-20.

Sudhakar Babu, N. 2003. Diversified Uses of Sunflower. Directorate of Oilseeds Research, Hyderabad. pp 1-24

Sukumara, S., Terayama, N., Kawasakatsu. S., Ichiara, A. and Saita, H. 1980. Conjugated Serotinine and Phenolic Constituents in Safflower Seeds. Journal of Agricultural Biologists and Chemists. 44:2951-2954.

Suman Kumar, Dhruv-Srivastava, and Khan, M.A. 2002. Physico-Chemical Characteristics of some Cultivars of Safflower (*Carthamus Tintorius* L) Seeds. Indian Journal of Agricultural Biochemistry. 15: 29-32.

Surendranath, M.R., Azeemoddin, G., Achutaramayya, D. and Thirumala Rao, S.D. 1985. Processing of Sesame Seed and Storage of Oil and Cake. Journal of Oil Technologists' Association of India. 17:4-6.

Surendranath, M.R., Gangi Reddy, N.C., Azeemoddin, D. and Thirumala Rao, S.D. 1982. Processing of Sesame Seed and Storage of Oil and Cake. Oils and Oilseeds Journal. 35:43-48.

Surendranath, M.R., Gangi Reddy, N.C., Azeemoddin, D. and Thirumala Rao, S.D. 1984. Preparation of Low Fat High Protein Sesame Seed. Journal of Food Science and Technologists' of India, 21:425-426.

Stefansson, B.R. Hongen, F.W. and Downey, R.K. 1961. Note on the Isolation of Rape Plants with Seed Oil Free from Erucic Acid. Canadian Journal of Plant Science. 41:218-219.

Tairo, H. 1985. Oil Content and Fatty Acid Composition of Peanuts Imported in Japan. Journal of the American Oil Chemists' Society. 62(4): 699-702.

Tao-HaiRong: Gao-XiuKu; Xiaowu-KaiTi; Li-WenSheng; Li-He. 2000. Analysis of the Compents of Unsaponifiables in Safflower Oil. Chinese Journal of Oil Crop Sciences. 22(4):40-42.

Tashiro, T., Fukuda, Y., Osawa, T. and Namiki, M. 1990. Oil and Minor Components of Sesame (*Sesamum Indicum L.)* Strains. Journal of the America Oil Chemists' Society. 67(8): 508-511.

Taha, F.S., Fahmy, M. and Sadek, M.A. 1987. Low Phytate Protein Concentrates and Isolates from Sesame Seed. Journal of Agricultural and Food Chemistry. 35: 1289-92.

Taha, F.S., Helmy, E.D., Fahmy, M. and Ali, M. 1989. Nutritional Evaluation of Sesame Seed Protein Products. *Grasses Y. Aceites.* 40(2): 109-113.

Taneja, V.K. Sharma, N.K. and Warle, R.D., 1986. Partial Hydrogenation of Indian Soybean Oil. Journal of the Oil Technologists' Association of India. 18(3):79-81.

Taranathan, R.N., Wankhede, D.E.B. and Rao, M.R.R. 1975. Carbohydrate Composition Of Groundnut (*A. Hypogaea* L.). Journal of the Science of Food and Agriculture 26: 749-754.

Tashiro, T., Fukuda, Y., Osawa, T. and Namiki, M. 1990. Oil and Minor Components of Sesame (*Sesamum Indicum* L) Strains. Journal of the Amrican Oil Chemists' Society. 67: 508-511.

Thayumanavan, B. 1981. Keeping Quality of Oils from Dehulled and Whole Seeds. Oilseeds Journal. 12:54-56.

Tkachuk, R. and Irvine, G.N. 1969. Amino Acid Composition of Cereals and Oilseed Meals. Cereal Chemistry. 46:216-218.

Toliwal, S.D., Lokhande A.R., Raval D.A and Kulkarni A.S, 1993. Chemical Composition of Oils from Some Varieties of Groundnut (*Arachis hypogaea* L.). Journal of the Oil Technologists' Association of India. 25: pp.31-33.

Toma, R.B., Tabekhia, M.M. and Williams, J.D. 1979. Phytatae and Oxalate Contents of Sesame Seed(*Sesamum Indicum* L).Nutritional Reports International.20:25-31.

Tyagi, P. and Vasistha, A.K. 1983. Component Fatty Acids and Glycerides of Oils from Different Genetic Varieties of *Sesamum Indicum*. Journal of the Oil Technologists' Association of India. 15: 28-32.

Tyagi, A. and Bhowmick D.N. 2001. Separation, Characterization and Estimation of Chlorogenic Acid from Aqueous Extracts of Sunflower Meal. Journal of the Oil Technologists' Association of India. 33(3): 108-113

Uppstrom, B. 1995. Seed Chemistry. In *Brassica* Oilseeds-Production and Utilization Ed. D.S.Kimber and D.I.McGregor, CAB International. Wallingford, U.K. pp. 217-242.

Urie, A.L. 1986. Inheritance of Partial Hull in Safflower. Crop Science, 26.

Urie. A.L. Perterson, W.F. and Knowles, P.F. 1979. Registration of Oleic Leed Safflower Reg. No. 8. Crop Science. 19(5):747.

Uzo, J.O., Adedzua, D.K. and Onwakwe, R.O. 1985. Yield, Yield Components and Nutritional Attributes of Cultivated Sesame (*S. Indicum*) and its Endemic Wild Relatives in Nigeria. FAO Plant Production and Protein. 66:166-176.

Vaughan, J.G. 1970. The Structure and Utilization of Oilseeds. Chapman and Hall, London, UK. pp. 41-43.

Verma, N.S., Dumka, P.K. and Tyagi, S.N. 1977. Biochemical Studies. In. Research on Sunflower, Experimental Station Technical Bulletin, G.B.P. University Agriculture and Technology, Pantnagar, India. pp. 44-46.

Venkat Rao, S., Nagaraj, G., Misra, R.C. and Halder, J. 1988. Yield ad Quality Characteristics of Nine Niger Genotypes Developed in Orissa. Journal of Oil Technologists' Association of India. 28(2):57-59.

Vermorel, M., Heaney, R.K. and Fenwick, G.R. 1988. Antinutritional Effects in the Meals from Cultivars Darmor and Jet Neuf, and Progoitrin Together with Myrosinose in Growing Rat. Journal of the Science of Food and Agriculture. 44:321-334.

Vibhakar, S., Rajalakshmi, S., Nagaraja, K.V. and Kapur, O.P. 1981. Physico Chemical Characteristics of Safflower Oil. Industrial Food Packer. 35(1): 11-12.

Vidyarthi, N.L. and Mallya, M.V. 1940. Component Glycerides of Niger Seed Oil. Journal of Indian Chemical Society, 17:22.

Vidyasagar, K., Arya, S.S., Premavalli, K.S., Parihar, D.B. and Nath, M. 1974. Chemical and Nutritive Changes in Refined Groundnut Oil During Deep Fat Frying. Journal of Food Science and Technologists, India. 11: 73-75.

Villegas, A.M. Gonzalez, A. and Calderon, R. 1968. Microbiological and Emzymatic Evaluation of Sesame Protein. Cereal Chemistry. 45: 379-83.

Vles, R.O. and Gottenbos, J.J. 1989. Nutritional Characteristics and Food Use of Vegetable Oils. In. Oil Crops of the World, Eds. Robbelen, G,. Downey, R.K and Ashri, A. Mc Graw Hill Publishing Co. pp. 63-66.

Vogel, H.C.A. 1989. The Nature Doctor, p. 247-248, 258-259,534.

Vohra, P 1989. Carbohydrate and Fibre Content of Oilseeds and their Nutritional Importance. In. Oil Crops of the World Ed. G. Robbelen, R.K. Downey and A. Ashri, Mc Graw Hill Publ. Co. pp. 289-225.

Walker, M.E. and Carter, M.L. 1971. The Effect of Fertilization and Storage Tempeature on Percent N, Oil and Germination of Spanish and Runner Peanut Seed. Research Bulletin. College of Agricultural Experiment Station, Georgia University. U.S.A.88: pp.17.

Wankhede, D.B. and Tharanathan, R.N. 1976. Sesame Carbohydrates. Journal of Agricultural and Food Chemistry. 24: 655-58.

Weiss, E.A. 1971. Castor, Sesame and Safflower. Barnes and Noble, Inc., New York. pp.529-744.

Weiss, E.A. 1983. Oilseed Crops. Longman, London.

Were, B.A., Lee, M. and Stymne, S. 2001. Variation in Seed Oil Content and Fatty Acid Compostion of *Sesamum Indicum* L. and its Wild Relatives in Kenya. *Sveriges Utsadesforenings Tidskrift.* 111(4):178-183.

Westphal, A. and Kelber, E. 1973. Performance, Oil Content and Quality of Niger and Sunflower. Proceedings of Annual Agricultural Seminar, Addis Ababa. Pp.40-76

Wichman, D. 1996. Safflower for Forage. Proceedings of North American Safflower Conference, Montana, Canada. Ed. H.H. Mundel, J. Braun and C. Daniels, A.B.Lethbridge, Jan. 17-18, pp. 11

Wilson, L.A. 1999. Current Developments in Soyfood Processing in North America. Proceedings of the World Soybean Research Conference VI, Ed. H.E. Kauffman Chicago, USA, August 4-7, pp. 394-402.

Williams, K.A. 1966. Oils, Fats and Fatty Foods, Their Practical Examination. J.A. Churchill Ltd. London. pp. 3-7.

Woodroof. J.G.1983. Peanuts Production, Processing, Products, Avi Publishing Company. Inc., Westport, Connecticut.

Worthington, R.E. 1977. The Linolenic Acid Content of Peanut Oil. Journal of the American Oil Chemists' Society, 53(4):167-169.

Worthington, R.E. and Holley, K.T. 1976. The Linolenic Acid Content of Peanut Oil, Journal of the American Oil Chemists' Society. 53: 515-516.

Worthington R.E., Hammons, R.O. and Allison, J.R. 1972 Varieal Difference and Seasonal Effects on Fatty Acid Composition and Stability of Oil from 82 Peanut Genotypes. Journal of Agricultural and Food Chemistry. 20 (3):727-730.

Yarosh, N.P. and Ivanenko, E.N. 1984. The Fatty Acids and Oil Percentages in Seeds of Different Ecological Types of Sesame. *Sbornik Nauchmykh Trudor Po. Prikladnor Botanike Geneitike I Selektsi*. 88: 48-54.

Yen, G.C. 1990. Influences of Seed Roasting Process on the Changes in Composition and Quality of Sesame (*Sesame Indicum* L) Oil. Journal of the Science of Food and Agriculture. 50: 563-70.

Yen, G.C., Shyu, S.L. and Lim, J.S. 1986. Studies on Protein and Oil Compostion of Sesame Seseds. Journal of Agricultural and Food Chemistry, 35: 177-81.

Yermanos, D.M., Hemstreet, S., Saleeb, W. and Huszar, C.K. 1972. Oil Content and Composition of the Seed in the World Collection of Sesame Introductions. Journal of the Amrican Oil Chemists' Society. 49: 20-23.

Yoshida, M. and Kashimoto, T. 1979. Determination of Sesamolin and Sesamin in Sesame Oil. Proceedings of Osaka Perfectural Institute of Public Health. Food Sanitation. 10: 41-46.

Young, C.T. and Hammons, R.O. 1978. The Amino Acid Content of U.S. Commercial Peanut Varieties, Proceedings of the American Peanut Research and Education Association. 10:75.

Young, C.T. and Horis, A.R. 1990. A Method for Rapid Analyses of Head Space Volatiles-Green and Roasted Peanuts. Journal of Food Science, 55:279-280.

Young, C.T., and Mason, M.E. 1972. Free Arginine Content of Peanuts (*Arachis Hypogaea* L.) as a Measure of Seed Maturity. Journal of Food Science. 37:722-725.

Young, C.T., Mason, M.E., Matlock, R.E. and Waller, G.R. 1972. Effect of Maturity on the Fatty Acid Composition of Eight Varieties of Peanuts Grown at Perkins, Oklahoma in 1968. Journal of the American Oil Chemists' Society. 49:314-317.

Young, C.T., Matlock, R.S. Mason, M.E. and Waller, G.R. 1974. Effect of Harvest Date and Maturity upon Free Amino Acid Levels in Three Varieties of Peanuts. Journal of the American Oil Chemists' Society. 51:269-273.

Yousuf Ali Khan, R., Lakshminarayana, J., Azeemoddin, G., Achutaramayya, D. and Thirumala Rao, S.D. 1980. Shelf Life of Groundnut and Sunflower Oils. Abstracts of the Indian Congress on Oilseeds and Oil. pp. 33-34.

Zang, L.P. 1997. Safflower: A Versatile Plant. Proceedings of IV International Safflower Conference, Bari, Italy, June 2-7. pp 329-331.

Zarkadas, C.G., Gagnon, C., Gleddie, S., Khanizadeh, S., Cober, E.R. and Guillemette R.J.D., 2007. Assessment of the protein quality of fourteen soybean [*Glycine max (L)* Merr.] Cultivars Using Amino Acid Analysis and Two-Dimensional Electrophoresis. Food Research International. 40(1): 129-146.

Chapter 3

Tree Borne and Minor Oil Sources

In addition to cultivated oilseeds, many other seeds and related plant products serve as sources of vegetable oils. They are obtained from large trees, shrubs or from different crops and their products. It may be the seeds or the husk, or the germ or other parts of the fruit or seed which are rich in oil. These are in general, referred to as non-conventional vegetable oil sources. They are highly important for the food/feed and non-food industries. Non- food and industrial uses are plenty.

But for the availability of these non- traditional oilseeds and their products, there would have been great pressure on the cultivated seeds. The products, mainly oil, help run our machines and protect them from getting rusted. Some of them protect our crops from pests and deseases. They even serve as beauty aids in the form of cosmetics. Some of them serve as sources of medicine for men and animals.

In recent years, some non-conventional/ non-edible oilseeds have become the raw material for biofuel/bioenergy. They have gained prominence, due to shortage of conventional oils. Seeds of more than 100 species occurring in the wild or cultivated are known to contain oil in considerable quantities. There is a vast scope for production and utilization of oil from these sources (Ambasta, 1986, Bhattacharya, 2002 and Vimal and Naphade, 1980).

In this chapter, the major non-conventional oilseeds of tree origin, by product oilseeds or other minor sources, are being dealt with. They are ricebran, corn, *sal, mahua, jatropha, kokum, melia,* rubber, mango kernel, *kusum, simarouba,* neem, *karanj, crambe, , moringa, tumba,* jojoba and tobacco seed etc. Emphasis is given to the composition and quality of oil, cake and other plant parts. Their food, non-food and medicinal uses are presented. Oil extraction and related post harvest technological aspects are also discussed.

Palm Oil

Oil palm (*Ealeis guineensis* jacq.) is a tree which can yield 4-6 tonnes of oil/ha. Palm oil is widely consumed all over the world and is next only to soyabean oil. Oil palm has versatile uses. Oil palm has originated in humid tropical regions of Africa. It can thrive well in regions upto 15^0 on either side of the equator. It belongs to *Spadiciflorae* order of the *Palmae* family. It is a member of the *Cocoineae* tribe. It bears fruit within three years of it's planting. The oil palm fruit, which is the item of economic value can produce two different kinds of oil. Palm oil is extracted from the flesh (mesocarp), while the seed or kernel yields palm kernel oil. Each fruit yields about nine parts of palm oil to one part of palm kernel oil.

The total area under oil palm plantation in the world is 9.16 million hectares. Production of the palm oil stands at 33059 million MT (over 20% of the global vegetable oil), with Malaysia and Indonesia, accounting for lion's share. While the production has been growing at the rate of 7%, its consumption has been growing slightly faster at 7.15%. Acreage under the plantation has been growing at the rate of 6.47%, which is much higher than that of soybean (3.36%). The productivity of palm oil in the global scenario is 3.67 MT per ha.

Table 74: Country-wise details of area, production and productivity of oil palm.

Country	Area harvested (m.ha)	Oil production (m Mt)	Productivity (Mt/ha)
Indonesia	3.69	13.92	3.77
Malaysia	3.55	14.96	4.21
Colombia	0.17	0.66	3.90
Thailand	0.28	0.69	2.45
Other countries	1.43	3.36	2.35
World	9.16	33.59	3.67

The palm trees are grown in plantations. The palm fruits grow in bunches on palm trees. Each bunch wieghing 4-20 kg contains 200-2000 spherical to ovoid fruits. High mesocarp yielding trees are selected for obtaining higher oil. Harvesting the bunches is easy from young trees. Sickles mounted on metal rods are used to harvest bunches from tall trees.

Palm Oil

Palm oil is a fruit coat oil similar to olive oil. It is obtained from the large bunches of small fruitlets. Palm oil is present in the fibrous mesocarp of the fruits at a concentation ranging from 20-24% (Gascon *et al.*,1989). The mesocarp or the pulp of the palm fruit contains around 45-56% oil (Purseglove, 1975 and Godin and Spensley, 1971). The oil content on dry weight basis in the mesocarp may go upto 75%. The fruits contain very little oil after anthesis which increases gradually and suddenly during fruit development. Initially higher levels of phospholipids and polar lipids are present which decrease with an increase in nonpolar lipids (glycerides) till final maturity. The fruit bunch also yields 2-3% oil. An average yield of 4.0- 5.0 tonnes per ha oil is obtained from palm plantations. The palm oil yield can be as high as 8 tonnes/ ha, the highest for any oil crop.

Freshly extracted oil has some disadvantages. Firstly, the free fatty acid content is higher in the oil. This is because the flesh contains a very active lipase which hydrolyses the glycerides. Sterilisation of the fruit bunches is the first processing step, which stops or prevents this process, inspite of a little hydrolysis that might have already occurred. During refining the free fatty acids are removed, and also a part of the monoglycerides, but the diglycerides remain in the oil.

Secondly, crude palm oil has a very high carotene content, giving it its red colour. This color has to be removed in refining, requiring somewhat special conditions. Thirdly, there is a relatively high content of natural antioxidants (tocopherols), and along with a high proportion of tocotrienols. These constituents impart a high degree of oxidative stability to palm oil. They persist after refining, although at a somewhat reduced level.

Oil Palm Processing

Fresh fruit bunches (FFB) harvested from oil palm plantations are highly perishable and require processing within 24 hours. Delay in processing adversely affects the quantity and quality of palm oil. The extraction of crude palm oil (CPO) from the fruit mesocarp and extraction of palm kernel oil (PKO) from the kernel have to be carried out one after the other.

The FFB are to be processed within 24 hours of their harvest. The received bunches should not be stored for long. But in peak seasons, the FFB may need to be stored in open platforms for several hours until the mill is ready to process them, by providing a buffer between transportation and processing. In the oil mill the fruit is sterilized, stripped, and pressed to obtain the crude oil while refining results in a bland, golden yellow oil suitable for edible applications. The different processes involved in palm oil extraction are detailed below.

Sterilization: Sterilization prevents the build up of FFA and loosens the fruit to aid easy stripping. The FFB are exposed to a steam environment at a pressure of 3kg/cm^2, which corresponds to a temperature of 120^{0}C for 1 hour. Sterilized bunches are fed continuously into a rotary drum stripping machine which strips and separates fruit from bunches by a rolling cum dropping action inside a rotating perforated drum. Loose fruit passes through gaps between channel bars running longitudinally along the drum, while empty bunches are eventually discharged at the end of the drum.

Digestion: Digestion loosens the mesocarp from the central nut and facilitates rupture of fat globules to release oil and it gives a pulpy mass. Proper digestion is essential for efficient extraction of oil. The digested mixture of fiber, nut and oil is forced through a perforated horizontal cylinder by the rotation of one screw or counter rotating screws. Sufficient pressure develops inside the cylinder. The pressure should be such that it should not break the kernels. Kernel breakage causes lauric acid intrusion into the resulting oil.

Oil recovery and drying: Crude oil from the press consists of a mixture of oil, water and non-oily solids. Through clarification, non-oily solids and water are removed through a settling process which takes several hours. The oil is further clarified through centrifugation. Finally, the oil is dried, using vapour extraction units or vacuum driers to prevent FFA formation. Crude palm oil is generally stored in welded steel tanks.

Kernel oil extraction: The kernels are separated from the cake left over after extraction of palm oil in the press. The press cake is conveyed through a paddle conveyor wherein, it is loosened and moisture is removed partially. Cyclone separator in the next stage sucks the low-density fiber and leaves the nuts down. Nuts are then cracked in nut cracking units and shell is separated in cyclone separators. The kernel is dried to safe moisture levels usually by employing bin drier and is processed in expellers to extract kernel oil. The expelled oil is filtered in cloth filters and is stored in steel welded tanks. The kernel oil is lauric rich and is broadly similar to coconut oil.

Properties of Palm Oil and Palm Kernel Oils

The product of a palm oil mill is called Crude Palm Oil (CPO), which is blood red in colour and consists of liquid phase olein and solid phase stearin. Through a simple fractionation process, liquid olein and solid stearin can be separated, making the the palm oil more versatile from the point of view of utility. The extraction of palm oil by mechanical and physical means, without the need for any solvents, makes it an absolutely valuable natural vegetable oil.

Table 75: Characteristics of crude palm and palm kernel oils.

Characteristic	Palm oil	Palm kernel oil
Density	0.891	0.908
Refractive index, n_D 40^{0}C	1.453-1.459	1.4495-1.4515
Melting range, ^{0}C	27-50	25-30
Titer, ^{0}C	40-46	20-28
Acid value, mg KOH/g oil	2-15	3-17
Saponification value, mg KOH/g oil	196-209	244-264
Iodine value	46-60	14-20
Thiocyanogen value	44-48	13-18
Hydroxyl value	7-24	-
Unsaponifiable matter,%	0.3-1.0	0.2-0.8

Source: Cornelius, 1977.

Palm oil has very good smell and taste. It's colour ranges from yellow to deep orange. It is highly stable at ambient temperatues as well as during frying. Palm oil melts over a wide range of temperatures namely, 25-50° C, while it is 20-30 ^{o} C for the kernel oil. Due to presence of higher levels of saturated fatty acids especially, palmitic acid, the iodine value of palm oil is low. The palm kernel oil has still lower I.V. indicating higher saturated fatty acids in it. The S.V. of palm oil is around 200 while that of the kernel oils is higher indicating the presence of low molecular weight fatty acids. On an average, the total saturated fatty acid content of the crude palm oil is around 45-50% with equal quantities of unsaturated fatty acids. The unsaponifiables are low at 0.3-1% in the oil. The free fatty acid content of the oil is also normally low. The oil from over ripe fruits contains higher FFA. Both the oils have non-drying characteristics.

Fatty Acid Composition

The fatty acid composition of palm oil is similar to that of the adipose tissue in most people on an ordinary diet. Palm oil contains equal proportions of saturated fatty acids and unsaturated fatty acids. The saturated fatty acids are made up of 44% palmitic acid and 5% stearic acids. The unsaturated fatty acids consist of 39% oleic acid and 10% linoleic acid. Palm oil triglycerides carry linoleic acid predominately in the 2-position, which favours absorption and availability for use in the body.

Table 76: Fatty acid composition of palm oil and palm kernel oil.

Fatty acid	Typical Range	Immature fruit	Elaeis oleifera	Palm kernel oil
Carprylic acid	–	–	–	2.4-6.2.
Capric acid	–	–	–	3.0-6.3
Lauric acid	–	–	0.05	44.5-52.0
Myristic acid	0.5-5.9	trace	0.3	14.1-18.6
Palmitic acid	32-47	57.5	25.0	6.5-10.4
Stearic acid	2-8	7.0	1.2	1.3-3.5
Arachidic acid	–	trace	0.1	0- 1.9
Palmitoleic acid	–	–	1.4	0-0.6
Oleic acid	40-52	0.9	68.6	10.5-18.5
Linoleic acid	5-11	29	2.1	0.7-2.5
Linolenic acid	–	2	0.9	–

Source: Cornelius, 1977.

Palm oil can be fractionated (through winterisation) into liquid olein fraction (80-90%) and a solid stearin fracrion(10%). The liquid fraction has a cloud point of 5-8°C. The solid stearin faction has a melting point of 35-40°C Palm olein contains higher proportions of oleic acid ranging from 60-65% and linoleic acid upto 25% with about 10% of palmitic acid. The solid palm stearin contains 65% palmitic, 10% stearic, 20% oleic and 5% linoleic acids.

Palm kernels contain around 52% oil, along with 9% protein, 5% fibre, 2% ash and 23% carbohydrates. Mostly expellers are used to extract oil from the dried kernels. Palm kernel oil contains low molecular fatty acids like capric and caprylic acids in the range of 2-6% each. The major fatty acid is lauric acid at 40-55% level. The other important fatty acids are myristic(14-20%), oleic (10-20%) and palmitic (6-11%) acids. Linoleic and stearic acids are present in the range of 2-4% each in the kernel oil. In composition, this oil is almost similar to coconut oil.

Table 77: Palm Kernels-proximate composition.

Constituent	%
Dry matter	92.0
Crude protein (N x 6.25)	8.5
Oil (either extract)	49.0
Crude fibre	5.8
Carbohydrate (nitrogen-free extractives)	26.9
Ash	1.8

Source: Cornelius, 1977.

Minor Components of Crude Palm Oil (CPO)

Palm oil has some very important minor constituents. Mainly they work as antioxidants and help in free radical quenching and as anti-aging and anti-caner agents. Vitamin A and E have their normal physiological roles as vitamins. Unrefined palm oil and or red palm oil are the richest sources of beta-carotene, which is widely regarded as an anti-cancer agent of great promise. Beta-carotene, a major carotene in crude palm oil is a potent antioxidant and precursor of vitamin A (Krinsky, 1993).

Table 78: Unsaponifiable matter of palm oil.

Constituents	Content
Carotenoids	500-700 mg/kg
α-Carotene	36.2%
β-Carotene	54.4%
γ-Carotene	3.3%
Lycopene	3.8%
xanthophylls	2.2%
Tocopherols	500-800 mg/kg
α-tocopherol	35%
γ-tocopherol	35%
δ-tocopherol	10%
Sterols	about-300 mg/kg
Cholesterol	4%
Campesterol	21%
Stigmasterol	21%
β-Sitosterol	63%
Phosphatides	500-1000 mg/kg
Total alcohols	about-800 mg/kg
Triterpenic alcohols	80%
Aliphatic alcohols	20%

Source: Loncin *et al.,* 1970.

Palm oil and palm oil products are naturally occurring sources of the antioxidant vitamin E constituents, tocopherols and tocotrienols. These natural antioxidants act as scavengers of damaging oxygen free radicals and play a protective role in cellular aging, atherosclerosis and cancer. (Walton, and Packer, 1980; Hirai, *et al.*, 1982; Cross, 1987; Elson, and Qureshi, 1995). Due to higher SFA and MUFA, palm oil is normally a stable oil with good frying qualities. Presence of tocopherols especially the γ-tocopherols, further strengthens the stability characteristics including its frying quality.

Studies on health aspects of palm oil confirm that its impact on blood cholesterol and lipoprotein are bebeficial. Studies have shown that palm oil enriched diet does not raise blood cholesterol, but in fact reduces the levels of blood cholesterol and LDL-cholesterol compared with a diet containing coconut oil, butter fat, vegetable ghee or hydrogenated cottonseed oil. This must be mainly due to the presence of carotenoids, tocopherols and the phytosterols.

Food Uses of Palm Oil

Palm oil has a wide range of applications. About 80% of the world production is used for food applications, the rest is feedstock for a number of non-food applications. Refined, bleached and de-odorized (RBD) palm oil, is obtained by removing the free fatty acids, colour, and unwanted flavour through refining. RBD palm oil is a neutral and near white oil which can be separated into (i) palm olein, a liquid fraction and (ii) palm stearin, a solid fraction. The average stearin: olein ratio in CPO is about 25: 75. Palm olein is commonly used for home cooking, in restaurants and fast food restaurants, and also in snack foods and instant noodles manufacture. Palm oil can be reused several times, because it has a good shelf life and frying properties.

Red palmolein: During the process of refining and bleaching, palm oil loses one of its valuable components, carotenoids. To retain the carotenoids, CPO is refined under special conditions involving mild treatments. "Red Palm Oil" contains more than 80% of the carotene originally present in crude palm oil and also the tocotrienol (vitamin E).

Shortening: Presently palmolein, palm stearin and sometimes the palm kernel oil are the sources of bakery shortenings. They are used without any hydrogenation and different grades of shortening can be made with different proportions of blending. Palm oil does not produce trans-fatty acids on hydrogenation. It can be used solely or for blending in different proportions with other vegetable oils. Refined palmolein can be used as such, or with mild hydrogenation. Fatty acid composition of palm kernel oil is similar to butter, which gives the butter like feeling while eating.

Confectionary fats: Unique taste of chocolate is due to the coco-butter, which is one of the main ingredients of chocolate making. Palm oil mid fraction and different blends of it can be used as coco-butter equivalents.

Palm vitee: In recent years, the palm vitee, a new health product from palm oil has been introduced. Palm Vitee is rich in Vitamin E, particularly the tocotrienols which act as more effective antioxidants in the biological system and may play a protective role in cellular ageing, heart disease and cancer. The palm Vitee promises great potential for the marketing of value-added products derived from palm oil.

Oleo Chemical Products for Non Food Uses

The techno-economic superiority of palm oil makes it a much favoured ingredient in the manufacture of various non-edible products like soaps, detergents, cleansing agents such as surfactant. Palm oil also is an excellent raw material for the production of oleochemical products like fatty acids, fatty alcohols, glycerols and other derivatives. These are widely used in the manufacture of cosmetics, pharmaceuticals, household and industrial products. Oleochemicals manufactured from palm oil and palm kernel oil are now popular ingredients in the manufacture of environment friendly detergents as they are biodegradable.

Palm oil and kernel oil play an important role as feedstocks for the production of oleochemicals. These oleo chemicals can be used for the production of candles, cosmetic products, soaps, pharmaceuticals, textiles and in rubber/plastic processing.

Palm oil based bio-lubricant can be synthesized by esterification of palm oil. The bio-lubricant might be used in agricultural and food industrial machineries. Palm bio-emollients are used in cosmetics as they function as tenderizer and moisturizers of skin. The advantge of palm bio-emollient over petrochemical based emollient is that, the bio-emollient is safe and non-toxic to human body, degradable, and has no side effects to skin. Technology is available for producing palm oil based plasticizer. Addition up to 30% palm oil based plasticizer may improve plasticity and elasticity of PVC. Similarly bio-candles can be produced using palmoil fatty acids. The quality of the palm-based soaps is comparable to soaps derived from tallow.

Oil Palm Plantation and Industry Waste Utilization

Fibres from fronds and empty fruit bunches are reconstituted to make materials such as medium density fibre boards and chip boards. Fibres from trunk are transformed into beautiful furniture. The empty bunch can be utilized as medium for mushroom growth. The mesocarp fibre that comes out of the mill has 10-12 per cent oil that can be extracted through solvent extraction technology.

Palm fibers: Empty Fruit Bunch (EFB)is an abundant source of natural fibre with 25% lignin, 60% cellulose and 2% hemi cellusose content. EFB fibres are clean, bio-degradable and compatible than many other fibres from other wood species. EFB fibre is suitable for the manufacture of mattresses, car seats, insulation, composite panels, medium-density fibreboards, thermal insulating materials, packaging materials, rubberized mattresses, coir geo-textiles etc.

Compost: The waste derived from oil palm industry is basically cellulosic and organic in nature with high nutrient content. This can be converted into high quality organic fertilizer, rich in potassium. The harvesting operation produces enormous quantity of fronds, which can be shredded and used for making compost in the plantation itself. The compost can be produced in a palm oil mill from EFB and Palm Oil Mill Effluent (POME).

Pulp and papers: Owing to their high cellulose content, both EFB and palm fronds can be processed into pulp using various pulping processes. The pulp is then blended with other pulp material for the production of different types of papers.

Plastic composites: The incorporation of oil palm trunk (OPT), oil palm frond (OPF) and empty fruit bunch fibres as fillers in plastic mattresses such as polypropylene (PP) and polyethylene (PE) results in the production of versatile composites that have many industrial applications.

Briquetted charcoal: Briquetted charcoal can be produced from EFB also with good burning properties. The EFB wastes can be formed to briquetted blocks by pressing at high pressure. The temperature produced by friction melts the lignin in the fibers, which act as a binding material.

Palm lumber: The amount of palm wood from every hectare of oil palm plantation at its felling is about 70 tonnes. However, the quality of the wood is not as good as conventional woods. Only the first 4 meters of the trunk can be used for lumber production. However, with proper treatment and processing, a high quality lumber can be produced from old palm tree. The wood has a typical structure resembling tiger skin and therefore it is called "tiger wood".

Cattle feed: Palm kernel cake from the mills after solvent extraction of residual oil containing 12-15% of protein can be used for making cattle feed.

Palm oil mill effuent (POME): Palm oil mill effluent contains more than 95% water and the remaining is made up of fibrous fruit residues and a little amount of palm oil which can also be utilized for the formulation of animal feeds.

Vitamin concentrates: Vitamin concentrates can be produced from palm fatty acid distillate (PFAD), which is a by-product of palm oil refining. PFAD contains both tocopherols and tocotrienols. It may find application as antioxidants in food industry. It may also serve as an active ingredient in pharmaceutical products.

Energy generation from biomass: Tremendous amount of energy can be generated from the oil palm biomass by controlled combustion processes like gasification. This biogas is easily combustilble and can be utilized for heat application as well as for power generation.

Oil Palm for Health and Nutrition

Palm oil has excellent health attributes. The important advantages of palm oil are:

1. Palm oil, like all other types of vegetable oils, is cholesterol-free.
2. A palm oil diet helps in raising the level of good HDL cholesterol and lowering the bad LDL cholesterol.
3. Palm oil reduces the tendency for blood to clot, therey reducing the risk of heart diseases.
4. Red palm oil is a rich shorce of beta carotene, an antioxidant and precursor of Vitamin A and tocotrierols (Vitamin E). Red palm oil contains 500-700 mg/g of tocotrienols and Vitamin E. Carotenoids in RPO can also serve as powerful antioxidants and anti-cancer agents and is good for heart. The vitamins also have a protective role in cellular ageing.
5. Palm oil has powerful antioxidants which can keep cancer at bay.
6. Solid fats made from palm oil are better for health as they do not have the unhealthy trans fatty acids present in the hydrogenated fats.

Palm Kernel Meal

Palm kernel cake is a product of palm kernel oil extraction. Palm kernel cake, in the powdered form is called meal or flour. Palm kernel meal contains about 20% protein being the lowest among the oilseed meals (Hartley 1967). The carbohydrates are present at around 50% with 15% crude fibre. Palm kerkel proteins have a poor amino acid balance. Lysine is the limiting amino acid as in any other vegetable protein. Methionine content is better than that in soybean.

Palm kernel meal contains a good amount of minerals, particularly calcium, phosphorus and iron (Cornelius, 1977). Morrison (1949) has reported 0.69% phosphorus, 0.42% potassium and 0.017% iron in palm kernel meal. The ratio of calcium to phosphorus in palm kernel meal is more favourable than in other oilseed meals (Mc Donald *et al.,* 1982). The digestibilities of nutrients in palm kernel meal are relatively low (Morrison, 1949).

Palm kernel cake and meal are not suitable for human consumption. It lacks palatability and has to be used mostly as animal feed. The palm kernel cake is dry and gritty. The solvent-extracted palm kernel cake is less palatable than the expeller-extracted palm kernel cake (Hutagalung *et al.,* 1983). The cake is therefore pelleted and mixed with well-liked feeds like molasses, cassava, or broken rice to improve it's intake by the animals. Palm kernel cake has been observed to be good feed for dairy cattle as well as pigs and it is rarely used in poultry rations (Gohl, 1975).

Rice Bran Oil

Rice grain obtained after harvesting and threshing is known as paddy. It has a protective layer of husk, consisting of two parts, the lemma and the palea. The husk accounts for about 20-22% by weight of paddy. Dehusked kernel is nothing but brown rice, which again has some covering layers.They are pericarp, testa (or tegmen or seed coat) and aleurone layer. The outer layers are fibrous and also contain phytin. Fibre and phytin both interfere in the digestion process and absorption of some minerals(calcium, iron) and protein. The covering layers also interfere with cooking process. Hence these layers are removed by milling (called debranning, polishing or whitening) to prepare rice as food. Such polishings are called rice bran. Husk, bran and broken rice are the three by-products of rice milling, all of which have many uses.

Composition of Rice Bran

If rice is parboiled, the aleurone layers of the grain are broken up and the oil is pushed outward. More oil comes into bran during milling. The vitamins and minerals in the outer layers are retained in the milled rice after parboiling and bran from parboiled rice contains less vitamins and minerals.

Carbohydrates (nitrogen-free extract) constitute some 30-50% of bran. The carbohydrate portion consists of 10-25% starch, 9-17% hemicelluloses, 5-10% cellulose, 3-10% lignin and 8-15% other sugars. Rice bran contains 15-25% oil. The bran from parboiled paddy contains more oil than the bran from fresh paddy. The bran also contains 15% protein. Non-protein nitrogen accounts for about 16% of total nitrogen; most of it is free amino acids. The bran contains 7-15% ash, which has sufficient levels of Ca, Fe, P, K, Mg and Si etc. Among the minerals, the predominant components are phosphorus and potassium, accounting for 1-3% each. Calcium followed by iron are the next major minerals in rice bran at 30-130mg and 10-50 mg/100g bran. It is rich in B vitamins and tocopherols (Bhattacharya and Shankara,1988). Nicotinic acid is the major B vitamin at 25-30mg and tocopherols at 3-13 mg/100g bran.

Table 79: Oil content of bran from different mills.

Mill type	Grade	Oil content %
Huller	Raw	4-6
	Parboiled	4-6
Sheller	Raw	12-15
	Parboiled	15-20
Modern	Raw	15-20
	Parboiled	25-30

Source: Gopala Krishna *et al.*, (2005)

Rice Bran Oil (RBO)

Rice bran is an important source of nutritious oil. At the rate of 5% bran in paddy and 15% oil in the bran the potential availability of rice bran works out to 3.9 million tonnes in the world. Rice bran contains not less than 15% and even upto 25% oil. A factor, which influences the oil content of bran is parboiling. The bran from parboiled rice contains oil up to 30%. The oil present in rice is intended for supply of energy and nutrition to the growing plant after germination. For this purpose the enzyme namely lipase is present in the rice bran. During milling of rice the enzyme and oil get mixed up and the oil begins to get hydrolyzed. Because of this the FFA content of the oil rises to 10-20% in a day and may reach a level of 70% in a month. Oil with more than 3% FFA is not fit for edible purpose.

Crude RBO contains 90-96% saponifiable lipids. The triglycerides account for 83-86%, with 3-4% diglycerides and 6-7% monoglycerides. The unsaponifiables account for 4-5% of the oil. Significant quantities of FFA, waxes, glycolipids and phospho lipids are also present in the oil. Phytosterols account for about 40% of the non-saponifiables, followed by 28% triterpene alcohols, 17% hydrocarbons, 10% sterol esters and 2-5% tocopherols. The high amount of waxes in the RBO increase the viscosity of the oil. Ricebran oil is twice more viscous than other vegetable oils. The diglycerides present in the oil also add to the viscosity. The wax contains about 15% of resinous matter. On saponification of the waxes, fatty alcohols are liberated at about 35%. Waxes hold a very large quantity of water in emulsification. Crude rice bran oil holds its own weight of water without the water separating out of the oil.

Table 80: Composition of crude ricebran oil.

Constitutent	Content
Saponifiable lipids	90-96%
Triglycerides	83-86
Diglycerides	3-4
Monoglycerides	6-7
Free fatty acids	3-5
Waxes	3-6
Glycolipids	6-7
Phospholipids	2-4
Unsaponifiable lipids	4-5

Crude ricebran oil contains about 600-700 ppm phosphorus, If phosphatides remain in the oil during deodorization or physical refining, the resulting oil will be darker in colour and poor in initial flavour and flavour stability. P content of the pretreated oil prior to physical refining should not be greater than 5ppm.

There are three ways to obtain low FFA oil from rice bran (1) extract oil immediately after the bran is separated from rice (2) to stabilize the rice gain and (3) to deactivate lipase enzyme. Rice bran can be stabilised by simply drying to low moisture or by subjecting to heat or by treating with chemicals such as an acid. The oil content of the rice bran is low (10-15%) and hence the oil is normally extracted through solvent extraction. Modern solvent extraction plants can reduce the proportion of residual fat in the meal to 1-2 per cent giving more economical yields and a more stable defatted bran. The rice bran oil is refined, degumed, dewaxed and bleached to obtain good quality oil. The RBO is a semidrying oil with higher level of upto 5% unsaponifiables. It's S.V. is similar to that of other vegetable oils. Normally, it is darker in colour than other edible oils. However, it can be refined to a pale yellow (golden) colour. It's I.V. is medium, indicating semidrying characteristics.

Table 81: Characteristics of rice bran oil.

Chatacteristic	Value
Colour in 1″cell	Y+5R=20
RI at 40°C	1.4600-1.4700
SV	180-195
IV	90-105
Acid value	1.5
Unsaponifiables%	3.0-5.0
Flash point	250° C

Fatty Acid Composition

The chemical composition of RBO is very close to that of groundnut oil. Both the oils contain about 35% linoleic acid. The major fatty acid in rice bran oil is oleic acid followed by linoleic acid, which are in the range of 40-45 and 30-37% respectively. Palmitic acid is present at 15-20% followed by about 2% each of stearic and arachidic acids. Myristic, gadoleic, behenic and lignoceric acids are present at less than 1% each.

Since linolenic acid is present at a lower level of 2%, flavour reversion associated with soyabean oil does not occur in RBO. The oils extracted from the different rice varieties viz., *Indica* and *Japonica* are almost similar in the fatty acid composition. The palmitic acid levels are lower in *Japanica* oil while oleic acid is lower in *Indica* oil. The linoleic acid is higher (40-50%) in *Japonica* oil than *Indica* oil (38-47%).

Table 82: Fatty acids in the oil of *Indica* and *Japonica* types of brown rice.

Fatty acid	Fatty acid content (%)	
	Indica type (Indian varieties)	*Japonica* type (Japanese varieties)
Myristic	0.2-0.4	0.2-0.4
Palmitic	18.0-20.0	14.9-16.7
Palmitoleic	0.2-0.3	0.2-0.3
Stearic	1.6-2.2	1.6-2.1
Oleic	38.3-46.5	40.9-49.5
Linoleic	28.8-35.6	28.5-36.7
Linolenic	1.1-1.8	1.2-1.5
Arachidic	0.8-1.1	0.7-0.9
Gadoleic	0.5-0.7	0.6-0.9
Behenic	0.4-1.0	0.4-0.5
Lignoceric	0.6-1.0	0.7-1.1

Unsaponifiables

RBO has a high unsaponifiable fraction and is rich in minor constituents such as phyto sterols (1.8%), oryzanol (1.6%) and tocopherols (81.3 mg/100g). Tocopherols and tocotrienols which are present in high concentrations (550ppm), provide oxidative stability to RBO and improve its shelf life. RBO has higher content of squalene (320mg/100g). The beneficial effects of these unsaponifiable components are well known. Tocopherols and tocotrienols act as potent antioxidants and anti-cancer agents while squalene has been shown to have antioxidant, antidermatitic and anti-tuberculosis properties. The unsaponifiable matter of the oil has reduced serum cholesterol in experimental animals.

The sterol composition of unsaponifiable matter is higher in rice bran oil (2.2%). Plant sterols such as stigmasterol, beta-sitosterol and compesterol have cholesterol lowering activity. The esters of phytosterols with long chain fatty acids as well as free sterols are equally effective in their cholesterol lowering properties.

Five classes of lipids namely neutral lipids, glycolipids, phospholipids, ceramides and wax lipids are present in rice-bran. Neutral lipids are very much similar to groundnut lipids. The unsaponifiable lipids of rice bran oil consists of four classes of compounds viz., sterols, 4-methyl sterols, triterpene alcohols and less polar compounds. The rice bran oil, in general, has higher amounts of unsaponifiable matter. But within the rice bran oil, the physically refined oil has higher amounts of the four classes of unsaponifiable matter than the chemically refined oil.

Table 83: Unsaponifiable lipids of RBO.

Phytosterols	**% in oil**
Sterols	1.80
β-sitosterol	0.88
Campesterol	0.51
Stigmasterol	0.27
4-Methyl sterols	0.40
Citrostadienol	0.17
Gramisterol	0.16
Obstusifoliol	0.03
Triterpene alcohols	1.20
24-methylene cycloartanol	0.49
Cycloartenol	0.48
Cycloartanol	0.11
Less polar compounds	0.80
Squalene	0.12
Tocopherols	0.04
Tocotrienols	0.07

Eight sterols have been identified by Japanese workers (Fujino *et al.*,1974). The three main sterols are, β-sitosterol (53-59), campesterol (16-26) and stigmasterol (10-13%). Glycolipids of rice-bran deserve special mention as they contain sterylglycosides along with glyceroglycolipids. Sterylglycosides are generally not found in other oilseeds. α-glycosyl α-steryl, β-acyl glycerides, may be considered as markers for rice-bran cake. Moreover many other glycolipids like tetra-glycosyl diacyl glycerol, monoglycosyl monoglyceride, diglycosyl diglyceride, triglycosyl diglyceride and tetraglycosyl diglycosyl diglyceride, are also present, which contribute to the sweetness of rice. Ceramide monohexoside, ceramide dihexoside, ceramide trihexoside and wax lipids are also present in appreciable amounts (Nasirullah and Nagaraja, 1987).

Tocopherols/Tocotrienols

Rice bran oil contains 45-137 mg/100g of tocopherols and tocotrienols of which about 57% is the latter. Alpha-tocopherol and the related compounds have a powerful antioxidant activity. Vitamin E is useful in biological system for alleviating a multiplicity of deficiency syndromes, like antisterility, nutritional myopathy, decreased immune function, cardiovascular problems, cancer and aging. Rice bran oil is an excellent salad and frying oil with very good oxidative stability (Gunstone, 2005), due to the presence of tocopherols and tocotrienols (860 ppm) and oryzanols (ferulic acid esters of sterols and triterpenols).

Table 84: Tocopherols and tocotrienols in rice bran oil.

Constitutents	mg/100g oil (1)	mg/kg oil (2)
Alpha tocopherol	61	347
Beta tocopherol	Traces	-
Gamma tocopherol	11	89
Delta tocopherol	9	42
Alpha tocotrienol	49	126
Gamma tocotrienol	29.2	301
Dalta tocotrienol	2.8	10

Source: (1) Goenka (2007) and (2) Gunstone (2005)

Tocopherols have been isolated from rice bran oil by molecular distillation and treatment with anion exchange resins. An enrichment procedure for tocotrienols involves treatment of the methyl ester of the residue from physical refining of rice bran oil with anion exchange resin to remove any free fatty acid fraction by chromatography on C-18 alkyl bound silica gel. Partial concentration of the tocopherols is achieved by reacting the deodourizer distillate from rice bran oil which is rich in fatty acids, with calcium hydroxide or sodium carbonate in the presence of a suitable solvent and concentrating the filtrate. The concentrated tocopherols are mixed with hardened soya oil to give 40% tocopherol containing food additives.

Rice bran oil is the only edible oil apart from palmolein witch has significant amount of tocotrienols. Tocotrienols apart from being powerful antioxidants, are known to inhibit HMG-CoA reductase, a key enzyme in the synthesis of endogenous cholesterol with a resultant decrease in blood as established in human subjects.

Edible Uses of Rice Bran Oil

RBO oil is toxicologically safe for human consumption. Two advantages which rice bran oil has over other oils are its very low content of linolenic acid and its high content of tocopherol, both important for oxidative stability. Linolenic acid is suspected of being a key factor in the flavour reversion found with soyabean (Hoffman, 1961), which limit its use in certain commodities. Flavour reversion has not been noted in rice bran oils and their performance in accelerated storage tests has been excellent (Feuge and Reddi, 1949; Swift *et al*, 1950) from both the point of view of peroxide formation and the organoleptic quality.

The keeping and the frying qualities of RBO oil are better than other edible vegetable oils. Foods deep-fried in rice bran oil absorb less oil when compared with foods fried in groundnut oil. The organoleptic studies of the foods fried in rice bran oil revealed the good acceptability of the products (Rukmini, 1987a). Rice bran oil lowers serum cholesterol level by reducing LDL and VLDL cholesterols, without in any way influencing the HDL cholesterol (Gunstone, 2005).

Rice bran oil performs well in frying tests giving low peroxide, foam, FFA and polymer formation (Lynn *et al.*, 1968), as well as an adequately high smoke point (213^{0}C) and fire point (352^{0}C) comparable with or superior to competitive oils in common use (Feuge and Reddi, 1949). There have also been reports of reduced losses (Akiya, 1967), decreased sticking and a cleaner odour and flavour (Lynn *et al.*, 1968) than with other oils. It has been evaluated in a wide range of commercial household applications with promising results (Thirumala Rao and Murti, 1961; Lynn *et al.*, 1968).

Salad Oil

The most important features of a good grade salad oil are cold stability, flavour stability, good colour and the ability to form stable emulsions when used in mayonnaise or emulsified salad dressing product. Emulsifiers and crystallisation inhibitors can improve some of these characteristics. Refined bleached deodorised rice bran oil can claim good cold stability as measured by the normal methods (Feuge and Reddi, 1949). Its performance can be improved through winterisation. (Feuge and Reddi, 1949; Swift *et al.*,1950 Lynn *et al.*, 1968; Lynn, 1969). The cold test, is a measure of the length of time before a bottle of oil, immersed in water at 0^{0}C, will develop cloudiness. A high yield of liquid on winterising rice bran oil is also of considerable benefit in the economics of salad oil production. Liquid oils of this type are also used extensively for fish canning.

Margarine and Shortenings

Margarine is a water-in-oil emulsion, the fat trapping the water phase within an interlocking network of crystals. The amount, nature and size of the crystals are important and determine the plastic properties of the material. The melting range must be wide so that over a range of temperatures the mixture is form-retaining although plastic. RBO is a very useful raw material for making margarines.

Shortenings and other compound cooking fats, extensively used in the home and in the food industry, consist of a blend of various fats hydrogenated to the desired melting point, together with a small amount of emulsifier. The behavior of rice bran oil on hydrogenation appears to present no unusual difficulties. The iodine value, refractive index and colour have been found to decrease readily and evenly becoming similar to other oils (Feuge and Reddi, 1949; Takeshita and Oba, 1962). In the hydrogenated state, rice bran oil is exceptionally stable and resistant to oxidation and appears better than hydrogenated cotton seed and groundnut oils (Feuge and Reddi, 1949).

By-Products from Rice Bran Oil

Crude rice bran oil contains gums (1-2%), waxes (3-6%), free fatty acids, oryzanol (1-2%), tocopherols (<1.0%), tocotrienols (<1.0%), sterols (1-0%), squalene (<1.0%), steryl esters (1-2%) and glycolipids (4-6%) depending on the quality and processing conditions of the bran. During rice bran oil refining most of the components are removed partially or totally from the oil at different processing stages as by products. These valuable components can be recovered or purified in the form of by products by adopting suitable processes.. Soap stock produced during alkali refining is a good source for fatty acids and the only commercial source for gama oryzanol. Bleaching operation produces significant quantities of spent bleaching earth which contains around 25% of oil.

Lecithin

Lecithin is a by-product of soybean oil industry. Lecithin usage in foods can bring about a reduction in the fat which is desirable in the present context. The advantage of rice bran lecithin over soya lecithin is that the former has a lower amount of linolenic acid (<2%), (fishy odour causing fatty acid) than the latter (>5%). Thereby the stability is much better. In order to recover this valuable by-product from rice bran oil industry, a method using 10-20 kg of wet-gum sludge has been developed.

Water or acid degumming of rice bran oil produces lecithin which contains higher amounts of phospholipids which are used as emulsifiers, dispersing agents, release agents in the baking, confectionery and other industries. Rice bran lecithin exhibits more oxidative stability as it contains a smaller amount of linolenic acid than soya lecithin. Lecithins may also be chemically or enzymatically modified to enhance their surfactant properties and functionality. Enzymatic degumming directly produces lyso-lecithin which has superior surface active eproperties.

Oryzanols

Oryzanol is ferulic acid ester of sterols and triterpene alcohols. It can be extracted from rice bran soap stock, which however does not have any food value (Gunstone, 2005). Rice bran oil contains 1.5-2.9% of oryzanol and during refining, about 90% of the oryzanol goes into the soap-stock as a refinery waste. The crude rice bran oil has an oryzanol content of 1.8-2.0% while it is 0.19-0.22% in the refined oil and 2.21-6.71% in the soapstock. The bran from different varieties of rice contain oryzanol to the extent of 1.6-2.6%. Oryzanol needs to be recovered from the soapstock of rice bran oil by using solvent partition technique. Oryzanol is being produced commercially in rice growing countries like Japan and India.

Oryzanol (1-2% of unsaponifiables) and phytosterol content in rice bran oil help reduce cholesterol and triglycerides and elevate HDL cholesterol significantly. Oryzanol has been shown to increase fecal excretion of bile acids. Oryzanol contains cycloartenols which have a structure similar to cholesterol and which is a precursor of phytosterols. It may compete with binding sites of cholesterol in the liver, causing greater amount of cholesterol to be excreted and metabolised to bile salts and pigments, compared to other edible oils.

Ricebran and ricebran oil derives it's beneficial nutritional properties from the ferulic acid (4-hydroxy-3methoxy cinnamic acid)esters of triterpene alcohols and plant sterols, collectively called oryzanol or γ-oryzanol. Recent investigations have demonstrated a multitude of beneficial physiological effects associated with oryzanol intake. These include: hypocholesterolemic activity, reduction of wrinkles in aged women, melanin formation (conversion of grey hair to natural black), preventing discoloration of nails, controlling perspiration odour, preventing motion sickness, correcting nervous system disorder, serving as antioxidant and preservative for foods and cosmetics, treatment of atopic dermatitis and senile xeroderma, etc.

A plethora of oryzanol containing transdermal and moisturizing cosmetic preparations have appeared in patent literature. Ricebran is the most abundant source of oryzanol and most readily accessible.

Table 85: Rice bran oil by-products and their utility.

Material	Amount (% in oil)	Use
Wax	2-6	In polish, food wraps, cosmetics, leather, etc.
Vitamin E	0.1	Medicinal as antioxidant
Squalene	0.3-0.4	Medicinal
Oryzanol	1-2	Medicinal, as antioxidant
Fatty acids	Variable	Industrial
Soap	Variable	Industrial
Gum (phosphatides)	1-3	Emulsifying, wetting, dispersing agent

Rice Bran Wax

Upgraded and bleached rice bran wax has potential applications in various industries like paper coating, candles, water proofing polish, cosmetics, carbon paper, printing inks, adhesives, fruit and vegetable coating, pharmaceuticals etc. and may substitute waxes like carnaube to some extent. The wax is also a useful source to prepare high value products like triacontanol and octacosanol, which are gaining importance in agriculture and medical fields as plant growth stimulant and cholesterol reducing agent respectively. Deodoriser distillate is rich in high value products like tocopherols, tocotrienols, squalene, phytosterols and triterpenols.

Ricebran oil contains upto 6 percent of waxes. About 20,000 tonnes of such waxes are produced by the ricebran oil processing industry for which a suitable outlet has to be found. Rice bran wax has various uses in areas such as: paper coating, candles, moulded novelties, electrical insulation, textile and leather sizing, water proofing, polish (floor, furniture, shoe), cosmetics, carbon paper, typewriter ribbons, printing inks, greases, lubricants, matches, explosives, crayons, adhesives, chewing gums, fruit and vegetable coatings, pharmaceuticals, etc.

A method has been standardized for upgrading commercial crude wax from ricebran oil. The wax 1has a high content of oil and fatty acids. These are first removed by solvent treatment. The defatted wax still contains about 3 per cent of soft wax and about 15 percent of resinous matter. These are removed in the subsequent steps. The upgraded wax thus obtained has a tan to off-white colour depending on the quality of starting wax. The upgraded wax has physico-chemical characteristics similar to carnauba wax. Potential applications of this upgraded wax are in cosmetics where natural ingredients find more favour compared to synthetic counterparts. It also finds use in food applications and in pharmaceuticals. However, the waxes obtained from commercial sources retain a bad odour even after upgradation.

Triacontanol

The major components of ricebran wax are long chain fatty alcohols and long chain fatty acids. A major component of these fatty alcohols is n-triacontanol which is known to have plant growth promoting properties. Ricebran wax constitutes a cheap source for this valuable chemical. There have been sporadic attempts to isolate this component from rice bran wax. However, the methods described in literature are all cumbersome or are not economical to operate on a commercial scale. Maximum yields reported in these methods are low at 12 percent of the starting material. Methods to upgrade yields upto 35 percent are being developed.

Conclusion

Rice bran oil due to its better composition of fatty acids and health promoting components in its unsaponifiables is a very good edible cooking medium. It can serve as salad oil and also as a good frying oil. It is gaining popularity all over the globe.

Corn Oil

Corn *Zea mays*, is native to both North and South America and has been the staple food grain of the local Indians. It belongs to the *Graminae* family and *Maydeae* tribe. United States, being the major cultivator contributes to about 50% of the world production. Other important countires are China, Brazil, Mexico, Romania, Argentina, and France. Corn is next only to wheat and rice, as a food grain, grown all over the world. More than 90% of the corn grain is processed and fed to animals in the Western world. Asia and Africa utilize the entire produce for human consumption by traditional processing, without separating the germ (Salunkhe *et al.*, 1985).

Only a small fraction of the corn is used for obtaining corn or maize oil. Corn oil is a by-product of the starch indusrties and the corn meal industry. Corn grain contains 12-25% germ, which contains upto 50% oil. It is obtained with a wet de-germination process from the sweetener, starch, and alcohol processors. In the corn milling industry, the germ is obtained with a dry de-germination process. There is a great potential for production of highly nutritious corn oil in view of significant corn grain production and a larger proportion of germ in the grain corn.

Corn Composition

The chemical composition of corn has been reviewed by Watson (1982) and Allen (1979). The oil content in different corn varieties ranges from 1.2 to 5.7% (Qucken-bush *et al.,* 1961). However, most field corns contain about 4 to 5% oil. The high-oil types contain up to 19.5% oil (Aquilera and Lusas 1986). Sweet corn and high-oil corn are relatively rich in oil while the popping corns are rich in crude fiber. Most of the oil is concentracted in the germ. The germ of regular corn varieties contains about 35% oil while that of high-oil corn can contain up to 50% oil.

The germ of regular corn varieties is separated through wet or dry milling and used for commercial oil extraction. The high-oil corn can, however, be directly utilized for commercial oil extraction. Certain hybrid seeds with good yield potential contain about 7 to 8% oil in whole kernel (Weber and Alexander 1975). The high oil and protein in these varieties is due to the large size of germ (Dudley 1974). The high oil corn exhibits superior protein quality over normal corn (Weber and Alexander 1975).

Corn Oil Extraction

The germs are separated through milling. Oil is recovered from dry milled germ through direct solvent extraction or by continuous mechanical screw pressing (expelling) methods. Solvent extracted germ oil is lighter in colour than expelled oil. Refining process is used to produce a light coloured oil. Wet milled germ is of high oil content (about 50%) and is best processed by combination of both expelling and solvent extraction methods. The recovery of maize oil from either dry milled or wet milled germ does not involve any specific processing problems. Crude corn oil, like other vegetable oils, can be strored for longer periods without deterioration. At temperatures above 60^0C, deterioration is fast and refining properties get adversely affected when the oil is stored for 5 days or more at this temperature.

Refinining of Corn Oil

The crude corn oil contains about 1.5 per cent phospholipids, 7.5 per cent of which can be precipitated by the degumming (using small quantities of water or through steaming of the oil). Phospholipids and some nontriglycerides also get removed to some extent during this process. The precipitated maize lecithin can be utilized after dehydration. Refining or neutralization of crude corn oil with aqueous sodium carbonate solution or alkali solutions or by both the procedures, reduces, the degree of colour reversion and increases stability.

Colour reduction during refining can be an index of refining effectiveness. A colour of 3.5 red before deodorization is a useful measure of adequacy of refining of crude corn oil. Corn oil contains traces of high melting waxes which precipitate at normal temperatures by winterization. The low amount of wax(0.05%) in corn oil does not warrant any winterization procedure, especially, when the oil is to be hydrogenated or used as a cooking-oil. However, to ensure complete clarity under cool conditions winterization can be carried out. This process also, removes the traces of di-saturated triglycerides that may be present.

The wax-free or dewaxed corn oil withstands a cold test (remaining clear at 0^0C) for 98 hours or more. Well-refined corn oil is placed for deodourization to produce a quality maize oil with low free fatty acid content, light colour and stability against oxidation (Reiners and Gooding, 1970).

Corn Oil Characteristics

Crude corn oil has a darker reddish amber colour. It can easily be processed to a light coloured oil. Wet milled oils are somewhat difficult to bleach to a light colour. This is mainly because of the colour of the germ that develops under wet conditions. In general, corn oil is mostly underbleached and deodorized to retain a darker color for appeal to the consumer.

Corn oil composition depends upon the seed type, climatic conditions, and growing season. Due to higher unsaturation corn oil has higher iodine value. It's melting point is higher compared to other vegetable oils, due to which it can become cloudy at temperatures around 12^0C. It's S.V., specific gravity, refractive index and unsaponfiables are similar to other edble oils(Sonntag, 1979). The low melting point, low smoke point, pale yellow colour, and bland flavour make it a favourite food oil. Indian corn oil has been observed to have the following characteristics: density=0.91-0.92, refractive index at 28^0C=1.4720-1.4735, acid value=3-16.1, saponification value=192-197 and iodine value=105-113 (Vasishtha *et al.*, 1991).

Table 86: Characteristics of corn oil.

Characteristics	Value
Specific gravity at 25/25^0C	0.915 to 0.920
Refractive index at 25^0C	1.470 to 1.474
Iodine value	103.0 to 128.0
Saponification value	187 to 193
Unsaponifiable matter, %	2.0 max.
Titer, ^{0}C	14.0 to 20.0
Melting point ^{0}C	12.0 to10.0

Fatty Acid Composition

Linoleic acid is the major fatty acid, followed by oleic and palmitic acids. Their concentractions are, 50-60, 25-30 and 10-15 percent respectively. Stearic acid is present at 2%. Linolenic and arachidic acids are present only in traces(Gupta *et al.*, 1979). Maize oil inspite of it's richness in linoleic acid, is superior from storage and nutritional points of view. It is also highly digestible. The low level of linolenic acid imparts it a good keeping quality (Gupta *et al.*, 1979). The production of corn oil is dependent on the demand for the maize starch as the oil is a by-product of the milling industry. (Weber, 1979). Development of hybrids having high oil content can go a long way in increasing oil production with desired level of PUFA (Weber and Alexander, 1975). Corn oil from different Indian genotypes have the following fatty acid composition: myristic=0.1-5.7, palmitic=8.4-23.4, stearic=0.4-8.3, oleic=24.2-40.1, linoleic=37.1-49.7, linolenic=0.8-6.3 and arachidic acid=0.2-4.9% (Vasistha *et al*, 1991).

Table 87: Fatty acid(%) composition of corn oil.

Fatty acid	% composition
Myristic	0.1
Palmitic	10.9
Palmitoleic	0.2
Margaric	0.1
Stearic	2.0
Oleic	25.4
Linolenic	59.6
Linolenic	1.2
Arachidic	0.4
Behenic	0.1

Most of the lipids in corn are in the free form. Bound lipids constitute 0.3 to 0.9% of the kernels, and most are bound to starch (Morrison 1978). Triglycerides comprise about 80% and phospholipids about 8% of the total lipids (Watson 1982). Other minor lipids include diglycerides (1-2.9%), sterols (1.3-5.5%), sterylesters (1.1-2.9%), glycolipids (2-5.4%), free fatty acids (0.3-0.9%), and waxes, pigments and odour-active compounds. The germ lipids are relatively rich in neutral lipids and low in phospholipids and glycolipids as compared to endosperm lipids (Weber 1979). Similarly, the germ lipids are low in linolenic acid and in the level of free fatty acids as compared to endosperm lipids. Thus, germ oil has a more desirable oil composition than endosperm oil from the view point of keeping quality(Watson 1962).

Unsaponifiables

Corn oil contains traces of waxes (0.05%) that are esters of myricil and ceryl alcohols with tetracosanoic acid. The melting point of these waxes is 81 to 82^{0}C. The waxes cause the oil to cloud when cooled to a low temperature unless removed by a dewaxing process. Dewaxed oil is important for salad oil production but is not required when corn oil is to be hydrogenated to margarine or shortening.

Corn oil has a higher tocopherol content (about 0.1%), and related antioxidants which account for the oxidative stability. The reverted flavour of corn oil is somewhat pleasant, usually characterized as popcorn-like or musty flavour (Weiss, 1983).

Carotenoid pigments such as xanthophylls (0.2-5.0 mg/kg) and carotenes (0.2-7 mg/kg) are present in corn oil. Of the sterols, β-sitosterol is the major component (80%), while campesterol (12%) and stigmasterol (8%) are present in minor amounts. Triterpene alcohol and 4-methyl sterols are also present. Polar lipids include monogalactosyl and digalactosyl diglycerides. Phosphatidylcholine is the major component (64%) of the seven identified phospholipids (Weber, 1978; 1979). The glycolipids include cerebroside, sterylglycoside and sulfolipids. The carotenoid pigments vary quantitatively more than the other components because they are slowly oxidized to colourless compounds during storage.

Food Uses of Corn Oil

Corn oil is popular as a food oil mostly due to its flavour stability during storage and cooking. About 60 to 65% of the corn oil is consumed as salad or cooking oil. Its bland flavor and high smoke point have made it a popular deep-frying fat. It is used for repeated fat frying purposes as it does not absorb flavours from one food and transmit them to the others. The high content of unsatured fatty acids in corn oil makes it a good dietary agent of reducing blood cholesterol levels. The larger use, about 30% or so, of corn oil is in margarine. Other food uses are as a liquid fat, or as a blended fat with harder fats. Bakers use the oil for greasing tins or for brushing baked goods.

Non-Edible Uses of Corn Oil

Corn oil to a limited extent is used in pharmaceuticals and cosmetics, coatings for tablets and pills, vitamins, antibiotics, face and body powders, etc. (Rankin 1982). The hydrolyzed fatty acids are recovered from crude oil during the refining step (Reiners, 1978). It contains 92% fatty acids, and is high in unsaturated fatty acids. It is used as a high energy source, and to control dust in beef and poultry feeds (Shroder and Heiman 1970). The soap stock, a by-product of corn oil refining, is used in the manufacture of soap and associated products. Smaller quantities of either crude or refined corn oil are also used in ammunition, chemicals and insecticides, paint and varnish substitutes, rust preventive compounds and textiles. Sulfonated corn oil has important applications in tanning and processing of leather.

Defatted Germ Flour

The defatted corn germ flour is presently used almost exclusively as an animal feed. The demand for low-cost vegetable protein supplements to increase the nutritional value of cereals and other foods has encouraged research on improved products from the defatted corn germ meal. It is rich in proteins (20-25%), starch (25%), crude fibre (10%) and minerals (3-5%). It contains higher levels of copper and manganese. It is rich in choline, niacin, pantothenic acid and pyridoxine. It has a balanced amino acid composition, with sufficient levels of of lysine, sulfur-containing amino acids and tryptophan as proposed by WHO/FAO for reference protein. The food-grade germ meal can be used as a protein supplement in a variety of bakery, macaroni and traditional products, and to prepare protein isolates and concentrates.

The addition of only 25% corn germ flour to cookies raises the protein content from 5.4% to only 5.7%, but doubles the lysine content (Blessin *et al.*, 1974). Corn germ flour when added to non-vegetarian items and bread improves the nutrition and food quality(Sutescu *et al* 1972) especially texture and flavor (Tsen *et al.*, 1974). Animal feding studies using bread with 6% corn germ improves the protein quality (Zaharia and Sutescu 1972).

Defatted corn germ flour supplemented with corn starch and milk proteins has been extruded into a product suitable for use as a nutrient snack. Macaroni supplemented with up to 20% defatted germ flour is highly acceptable (Lucisano *et al.*, 1984). Germ meal with whiter colour and free from residual lipids and bitter compounds has better functional food characteristics. The defatted corn germ meal obtained in SC-CO_2 extractional process is reported to exhibit superior sensory, functional and storage properties for food applications than that produced in expeller press and hexane extraction.

Attempts have been made to prepare protein isolates from defatted corn germ flour that can be used as a food ingredient to impart favourable changes in structure, texture and composition of food products during preparation, processing and storage (Kinsella, 1982; Nielson *et al.,* 1973). The isolate has upto 85% protein with a good amino acid balance. It has mild flavor, high solubility and oil emulsifying properties for use as a food additive. Similar characteristics of protein isolates of corn germ have been reported by Banu *et al.,* (1974).

Cotton Seed (*Gossypium*) Oil

Cotton is grown primarily for its textile fibre from times immemorial. Cottonseed, a by product has been used largely as an animal feed till the fifties in the 20th century. Deficiency in edible oil supplies, and the good quality of cottonseed oil have made it an important source of oil in recent times and has become the number two vegetable oil world over. Cottonseed forms two thirds of the *kapas* produced, the remaining 1/3 being the fibre. The seed contains 14-23% oil.

Cotton originated in the tropical regions, but 50% of its present production is in the temperate regions. China, Russia, U.S.A, India and Pakistan account for 80% of world production There are four cultivated species namely *G.Hirsutum, G.Barbadense, G. Arboreum* and *G. Herbaceum.* The seeds of the latter two species are small and have lower oil content than the former two species.

Cotton plants are perennials, but behave as warm season annuals. The cultivated plants are day neutral in flowering response. Cotton grows well on moderately fertile soils ranging from sandy to heavy clay soils. Cotton plants are heavy users of moisture and good rainfall is needed during its growing period especially blooming onwards. It also needs good sunshine.

The plants begin to flower within 60 days of sowing and takes another 50-55 days for pod maturity. Cotton has been an important article of commerce for production of fibre. However, recently, cotton seed has also assumed industrial importance as a potential source of vegetable oil. Cottonseed linters for various industries and cottonseed cake as cattle feed are the other important by- products of cottonseed crushing industry. The fried seeds are sometimes eaten after removal of the hulls through biting between the teeth.

Seed Composition

The seeds have 35-55% hull with 65-45% kernel. The seed oil content ranges from 14-25% with a seed protein content of 15-22%. American type seeds are generally higher in oil and linter contents and are generally preferred for crushing. High oil and low linter content seeds are preferred for oil extraction purposes. Similarly low hull types are better for higher oil recovery. The Bangladesh cotton seeds have 19-20% oil and 20-22% protein (Uddin, 2005). The different species of cottonseed namely *G.Hirsutum, G.Barbadense, G. Arboreum and G. Herbaceum* have a similar oil content of 33-35% (Vigneswaran *et al.*, 2007).

Cottonseed Oil

Cottonseed oil termed as "Heart oil" is an unsaturated edible oil. It has a 2:1 ratio of polyunsaturated to saturated fatty acids and generally consists of 73% unsaturated fatty acids including 18% monounsaturated (oleic) and 55% polyunsaturated (linoleic) and 27% saturated (palmitic and stearic) fatty acids (Table. 89). Cottonseed oil is cholesterol free, as are all oils extracted from plants. An additional benefit from cottonseed oil is that high level of antioxidants are retained at high levels in fried products preserving their freshness and creating longer shelf life.

Cottonseed oil is a vegetable oil extracted from the seeds of the cotton plant after the cotton lint has been removed (Pandey, 1987). The oil must be refined to remove gossypol, a naturally occurring toxin that protects the cotton plant from insect damage. Therefore, unrefined cottonseed oil is sometimes used as a pesticide. In its natural unhydrogenated state, cottonseed oil, has neither cholesterol, nor trans fatty acids. However, it does contain over 50% Omega-6 fatty acids and only trace amounts of Omega-3 fatty acids, and the imbalance is considered unhealthy, if not used in moderation or balanced elsewhere in the diet. Further, these polyunsaturated fats can potentially go rancid during the extraction process.

Cotton oil is generally clear with a light golden colour. The degree of colour depends on the amount of refining. Clear, colourless oils are not necessarily better oils, but may have been refined more severely. Cottonseed oil's light, non-oily consistency and high smoke point make it the most desirable oil for cooking (frying). The characteristics of cotton seed oil are given in Table 88.

Table 88: Typical analytical values for cotton seed oil products.

Characteristic	Cotton seed cooking oil	Hydrogenated oil
Lovibond colour (Red Max.)	2.0-6.0	2.0-2.6
Free fatty acid (as oleic % max.)	0.05	0.05
Peroxide value (meq/kg max.)	1.0	0.5
Iodine value	103-116	50-70
AOM stability (hrs.)	15	100-200+
Cloud point (⁰C)	1-3	-
Melting point (⁰C)	10-15	40-45
Pour point (⁰C)	-	40-55
Smoke point (⁰C)	220	-
Flavour	Bland	Bland

The refined ol is pale yellow in colour. It is low in FFA. Due to higher unsaturation it's I.V. is higher. The smoke point is 220°C. It's M.P. is around 15° C and can become cloudy during winter seasons. It is in general similar to other edible oils in its characteristics.

Fatty Acid Composition

Cottonseed oil is rich in palmitic acid (22-26%), oleic acid (15-20%), linoleic acid (49-58%) and 10% mixture of arachidic, behenic and lignoceric acids in the crude oil. It also contains some quantity of cyclopropene fatty acids. These are undesirable components and are largely removed while refining, particularly during deodorization, and also during hydrogenation. They do not pose any health hazard, when cottonseed oil, is used as a cooking medium. It is a nutritious oil due to higher 18:2 levels.

Table 89: Fatty acid composition of cottonseed oil.

Fatty acid	Cotton seed cooking oil	Partially hydrogenated oil
Myristic (14:0)	0.8	0.9
Palmitic (16:0)	24.4	22.5
Palmitoleic (16:1)	0.4	0
Stearic (18:1)	2.2	5.5
Oleic (18.1)	17.2	50.0
Linoleic (18:1)	55.0	20.3
Linolenic (18:3)	0.3	0.3
% Saturates	27	29
% Monounsaturates	18	50
% Polyunsaturates	55	21

The different species of cottonseed have some differences in fatty acid composition among them. *G.hirsutum* has the lowest level of linoleic acid (47%) against the highest (53%) level in *G. herbaceum*. The hybrid as well as *G. barbadense* have around 52% linoleic acid in their oil with *G. arboreum* having 49%. *G. hirsutum* has the highest palmitic acid content of 27% with the others having around 24-25%. The hybrid has higher stearic acid (3.4%) and lowest oleic acid (20%).

Table 90: Fatty acid composition of different species of cotton.

Species/varieties	Oil%	Myristic	Palmitic	Palmitoleic	Stearic	Oleic	Linoleic
G. arboreum	33	0.4	23.7	0.7	1.7	24.5	49.1
G. herbaceum	34	0.4	22.9	0.8	1.7	21.3	53.0
G. hirsutum	34.	0.8	26.5	0.5	1.4	23.4	47.4
G. barbadense	35	0.6	24.5	0.9	0.8	21.0	52.2
Hybrid 6	35	0.7	23.4	0.5	3.4	19.6	52.2

Source: Vigneswaran, *et al.*, (2007)

Cottonseed Oil Uses

Cottonseed oil normally, has bland, neutral flavour that does not mask food flavours. It is ideal for frying and good for cooking, salad dressing and other applications. About 56% is consumed as cooking and salad oils while about 36% goes into baking and frying fats, and a small amount into margarine and other purposes. In India nearly entire cottonseed oil produced is utilized for edible purposes, mostly for vanaspati. A small quantity (5-10%) is used for manufacturing soap. Cottonseed oil can be used for all kinds of culinary applications and can be customized to fit into a number of unconventional uses. It also has a lot of industisal uses. It can serve as a wetting agent as a brominated oil, as an emusifier in medicine and in agricultural sprays. It serves as a tallow substitute after hydrogenation.

Cottonseed oil is commonly used for manufacturing snack foods like potato chips and others. Along with soybean oil, it is very often partially or fully hydrogenated. The important quality of cottonseed oil is the high smoke point. This means that the oil can get super hot and still not smoke-a huge plus for deep-frying. However, hydrogenated cotton seed oil is considered unhealthy due to the presence of trans fatty acids. Cottonseed oil is the first oil to be hydrogenated in mass production. Hydrogenated castor oil has mainly been used for candle manufacture. Only recently, it has become a popular cooking medium.

Table 91: Cottonseed oil edible applications.

Culinary uses	Remark
Deep frying	Cottonseed oil is an excellent flavour carrier and does not mask the fresh flavours of foods.
Baking	Blending of fully hydrogenated cottonseed oil with a partly hydrogenated base stock creates a shortening with a solid fat index that achieves an optimum plasticity for baked foods.
Margarines, icings and whipped toppings	Cottonseed oil is noted for its ability to form beta prime crystal, which helps promote the desired consistent texture and smooth, creamy appearance in shortenings and spreads.
Salad dressings	Excellent choice for salad dressings owing to its neutral flavour profile and the fact that it can be winterized

Table 92: Non conventional/industrial uses.

Cottonseed oil	Uses
As Brominated oils	Used as wetting agent in production of cloudy flavoured soft drinks.
Fat for Intravenous use	Emulsified cottonseed oil can be used as intravenous nutrition for patients requiring high calorie diet
As Tallow substitute	Cottonseed oil could be hydrogenated to tallow like product for soap making and textile sizing.
As Emulsifier in agricultural pesticides	Cottonseed oil is ideal for low volume and ultra low volume spray technology as it helps in increased chemical adherence

Cotton Seed Meal and Peptones

Cottonseed meal is the end product of processing of cottonseeds. It is obtained through the scientific process of deliniting, dehulling and oil extraction through expeller and solvent extraction systems. Cottonseed meal contains negligible oil but very high protein content of about 40-42 per cent. The cottonseed meal composition is given in Table 93 and the nutritive value of cottonseed meal is given in Table 94.

Table 93: Chemical composition of cottonseed meal (Solvent Extracted).

Species	Nitrogen (%)	Gossypol (%)	Ash (%)	Carbo-hydrate(%)
G. arboreum	8.5	1.55	6.05	11.
G. herbaceum	8.0	2.02	8.5	15.
G. hirsutum	8.8	2.0	7.3	12.8
G. barbadense	8.2	2.2	8.7	13.3
Hybrids	8.9	2.25	8.6	11.75

Source: Vigneswaran, *et al.*, 2007.

The nitrogen content of the cake ranges from 8-9% indicating 50-55% protein in the solvent extracted cottonseed meal. The carbohydrate content ranges from 11-15% with an ash content of 6-9%. Hybrids have higher nitrogen (protein) and ash contents while *G. herbaceum* meal contains the highest sugars. The composition reveals the usefulness of cottonseed meal as a good animal feed. Cotton seed meals from Bangladesh have 25-26% protein, 4.5% ash and 12% fibre (Uddin, 2005).

Decorticated cottonseed meal is a rich source of protein which increases ration efficiency and supplies highly digestible protein for tissue development, healthful activity of feed digesting rumen microorganisms and adequate production of hormones and enzymes. It stimulates appetite and can be used in unrestricted amount without fear of digestive disturbances or other adverse effects. Cottonseed meal with gossypol content less than 0.04 per cent and acceptable levels of aflatoxin can be fed to animals without any problem. The best practical approach for detoxification is to treat the meal with calcium hydroxide and ferrous sulfate.

Table 94: Composition of different grades of cottonseed meal (%).

Commodity	Crude Protein	Ether extract	Crude fibre	Digestible protein	TDM	Ash	Calcium	Phosphorus
Cottonseed	18.0	18.50	25.0	12.5	66.0	4.3	0.24	0.58
Undecorticated Cottonseed meal	21.0	7.50	29.0	15.8	72.0	6.5	0.32	0.56
Decorticated Cottonseed meal	39.0	1.0	14.8	33.2	73.4	8.1	0.49	1.2

Peptones are partially hydrolysed proteins obtained on controlled hydrolysis of high molecular weight proteins by acid, alkali or enzymes. Generally peptones prepared from meat, soybean and groundnut are used in microbiological work. Peptones can also be prepared from several other sources namely, fish, casein, gelatin, keratin, sunflower, etc. Casein, a milk protein also yields a good quality peptone on either acid or enzyme hydrolysis. Cottonseed meal has certain advantages over other protein sources. Cottonseed protein has good profile of amino acids. Cottonseed meal is cheap and available in abundance and has higher protein, similar to other oilseed cakes.

Gossypol

Gossypol is a polyphenol $C_{30} H_{30} O_8$ derived from the cottonseed plant. It decomposes at about 177-182°C. It is a yellow pigment. Gossypol is a polyphenolic aldehyde that permeates cells and acts as an inhibitor for several dehydraogenase enzymes. It reversibly inhibits calcineurin to calmodulin. It also inhibits replication of the HIV-1 virus. It is an effective protein kinase C inhibitor. It appears to be involved in low fertility in males. Thus gossypol appears to have a potential as a contraceptive.

The gossypol content of Indian genotypes range from 0.56-1.76%. *Arboreum* group contains the lowest while the highest is in hybrids (Varghese *et al.*,2004). The different species have been analysed to contain 1.6 to 2.25% gossypol. The cultivated hybrids have the highest level of 2.25% while the seed meal of different specices have 1.6-2.2% (Vigneswaran *et al.*, 2007).

Gossypol, the yellow pigment is present in cottonseed in "free" and "bound" (with protein) forms, It is the "free" gossypol that is considered as toxic, specially for non-ruminant animals like poultry, swine, etc. To prepare edible flours for human use, the free gossypol of cottonseed kernels and or meal should be quite low. Cooking the cotton seed reduces the free gossypol considerably. Expeller cakes generally contain low free gossypol. Free gossypol is considered non-toxic to ruminant animals.

Gossypol is a highly reactive compound and has unfavourable physiological effects on non-ruminant animals like swine and poultry. During seed processing, part of gossypol goes into the oil and the rest is retained in the meal. On storage of oil, gossypol degrades into dark coloured bodies, which are difficult to remove. Food and animal feed industries must manage the gossypol levels to avoid toxicity. Some ruminant microflora can digest gossypol, but only to a certain level. Gossypol present in the fresh raw oil can be easily removed with alkali. Cottonseed oil after refining is mostly free from toxic constituents. Genetically engineered cotton plants that contain very little gossypol in the seed, but still contain the compound in the stems and leaves have been developed. This provides protection against pests and diseases, while allowing the seed to be used for oil and meal for human consumption.

Coconut

Coconut is a perennial tree that grows in the tropical regions. Philippines, Indonesia, India, and Srilanka are the major producers of coconut. The tree grows under a rainfall ranging from 130-230 cm with a mean annual temperature of 27^0C and a very good sunlight with a sunshine of 2000 hrs, at least 120hr/month. Coconut is widespread in the coastal regions of the tropics, mostly in between 20^0 N and 20^0 S latitude. India accounts for about 20% of world coconut production.

The centre of origin of coconut is not very clear. Different theories suggest it to be different places like tropical America or central America or South Asia or Pacific. Coconut belongs to the family *Palmae*. The fruit weighs more than 1kg. It is ovoid in shape with green or yellow colour. The nut has smooth epidermis over a fibrous mesocarp (husk). The actual coconut lies inside with a hard endocarp (shell). Inside it there is a 1-2cm thick endosperm (kernel or meat) usually called *copra,* which is edible and a source of oil. A thin brown layer known as testa lies between the kernel and shell. Inside the *copra,* water upto 300ml may be present.

Coconut tree enjoys a unique position as a source of food, oil, beverage, fibre, fuel, shelter and timber. About, 60% of coconuts produced are consumed raw as direct food or in preparation of food products. The rest namely 40% is utilized for oil extraction. Coconut is essentially an oil crop though every part of the palm is useful to mankind in some way or the other. The oil is extracted by crushing the *copra* in rotary mills, expellers and hydraulic presses. Coconut oil has high saponification value, a low iodine number, and is used extensively for edible and industrial purposes.

The *Copra*

The meat of the ripe coconut is used in cooking, preparation of sweetmeats and for house hold oil production. The milk or cream obtained from grating of the meat is used in the preparation of delicacies and also as a substitute for cow's milk. Coconut syrup and coconut honey, which is an excellent substitute for rural honey, are obtained from coconut milk. Extraction of edible quality protein, oil and flour from fresh coconut meats are feasible through different processes. Desiccated coconut, which is the dried, shredded coconut meat, is used in the confectionery and food industries. *Copra* should be dried to 5% moisture for safe storage.

Coconut Water

The liquid or coconut water in the tender nut, is a refreshing drink, is often recommended as a medicine in cases of gastroenteritis where it is considered as substitute for saline glucose. Coconut water is also the basic ingredient of *Nato De Coco*, a delicious Philippino food, as well as coconut vinegar, coconut sauce and coconut lemonade. Industrially coconut water can also be used as a rubber coagulant in the production of high quality rubber. Sweet toddy, obtained by tapping the unopened inflorescence of the palm, is an excellent beverage. Palm sugar or jaggery can be obtained by boiling and evaporating fresh toddy. An alcoholic drink, arrack is obtained by distillation of fermented toddy. On storage, the continuing fermentation converts the alcohol into acetic acid or vinegar.

Table 95: Compostion of coconut kernel.

Constituent	Per cent
Moistrue	50
Oil	34
Ash	2.2
Fibre	3.0
Protein	3.5
Sugars	7.3

Source: Banzon and Velasco (1982).

Extraction Methods and Quality of Oil

Coconut oil is extracted by dry process and wet process. Dry process is more popular. The wet process of fresh coconut represents both the oldest and most modern method of recovering coconut oil. Coconut proteins obtained by wet process are deliciously well suited for human consumption. The yield of coconut oil from coconuts and copra depends on the cultivar, method of oil extraction, oil content of the *copra* crushed and the efficiency of the milling equipment employed and the quality of the coconut oil depending on the initial quality of the *copra* used for extraction and also on the processing methods. The quality of coconut oil is judged by its colour, flavour , clarity and appearance.

Copra (5% moisture) is disintegrated and cooked at 120^0 C. The oil is extracted in an expeller. The filtered oil is collected in stainless steel vessels. The extracted oil will have 0.2% moisture and the natural aroma / roasted flavour of *copra*. It has a very good keeping quality.

Physico-Chemical Properties

The cultivar, season, method of oil extraction and refining influence the physico-chemical properties of coconut oil. The oil content ranges from 34 to 45 per cent in the ripe endosperm of coconut and from 60 to77 per cent in well dried *copra*. The kernel contains 4-4.5% proteins. The oil content is influenced by the water content in the nut, stage of maturity and the type of coconut variety (FAO, 1984). Maximum oil and protein content is found near the testa with decreasing levels towards inner cavity. The per cent total soluble sugars, reducing sugars, water and minerals are maximum in the inner portion of the endosperm and decrease towards the testa portion. The fiber is almost evenly distributed throughout the endosperm. The testa is rich in fiber content. Aging affects the oil content showing decreasing trend as age increases.

The quality of the oil in immature nuts deteriorates to a marked extent on storage due to increase in acid value. There is only a slight increase in the acid value of the oil extracted from the copra of matured nuts. The pleasant taste, mild and sweet aroma of raw coconut oil is due to the presence of minute quantity (about 150 ppm) of δ- and γ- lactones resulting from oxidation. In addition, twenty different pyrazines, furans, esters, alcohols and pyrrols also contribute greatly to the nutty flavour of the roasted coconut. Heat treatment of distintegrated *copra* at 160^{0}C for 15 min gives intensive roasted aroma and a better keeping quality to coconut oil. The effect is ascribed to the antioxidant activity of the products of non-enzymatic browning due to heating rather than the change in the moisture content of the oil. It is consumed in its natural form in many countries. There is an increasing demand for the colourless, bland, refined coconut oil.

Table 96: Characteristic of coconut oil.

Characteristic	Range
Specific gravity at 5.5°C	0.869 to 0.874
at 15.5°C	0.917 to 0.919
Refractive index at 40°C	1.448 to 1.450
Iodine value	7.5 to 10.5
Saponifiacation value	250 to 264
Unsaponificable matter, %	0.5 max
Titer, °C	20.0 to 24.0
Setting point, °C	21.8 to 23.0
Melting point, °C	23.0 to 26.0

Source: Sonntag, (1979).

The specific gravity and I.V. are lower while S.V. and M.P. are higher for coconut oil compared to other vegetable oils. This is mainly because of the saturated fatty acids, mainly, lauric acid, in coconut oil.

High grade coconut oil is low in pigments, totally free from phosphatides, gums and other non-glyceride substances commonly present in oils from seeds of annual plants. It has the lowest turbidity, the lowest iodine value and the highest saponification value among the vegetable oils. The percentage of non-saponifiable matter is less than 0.5 per cent consisting of about one-third sterols, some tocopherols and squalene. The changes in the crystallization of coconut oil on refining and bleaching is due to the removal of free fatty acids and other minor components such as natural phospho-lipids in crude coconut oil.

Fatty Acid Composition

The fatty acid composition of coconut oil from progressively maturing bunches shows a large variety of fatty acids which differ mainly in molecular weight rather than in degree of unsaturation. About 90 percent of fatty acids of coconut oil are saturated. They consist of C_{12}, C_{14} and C_{16} acids with C_{12} predominating. The stage of maturity at harvest is known to influence the fatty acid composition of coconut oil. It is extremely resistant to the development of rancidity. The crude coconut oil is relatively stable but major reductions in oxidative stability occur during the processes viz., degumming, refining, bleaching and deodouration. Addition of citric acid at the end of deodourisation process when the oil has cooled to 100^0C is effective in regaining considerable oxidative stability in the oil.

Table 97: Fatty acid composition of coconut oil.

Fatty acids	%	Composition
6:0	0.4	-0.6
8:0	6.9	-9.4
10:0	6.2	-7.4
12:0	45.9	-50.3
14:0	16.8	-19.2
16:0	7.7	-9.7
18:0	2.3	-3.2
18:1	5.4	-7.4
18:2	1.3	-2.1

Coconut oil also contains C_6 to C_{10} saturated fatty acids, not commonly found in other vegetable oil. Due to the higher proportion of C_6 to C_{10} lower chain saturated fatty acids coconut oil yields the highest amount of glycerol upon processing than any other common vegetable oils and maximum glycerine can be recovered by high pressure splitting of coconut oil or by converting the oil into methyl esters which can be fractioned at low temperature and transported easily compared to fatty acids which are corrosive (Bhat, 1988). Coconut oil can yield upto 14% glycerol on hydrolysis as against 8-10 % by other vegetable oils.

Coconut oil on storage is liable to a type of rancidity known as ketone or perfume rancidity. This is due to β-oxidation of fatty acids by bacteria or fungi. The rawe oil can be storted for upto one year without any deterioration in quality.

Application

Coconut oil finds extensive use in the food industry. Due to its high content of saturated fatty acids, it is resistant to oxidative rancidity and retains a pleasing flavour. It is therefore used as a fat source in the preparation of filled milk, infant milk powder and other bakery products. The short chain fatty acids can be converted into value added products like artificial fruit flavours, useful chemical intermediates and plasticizers. Coconut oil is partially solidified and filtered to get a solid hard fraction with a melting point of 32^{0}C called coconut stearin, which is used in confectioneries. The high levels of low molecular weight saturated fatty acids in the oil impart to it a melting point of 24-27^{0}C.

Coconut oil is used in the margarine, biscuit and cooking oil industries, and also in the production of toilet articles such as hair oil, shaving creams, shampoos and cosmetics. The major industrial use of coconut oil is in soap manufacturing by saponification. The resulting glycerol is used mainly in the pharmaceutical industry and in nitroglycerine manufacture for explosives. Coconut oil is also used in the production of plasticizers, the manufacture of safety glass, and synthetic resins as well as rubber substitutes and detergents. The oil can also be used as a lubricant and fuel. Production of petrol, heavy oil and kerosene oil substitutes is possible from coconut oil. The cake obtained after oil extraction can be used as fertilizer for field crops or fed to cattle, poultry or young growing swine.

Clinical and Pharmaceutical Uses

Coconut oil is a good and ready source of energy. It is used widely in hospitals for patients with pancreatic lipase deficiency (malabsorption syndrome), obesity, post-surgical patients and premature infants. Since coconut oil is rich in medium chain fatty acids, they are easily absorbed by the body. Medium chain triglycerides are utilized in infant formulations.

The high contents of medium chain triglycerides in coconut oil are beneficial in lowering serum and liver cholesterol levels. The presence of high amounts of short chain fatty acid content in coconut oil refutes the claim on the bad cholesterol present in coconut oil. The bad cholesterol is attributed to the high molecular weight saturated fatty acids (C_{16} and C_{18}) which have been found to be only about 10 per cent in coconut oil.

Coconut oil is used in pharmacy as an ointment base, on account of its ready penetration into the skin and appreciable water absorbing property. Specially prepared coconut oil is used as a good eye lotion. A mixture of coconut oil and lime water is used to treat burns.

Industrial Uses

The highest saponification value of coconut oil coupled with the lowest iodine value gives it a very high hardness number. Therefore, the soap made with it, is extremely hard. Coconut oil gives the soap a body and stability. Coconut oil soaps are remarkable for their excellent lathering qualities due to the relatively high proportion of low and medium chain fatty acids. These soaps lather freely in hard and saline waters and hence coconut oil is used widely in the manufacture of marine soaps. Coconut oil can be easily saponified in cold conditions due to the presence of lower molecular weight fatty acids. Hence, it is widely used for the manufacture of cold pressed soaps.

The soaps made with oconut oil are capable of taking filling materials (fillers) like silicates, soda ash and earth etc. The oil blends well with other oils and fats to give soaps of varying degrees of hardness. Detergent industry depends mainly on coconut and palm kernel oils for lauric acid. Coconut oil is a potential fatty raw material in modern chemical industry due to it being biodegradable, its high lauric acid content and other properties. The amides of coconut oil fatty acids with ethylene oxide and pyridine hydrochloride give products which are used as demulsifiers in petroleum industry.

A new fuel, cocohol, produced by a chemical combination of ethyl alcohol and coconut is useful to operate the high speed diesel engines at a greater thermal efficiency than diesel oil. Coconut oil itself can substitute diesel fuel but suffers drawbacks, particularly, the production of toxic acrolein (CH_2=CH-CHO) due to incomplete combustion. To get rid of this toxic chemical, one can convert coconut oil into ethyl and methyl esters which produce glycerol, a by-product of high cash value. The heat of combustion of ethylester of C_{12} acid approaches that of gasoline (petrol).

Coconut oil serves as a raw material for many oleochemicals. It is the richest source of glycerol among vegetable oils. Fatty acids ranging from 6:0 to 18:0 can be obtained, the major one being lauric acid. Fatty acids can be converted to fatty alcohols, fatty amines etc. through various chemical reactions. Metallic salts of fatty acids serve as lubricants. Low molecular weight fatty acids and their esters are utilized by the cosmetic industry. Polyesters are useful as agricultural pesticides. Fatty amines and amides are utilized by the textile and plastics industry. Medium chain triglycerids are speciality dietary food items. Some products are utilized by the pharmaceutical industry also.

Disadvantages of Coconut Oil

1. Higher oil absorption by the food during frying
2. Refined coconut oil gets hydrolysed 2-10 times more rapidly than normal oils and they possess soapy off flavour. Upon hydrolysis, the released short chain fatty acids often cause undesirable flavours.
3. Higher susceptibility of triglycerides to hydrolysis that cause rapid development of free fatty acids in the oil and simultaneous rapid decrease in the smoke point, as compared to other higher molecular weight fatty acid oils.
4. The oil extracted from *copra* contaminated with sulphurous materials may impart a rubbery flavour to the oil upon exposure to sunlight. A

brief contact of the oil with N catalyst and H_2 gas before refining can eliminate all these undesirable flavours.

5. Admixture of coconut oil with other oils produces foaming. Addition of 1-2 ppm of antifoaming agents to this mixed oil or blend for frying improves its stability significantly.

Coconut Shells and Stem

The hard shells of coconut are used in the manufacture of shell charcoal, activated carbon and shell flour. Shell charcoal is used as a refining agent, both as a deodourizer and as a decolourizer. Activated carbon is used for purifying, refining and bleaching of volatile oils and chemical solutions and also as an adsorbent of gases. Coconut shell flour is used extensively as a compound filler for synthetic resin glues and as a filler in thermoplastics. It gives a smooth and lustrous finish to moulded articles and improves their resistance of moisture and heat. The fibrous husk of coconut yields a fiber known commercially as coir which is used in making mats, matting, ropes, cordages, nets and bags as well as hardboard and rubberized coir products. Coir dust is used as an insulating material as it has a low thermal conductivity.

The stem of the palm, after immersion in saline water which increases it's durability, is used for the construction of houses and cattle sheds and in the manufacture of particle boad. The plaited coconut leaves are used for thatched houses, fencing and for making baskets. Thampan, (1981) has described in detail the food and commercial products that can be obtained from the palm.

Olive

Olive plant is native to Asia Minor. It belongs to the *olea* family and genus *oleaceae*. It is an ever green tree thriving in temperate climates (mainly Mediterranean region of Europe). Fruit yield ranges around 12,500kg/ha. The ripe fruit contains 35-70% oil on dry basis. World production of olives is around 8-9m tonnes of which 0.6m tonnes are table olives. Around 1.6m tonnes of olive oil is produced annually. Olive oil though costly has a good name as heart friendly oil. Many plant breeding problems work against its wider cultivation.

Green Olives

All the industrial processes involved in preserving green olives are generally divided into three main processing stages or steps; a) soaking the olives in a caustic soda solution to remove their bitterness, b) washing the olives and c) preserving them in brine, with or without fermentation. In the preparation of Mediterranean black olives, the first stage i.e, the removal of bitterness, is omitted. However, black olives like green olives can be preserved in brine as well as a variety of other ways such as in salt, vinegar, oil, etc. Edible olive preparation is carried out to suit mostly the regional preferences.

Olive fruits are harvested when they turn green-purple to black. Over ripe fruits have lower oil content. Olive fruit contains 50% water, 1.5% protein, 22% oil, 19% carbohydrates, 6.0% cellulose and 1.5% minerals. The oil is concentrated in the pericarp.

Olive Oil

Oil from the olive fruit has been extracted for more than 6000 years by crushing the olives, followed by kneading and pressing of the crushed material. The crushed material is then allowed to settle and the oil is decanted. Today, better techniques and modern equipment make it possible to extract quality oil at a much lower cost. The early extraction methods used millstones and standard presses. These have now given way to mechanical crushers and kneaders as well as to super press equipped with centrifuges. Oil yield between 15 to 35% is normally obtained and the amount depends upon the cultivar harvested, the degree of fruit ripeness and the extraction method employed.

Olive oils are marketed under different names. Virgin olive oil, is one type, which can be extra-fine or semi-fine according to the acidity level of the oil. The virgin olive oil is a natural oil which is obtained using mechanical processes. It is not mixed with oils of a different type or olive oil obtained by any other means. Pure olive oil is a mixture of virgin oil and olive oil obtained by refining. When virgin oils have high acidity levels, they are refined and mixed with other grade olive oils. The quality of olive oil depends on the specific processing technique.

Properties of Olive Oil

Olive oil usually has a greenish yellow colour and a characteristic olive flavour and odour. Food grade of olive oils are consumed without the usual edible oil processing after extraction. The high levels of free fatty acid in olive oil may be attributed to the high moisture content in olive fruits, which is favourable to enzyme action. Another cause for high free fatty acid content may be the bruising of the olive fruits at harvest. Olive oil remains liquid even at freezing temperatures due to it's low cloud point. The unsaponifiables in the oil are low. It's saponification value is around 190. Iodine value is low indicating it's non-drying nature and lower unsdaturation. The oil also has very good sensory and taste properties. It's physical properties are mostly similar to that of groundnut oil.

Table 98: Characteristics of olive oil.

Characteristic	Value
Specific gravity at 25/25°C	0.909 to 0.915
Refractive index at 20°C	1.4680 to 1.4705
Saponification value	188 to 196
Unsaponifiable matter, %	1.8 max
Iodine value	80.0 to 88.0
Titer, °C	17.0 to 26.0
Melting point °C	0
Cloud point °C	-5.6
Pour point, °C	-10.0

Source: Sonntag, (1979)

Fatty Acids of Olive Oil

The major fatty acid is oleic acid followed by linoleic, palmitic and stearic acids. Traces of myristic, palmitoleic, linolenic, arachidic, behenic, lignoceric and eicosenoic acids are also present. Oleic acid upto 80% is present in the oil with 20% of other fatty acids. The proportions of oleic and linoleic unsaturated fatty acids and palmitic saturated fatty acids vary, depending upon the origin of the olive oil. Like most other vegetable oils, olive oil develops more unsaturated fatty acids in cold climates and with advanced maturity of the fruit. The glyceride composition of olive oil is as follows: triolein, 40-60%, linoleodiolein, and saturated diolein 12-20% each, and saturated, oleo linolein, 7% (Gunstone, 2005).

The olive oil is a good source of vitamin E (100mg/kg oils) and polyphenols (300 mg/kg oil). The major constituent of tocopherols in olive oil is α tocopherol (88%). The phenolics present are hydroxy tyrosine, tyrosol, vanillic acid, caffeic acid and lignans. The unsaponifiables range from 0.5 to 1.5% out of which 40% is constituted by squalene. The actual level of squalene ranges from 150-170mg/ 100 ml oil. This is against a range of 5-50 mg/ 100ml in other oils (Gunstone, 2005).

Table 99: Fatty acid composition of olive oil.

Fatty acid		**% Composition**	
		Mean	**Ranges**
Myristic	14:0		0.1 to 1.2
Palmitic	16:0	9.0	7.0 to 16.0
Palmitoleic	16:1	0.6	-
Stearic	18:0	2.7	1.0 to 3.0
Oleic	18:1	80.3	65.0 to 85.0
Linoleic	18:2	6.3	4.0 to 15.0
Linolenic	18:3	0.7	-
Arachidic	20:0	0.4	0.1 to 0.3

Sorce: Sonntag, (1979).

Olive oil is widely used as a table and cooking oil due to its characteristic flavour and stability. Only 10 to 18% of the olive oil fatty acids are saturated, but the majority of the unsaturated fatty acids are monounsaturated. The high oleic and low linoleic fatty acid content help make olive oil more resistant to oxidation than most liquid oils. Additionally, virgin olive oil has low tocopherol levels, but the chlorophyll's degrade into pheophytins for oxidative stability. The polyphenols are usually removed from the liquid oils during processing (Sonntag, 1979).

Light causes significant deterioration in olive oil quality in the presence of air. The oil develops an off-flavor from oxidation and becomes colourless due to the loss of chlorophyll and carotene. Packaging studies have shown that glass provides better protection than polyethylene plastic containers. A storage life in excess of 2 years has been demonstrated by olive oil evaluations of hermetically closed glass bottles stored in the dark (Kiritsakkis,1984).

Benefits of Olive Oil

Olive oil is highly stable due to higher levels of oleic acid and vitamin E that are present in the unsaponifiables. The oleic acid has a role in reducing cholesterol and increasing the good cholesterol namely HDL cholesterol. Olive oil also lowers triglycerides and low-density-lipoprotein (LDL) cholesterol- the "bad one". Hence olive oil is considered heart friendly. The Mediterranean people who consume olive oil have low incidence of heart attacks. This is because of high level of fruits, vegetable grains, legumes and low meat in addition to the heart friendly olive oil.

The important antioxidant functions of the minor components of olive oil have major implications for the prevention of arteriosclerosis and its chief complication, namely, the heart attack. Fatty and non-fatty components, have been demonstrated experimentally in animal models and in human cell lines to have cancer prevention properties.

Olive oil enriched diets also have beneficial effects in treating insulin resistance, obesity and in lowering blood pressure. Virgin olive oil, in the context of Mediterranean diet, prevents age-related congnitive decline and dementia.

Neem (*Azadirachta indica*)

The margosa or neem tree is native to India. Neem is known as "*aristha*" in *Sanskrit* meaning "sickness reliever". The tree belongs to *Meliacea*. Neem is a large evergreen glabrous tree. It is tolerant to soil salinity, alkalinity, and acidity and hence can grow well in all types of soils. It can even grow under drought conditions. Neem plantations help in arresting desertification. Neem trees serve as wind breaks. They are grown as avenue trees and provide shelter and shade to humans and animals. The tree purifies the air by releasing large amounts of oxygen. Every part of the tree is useful. Its twigs, leaves, timber, seed, seed oil, and cake have widespread uses at home, in agriculture, in industry and in medicine. (Axtell,1992). Neem is sacred in India, as it is health giving with medicinal and insecticidal properties.

Neem is reproduced by seed. Neem tree can grow up to a height of 40-60 feet, with a girth of 6-8 feet. It has a straight trunk with spreading branches. The wood is relatively heavy with specific gravity varying from 0.56-0.85. The tree starts flowering from 6th season onwards. The flowering season is from January to April. The fruits ripen during June-August which is the monsoon season in India. The green fruit becomes yellow on ripening. The overipe fruits fall down on their own. They can also be harvested by beating with sticks.

The fruits contain a small amount of pulp and a seed with brittle testa. The ovoid neem fruit weighs about 0.5g with a size of 2 cm x 1 cm. The composition of neem fruit is 51% pulp and skin, 40% seed and 17% kernel (25-28% of dry fruit). The fruit yield per tree ranges form 30-60kg. The kernel constitutes 45% of the seed. The seed also contains 5.5% bitters. The sun dried seed with 5-8% moisture can be stored in gunny bags, safely for a period ranging from 3-8 months.

Neem Seeds

The seeds are off white to cream in colour. The brittle testa can easily be removed to obtain a greenish brown kernel. The kernel contains 40-45% oil. It's protein content is 8-15%, carbohydrates 16-35%, crude fibre 5-16% and ash 3-12%. Depulping and decorticating are prerequisites for oil extraction of neem kernels. Oil from neem kernels are obtained using *ghanies* or oil expeller. Yield of oil is 30-40% from kernels. The cake retains around 10% oil which can be recovered through solvent extraction.

Forty kg of fresh fruit can yield 24kg of dry fruit (60%), 1.10kg seed coat (4.5%), 6.0kg husk (20%) and 5.48kg kernel (23%). The 5.48 kg kernel can yield 2.48kg neem oil (45%) and 3.02 kg neem cake (55%) (Ketkar,2004). Neem oil contains non-lipid components and sulfur containing compounds which impart bitterness and pungent disagreeable smell to the oil.

The Neem Oil

The oil is greenish yellow to brown with a strong disagreeable garlic like odour and an acrid bitter taste. The bitter taste is due to the presence of sulphur containing compounds like nimbin, nimbidin and nimbosterol. Cold pressed oil is more odoriferous than refined oil. Better grade oil without pungency is obtained from kernels by initial extraction with alcohol followed by hexane extraction. Such extraction is also useful in obtaining better grade cake. The oil with a specific gravity of 0.92 has a saponification value of 193 and iodine value of 75. It is a non-drying oil. It's unsaponifiables are slightly higher at around 4%.

Table 100: Properties of neem oil.

Property	Value
Specific gravity	0.9189
Saponification value	193
Iodine value	75
Acid value	4-6
FFA%	0.2
Refractive index 30^{0}C	1.4612
Unsaponifiables%	3.8
Tocopherols	1.17mg/g

The oil mainly contains oleic acid (45%) followed by stearic(20%), palmitic (20%) and linoleic (15%) acids with around 1% of arachidic acid. The saturated fatty acids account for about 42% while the remaining is accounted for by the unsaturated fatty acids. Due to low level of PUFA(18:2), the crude and refined oils are normally stable and do not become rancid quickly.

Table 101: Fatty acid composition of neem oil.

	Crude oil	Refined oil
16:0	19.4	19.4
18:0	21.2	18.3
18:1	42.1	45.7
18:2	14.9	14.4
20:0	1.4	1.4

The glyceride composition of neem oil is as follows: palmito-distearin, 0.2; oleo-palmito-stearin,20.3; oleo-distearin, 1.6; palmito-oleo-linolein, 6.6; palmito-diolein, 26.3; stearo-oleo-linolein, 3.6; stearo-diolein, 24.9; and linoleo-diolein, 16.5%.

To obtain neem oil and deoiled meal free from bitters and odoriferous compounds, a three-stage counter-current pretreatment of neem seed kernels, with aqueous ethanol (water : ethanol 15 : 85 v/v) has been employed. It has been found effective and has resulted in the oil and meal with almost negligible quantities of bitters and odoriferous compounds (Vasishtha *et al.*, 1992). Such a debitterized oil has been found to be nutritionally as well as chemically safe like any other edible oil(Rukmini,1987).

Isolation of Active Ingredients

By treating the oil with dilute sulphuric acid we get about 2-4% of a sticky gum which contains the active ingredients. They can be precipitated by milder treatments (namely water, buffer or citric acid). Bitter triterpenoids can be extracted directly from oil or hexane solution of oil into alcohol. Kernels also can be directly extracted with 95% alcohol. By following various steps one can obtain better quality cake suitable as animal feed and a pesticide concentrate (3% yield) containing upto 8% azadirachtin.

Bitter Principles

Medicinal, pest control and other properties of neem have been known to Indians from ancient times. All parts of the tree are useful and offer great potential for agricultural, industrial and medicinal exploitation. The most valuable part of the neem tree is seed. The crude oil obtained from the seed contains a large number of non-lipid constituents which exhibit exceptionally powerful anti-feedant, pesticidal and medicinal properties. Neem oil is unique in containing terpenoids which are loosely termed as "bitters" and some organic sulphur compounds that impart a pungent and rather disagreeable odour. Because of bitter taste, dark colour and unplesant odour neem oil is largely being used for making soaps and a few pharmaceutical preparations.

Nine limonoids have been isolated from neem. The important ones are azadirachtin, salanin, meliantriol and nimbin. Azadirachtin is the first active ingredient isolated from neem. It is present between 2-4 mg/g kernel. The highest reported has been 9 mg/g kernel. It does not kill insects, but repels, disrupts growth and reproduction. It works as feeding deterrent and or growth regulator of insects and nematodes. Meliantriol and salanin also act as anti-feedants. Nimbin and nimbidin (2% of kernel) have antiviral properties. Other active constituents of neem paralyze the swallowing mechanism.

The bitter principles of the oil (2%) can be extracted with alcohol. Nimbidin (yield, 1.2-1.6%), the main constituent, is highly bitter and contains sulphur. On hydrolysis it gives neutral nimbidinin and nimbidic acid. Along with nimbidin, two more bitter constituents, free from sulphur, namely, nimbin(0.1%) and nimbinin (0.01%) are also present. Other minor constituents, namely, desacetylgedunin, salannin, azadirone, epoxyazadiradione and vepinin ($C_{28}H_{36}O_5$; yield, 0.15%) have also been identified in the oil. The oil also contains nimbinene, 6-0-acetylnimbandiol ($C_{28}H_{32}O_7$, mp 121^0), and 1,3-diacetylvilasinin ($C_{30}H_{40}O_7$, mp 157^0).

Nimbidin has been found toxic at 0.25 mg/g body weight of frogs. Nimbidin has been observed to be toxic to many common fungi and nematodes. It is highly useful as an external application for the skin disorders. Nimbin is antipyretic and non-irritant. Sodium nimbidinate, obtained from nimbidin, has potent diuretic and anti-inflammatory properties. It is highly active when injected. At larger doses, it can serve as an oral drug. When administered intramuscularly in congestive cardiac failure with anasarca, good response has been observed. It produces uterine contraction. It has also been observed to have a role in reducing blood pressure in experimental animals. Nimbidin and sodium nimbidinate kill *Paramoecium caudatum* in 1/500 dilution. Sodium nimbidinate, being water-soluble, is more active than nimbidin. Sodium nimbidinate and sodium nimbinate are reported to have spermicidal activity, indicating a useful role in birth control.

Nimbidol a fat-soluble, bitter and odorous fraction, obtained from the mother-liquor of the bitter principles, is a mixture of nimbidin, fatty impurities and the sulphur-containing constituents. Laboratory trials on rats have shown that the oil and nimbidol at a dose of 8 mg/kg body weight has anti-arthritic properties.

Azadirachtin

Neem seed, kernels and oil are good sources of active bitter principles, including azadirachtin which vary widely with trees from different geographical regions, soil, climate, seed maturity, length of storage and the method of processing of oil. Seeds from Nicaragua, Indonesia and Togo contain higher percentage of azadirachtin in kernel extracts as revealed by high performance liquid chromatography (HPLC) analysis (Srimannarayana, 1995).

Generally neem seeds from arid and dry regions are charecterised by high azadirachtin content while trees from coastal regions contain lower percentage. Indian neem seed samples have been observed to contain a maximum of 0.6 to 0.7% azadirachtin with a range of 0.2 to 0.5%. Azadirachtin decomposes rapidly under sunlight. The azadirachtin contents are lower if the extracts or seeds are kept in humid condition.

Table 102: Azadirachtin content(%) of neem from different localities.

Content	Azadirachtin%	Content	Azadirachtin %
Nicaragua	4 .79	Mauritius	3.29
Togo	3.87	Niger	1.53
India	3.50	Indonesia	4.75
Sudan	1.90	Myanmar	3.92
Maharashtra (India)	4.50	Andhra Pradesh (India)	4.40
Karnataka (India)	5.70		

Although more than 200 products have been isolated from neem, azadirachtin is the most active principle. The related meliacins notably nimbin, epinimbin, salannin and meliantriol also exert physiological activity although feeble in nature. The bioactivity of neem products can control more than 300 insect species, besides many plant nemadotes and pathogens. Neem products exert behavioral, physiological and biological effects on insects. Neem products are ideal in integrated pest management since they leave no residual toxicity and are environmentally friendly. The products are nontoxic to non-target organisms like beneficial predators, parasites, mammals, spiders and pollinators. One disadvantage with neem derivatives is that they are highly degradable and last only for a short while under field conditions when compared to conventional insecticides. Neem derivatives are better for prophylactic proposes. Cold pressed oil contains 1700 ppm azadirachtin.

Neem Cake

The cake left after the extraction of oil is used as animal feed, as organic manure, insecticide and nematicide. Though bitter in taste, cattle get accustomed to it on regular feeding. The fresh cake smells bad, but the odour is lost on sun drying. The animals that feed on this cake are healthy, with normal milk yield. For palatability jaggery can be added to the cake. The keeping quality is good and it is not easily spoiled on storage nor is it attacked by fungi. The processed cake can be employed as a good poultry feed. The bitter principles in the cake act as vermicides.

The chemical composition of a sample of expeller cake is as follows (dry matter basis): moisture, 12; carbohydrates, 26; crude protein, 36; crude fibre, 9; fat, 9; minerals, 8; calcium, 0.8; phosphorus, 0.8 and potash, 1.48%. The method of oil extraction and the residual oil content greately influences the the composition and quality of the cakes.

Table 103: Composition of neem cake.

Constitutes	Content
Crude Protein	13-35%
Carbohydrates	26-50%
Crude fibre	8-26%
Fat	2-13%
Ash	5-18%
Acid insoluble ash	1-17%

The residual bitter substances and odoriferous principles in the defatted meal can be removed by extraction with ethanol or methanol. The meal on further processing and drying is a light straw-coloured granular powder of agreeable taste and flavour. It is suitable as feed for cattle and poultry. Composition of this meal is as follows: moisture, 10.7; nitrogen, 10.1;protein, 63.0; carbohydrates, 10.4; other organic matter, 8.6; ash, 7.4; P_2O_5, 4.9 and iron, 0.2%. Neem cake is rich in calcium, magnesium and phosphorus.

Table 104: Mineral composition of neem cake.

Constituent	mg/100 g
Phosphorus	89
Calcium	399
Magnesium	190
Iron	8.7
Zinc	8.0
Copper	2.4
Manganese	2.2

The amino acid make up of the meal is as follows: glycine, 9.4; alanine, 5.3; aspartic acid, 6.8; glutamic acid, 11.8; serine, 7.6; threonine, 5.3; valine-methionine, 4.5; nor-leucine-leucine-isoleucine, 8.0; arginine, 6.3; histidine, 3.4; lysine, 3.9; proline, 5.9; tyrosine, 3.4; phenylalanine-tryptophan, 5.3; cysteine, 3.1; glutamine, 1.6; & two unidentified ones, 3.9, 4.6%. The cake is rich in most of the essential amino acids except valine and tryptophan which are in negligible quantities. The arginine content is higher indicating would healing properties. Protein fibres have also been manufactured from the cake.

Neem Products and Agriculture

In agriculture, neem plays an important role in plant protection. The bitter principles of the neem seed kernels are highly inhibitory to insect feeding and development. The neem products have been found to be effective in controlling more than 2000 insect species. Basal application of neem cake and urea mixture at 1:5 ratio reduces many insects. Such blending also helps in slow release of nitrogen due to slower nitrification and reducing nitrogen losses. The crops have the double benefit of better nitrogen availability and less infestation.

The cake is mostly utilized as organic manure. It has higher N, P, K, Ca and Mg than farm yard manure. The cake also serves as insecticide and nematicide when used as manure. The doses of neem cake differ for each crop. The alcohol-extracted cake is a better nitrogenous manure than unextracted cake and by proper processing, manurial value can be improved. Application of the cake increases the yield of crops, in general. In conjunction with superphosphate, the yield is much higher. The cake is not only an organic nitrogenous manure, but is an inhibitor of nitrification. This property helps in efficient use of the nitrogenous fertilizers. When mixed with urea, neem cake reduces conversion of urea to nitrogen gas. Hence the release of nitrogen to crops becomes slow and fertilizer availability and efficiency to crops increase.

The cake gives protection from grain moth, lesser grain borer and red flour beetle. It reduces gall formation and also helps in the reduction of certain parasitic fungi. It also controls nematodes on the root tissue. The cake inhibits the egg-laying capacity of the female nematodes. The aqueous extract of the cake inhibits larval emergence of *Meloidogyne incognita, M. Javanica* and *Pratylenchus spp.* Black scurf of potato can be significantly reduced by adding neem cake to the soil. In combination with nemagone, the cake gives better results than either of them alone. The cake reduces the occurrence of nematodes in wheat and increases the yield considerably. When applied to sugarcane, potato, cotton and paddy there is a similar increase in yield. It protects crops from termites. Tobacco seedlings are protected from ground beetle when neem cake is applied at the rate of 50 kg/ha to the crop. Azadirachtin and possibly other constituents present possess systemic action, spreading to all parts of the plant to which the cake is applied

Urea coated with an extract obtained from the oil regulates the rate of nitrification and increases the yield of grain and straw of wheat. The pulp water, obtained during the depulping of fresh or dry fruits, can be used for manuring and for composting, and can increase crop yields. The pulp water contains; nitrogen, 1.2; and phosphoric acid, 0.27%. A paste made from the cake or the oil-emulsion, when applied to the trunk of orange or other Citrus spp., protects them from pests.

Neem oil (2-3%) controls rice brown plant hopper, leaf folder, gall midge and ear head bug. Neem products also show efficacy in control of fungal diseases, viral infections and nematodes. Neem products are non-toxic to beneficial organisms like predatory spiders and mirid bugs and hence are compatible with integrated pest management. Neem leaves also are incorporated into the soils to improve their physical structure like aggregate stability and bulk density, apart from improving fertility levels.

Uses of Other Neem Tree Parts

The shell from the seeds can be used for the production of activated carbon and tooth powder. It can be used as fuel and for manufacturing briquettes or hardboards. The powdered shells are used as fillers in thermoseting-moulding compositions. The flowers, generally, dried ones, are eaten either raw or in curries and soups, or as a fried dish in South India. They are useful in some cases of atonic dyspepsia and general debility. The flowers provide plenty of nectar. The air-dried blossoms contain β-sitosterol and its β-D-glucoside, kaempferol, thioamyl alcohol (7.6%), benzylalcohol (9.67%), benzylacetate (8.2%), an un-identified alcohol (3.9%), and an essential oil (0.025%). Two of the three isomeric sesquiterpenes isolated from the essential oil are azadirachtene and margosene. The component fatty acids present in the waxy material are: behenic, 0.7; arachidic, 0.7; stearic, 8.2; palmitic, 13.6; oleic, 65.3; and linoleic, 8.0%. A glycoside of myricetin (melicitrin) is also present.

Some trees, especially near the water courses exude a sap naturally from the stem-tip. Neem gum is useful for preparing colours in silk dyeing. It is also useful as a general purpose adhesive at rural household levels. The sap is considered refrigerant, nutrient and tonic. It is also useful in skin diseases, constipation, atonic dyspepsia and general debility. Sometimes, it is made into toddy. The fresh exudate has a strong smell of fermented liquor with the characteristic odour of the tree. It is slightly sweet and contains: total solids, 12.5%; reducing sugar, 0.5%; and total acid (as acetic acid), 0.8%. The amino acids present in the exudate are: arginine, 15.2; tyrosine, 65.0; tryptophan, 333.3; proline, 58.5; glutamic acid 38.8; serine, 17.0; and phenylalanine, 154.5 mg/100ml. Tryptophan content is fairly high. The pollen-grains of margosa cause pollinosis to some extent.

An average tree can produce 350 kgs of green leaves annually. Neem leaves contain upto 8% protein with 15% crude fibre, 12% ash, 5% lipids and 60% carbohydrates. The leaves and twigs serve as manure and also as cattle feed. Neem twigs are used as tooth brush in many parts of India. The leaves are utilized as antibacterial and antiviral. Its timber has uses in home and farm. The seed and its components namely cake and oil have uses in agriculture and medicine.

Neem Oil as a Suckericide on Tobacco

When the apical dominance is removed to improve the quality of tobacco, auxiliary shoots-known as suckers-develop. Neem oil emulsion at 10 per cent is an effective suckericide for tobacco (Bangarayya *et al.*, 1982). It is superior to manual desuckering, and other oil emulsions. The 10 per cent neem oil emulsion is more economical than other oil emulsions.

Medicinal Uses

Medicinal uses of neem have been known to Indians since ancient times. Neem products have been part of *ayurvedic* medicine. Recent studies have indicated many medicinal properties. Neem products act as anti-malarial, anti-diabetic, anti-inflammatory and anti-ulcer agents. They also possess contraceptive properties and have application in skin and dental care. Neem oil and neem extractions serve as antibiotics. They are also useful in veterinary medicine. The compound nimbidin has anti-ulcer potential besides acting as anti-inflammatory, analgesic, and anti arthritic agent. Normally anti-inflammatory agents are ulcerogenic, while nimbidin with anti-ulcer properties, can be a better medicinal compound. More research needs to be carried out to identify natural eco-friendly and side effect free medicines from neem.

The oil has many therapeutic uses. Medicinal properties of the oil are attributed to the presence of bitter principles and odorous compounds. The bitter principles are used in the pharmaceutical industry. The alcoholic extract, after removal of the bitter principles, is used as an agrochemical for the preparation of water-based pesticidal spray against agricultural and horticultural pests. The oil is a useful remedy in some chronic skin diseases and ulcers. It is a common external application for rheumatism, leprosy and sprain. Intrauterine medication of oil controls different types of metritis. The warm oil relieves ear trouble. A few drops of oil taken in betel-leaf provides relief in asthma. The oil is reported to have anti-fertility properties. It possesses antiseptic and anti-fungal activity and is found to be active against both gram-positive and gram-negative organisms.

Neem leaves have many medicinal uses. A strong decoction of leaves can be used as a mild antiseptic. The hot infusion is used as anodyne for fomenting swollen glands, bruises and strains. A paste of neem leaves is applied for reducing cowpox and in general for teating all types of viral infections. The fresh and mature leaves of neem and seeds of *Psoralea corylifolia* are used to prepare medicine for leucoderma. The Khasi and Jaintia tribes of Meghalaya use the neem leaves for diarrhoea, dysentery, tuberculosis and heart disease. The leaves are also used as an insect repellent and in control of nematodes. Fresh mature leaves yield an odourous viscous essential oil which exhibits antifungal activity. The tender neem leaves in combination with *Kali mirch* or pepper (*Piper nigrum*) are effective in getting rid of intestinal worms. Fresh juice from the leaves mixed with honey is good for Jaundice and skin diseases.

Industrial Uses

Neem oil has since long been used as lamp oil and lubricant. Since it is a non-edible oil with medicinal properties it is mainly being used in soap and facial cream manufacture. Since neem oil is non drying it resists degradation better than other oils and hence it serves as a lubricant. In villages it is used as cart wheel lubricant. The powdered leaves and purified oil are used as ingredients of cosmetics.

Neem oil has been experimented as biodiesel. Its cetane number is 47 against 46 for diesel. The calorific value of neem oil is 9400 as against 10,500 Kcal/ Kg for diesel. Higher viscosity (30 vs 5), fire point (210 vs 107) and flash points (180 vs 80) are its disadvantages (Mathur and Das, 1985).

The oil is widely used in cosmetic preparations such as creams, hair-lotions, medicated soaps, washing-soaps and tooth-pastes. Either solely or in combination with other oils, it is largely used in the manufacture of soaps. It also serves as a source of olein and stearin, and high-melting stearin after hydrogenation. The oil is also used in disinfectants and emulsifying agents for insecticides. Neem oil when applied to hair is reported to prevent baldness and greying of hair. The essential oil present in the seeds has anti-microbial activity.

Pyrolytic degradation of the oil yields 'pyronimin' a denaturant for alcohol. The oil can be converted into 'polyol' to substitute polypropylene glycol, used as a rocket-propellent fuel. It can also be used in the textile and rubber industry. Neem oil is also useful in curing raw hides and skins. This method dispenses with the use of common salt, a major pollutant in tanning effluents. The leathers prepared by this method do not deteriorate in quality during storage, and compare well with the leathers obtained from wet-salted stock.

Timber

Neem is a member of the mahogany family, and the properties of its wood resemble mahogany. It is relatively heavy, with a specific gravity varying from 0.56 to 0.85 (average, 0.68). When freshly cut, it has a strong smell. Although easily sawn, worked, polished, and glued, it must be dried carefully because it often splits and warps. It also splits easily when nailed, hence holes must be pre-bored. Nevertheless, it is a good construction timber and is widely used in carts, tool handles, and agricultural implements. In South India it is a common furniture wood.

The heartwood is red when first exposed, but in sunlight it fades to reddish brown. It is aromatic, beautifully mottled, narrowly inter-locked, and medium to coarse in texture. It is subject to only slight shrinkage and can be readily worked by hand or machine. Although it lends itself to carving, it does not take a high polish. The timber is widely used as building material. The timber is durable even in exposed situations. It is seldom attacked by termites. It is resistant to woodworms, and it makes useful fence posts and poles for house construction. Pole wood is especially important in developing countries. The tree's ability to resprout after cutting and to regrow its canopy after pollarding makes neem highly suited to pole production. Neem wood also serves as fuel in rural areas.

In conclusion, it should be mentioned that neem and its products have great potential. In these days of environmental pollution, there is great scope for the exploitation of neem. The tree and its products not only serve to purify the environment but also improve the health standards of living beings when utilized properly.

Jatropha

Several species of *Jatropha* are cultivated world wide for their ornamental characteristics as well as for their economic uses. Nine species have been recorded to have importance. They are *Jatropha curcas, J. gossipifolia, J.*

integerrima, J. multifeda, J. podogarica, J. heterophylla, J. glandulifera, J. nana and J. wightiana. Among these *J. curcas and J. gossypifolia* have potential for exploitation (Sujatha and Mukta, 1995). *Jatropha curcas* is a semi-wild large shrub or small tree (Anonymous, 1959). It grows in any type of soil in the waste lands of semi arid tropics, with rainfall ranging from 30-100cm. It is sensitive to frost and cannot grow in cool climates. It grows wild in tropical climates. It is grown commercially for its oil in some countries. It is native to Mexico and central America. It has been introduced into other parts of the world, especially Asia and Africa by the Portuguese. It occurs almost throughout India and in Andaman Islands.

Jatropha curcas Plant

Jatropha belongs to *Euphorbiceae*. Being a shrub, *curcas* plant grows up to a height of 3-5m but can reach heights of 8-10m. Animals normally do not eat the tree due to its bad odour and taste. Hence to protect against free roaming animals it is used as a fence tree. *Jatropha* plant also serves to reduce wind erosion. When planted parallel to slopes to fix small earth or stone dams, *Jatropha* plants help control water erosion. The roots of the plant grow close to the ground, anchoring the soil like small dikes or earthen bunds. These dikes effectively reduce the surface run off during downpours, thus helping more water to penetrate into the soil and boosting harvest.

Since the plant also is not much afflicted by pests or diseases, it has good scope for wider cultivation. *J. curcas* is a fast growing plant and can be propagated using either seeds or cuttings, in the rainy season. Plants raised from seeds are more robust and live longer. They are grown at spacings like 1m x 1m or 1.5m x 1.5m or 2m x 1.5m. NPK at 50:50:25 kg/ha/annum can be applied. The plants start bearing within the first year itself, but yield stabilises after five years and continues upto 50 years. It has pale green leaves and unisexual flowers. Pollination is carried out by insects and moth because of the sweet perfume and copious nectar of the flowers. Pistillate flowers open before staminate flowers in the same inflorescence. After pollination a trilocular ellipsoidal fruit is formed. Fruits formed in winter mature after more than 50 days. The fruits are harvested when they turn from green to yellow to black. The fruits contain 3 oval seeds. Seeds, 1.5 to 3cm in diameter, weigh 1.5 to 3g.

Seed Composition and Quality

The *Jatropha* fruit is a capsule and contains three seeds. The seed contains 40% hull with 60% kernel. The seed is similar to castor seed but smaller in size. The seeds are dark brown in colour. The seed yields range from 200-2000kg per ha at 2x2m spacing. The ripe fruits are collected from the short trees, dried and separated. Later seeds are separated and decorticated. The fruit contains 50-60% seed with around 60-70% white kernel. The whole seed oil content ranges from 28-35% while the kernel oil content is higher at 40-50% (Chatakanonda, *et al.*,2005). Lot of variation in oil content with respect to location and genotype exists in *jatropha* seeds. The whole seed oil content varies between 33-39% and kernel oil from 47-58% (Ginwal *et al.*, 2004). Oil is expelled using an expeller. The oil yield is around 25%. Double crushing in expeller yields oil at 28-30%.

Table 105: Composition of *jatropha* seeds.

Constituent	%
Moisture	6.2
Protein	18.0
Oil	38
Sugar	18.0
Fiber	16.0
Ash	5.3

The seed also contains some quantity of starch, sucrose, glucose, gluten and lipase. The oil and protein contents of Indian seeds have been reported as 33 and 23% respectively (Nagaraj and Muktha, 2004). The kernel has 26-30% protein, 55-62% oil and 3.6-5.2% minerals. The seed possesses poisonous and purgative properties though it is rarely used as a purgative. The seeds contain two toxic principles curcine, similar to ricin and curcasin, a resinous substance (resino-lipid), having purgative properties. *Jatropha* seed oil is normally colorless and odourless. But on standing turns yellow to yellowish brown in colour.

Table 106: Characteristics of *jatropha* oil.

Characteristic	Value
Specific gravity 15°C	0.918-0.923
Refractive index 40°C	1.462-1.465
Saponification value (S.V.)	188-196
Iodine value (I.V.)	93-107
Hydroxyl value	4-20
Unsnaponifiable matter	0.4-1.1%
Viscosity	7.1 centipoises

The oil is semidrying with I.V. of 93-107 and S.V. of 188-196. It has a hydroxyl value of 4-20, being very low in comparison to castor oil. *Curcas* oil is low in viscosity, slightly soluble in alcohol but easily soluble in petroleum ether(hexane) and is optically inactive. The toxic principle goes into the alcohol soluble portion which property is normally used to detoxify the oil. The oil can be stored for longer periods without much deterioration. Polymerisation reactions may sometimes take place when the oil is exposed to air which is normally not a problem.

The oil contains mainly oleic and linoleic acids. These two unsaturated fatty acids total upto 80% with the saturated fatty acids namely myristic, palmitic and stearic acids together accounting for the remaining 20%. Nagaraj and Muktha (2004) have reported 12.6% palmitic, 3.9% stearic, 41.8% oleic and 41.8% linoleic acids for Indian *jatropha* seeds. *Jatropha* seed oil from Thailand has 42.4% oleic, 35.2% linoleic, 14.7% palmitic and 6.9% stearic acids (Chatakanonda, *et al.*,2005).

Table 107: Fatty acid composition (%).

Fatty acid	Range	Mean
C14:0 (myristic)	0.5-1.4	1.4
C16:0 (palmitic)	12-17	15.6
C18:0 (stearic)	5-9.7	9.7
C18:1 (oleic)	37-63	41
C18:2 (linoleic)	19-41	32.1

Source: Bringi, (1987)

The oil from *J. curcas* seeds has been analysed using HPLC. The lipids detected are triacylglycerol (91.90%), 1,3-diacylglycerol (3.80%), free fatty acids (3.20%), gamma-tocopherol (0.02%), phytosterol (0.27%) and sterol fatty acid esters (0.22%). The polar lipids are phosphatidylcholine (0.78%) and phosphatidylethanolamine (0.32%), with digalactosyldiacylglycerol (DGDG) and monogalactosyldiacylglycerol (MGDG) present in trace amounts. The neutral lipids account for about 98.0% with 1.10% phospholipids and trace amounts of glycolipids. MGDG and DGDG are the two major glycolipids in *J. curcas* seed oil (Abigor and Uadia, 2001).

Utility of *curcas* Oil

The oil is semidrying and can be used for preparation of semidrying alkyds and varnishes. The oil serves as lamp oil and lubricant. It can be used for making soap, cosmetics and candles. Hardened oil serves as a substitute for tallow. A varnish can be made by boiling the oil with iron oxide. It can be used for wool spinning. It also serves as hair growth stimulant. The oil also has commercial potential in the industry for manufacturing soaps, varnishes and paints.

Nutritional as well as non- food applications are restricted due to the toxicity of *J. curcas*. The seed oil contains phorbol esters, a family of compounds known to cause a large number of biological effects such as tumour promotion and inflammation. Therefore it is necessary to find feasible routes for detoxification of the oil. *J. curcas* seed oil has been treated by traditional oil refining processes and examining the content of phorbol esters. Parameters of several refining steps has been varied to optimize the grade of detoxification. Almost no effect could be observed with degumming and deodorization, whereas the steps of deacidification and bleaching could reduce the content of phorbol esters by up to 55% (Hass and Mittelbach, 2001).

Utility of the Plant

All parts of the plant exude a sticky, semi-solid, astringent and pungent latex, containing the resinous substances (14.6% in the latex coagulum) without any rubber. The latex appears to have some anti-cancerous properties. The latex contains an alkaloid known as 'Jatrophine' which is believed to have anti-cancer property. The latex on drying turns to a bright reddish brown, brittle substance, similar to shellac. It stains linen indelibly and is useful as marking ink. The bark contains tannin (37% on dry weight basis) alongwith wax, resin, saponins, reducing sugars and traces of a volatile oil. The wax is a mixture of melissyl alcohol and melissyl melissate.

A dark blue dye used in the Philippines for colouring cloth, fishing nets and lines is obtained from the bark. A dye can also be extracted from the leaves and tender stems which can be concentrated to a yellowish syrup or dried to a blackish brown lumpy mass; the dye gives fast colours like tan and brown to cotton. Tender twigs of the plant are used for cleaning teeth; the juice has a role in relieving toothache and strengthening gums. Tender twigs and leaves serve as manure for coconut trees. In Assam region of India,, leaves have use as feed for the *eri* silkworm.

In some parts of the world like Java and Malaya, tender leaves are eaten after steaming. Even cooked nuts are eaten after stewing. The ash of the tree (roots and branches) is used as salt. Fruit hulls, seed shells and wood are used as fuel. Seeds dipped in oil serve as torches. Leaf juice stains red and indelibly marks linen cloth black. The plant serves as fish poison. Extracts of the seed and other plant parts inhibit fungal growth and hence serve as fungicide and pesticides. Seed powder is used to kill rats. The wood is useful as fire wood. The wood is very flexible if thinly split and is used for basket making. A water extract of the whole plant has molluscicide effects against various types of snail, and has insecticidal properties (Ecky, 1954).

Medicinal Uses of Jatropha

A decoction of bark is given for treatment of rheumatism and leprosy. Seed oil is strongly purgative and is used as an antiseptic for cough, skin diseases and as a pain reliever from rheumatism. It contains phorbol, curcin (taxalbumin), alkaliods, flavones, glycosides and sterols (stigmasterol, β-sitosterol, campesterol and cholesterol). Water extract of the branches have some role in HIV and tumor treatment. Plant sap is useful in alleviating dermatomucosal diseases.

The plant juice is used as purgative and haemostatic and for stupefying fish in Java and Philippines. The leaves have rubefacient, lactagogue, and insecticidal properties. In Ghana, the leaves are used for fumigating houses against bed-bugs. The leaf juice is used as an external application for piles and tongue inflammations of babies. The twig sap stops bleeding and hence used for dressing wounds and ulcers. Sap emulsion with benzyl benzoate is effective against scabies, wet eczema and dermatitis. A decoction of leaves and roots is given for diarrhoea. The yellow root oil is a good anthelminthic. The root bark is used in external applications for sores. In Konkan region of India, the bark is rubbed with asafoetida and buttermilk and the paste given in cases of dyspepsia and diarrhoea. A decoction of the bark is useful in treating rheumatism and leprosy.

Seeds have purgative properties. One to two roasted seeds are sufficient to act as purgative, while larger doses may become unhealthy. The seeds and seed extracts have use in traditional medicine for treating arthritis, gout and jaundice. The oil is useful in treating skin diseases and to obtain relief from rheumatism. The oil also serves as a hair growth stimulant. The seeds are useful in treatment of syphilis. Juice and latex serve as antibiotic agents and hence have widespread uses in medicine and veterinary fields. They serve as diuretics and anti-cough agents etc. Leaves are used to treat jaundice.

Other Species of Jatropha

Along with *J.curcas* other species namely , *J. gossipifolia, J. integerrima, J. multifeda,* and *J. podogarica,* have been analysed for their oil, protein and fatty acid profile. *J. podogarica* has the highest seed oil content of 39.9 % followed by 33% in *J. gossipifolia* and *J. integerrima and* 32% in, *J. multifeda* (Nagaraj, 2007). The seed protein content in all species is around 22%. The linoleic acid is maximum (75%) in *J.integerrima*, followed by *J.Podogarica* (69%) and *J.Gossipifolia* (61%) with the lowest content in *J.multifeda* (50%). Oleic and palmitic acids are higher in *J.Multifeda* (27 and 18% respectively) followed by *J.gossipifolia* (23 and 11% respectively). Stearic acid is present at around 5% in all the species.

Table 108: Fatty acid composition of some *jatropha* species.

***Jatropha* species**	**Fatty acid profile %**			
	16:0	**18:0**	**18:1**	**18:2**
J. Gossipifolia	11.3	5.1	23.0	60.8
J. Integerrima	8.6	4.4	12.0	74.8
J. Multifeda	17.9	5.4	26.8	50.1
J. Podagarica	9.4	4.5	16.9	68.8

Source: Nagaraj, (2007).

Curcas oil as Bio-Diesel

The oil can be used directly in diesel engines or mixed with diesel or can be converted to methyl esters(bio-diesel) and utilized. Wherever, the seeds and oil are available cheaply, the bio-diesel can serve as an alternate and ecofriendly fuel. The yield of methyl esters (bio-diesel) is around 81% from the oil when reacted with methanol and sodium hydroxide(Chatakanonda, *et al.*,2005). The cetane value of biodiesel made from *curcas* oil is comparable to diesel oil. It's energy content is 3% less than that of diesel fuel. The carbon residue is less while sulfur residues are practically zero and hence better than fossil fuels. Because *jatropha* plant is renewable, ecofriendly and the oil, biodegradable with less sulfur emissions, great future exists for its widespread use.

Table 109: Physical and chemical properties of diesel fuel and *curcas* oil.

Property	***Curcas* oil**	**Diesel oil**
Viscosity (cp) (30°C)	52.6	3.60
Specific gravity (15°C)	0.917-0.923	0.841-0.85
Solidfying Point (°C)	2.0	0.14
Cetane Value	23-41	47.8.59
Flash Point (°C)	110/340	80
Carbon Residue (%)	0.64	0.05-0.15
Distillation (°C)	284.295	350-370
Sulfur (%)	0.13.0.16	1.0-1.2

Jatropha Cake

The oil cake, which is a by product, is useful as organic fertiliser. The cake contains about 50% protein. It has sufficient levels of minerals, N= 5.7-6.5%, P_2O_5 = 2.6-3.0%, K_2O=0.9-1.0%, CaO=0.6-0.7%, and MgO=1.3-1.4%. Composition-wise it is next to neem cake , but better than FYM and *pongamia* cake. Because of toxic constituents, it has limited use as animal feed. Some antinutrients like trypsin inhibitor, lectin, saponin and phytate have been identified in the meal. Roasting does not have much influence in reducing the antinutrients except for slightly increasing protein digestibility (Aderibigbe, *et al.*, 1997). Oil cake has also scope for production of bio-gas. It can supplement energy requirements in the rural set-up.

Karanj (*Pongamia glabra*)

Pongamia glabra also known as karanja is found in India, Malaya, Seychelles, East Indies, Philippines, and Pacific Islands. In India, it grows well in the states of Orissa, Karnataka, Maharashtra, Uttar Pradesh, Madhya Pradesh, Andhra Pradesh and Tamil Nadu. The karanj tree is a hardy and adaptable tree. It is native of Western ghats in India. *Pongamia glabra* belongs to *Leguminosae.* It helps in atmospheric nitrogen fixation and inproves soil fertility. It is an evergreen tree with a bole and a fairly spreading crown of 18m height and 1.5m girth. It is drought tolerant and grows well in all types of soils including saline soils. It is an eco-friendly tree providing shade to humans and animals. This tree is useful in afforestation programmes and as an avenue tree.

The tree comes to fruiting in 4-7 years. The flowering season varies from April to September depending on location. It fruits twice in a year-during September/October and March/April. Green pods appear which mature after 10 to 11 months. The harvesting also gets spread out. The pods are beaten down from the trees using sticks. The dried pods are collected and shells removed using mallets and stones to separate the kernels.

The ripe pods are flat and elliptic, about 5-7 cm long with one to two kidney shaped reddish kernels. The size of the seed is 1.7-2.0 cm in length and 1.2-1.8cm in breadth and weighs about 1-1.2 g. The pods contain about 50% of its weight of seeds. The kernels are white and covered by reddish brittle skin,called testa. The kernel content is 95% of the seed. A mature tree yields 40-60 kg of pods and the kernel yield is around 8 to 24kg per season. The seed oil is non-edible and has a lot of industrial uses.

Seed Composition

The seed contains 25-30% oil and 15-18% protein. It also contains 7% each of starch and crude fibre. The starch is mostly present in the kernel while the fibre is in the testa. The kernels contain 27-39% oil and 13.5% mucilage.

Table 110: Seed composition of karanj.

Constituent	% Composition
Moisture	19.0
Oil	28.0
Protein.	17.0
Starch	7.0
Carbohydratres	6.6
Crude fibre	7.0
Ash	2.0

The Karanj Oil

In ghanies 18-22% oil yield is obtained while in an expeller the yield is 24-27%. The oil is dark brown in colour with a repulsive odour and a specific gravity of 0.839. The iodine and saponification values vary in the range 83-90 and 183-195 respectively. The yellowish orange brown oil slowly darkens on storage. It has a bitter taste and disagreeable smell which is due to non-lipid associates. Extraction with alcohol or alcoholic alkali is useful in non-distructively separating the non-lipid constituents and improving the quality of oil. Simpler procedures of treatment with sulfuric acid also furnish the same result.

Table 111: Characteristics of karanj oil.

Characteristic	Range
Specific gravity	0.925-0.940
Refractive Index 40°C	1.4734-1.4790
Saponifiable Value	185-195
Iodine Value	83-90
Unsaponifiables	3.0-4.5%
Titre	34°C
Colour (Y+5R)	30

Freshly extracted oil is yellowish orange to brown and rapidly darkens on storage. It has a disagreeable odour and bitter taste. These factors and the ineffectiveness of conventional refining, bleaching and deodourisation make the quality of karanj oil unacceptable. Hence alternative upgradation methods have been attempted. Karanja soap which is initially yellow orange in colour turns green and then dark brown on storage. The non-lipid associates of the oil (3-5%) consisting mainly of flavanoids are responsible for the colour change in the alkaline medium of soap.

Fatty Acid Composition

Oleic acid is the main fatty acid (50-60%) of the oil, followed by linoleic acid (10-20%). Palmitic, stearic, arachidic and eicosenoic acids are the minor fatty acids. It also contains around 5% dihydroxy stearic acid which imparts high viscosity and hence lubricating properties to the oil. A commercial sample has been observed to contain a higher level of palmitic acid (22%), with 55% oleic and 15% linoleic acids (Basu et al.,1988). Another Indian karanj oil sample has oleic,54.5; linoleic,17.0; palmitic,11.3; stearic, 6.4; linolenic,4.5 and arachidic, 5.2% (Nagaraj and Mukta 2004).

The oil contains unsaponifiable matter ranging from 5-7%. Karanja oil being rich in oleic acid, is a valuable raw material for oleic acid and related oleochemicals. Tri unsatutrated glycerides constitute 37% with diunsaturated glyceride at 48% and monounsatutrated glycerides at 15% account for the total glycerides (Hilditch and Williams, 1964).

Table 112: Fatty acid composition of karanj oil.

F.A.	% composition	
	(1)	(2)
16:0	10.6	10.6-11.6
18:0	6.5	8.0-8.1
18:1	49.0	52.2-54.5
18:2	19.0	22.0-23.1
20:0	4.1	3.9-4.1
20:1	1.4	1.8-2.0
22:0	5.3	-
24:0	2.4	-

Source: (1) Hilditch and Williams, (1964), (2) Chakraborty and Bhattacharyya, (1988).

A large number of flavonoids have been isolated from the seeds of *pongamia pinnata*: karanjin (-) isolanchocarpin, pongapin, pongaglabrone, karanja chromene, (-) isolanchocarpin, pongamol, glabrine, and oleomethoxy-karanjin (Vimal and Naphade, 1988). Of 14 biosynthetic variants of flavones, isolated and characterised from the oil, karanjin is the major constituent. It does not appear to be responsible for the dark colour of the oil, but the real culprits seem to be the alkali labile minor constituents like isolonchocarpin (a flavanone) and pongamol (o-diketone) and two other chalkones are prone to oxidation (extended conjugation) leading to darkening of soap colour. An amino acid, glabrin along with the flavanoid karanjin, pongapin, karanjone and pongaglabrone and a diketone pongamol are present in the seeds.

Processing to Improve Quality of Oil

Selective removal of flavonoids, colour bodies and other impurities by alcohol or alcoholic alkali upgrades the fat. The extracted fat on bleaching and hydrogenation yield a reasonably acceptable material (m.p. 45^{0}C) for soaps. Upgrading efficiency is not consistent as some batches of dark commercial oils require multiple processing to give satisfactory results.

The soluble flavanoids can be rendered oil insoluble by careful sulphonation by treating with concentrated sulphuric acid. Treatment with 1-2% by weight of the acid at ambient temperatures produce a dark viscous gum. The decanted oil (92% yield) is washed with water to remove residual acid and bleached or alternately directly bleached, the earth acting as adsorbent for residual acid. The hardening rate of acid washed oil is considerably enhanced compared to raw oil probably due to removal of minor quantity of nitrogenous compounds present in the oil.

The oil can also be upgraded using a distillation process. The flavanoids are volatile and codistil with fatty acids in the conventional splitting and distillation process. The soap from distilled acid is yellow and acquires a mottled green colour on keping for 24 hrs. A prior high temperature deodorisation of the oil which removes fatty acids and volatile flavonoids yield an oil dark in colour but useful for manufacture of cheaper quality soap. Further splitting and distillation yield good quality fatty acids.

The quality of the oil can be improved through hydrogenation process also. Despite high requirements of hydrogen and sluggish rate of reduction, the crude commercial oil is partially hydrogenated under pressure to I.V. 45. This degree of hydrogenation is adequate for saturating alkali labile colour precursors to colourless, alkali stable compounds. The quality is acceptable for laundry soaps. Edible quality karanja oil has been obtained by mixed solvent refining process (n-hexane: 80% ethyl alcohol or 80% iso propyl alcohol 3:2 v/v)preceded by degumming in the oil phase with phosphoric acid and followed by silica gel bleaching and deodourisation (Chakraborty and Bhattacharyya, (1988).

Utility of Karanj and its Products

Karanj oil, has traditionally been used as lamp and lubricating oil. It has been used in rural medicine to cure skin infections. The oil is mostly used in soap manufacture and in leather tanning. The oil has medicinal uses in treatment of leucoderma, psoriasis, scabies and skin itches. Karanj seed, oil and cake have found a variety of uses based on their biological activity, attributable to the non-fatty minor compounds present. The oil besides being anthelmintic, has a curative value for diseases of the eyes, rheumatism, leucoderma and itching wounds (CSIR 1969).

In rural areas the leaves are used to prevent infestation of grains. The presence of a hypotensive principle and a substance producing uterine contraction has been reported in karanj (Bringi and Mukherjee,1987). Karanj seed and its isolates are known to possess piscicidal, insecticidal, nematicidal, bactericidal and antituberculin activities. Karanj oil, especially, karanjin, the major non-fatty physiologically active furan flavonoid, has synergistic activity for pyrethroids. Karanjin is also useful as a broad spectrum chemical in pest control. Both oil and karanjin have been found to be effecive against cockroaches and houseflies. Karanjin, however, does not prossess any contact insecticidal activity. Alcohol extract as well as some of the purer flavonoids possess aphicidal activity. Karanjin seems to be the acive ingredient. The water extract of karanj cake has some nematicidal activity which, however, is inferior to neem and mahua cakes.

Karanj oil synergises the knockdown capability of pyrethrins against housefly. It also synergises the action of lindane, endrin, heptacholor, and thiodan, but antagonised BHC and toxaphene against *Tribolium castaneum*. Combined use of Karanj oil and piperonyl butoxide as pyrethrum synergists against housefly has proved more economical than that of piperonyl butoxide alone (Vimal and Naphade, 1988). Karanj oil is becoming popular in recent years for making bio-diesel. FFA and unsaponifiables need to be reduced in the oil before it is converted to the methyl esters for use as bio-diesel.

Karanj Cake

The cake after oil extraction may be used as a manure. Karanj cake, like neem cake helps in slow release of nitrogen to plants when used in conjunction with urea and other nitrogenous fertilizers.

Karanja cake is bitter and has a pungent odour. The oil cake and deoiled meal are used mainly as manure, fungicide and insecticides. The cake contains N, 5.4%; P, 0.9% and K, 1.2%. Karanja cake also possesses nitrification inhibition property, which, can be used to advantage in blending with nitrogenous fertilizers. Though karanja cake is rich in proteins (30-35%), it is unpalatable and toxic to cattle even at 4% level because of presence of several toxic principles (karanjin, pongamin and glabrin).

Since the toxic principles are oil soluble, they move into the oil during solvent extraction. Deoiled karanja cake (digestible crude protein 26%, total digestible nutrients 71.4%) can replace 25-60% nitrogen of mustard cake in the feed of cows. It can also be incorporated into the compound feeds of broilers and layers in place of black sesame seed cake as protein source upto 30% and 40% respectively. Karanj cake is acceptable to rats as food.

Karanja cake is rich in glutamic and aspartic acids followed by leucine and lysine. It is also rich in serine, tryosine, histidine, methionine, arginine, and glycine. It is evident that the karanja protein is balanced with respect to all the important fatty acids. Unfortunately, due to many toxic constituents the cake is inedible. The detoxified cake can easily serve as a good animal feed.

Table 113: Amino acid composition of karanj protein.

AA	g/100g protein
Leucine	9.4
Lysine	9.3
Methionine	6.8
Arginine	6.2
Valine	3.7
Histidine	5.7
Isoleucine	5.0
Theonine	3.3
Phenylalanine	2.5
Aspartic acid	10.5
Serine	6.8
Glutamic acid	15.3
Glycine	5.5
Alanine	4.0
Tyrosine	6.0

Source: Bringi and Mukherjee (1987).

Karanj Wood/ Timber

Karanj timber is of low quality and is used as firewood (4,600 kcal/kg), and for cabinet work, cart wheels and posts. Leaves of the plant are used as fodder and green manure. The tree is generally grown along canal bunds, road sides avenues, seashore dunes, tidal and forest beaches (Karoshi and Hegde, 2002). Seed shells are useful for manufacture of paper pulp and as firewood.

Mahua (*Madhuca*)

Two major species of the genus *Madhuca* grow in India. They are *Madhuca Indica* or *Madhuca Latifolia* and *Madhuca Longifolia* . *Latifolia (indica)* is a medium sized large deciduous tree while *Longifolia* is a large semi-ever green tree. These two are so closely related that no distinction is made in the trade while dealing with their seeds or fat. These trees come to fruiting in 10-12 years. Mahua generally grows in rocky, stony and infertile lands. The tree is multipurpose (CSIR, 1962). It is valued for its flowers, fruits, seeds and wood. The flowering season is February to April. The flowers are useful in preparation of country liquor-*arrack* (alcohol). They are also eaten raw or cooked and used in making syrups.

The mature fruits fall on the ground during May-July or August – September. The orange brown ripe fleshy berry is 2.5 to 5cm long and contains one to four shining seeds. The seeds are ellipsoidal in shape, with 1.5-2.0 cm length and 1.3- 1.6 cm breadth. The yield of seeds per tree varies from 20-40 kg. The seeds contain 70% kernels which in turn contain 45% oil. Recovery of kernels is a village level activity and proper storage is necessary to maintain seed quality. Moisture content of 6-8% in the seeds is congenial for storage. The fruit shell is removed by hand or by beating with sticks.

Seed Composition

Mahua seeds contain 75% kernels and 25% hulls.The kernels contain oil ranging from 37-51%. The kernels also contain 8-12% protein , 28-41% sugars, 3-7% crude fibre and 2-4% minerals. The seeds also contain saponins ranging from 2-4%. The oil from mahua kernels is obtained initially by expelling in a *ghani* or expeller. The residual oil from the cake is recovered through solvent extraction. The oil is pale yellow in colour. Since it is in the form of solid at room temperature, it called a fat and is popularly known and valued as vegetable butter.

Mahua Fat

Fresh oil from properly stored seeds, is yellow in colour while commercial oils are greenish yellow with disagreeable odour and taste. Oil obtained from kernels has low FFA of 1-2% only. Poorly stored kernels can yield oil of up to 30% FFA. The absence of unusal fatty acids and or low level of unsaponifiables in mahua oil facilitate its refining to obtain good grade edibel oil/fat. Quality of the oil expressed from the seeds depends on the conditions under which they have been stored. Fresh Mahua oil is yellow in colour with an unpleasant taste. It is a good source of vegetable oil for edible purposes. Mahua oil is used for edible and cooking purposes in some rural areas in India. The oil is sometimes used as an adulterant for ghee. The Expeller and solvent extracted oils are bitter in taste and as such they are not edible. Refining and hydrogentation yield products similar to mutton tallow or cocoa butter(Subba Reddy, 2001).

Mahua oil can be refined and bleached by conventional alkali refining process and using conventional bleaching earths under vacuum. The oil needs hardly any degumming, or any special pretreatment before refining practice. High FFA oils can be satisfactorily refined by high vaccum steam stripping, mixed solvent refining and by miscella refining. The same can be acieved by a combination of ethanol or isopropanol extraction and alkali neutralisation.

Table 114: Characteristic of mahua fat.

Characteristic	Content
Melting point	23-31^0 C
Slip point	20-31^{0}C
Titre	38-48^{0}C
Saponificatin value	187-196
Iodine value	55-63
Refractive Index at 40^{0}C	1.4520-1.45620
Unsaponifables	1.0-3.0 %
Sterols	0.06-0.12 %
F.F.A	2.0-9.0%

Source: Bhattacharya, (1988).

Due to a melting point of 31°C, it stays as a solid at room temperature and hence it is referred to as mhua fat. It is a non-drying oil and has an I.V. of 55-63. It's unsaponifiables are somewhat low.

Oleic acid is the major fatty acid in mahua oil(32-38%). It is also rich in saturated fatty acids, namely, palmitic(22-37%) and stearic (19-24%) acids, which together account for about 45-57% of total fatty acids. Linoleic acid is present in the range of 14-18% in the mahua oil. Arachidic and linolenic acids are present in traces. In quality it is similar to shea butter, and hence it can easily substitute it.

Table 115: Fatty acid composition of mahua oil.

F.A	% Composition
16:0	22-37
18:0	19-24
20:0	0-0.1
18:1	32-38
18:2	14-18
18:3	0-1

Source: Bhattacharya, (1988).

Utility of Mahua Fat

Mahua oil is a substitute for ghee (anhydrous butter) fat in India. It is also used as cooking oil in some areas, especially the tribal regions. It can be used in the manufacture of vanaspati (hydrogenated oil). The oil can be blended with traditional vegetable oils like rice bran, cottonseed, or palm olein and by directed interesterification produce PUFA rich zero trans margarine fat base and vanaspati(Bhattacharya *et al.*, 1981). Mahua oil is a tallow substitute. It is used in chocolate manufacture and good quality toilet soaps. It also serves as a raw material for oleic and stearic acids and diesel substitute.

Kane *et al.*, (1966) have reported routes for the preparation of Cocoa butter substitute from mahua oil. Mahua oil can be fractionated to obtain its least soluble fraction (LSF) from acetone (1:7 w/v) at 4 ± 1^0C for 7 hrs and this fraction when blended with sal stearin (15:85) gives rise to a product of highly satisfactory solidification and dilatation(Ghosh Choudhury,1983). LSF of mahua oil can be blended with kokum fat in different proportions, namely, 7:3, 6:4 and 1:1, to obtain cocoa butter substitutes of different grades. Mahua oil when interesterified with kokum fat (3:2 mixture) and later fractionated from 7 volumes of acetone at 23^0C yields a good quality CBS.

Refined oil finds use in the manufacture of lubricating greases. The oil is also used for candles, as a batching oil in jute industry, in wool mills for smoothening or for easy spinning. Mahua oil has emollient properties and is reported to be used in treating skin diseases, rheumatism and headache. It is considered as a good laxative. It is useful in habitual constipation, piles and haemorrhoids. It prevents dry skin, wrinkles and restores skin flexibility. It serves as a galactogogue.

Mahua oil is an excellent oil for making premium quality toilet soaps. It can substitute tallow in soap manufacture. A mixture of sal oil and coconut oil (80:20 or 70:30) and mowrah oil with coconut oil (80:20 or 70:30) appear to be ideal tallow substitutes for toilet (mahua) soap (Bhattacharya,1988). Mahua oil is an important raw material for producing commercial stearic acid and oleic acid which are the basic fatty acids for oleochemicals. It is widely used in making laundry soaps, lubricants, creams, lotions, balms, and make-up foundations (3-100%).

Unsaponifiables

The unsaponifiables(0.8-1% of the oil) contain 0.4% phytosterols. Δ^{5}-Avenasterol, β-sitosterol, stigmadistanol and sitostanol account for the total sterols in almost equal proportions. The other constituents of the unsponifiables are the tocopherols, namely, α (0.038%), β (0.19%) and γ (1.7% of unsaponifiables). It is significant to note that Δ tocopherol, which imparts high temperature stability to oils is prersent in high concentrations. Mahua oil is normally a stable oil. The higher gamma tocopherol makes it more stable at frying temparatures also.

Mahua fat contains an unsaturated gum called illepe (0.5% IV= 280). It has polypropeniod structure. It has a role in rancidity. Partial hydrogenation is useful in solving the problem. Mahua fat contains 26% symmetrical glycerides of the SOS type and can be recovered by crystallization. It's ulitity in confectionery is limited due to low yield of stearin and presence of linoleic glycerides (Bringi. 1988a).

Mahua Cake

Mahua cake accounts for 50% of the extractions. The expeller cake contains 8-12% oil, 15-17% protein, 50-55% carbohydrates, 5-6% fibre and 6-7% minerals. The cake also contains 4-8% saponins. The solvent extracted meal is richer in all the constituents, except oil (Bringi and Mehta, 1987). The saponins in the cake need to be removed or detoxified, if it has to be used as an animal feed. Mahua meal with its saponins serves as a wormicide. It is also used in hair care (hair bath) due it's sapogenic activity. Seed cake also has insecticidal properties. Mahua cake is poor in nitrogen, but it contains 6% phosphatides, 11.8% sugars and some pectin bodies, in addition to the residual oil, proteins, fibre and other carbohydrates. The existence of phosphatides in the oilcake would make mahua seeds the richest source of phosphorus among oilseeds. It's industrial isolation from mahua cake would be one way of utilizing the cake for a better purpose.

The amino acid composition of mahua is as follows: aspartic acid 8.40, threonine 3.25, valine-methionine 7.33, serine 2.76, arginine 6.63, histidine 6.31, glutamic acid 19.0, lysine 6.63, alanine, 6.46, glycine 1.80, tyrosine-tryptophan 17.58, leucine-isoleucine-Phenylalanine 11.29 and Proline 2.50%. Mahua meal is used for the extraction of L-lysine (Govindarajan, 1952). Mahua meal is normally a poor nitrogenous fertiliser, but when boiled with alkali for a prolonged period upto 9 hours becomes more nitrifiable (Leley and Agashe,1971).

Mahua meal has found very little use as animal feed due to the presence of toxic saponin (7%) although it has 40-50% carbohydrates and about 14-16% proteins. At 10-20% level in the diet, mahua meal is reported to cause death within one month (Mulky,1976). Recently the structure of saponin has been determined. The hydrolysis products, sugars and the aglycone, protobassic acid, are not toxic and upgrading is basically a process related to hydrolysis. The cake with the toxic saponins serves as an orgzanic manure only.

Detoxification of Mahua Meal

The removal of saponin is therefore essential for the greater utilisation of mowrah meal. Ethanol extracted meal at 20-30% level in diet does not show any biochemical abnormalities in animals. However, this method causes a loss of solids and nutrients to a great textent. This can be preventeed by hydrolysis of saponin with 2% hydrochloric acid and then neutralizing with sodium hydroxide. The meal becomes non-toxic and there will be no solid matter loss. However, the saponin gets lost which otherwise could be used as a surface active agent.

An alternate approach is to examine ambient hydrolysis by enzymes. The conversion of saponin to sapogenin by the endogenous sapogenase in the presence of water (Mulky,1976) detoxifies the meal, but the enzymatic hydrolysis takes 10-14 days for optimum conversion. There is also a possibility of fungal infestation in the meal during this period. Another enzymatic hydroysis method, namely, incubation with β-D-glucosidase as magma with little water to hydrolyse the β-linked sugars at 2 and 3 position has been succesfully examined. This enzymatic hydrolysis is highly useful in detoxifying mahua meal (Bringi,1988b).

An elegant process based on the extraction of saponin by isopropanol has been reported by Bhattacharya *et al., (*1981). The process consists in refluxing mahua meal with 5 volumes of isopropanol at 80^{0}C for 2 hrs. There is 82% removal of saponin and the treated meal retains all the protein, with a loss of about 12% carbohydrates. Extraction of saponin is about 87% with ethanol with a loss of 3% protein and carbohydrates (Sen gupta *et al.,* 1981). The isopropanol treated meal is non-toxic and the meal could be exclusively utilised as animal feed. The saponin recovered has scope of utilisation as such in the non-ionic detergent formulations.

Mahua Flowers

Mahua flowers are a dietary ingredient of tribals, as well as feedstock. The flowers are extensively used in brewing country liquor. Mahua flowers as a dietary source does not show any toxic symptoms.. The flowers are available in March-April of the year and the seeds are available in May-August. The flowers are a good source of sugars (68%) and protein (6.67%). The flowers, with 2.6% ash, contain 137and 139mg/100g of P and Ca respectively (Jayasree, *et al.,*1998). The flowers also contain vitamin C and some B complex vitamins.

Table 116: Nutrient composition of mahua flowers.

Composition	Per 100 g
Moisture	11.61g
Protein	6.67g
Fat	0.09g
Total sugars	68.00g
Ash	2.60g
Crude fibre	1.90g
Phosphorus	137.00mg
Calcium	139.00mg
Ascorbic acid	17.00mg
Thiamine	28.00mg
Riboflavin	870.00mg
Niacin	4.80mg

Source: Jayasree, *et al.,* (1998)

Mahua flower protein has high levels of aspartic acid and glutamic acids (around 22% each). It is also richer in leucine (13%), glycine, alanine, isoleucine, serine, phenylalanine and threonine. Lysine content of the flower protein is higher than any cereal protein. It is also a good source of sulphur containing amino acids, namely, cysteine and methionine.

Fatty acid composition of the fat of mahua flowers is somewhat different from that of kernel oil. The flower oil has the following fatty acid composition: capric=11.3, lauric=14.8, myristic=4.0, palmitic=13.1, stearic=4.0, oleic=19.2, linoleic=24.0, linolenic=1.6, behenic=2.7 and lignoceric=4.5%. The flower oil is richer in low molecular weight fatty acids as well as some long chain fatty acids. It is richer in linoleic acid, an essential fatty acid, along with some quantity of the omega-3 fatty acid, namely,linolenic acid.

The flowers with very good composition of sugars and protein with low fat are good sources of calories. Nutritionally, the protein is endowed with very good make up of all the amino acds, especially lysine and sulphur amino acids. Though the oil content is low, it is also balanced with respect to all the fatty acids including EFA. Hence the flowers must be highly beneficial when eaten.

Table 117: Amino acid composition of desugared and defatted mahua flowers.

Amino acid	Mahua flowers (g/100 g ptotein)	Amino acid (g/100 g protein)	Mahua flowers
Lysine	4.67	Alanine	9.31
Histidine	3.63	Cysteine	3.35
Arginine	4.54	Methionine	1.80
Aspartic acid	20.28	Valine	7.25
Threonine	5.86	Isoleucine	7.91
Serine	7.29	Leucine	12.98
Glutamic acid	22.80	Tyrosine	3.94
Proline	7.10	Phenylalanine	6.40
Glycine	9.12		

Source: Jayasree, *et al.*,(1998).

Simarouba

Simarouba is an evergreen tree which thrives on most eroded and degraded gravel soils under sub-tropical climate. It appears to have originated in the tropical forest region of central America. *Simarouba glauca* tree is also called paradise tree or *aceituno.* This tropical tre1e grows well up to 1000 m above sea level in all types of well-drained soil with pH 5.5-8.0. However, a minimum of 1.0 m deep soil is preferred for its growth. It can easily grow in places with

250 mm to 2500 mm annual rainfall and mean maximum monthly temperature going up to 45^{0}C. *Simarouba* seedlings are raised during April-June in polybags. Seedlings older than two months are ready for transplanting. The grafts of elite lines or the apomictic seedlings with known sex should be preferred for planting to get higher and earlier returns.

In rainfed marginal lands/wastelands transplanting is done immediately after the onset of monsoon so that the plants can establish well by the end of the rainy season. If irrigation facility is available, planting can be taken up at any time of the year. The saplings/ grafts are planted at a spacing of 6.0-8.0 m in pits of 45 cm size half filled with top soil and 3-5 kg compost. Protective irrigation at weekly intervals during the first summer is essential. Regular weeding, interculture and related agronomic practices need to be practiced till the plants get thoroughly established. The tree begins to bear fruit when it is 6-8 years old.

For effective pollination and good bearing, planting of andromonoecious and female plants in a 1:20 ratio in the right geometry is advocated or some selected female plants may be grafted with a few branches of high-yielding andromonoecious scions. The low yielders can successfully be converted into high yielders by grafting them *in situ* with the scions of high-yielding elite genotypes. By top working and crown grafting, the male plants can successfully be transformed into females. The flowering is annual beginning in December and continuing up to February. The trees are polygamodioecious and the female trees are heavy bearers.

Pests and diseases are not a major problem on *Simarouba.* In the nursery the seedlings are affected by damping off (caused by *Pythium* sp.) and wilt. Drainage and application of commercial fungicides can take care of these diseases. The sooty mold (*Capnodium* sp.) is reported to occur on the 'honey' deposited by scale insects and aphids. The cattle generally do not prefer to browse on this plant. In view of these advantages it is easy to grow *simarouba.* The almond moth, *Ephestia cautella* occurs as a pest on stored decorticated seeds.

Simarouba Fruit

The drupelets turn black or greenish yellow when they are ready for harvest. For harvesting, the productive branches with ripe fruitlets are shaken or beaten mildly with a stick and the fallen fruitlets are gathered. These can be dried before or after depulping in direct sunlight. During April/May manually harvested drupelets are depulped, washed and sun dried, and transported/ stored. Each tree can yield 15 to 30 kg nutlets. The seeds contain up to 60% oil. The plantation can yield 1000-2000 kg/ha /annum oil. After decorticating, the oil is extracted from the kernels by expellers in the conventional oil mills. The extracted oil/ fat is refined packed and marketed.

Simarouba fruit is an etaero with one to five fruitlets. In a nutlet, the shell accounts for 40-58% and the rest is seed(42-50%). The weight of the nutlet is 0.7 to 1.2g and the seed is 0.2 to 0.4g. Manually harvested drupelets are depulped, washed and sun dried to less than 10% moisture and transported to the place of processing. The seeds with shell contain 27% oil while the kernels contain 60-75% oil. The seeds also contain 16.5% protein (Nagaraj and Mukta, 2004). The oil is normally extracted using expeller followed by solvent extraction. Soxhlet extraction yields upto 60% oil while chloroform: methanol (2:1) yields 67-68% lipids. Upto 85% of the oil can be recovered through expellers (Jena *et al.*, 2003).

Simarouba Oil

Simarouba oil is a non- drying oil with a low I.V. It is a solid at room temperature with a M.P. of 27°C. Unsaponifiables and FFA are low in the oil. It is an edible fat. Being free from trans - fatty acids, it can be used as a good margarine substitute. It has 62-65% unsaturated fatty acids and 34- 38% saturated fatty acids. It's stearic acid content is around 25% and palmitic acid is 10%. Due to higher levels of saturated fatty acids , *simarouba* oil remains solid at room temperature. Normally stearic acid is considered a good fatty acid from the nutritional angle, while palmitic acid is not so. It is very rich in oleic acid(60-65%), but deficient in the essential fatty acid, linoleic acid. Blending with high 18:2 oil would make it a more nutritious oil. Due to its good characteristics, it can serve as a good bakery fat.

Table 118: Physico-chemical characteristics of *Simarouba* Oil.

Characteristic	Value
Free fatty acids	0.06%
Iodine Value	53.8
Saponifiable Value	191.0
Unsaponifiables	0.4%
Refractive Index 25°C	1.4556
Titre	47°C
Smoke point	138°C
M.P.	27°C
Colour	10Y+0.8R
Solid	Snow white

Source: Armour, (1959)

The crude oil is refined, bleached, deodourised and fractionated. The high oleic liquid fraction is useful as frying oil while the solid palmito-stearin fraction is useful as cocoa butter substitute. This fraction is also usful in making ice cream and mayonnaise. The *simarouba* oil directly and the solid fraction serve as substitutes for vanaspati. It is free from trans fatty acids and hence is better than hydrogenated oil in quality.

The expeller and solvent extracted oils are bitter in taste and so are not edible. Both the oils are easily refined and bleached to get light coloured oils, which are totally free from bitter taste. Orgonoleptic studies have been conducted with refined and bleached *simarouba* seed oil by preparing etables viz. *Bajjis,* Potato Chips (deep fried), with groundnut oil as control. The above etables are acceptable and have been observed to be good in taste (Narasimha Reddy, 2001).

The non-edible uses of the oil include, manufacture of soap and some oleo-chemicals like glycerine, fatty acids, alcohols, fatty acid methyl esters, and fatty acid amines etc. It is also useful in making cosmetics, detergents, plasticizers, candles, lubricants etc.

Table 119: Fatty acid composition of Simarouba.

Fatty acid	(1)	(2)
16:0	11.3	10.9
18:0	23.0	25.6
18:1	64.9	59.6
18:2	1.6	3.3
18:3	-	0.4

Source: (1) Mukta and Nagaraj, 2004 (2) Armour, 1959.

Simarouba Cake

The cake is rich in proteins (50-60%). However, it is bitter and toxic to humans and animals and hence it is used as an organic manure (Armour, 1959). The cake has around 8-10 % N, 1%P and 1.2%K. There are also traces of calcium, magnesium and sodium. Defatted seed meal contains 50% protein and globulins are the major constituents upto 35% (Jayashri and Sunanda, 2001) The use of *aceituno* press cake as fertilizer has given good results on coffee, sugarcane, cotton and corn (Armour, 1959). Some attempts have already been made to detoxify the oilcake in El Salvador with a view to greater utilization of the proteins, especially in the livestock feed concentrates (Armour, 1959), without much success.

Medicinal Properties

S. glauca kernel cake extract has possible medicinal uses as an amoebicide. The active ingredient appears to be a crystalline glycoside and is referred to under the name of Glaucarubin. This product, under the trade name of "Glaumeba", has been introduced in most of the major tropical areas (Armour 1959).

The leaves, bark and wood of *Simarouba* have a long history of use as medicine by the natives in the tropics. The chemicals such as Ailanthinone, benzoquinone, canthin, dehydroglaucarubinone, glaucarubine, glaucarubolone, glaucarubinone, holacanthone, melianone, simaroubidin, simarolide, simarubin, simarubolide, sitosterol, tirucalla have been recorded from leaves, bark and seeds. The chemicals documented are known to possess medicinal properties. They can function as amoebicide, analgesic, anthelmintic, antibacterial, antidysenteric, antileukemic, antimalarial, antimicrobial, antitumorous, antiviral, astringent, cytotoxic, emmenagogue, ferbrifuge, skin hydrator, stomachic, sudorific, tonic and vermifuge (Joshi and Joshi, 2002).

The Shells

The shells (endocarp) can be used for making hard-boards. They can be pulverized and added to enrich the compost since they contain about 1.24% potash. They are also useful as fuel as they have high calorific value and are routinely used to heat the boilers. About 30% of the shells are mixed with the kernels and used for oil extraction in the oil mills. This increases the efficiency of the extraction process and hardly 7-8% oil is left in the oil cake after extraction through expellers which would be otherwise as high as 15%, if only kernels are processed.

Fruit Pulp

Aceituno fruits are similar in size and shape to the European olive. The somewhat astringent but semi-sweet pulp which surrounds the seeds of the *aceituno* is eaten mainly by children. The fruits are sometimes available for sale in the markets of El Salvador during the month of May (Armour, 1959).

The mesocarp or the pulp of the ripe fruit is generally 2-3 mm thick, contains 89.4% moisture and constitutes about 62% of the weight of the fruitlet. Its chemical composition is as follows: total soluble solids 15.2-17.20 Brix; total sugars 11.51%; reducing sugars 9.88% and ascorbic acid (vitamin C) 29.0-32.3mg per 100 g pulp. The pulp is used in the preparation of squash, beverage and jam (Srilakshmi, 1999; Ramjayam, 2000) which are very well accepted because of their attractive natural colour, flavour, and good taste. Microbial spoilage does not occur upto 105 days of storage period. The fruits can be a source of natural colourant. The pulp is perishable and requires prompt handling.

In case it is not possible to use pulp for preparing squash, beverage and/or jam, the pulp produced in huge quantities to the tune of 6000-8000 kg/ha can be mixed with leaf litter of the same tree or of some other plant and used in the production of vermicompost (about 4 tonnes per hectare) of excellent quality. The earthworms relish it very much and grow large in size and number and convert into vermicompost in a span of about one month.

Other Parts of the Tree

All parts of *Simarouba* plant are useful in some way or the other. However, most useful components are the seeds which can be used in the manufacture of *vanaspathi* and/or margarine. As an industrial oil it is well-suited for the manufacture of quality soaps, lubricants, paints, polishes, pharmaceuticals etc.

Leaf litter (about 20kg/tree/year) is a good feed for earthworms and it makes good manure and vermicompost. The leaf and the bark contain the chemicals useful in curing amoebiasis, diarrhoea and malaria. The bitter principle present in the bark and leaves finds an important place in native medicine preparation. A febrifuge is made by extracting the astringent juice of the bark. A crude remedy for diarrhea, particularly in children, is made by collecting the juices of the bark/leaf, adding them to the juice of mango bark, with the normal dose being one teaspoonful given three times daily (Armour, 1959).

Wood is light and generally insect-resistant, and hence useful in making light furniture, toys, packing material, pulp (for the paper industry) and matches. It also seves as fuel. The wood is taken from the male and low yielding trees and used in various ways. It is light and generally less preferred by wood eating insects, hence useful in making yoke for oxen. The oil tree is also important for sustainability of our cropping systems and soil health. This eco-friendly tree with well-developed root system and dense evergreen canopy can efficiently check the soil erosion and environmental degradation. It is well suited for soil and water conservation, wasteland reclamation and is ideal for watershed areas.

Kokum (*Garcinia Indica*)

Kokum trees are found in the ever green forests of Western Ghats of India and the states of Assam and West Bengal. They are small evergreen trees growing upto 18m. The leaves are oblong and reddish when young. Kokum tree flowers during November-February period. The fruits are collected during April – June. The ripe fruits are beaten down from the trees or pulled down using long sticks and the fruits collected from the ground. The fruits are 2.5-3cm in diameter. They are purple when ripe with yellowish tinge. Again the fruits are beaten with sticks to separate the seeds from the outer rind. The seeds are then covered with ash, dried and stored in heaps or in baskets. The outer rind which is called *amsul* is a culinary item. Seed constitutes 23-26% of the fruit. There are 5-8 seeds in a fruit embedded in an acid pulp. The yield per tree ranges from 60-80 kg fruits or 10-15 kg seeds.

The kokum seeds are decorticated and the kernels separated from which the valuable oil known as kokum butter (Mango Steen oil) is obtained. Normally the oil is extracted by boiling the pounded kernels. The floating oil is separated which on cooling solidifies. The lower quality kokum butter is obtained by a churning process. The kernels form about 60% of the weight of the seeds and yield about 33 percent of oil. On the whole seed basis the oil content is around 23 to 26 percent.

Table 120: Characteristicsof kokum oil.

Characteristic	Value
Melting point	39.4-43.0°C
Refractive Index (40°C)	1.4565-1.4575
Saponofication value	187-191.7
Iodine value	25-36.0
Unsaponifiable matter	1.2-2.3(%)

Source: Lakshmikantan (1978).

The kokum oil is solid at room temperature with a S.V. of around 190. It's I.V. is low and it is a non-drying oil. The unsaponifiable content is low at 1-2%. It is very rich in stearic acid (55%) followed by oleic acid (40%). Other fatty acids like palmitic myristic, and linoleic acids constitute about 5%. Being very rich in 18:0, it serves as its source for extraction and utilization.

Table 121: Fatty acid composition of kokum oil.

Fatty acid	% Composition
Oleic	39.4-41.5
Myristic	1.2
Palmitic	2.5-5.3
Stearic	52.0-56.4
Linoleic	1.0

Source: Kokum, ICOC, Bulletin, No.9.

Kokum butter when fresh has a faint, but unpleasant odour and a bland oily taste. It melts in the mouth like butter and leaves a sensation of coolness on the tongue. On keeping for long, the oil becomes rancid and acquires a brown colour. The oil is made up of 85% free lipids, 12.1% bound lipids and about 3% firmly bound lipids. (Thippeswamy and Raina, 1989). The glyceride composition of the oil has been calculated (Coleman, 1965) SSS= 3.6%, SUS=80%, SUU=14.3% and UUU=0.6%. The unsaponifiable matter (Kolhe, *et al.*, 1981) consists of 3-oxotriterpenes (12%).

Uses of Oil

Kokum oil or butter is used as lamp oil, as well as frying oil. It also has medicinal uses. It is used to relieve irritation, as a nutrient and in the preparation of poultices, fomentations etc. When administered internally, its action is similar to that of cod liver oil. As kokum fat is highly saturated, nutritional evaluation studies indicate it to be a good fat with lower digestibility. It is similar to any other hydrogenated vegetable oil with zero trans fatty acids. Around 500 tonnes per annum of kokum fat is being produced in India. It has a ready export market as the kokum fat is being used as cocoa butter substitute.

Kokum oil is also used in soap manufacture in admixture with other non-edible oils, such as neem, *mahua* and *khakan*. On account of its high content of stearic acid, it is highly suitable for candle making and as a source of stearic acid. Kokum oil is useful in manufacture of ointments, cosmetics, sizing of cotton yarn, suppositories, cocoa-butter substitutes and also as a carrier of medicines (Lakshmikantan, 1988).

Other Plant Product Uses

The fruit of kokum has long been used in South-West India as a semi-medicinal item of diet. It is acidic, slightly astringent and is considered to be superior to tamarind for the preparation of acid drinks. The fruit is also made into a *sherbet* (juice) which is useful in fever as a cooling drink. The seed of the ripe fruit is eaten raw as a delicacy. The dried outer fruit rind is known as *amsul* (am=sour, sul=rind). It is a popular culinary ingredient with taste and qualities similar to that of tamarind. The rinds contain 10% malic acid and small quantities of tartaric and citric acids and have lot of medicinal uses. The roots as well as fruits also have medicinal uses. The root is astringent. The fruit is useful as an anthelminthic and a tonic for the heart.

Seed Cake

The cake is mainly used as manure (CSIR,1956). The defatted seed meal is low in protein (16.5 percent). The amino acid composition of defatted meal (Bringi, 1987) is (in percentage) lysine (0.33), arginine (1.6), threonine (0.43), cysteine (0.13), methionine (0.15) valine (0.57), isoleucine (0.4), leucine(1.0) and phenylalanine(0.67). It is rich in arginine, the wound healing amino acid.

Melia – The Bead tree

Melia Azadarach L. is a native of West Asia, and is adapted to the warm tropics. The species is mostly cultivated for its shade value, due to its quick growing nature. It is sometimes grown as a shade tree in tea and coffee plantations. Under natural conditions, the plant regenerates freely from seeds during the rains. Artificial propagation is possible by direct sowing, transplanting seedlings from the nursery or by cuttings and root suckers. The plant grows fast in the early stages, the annual girth increment being 1.5-2.0 inches. It coppices well upto a girth of 3 feet. It can grow up to a height of 9-12 m, The spacing of trees can be 8x8m. It starts bearing fruit in 4-5 years.

The average life of Melia is 20 years and needs replacement around that time. It has a cylindrical bole and spreading crown. The flowering and fruiting occurs during March to July. Ripening takes longer time. Fruit bunches are collected from trees through plucking or by beating with sticks. Kernels constitute 9.5% of the fruit with an oil content of 30-40%. The whole seed oil is around 8-10%. The kernels contain 18% protein. The oil is a drying oil. The oil can be used in soap making, detergents and paints etc. It is sometimes used as hair oil. The saponification value of the oil is similar to other vegetable oils. It has higher iodine value indicating greater unsaturation. It is low in unsaponifiables (1%).

Table 122: Characteristics of melia seed oil.

Characteristic	Value
Density	0.9134
Refractive index	1.4691
Saponification value	190.8
Iodine value	134.7
Acid value	4.45
Unsaponifiables %	1.26

Table 123: The fatty acid composition of melia oil.

Fatty acid	%Composition
Palmatic	8.4
Stearic	4.0
Oleic	23.2
Linoleic	63.9
Arachidic	0.51

Source: Mukta and Nagaraj, (2004).

Melia oil has higher levels of linoleic acid (64%) with 23% oleic acid. It has around 8% palmitic and 4% stearic acids. Melia oil with around 87% unsaturated fatty acids is more unsaturated than neem oil. The oil can be used for making soaps, alkyd resins, protective coatings and greases etc., The unsaponifiables contain phytosterols and aromatic hydrocarbons. Seed oil cake can be used as animal feed.

Medicinal Uses

Various parts of the plant have therauptic values. The leaf juice is anthelminthic, antilithic (stone reduction), diuretic and emmenagogue (restores menstruation). A decoction of the leaves is regarded as astringent and stomachic. A poultice of the flowers is applied to eruptive skin diseases and for killing lice. The fruit is used in China as febrifuge and disinfectant. Seeds are prescribed in rheumatism. A gum collected from the tree is considered useful in spleen enlargement. An infusion of the bark is effective against ascariasis. The bark contains tannins, alkaloids, azaridine and paraisine. Aqueous extracts of heart wood are useful in asthma. Infusion of the bark has anthelminthic activity and is useful in treating intestinal worms. The leaves, bark and fruits have insect repellant properties. Leaf extracts are used for protection against bruises. An alkoloid and carotenoid are the active principles involved in the medicinal properties. The fruits are used as febrifuge (fever reducing agent).

Uses of Bark and Wood

The tree yields a valuable timber. The sapwood is yellowish white. The heartwood is red turning reddish brown with age and is mostly straight-grained and course textured. It is tough, moderately hard and heavy (specific gravity 0.56). It is highly durable and lustrous with dry feel. It seasons well with care and does not require antiseptic treatment. It is resistant to attack by white ants. It is easy to work, turn, peel and finish. The wood is used for toys, cigars, ammunition boxes, packs and museum cases. It is suitable for sports ware, roofing material and agricultural implements. It is generally used as fuel wood.

Uses of Leaves

The tree is lopped for fodder. Leaves are slightly bitter. They are occasionally eaten after boiling with vegetables. Leaves also serve as green manure. The leaves, bark and fruits have insect repellant properties.They have insect growtth regulation and antifeedant properties. Leaves are placed inside books and between folds of woolen garments to protect them against insect attack. Extracts of leaves are used as sprays to protect plants against infestation. A decoction of dry leaves (2-5%) is effective against locust. The active principle is reported to be an alkaloid soluble in hot water. A carotenoid, meliatin, present in aqueous suspension of leaves acts as a repellant to grass hoppers. Extracts of the plant are reported to be used in Ghana for protecting cocoa beans against infestation by *Ephestia* species. Alcohol and petroleum ether extracts of stem bark are toxic to carpet beetle larvae.

Uses of Fruits

The fruits of the plants possess an unpleasant odour and an intensely bitter nauseating taste. The bitter constituents are present in the pericarp and not the kernel. The bitter principle is bakayanin (m.p. 85-118^{0}C). The fruit contains an alkaloid azaridine (margosine), a brown resinous constituent, non-bitter acidic component, a sterol and some tannins. The fruit also contains glucose and starch which can be utilized for alcohol pruduction. A kind of whisky is reported to be made from the fruits. The fruits are useful as disinfectant. The stones of the fruits are used as beads in necklace, rosaries etc.

The fruits of *M. Azadarach* are poisonous to man and animals, though sheep, goats and birds have been eating the fruit without apparently any ill effects. Cases of severe poisoning, as a result of eating the fruits have been reported. The patients develop symptoms of colic, diarrhoea and vomiting. Symptoms of paralysis and narcosis have been produced is experimental cats, dogs and sheep by ingestion of the fruit.

Sal (*Shorea Robusta*)

The sal tree is the most famous of all the species of *Shorea*. It is a semi-deciduous, gregarious tree usually growing up to a height of 18-30m and girth of 1.8 to 2.1m. This tree has great capability for adaptability to varied climates and locations. It can grow under an annal rainfall ranging from 90-200 cm and varied temperature regimes. The sal is a tree known to Indians since ages and is mentioned in the epics like Ramayana. The tree is well known for its wood which is widely and extensively utilized. The timber is very strong with good mechanical strength. It remains free from insect and fungal attack due to its immunity to these organisms. The sal tree is a source of oleoresin. It is obtained through tapping and is used in manufacturing incense. The tree also serves as a host for tasar silk worms and lac insects.

The sal trees start bearing fruits, when they are 25-30 years old. The flower buds appear in mid- or late February and bear fruits in March/April. The ripening period ranges from eight to ten weeks. The wings of the fruit are green when immature and turn brown at maturity. The fruit with dry brown wings fall from the tree as such. The nut which remains green in the fruit gradually turns brown on storage. On artificial drying it turns brown quickly. The fruit consists of 23% wing, 30% pod and 47% kernel, which contains around 20% oil. The fruits which fall to the ground are collected from the second fortnight of May and June manually. They are sun dried for a couple of days. Clean kernels are obtained by removal of wings and pods.

Sal fruit is of the size of a pea with five segments of fruiting calyx enveloping it with three big wings. The dewinged seeds contain an outer thin, brittle seed. The pod breaks easily on pressure, opening up the hard ovoid kernel. The kernel weighs on an average 4 to 4.5g. The kernels constitute about 66.4% of the fruit, the rest being shells and calyx (12.8% and 20.8%) respectively. The dewinged, decorticated, dried kernels put in gunny bags are stored in the godowns near the extraction units, at moisture content ranging from 8-10%. High moisture and insect attack increase the FFA (free fatty acid) content of the oil in the seeds. The FFA value can go upto as high as 20% when the seeds are infested. Hence sufficient care must be taken to store the seeds properly. Treatment with Aluminum phosphide, is reported to protect the kernels from insect attack (Sethi and Maini,1970).

Seed Composition

Sal seed contains oil ranging from 10.5 to 18.6%. It's protein content ranges from 5.1 to 10.5% with a carbohydrate content of 50-63%. It has a mineral content of 1.6 to 3.0% and 1-10% cellulose. The seeds have tannins at a higher level of 8-15%. Sal oil is extracted from the seeds mostly by solvent extraction due to its low oil content. Food grade hexane is used for the solvent extraction. Sal kernels are broken and cooked in steam and flaked in roller mills before extraction.

The fat is obtained at the yield of about 12-13%. It has a greenish yellow colour and does not have an agreeable odour. It slowly gets bleached on exposure to light. The characteristics of Sal fat are given in Table 124. Refining is done by treatment with caustic soda to remove the FFA and bleached using Tonsil earth and activated carbon (Chand, *et al.*,1974). Deodourization is achieved by steam under vacuum.

Sal Oil

Sal oil is yellowish green to dark green in colour due to the presence of chlorophyll. The oil has a woody or terperoid flavour. Sal oil is called vegetable butter as it remains solid at room temperature. Sal oil has S.V. =175-192; I.V.=32-45; FFA=1-2; slip point=26-31 and refractive index at 40^{0}C= 1.4562-1.4619 (Nasirullah *et al.*,1981). It is a non-drying oil. It's unsaponifiables are low at around 2%. It is solid at room temperature & its M.P. is around 37^{0}C.

Table 124: Characteristics of sal oil.

Characteristic	Content
Melting point	34.5-37.0^{0}C
Slip point	*26.0- 36.8^{0}C*
Titre	48.0-53.4^{0}C
Saponification value	180-190
Iodine value	31-45.0
Refractive index at 40^{0}C	1.4560-1.4580
Unsaponifiables	0.65-2.2%
Sterols	0.06-0.43%
Phosphatides	0.9-0.1.0%
Free Fatty Acids	1.0-5.0%

Sal oil contains 50-60% saturated fatty acids and 40-50% unsaturated fatty acids. Stearic acid is the major saturated acid present upto 48% and oleic acid is the major unsaturated acid present upto 46%. Palmitic (4-8%), arachidic (3-8%), linoleic (1-3%) and linolenic (0-2%) acids are the minor fatty acids. Sal oil also contains a small quantity of epoxystearic acid (1.5-3%).

Table 125: Fatty acid composition of Sal oil.

Fatty acid	% Composition
Palmitic	4.0-8.0
Stearic	38-48
Arachidic	3-8
Oleic	37-46
Linoleic	1-3
Linolenic	0.2
Epoxy stearic	1-3

Sal oil is mainly used for making cocoa butter substitute. It is made through fractionation by selective solvent extraction. The oil also is used as cooking oil by tribals. It is permitted to be used in *vanaspati* (hydrogenated oil) manufacture. It can also be blended with traditional edible oils directly and through interesterification. *Ghee,* (butter oil) substitutes can be manufactured with sal oil. The oil can be used in the manufacture of soaps as tallow substitute. It is also a good raw material for stearic and oleic acids. Sal oil is used as rocket fuel. It (alkyl ester of the oil) has scope in manufacture biofuel.

Sal Glycerides

The glyceride composition of Sal fat has been studied by fractional crystallisation method. Enzymatic hydrolysis by pancreatic lipase followed by Thin Layer Chromatography and Gas liquid Chromatographic techniques have been used for this purpose (Mani and Lakshminarayana, 1968). The glyceride composition is given in table. 126.

Table 126: Glyceride composition of sal fat.

Glyceride	Mole percent
GS_3	1.0
Oleo-dipalmitin	-
Oleo-palmito-stearin	8.0
Oleo-distearin	42.0
Oleo-stearoarachidin	14.0
Palmitodiolein	5.0
Stearodiolein	25.0
Arachidodiolein	4.0
Triolein	1.0

Sal fat is a good substitute for Cocoa butter. A comparison of the properties of Sal fat and Cocoa butter are given in Table 127. Sal fat has characteristics which are in many ways similar to cocoa butter and hence has lot of scope in confectionary use. Modified products have been prepared from sal fat and Cocoa butter and their properties have been evaluated (Bhambhani, *et al.*, 1972). There is a large export market for Cocoa butter-like fats in countries like United States of America, United Kingdom and Japan. A number of plants have been set up for the extraction of sal seed fat.

Another factor associated with storage of kernels is the formation of epoxy glycerides due to the presence of enzymes. Analysis of fat extracted from kernels stored for different periods show the presence of a glyceride containing two moles of a saturated acid and one mole of 9,10-epoxy stearic acid and another glyceride containing 9, 10-dihydroxy stearic acid in place of epoxy stearic acid. The quantity of the epoxy and dihydroxy glycerides is variable. It is usually in the range of 0.7 to 3.8% expressed as the epoxy and dihydroxy fatty acid content of the weight of fat (Bringi, 1979).

The non-saponifiable matter constitutes about 0.90% of the oil. β- sitosterol (0.09%) has been identified as a constituent of unsaponifiables (Srinivasulu and Mahapatra,1970). Sal unsaponifiables contain 3-ketotriterpenes (15%). This appears to be a special feature in some fats containing high proportions of symmetrical glycerides with the exception of cocoabutter (Bringi,1988).

Sal Fat as Cocoa Butter Substitute (CBS)

The utilty of sal as a component for Cocoa Butter Substitute (CBS) by virtue of preponderant symmetrical glycerides content and export of large quantities for this purpose are well known. Sal fat resembles cocoa butter in its characteristics. Sal fat has slightly higher M.P. and 1-2% higher solid fat content. The quality of sal for CBS varies considerably. While early season fat is generally acceptable in terms of M.P., dilatation and cooling curve chaaracteristics, the quality invariably deteriorates as the season advances. Apart from partial glycerides, two minor constituents, 9, 10 epoxy stearic acid and its hydrolysed product 9, 10 dihydroxy stearic acid are responsible for the change in physical properties. The former is present at 3-7% while the latter is present at 2-3% (Bringi,1988).

Sal fat is used in soap making. The availability of large quantities of sal fat for soap making will reduce consumption of tallow fat. It is being used as an edible oil by certain sections of people in Orissa, Bihar and Madhya Pradesh in India. It's use as a polyol substitution product, which is used as fuel binder in space research, is being studied (Dhawan, 1976).

Table 127: Properties of cocoa butter and sal fat.

Properties	Cocoa butter	Sal fat
Saponification value	196.0	189.0
Iodine value	37.6	38.9
Melting point (°C)	34.5	36.5
Super cooling temperature (°C)	20.8	22.2
Rise in temperature on solidification (°C)	8.0	6.7
Plastic Range Index	2.5	4.25
Solid content Index		
30°C	61.9	59.5
33°C	16.9	39.5
36°C	0.0	1.1
38°C	0.0	0.0

Sal Cake

About 80-85% of the material would be left after solvent extraction which is the deoiled sal meal, composition of which is given in Table 128. The utilisation of this material would influence the economics of sal seed industry. The sal meals contain very high levels(65%) of carbohydrates and the starch alone accouts for upto 40%. The fibre content also is higher at 40%. The protein content is low at 8-10% and ash content is around 2-7%. The Ca and P contents vary around 2%. It may easily serve as organic manure, but has the potential to be used as animal feed.

It can be used by food and non-food industries. If the tannins, present at 8-12% giving an astringent taste, are removed, the cake can become a better grade animal feed. Sal starch has use in tablet manufacture. Sal tannin is useful in leather tanning .Sal meal is a good organic manure.

The deoiled sal seed meal contains about 8-9% proteins. The amino acid composition of the proteins has been studied. They are well balanced with a high threonine level. However, its use as a poultry or cattle feed has limitations because of the high tannin content. Use of sal seed meal beyond 5% level in poultry feed compositions results in discoloration of yolk. So it has been recommended that a safe level of 20% for cattle and up to 5% for poultry feed compositions may be maintained.

Table 128: Composition of deoiled sal meal.

Contents	Percentage
Moisture	5.0-12.9
Oil	0.7-0.4
Protein	8.7-10.0
Crude fibre	0.5-3.0
Ash	2.3-2.8
Acid insoluble Ash	2.1-6.5
Sand, silica	< 1
Ca	1.4
P	1.8
Total carbohydrates	64.7-64.9
Starch	30.0-40.0
Cellulose	41.7

Alcohol extraction of deoiled sal seed meal furnished ellagic acid in 1.12% yield (Srinivasulu and Mahapatra, 1970). Ellagic acid is gaining importance in Cancer Research. It effectively neutralizes the active form of the carcinogen benzo (a) pyrene *in vitro* (Bringi, 1988).

The deoiled sal seed meal contains 10-13% tannins. They constitute ellagitannins and gallotannins of the hydrolysable group. Studies have been made for the utilisation of these tannins. When blended with other tanning materials, sal tannins have given good quality leathers (Vijayramayya, *et al.*, 1979). Sal tannins have been observed to detoxify ricin and other antinutrients in castor meal (Gandhi and Mulky, 1994)

Extraction of starch from deoiled sal meal in the original state yields coloured products. When extracted after hydrolysis starch at 50% yield is obtained. The product is white in colour and resembles soluble starch in its properties. However, it has poor adhesion property. Deioled sal meal is suitable as an extender for urea formaldehyde resin adhesives for plywood.

Table 129: Amino acid composition of sal seed meal protein.

Amino acids	Percentage
Glycine	6.5
Alanine	6.3
Aspartic acid	6.6
Glutamic acid	10.3
Serine	6.0
Threonine	7.8
Valine-Methionine	7.6
Norleucine-leucine-isoleucine	12.0
Arginine	6.8
Histidine	4.8
Lysine	5.7
Proline	6.6
Tyrosine	4.3
Phenylalanine-Tryptophan	3.4
Cysteine	2.2
Unidentified	3.7

Potential of Sal Tree and It's Products

Out of the total forest area of India (about 75 m ha), approximately 30% is covered by tropically moist deciduous forests, Sal forests constitute roughly half of these accounting for approximately 10 m ha, located mostly in the states of Madhya Pradesh, Orissa, Bihar and their adjoining areas. It has been estimated that there is a potential of approximately 5.5 m tonnes of sal seeds per annum from these forests. At an oil content of 12-13%, the availability of sal oil will be of the order 700,000 tonnes per year. Assuming that only half of the seeds can be collected, India, can have about 3,50,000 tonnes of oil. This oil will be highly useful in meeting the food and non food requirements. It can help in sustituting the tallow requirements of the soap industry.

After taking out oil from the sal seed, about 2-3 m tonnes of deoiled cake will be available for making animal feed. Thus, sal tree would play an important role in rural developments (Dev, 1978), especially in a country like India where 80% people live in villages. An integrated project utilizing the various aspects of sal tree would help in economic uplift of the rural masses in the interior areas.

Rubber (*Hevea braziliensis*)

This tree belongs to *Euphorbiceae*. Rubber tree is a small evergreen tropical tree. It grows upto a height of 60-100 feet and matures in 15-20 years. The stem which has a girth of 8-12 feet is smoth and straight. The bark is greyish with trifoliate leaves. Rubber which is the main commercial product is obtained from the latex. The tree is outstanding due to its high latex and superior rubber quality. It flowers in February - March and the fruit matures during August-November.

Rubber Seed

The fruits are ellipsoidal and capsular with three carpels, each containing one seed. The weight of seed is 2-4g resembling castor seeds. The seed yield per tree ranges from 1 to 1.5 kg. The kernels constitute 5% of the fruit and 50-60% of the seed. The kernels contain 45-52% oil, 17.6% protein, 23% carohydrates and 2.5% ash. The seeds are quickly dried as they don't have good storage value. The kernel is also dried and stored. The oil is extracted through screw press and or solvent extraction. The seeds also contain linamarin or phaseolunatin due to which they are not edible. However, it is eaten by people living near rubber plantations, after soaking the seeds in water for 24 hrs and boiling them in water.

Rubber Seed Oil

About 4,000 tonnes of rubber seed oil are produced in India, chiefly from rubber plantations located in Kerala and Southern parts of Tamil Nadu. Cold hydraulic pressing yields higher oil than expeller pressing. Seeds deteriorate quickly and hence the commercial oil has high FFA and dark colour. The fresh oil is pale yellow while the commericial oil is dark in colour. It has low unsponifiable matter. Freshly extracted oil also has low FFA. It's digestibility is 97% as against 94% for groundnut oil (Gandhi *et al.*, 1990).

Table 130: Characteristics of rubber seed oil.

Characteristic	Value
Specific gravity (36^{0}C)	0.9163-0.9290
Refractive index (40^{0}C)	1.4660-1.4697
Acid value	0.74-3.0
Saponification value	181-196
Iodine value	132-139
Unsaponifiable matter	0.5-1.0%
Titre	28-30^{0}C

Physicochemical and instrumental characterization of rubber seed oil has been carrier out (Aigbodion and Bakare, 2005). The natural form of rubber seed oil is highly acidic (acid value= 43.6mgKOH/g). The saponification value of the oil is similar to other oils. The iodine value is higher indicating that it is a drying oil. It's unsaponifiable matter is low at 1%.

The oil contains significant quantities of linolenic acid (upto22%). It also has around 40% linoleic and 23% oleic acids. The total unsaturated fatty acids thus account for around 85%. The remaining 15% is accounted for by the saturated fatty acids, namely palmitic and stearic acids. Aigbodion and Bakare (2005) have reported 17.5%palmitic, 4.8% stearic, 25.3% oleic, 37.5% linoleic and 14.2% linolenic acids in rubber seed oil. The compositional variation is mainly due to the effect of temperature and location of the rubber plantations. The oil is effective against house flies and lice. It's insecticidal properties against stored grain pests also have been tested. It inhibits emergence of adult weevils (Nicholas and Daniel, 2005). It serves as a substitute of linseed oil. The oil aso can be used for soap making.

Table 131: Fatty acid composition of rubber seed oil

Fatty acid	Composition%
Palmitic	7-9
Stearic	7-8
Oleic	21-23
Linoleic	40-42
Linolenic	19-22

Rubber Seed Meal

The cake mostly serves as organic manure or animal feed. It contains 4-6% oil, 30-34% protein, 40-44% carbohydrates, 7-12% fibre and 6% minerals. The meal is similar to linseed meal. Staight meal is not liked by animals. However, it can be part of compound feeds. It increases the milk yield of animals, when consumed. At 10% level in compound feeds no detrimental effects have been observed (Ketkar, 1976). The seed also in combination with other feeds can be utilized for feeding animals. The seed shells also have similar uses.

Kusum (*Schleichera trijuga*)

Kusum is a dry forest tree. It grows up to a height 35-45 feet. It is the host for lac insects. The tree is a shade bearer and is frost and drought hardy. The fruit has a pulpy edible aril with one or two seeds (0.5 to 1.0g). The fruit pulp is edible and very tasty. Raw fruits are pickled. Ripe fruits have pleasant acidic taste. The seeds are obtained by plucking bunches of fruit and depulping. The seeds are eaten raw or roasted. The kernels constitute 60-65% of the seed. The trees which flower in February-April yield fruits in June-July, coinciding with the monsoon in India. The flowers are the source of a natural dye. The fruits are smooth, hard skinned berries, which contain one to two ellipsoidal seeds.

Kusum Seeds

Kusum fruit bunches are collected when they are slightly immature from the trees. They are heaped and dried. The pulp is removed by rubbing the fruit in water. The wet depulped seeds are further dried by spreading on the ground. Well dried seeds can be stored easily for more than 3-4 months in gunny bags. Dring storage some reduction in seed constituents occur (Srivastava and Pandey, 2001).

The oil content of the seed ranges from 50-62% . *Kusum* oil is obtained by pressing the seeds in an expeller or through boiling. A yield of 27-32% oil can be obtained through *ghanies* from the seeds. Expellers can yield 36% oil (Lakshmikantan, 1978). The yield of oil is low compared to the actual content of 51-62% oil in the seeds. The oil known as macasser oil in commerce is yellowish white to yellowish brown, semi-solid fat with a faint odour of bitter almonds.

The potential of the oil is estimated at 66,000 tonnes, but the collections are low and fluctuate between 4000 and 5000 tonnes per annum in India. The oil is composed of 37% glycerides and 58% of a cyanolipid.

Kusum Oil

The FFA varies widely and most commercial samples have FFA of 5-11%, I.V., 56-60 and S.V., 215-220. The major fatty acids are oleic 42, arachidic 20 and a C_{20} monoene acid 15%.The oil is unique among vegetable oils as it contains only 3.6-3.9% glycerine as against 9-10% in other oils and fats(Bringi, 1988). The oil on saponification gives 0.8 to 1.2% sodium cyanide expressed as HCN due to the presence of cyanolipids.

The characteristics of the oil are similar to other vegetable oils. It has higher S.V. indicating the presence of somewhat low molecular weigh fatty acids.

Table 132: Characteristics of *Kusum* oil.

Characteristic	Value
Specific Gravity- (32° C)	0.9099
RefractiveIndex (40°C)	1.4607
Acid value	31.3
Saponifiable Value	234.3
Iodine Value	60.2
Unsaponifiables %	2.1
Acetyl value	4.0
Hehner value	88.7

Kusum oil is rich in oleic acid(43%) followed by arachidic acid(21%) and eicosenoic acid (15%). It also contains 8% palmitic and 4.5% linoleic acids. Stearic, behenic and erucic acids are present at a lower range of 1-2% each. Mono-enoic fatty acids appear to be the major fatty acids in *kusum* oil.20:2 and 22:1 are also present at around 1-2%.

Table 133: Fatty acid composition of *kusum* oil.

Fatty acid	% composition	Fatty acid	% composition
10:0	0.2	18:1	42.6
12:0	0.3	18:2	4.5
14:0	0.3	20:1	15.2
16:0	8.0	20:2	0.6
16:1	1.3	22:1	1.9
18:0	2.3		
20:0	20.6		
22:0	1.5		

Uses of Kusum Oil

Kusum oil is not safe for edible purposes as the metabolism and biological effects of cyanoalcohols in the body are not known. However, the oil is a valuable raw material for soaps (91% fatty acids) and represents an additional source of scarce hard oil (Titer 46°C). But two major problems are encountered in saponification 1) the lye will contain 2.2% sodium cyanide, whose disposal is a problem requiring large dilution to bring CN to environmentally acceptable levels and (2) the glycerine in the lye arising by saponification of the 36%

glycerides cannot be recovered as the cyanogen containing alcohol/or the degraded product is highly corrosive to the equipment apart from the possibility of contamination of the distilled glycerol. *Kusum* soap is soft with more lather.

The cyanide in the oil makes it medicinal and is useful in treating itches and aches. The oil is used in indigenous medicine. In some areas it is used as edible oil and purgative. It can be used as a prophylactic agent against cholera, rheumatism and some skin diseases. It also serves as an illuminant, lubricant and hair oil. As a hair oil it has a stimulating and cleansing effect on the scalp. It is useful as a lubricant. The oil is a good source of the fatty acids oleic and arachidic acids.

Kusum oil has insecticidal activity. The cyanoester has also been synthesised with the same or different fatty acids using Wittig reaction. Cypermethrin, a synthetic analogue of natural pyrethrin, is an effective widely used insectide.

Technology for Removal of Cyanide

Since the oil is a mixture of glycerol esters and cyanogen containing alcohol esters, it is possible to preferentially hydrolyze the latter leaving the glyceride intact. This possibility has been realized and has formed the basis for developing a safe technology, free from environmental hazards associated with sodium cyanide.

Aqueous hydrolysis of *kusum* oil with ZnO at 120°-130°C (2kg/cm^2 pressure) preferentially hydrolyses cyanogen containing alcohol esters leaving glyceride unaffected. (The normal temperature for glyceride splitting is 230° C at 27 kg/cm^2). The fatty portion after hydrolysis contains the glyceride originally present in the oil and 65% fatty acids from cyano ester hydrolysis. This mixture on saponification gives less than 50 ppm cyanide as compared to 11000 ppm expected by saponification of the crude oil. Such treated fat can be mixed with other fat charges for soap making and the glycerol originally present in the oil can be recovered fully (Bringi, 1988).

Kusum Cake

The cake contain 5.5%N, 2.08%P and 1.13%K. It serves as a good organic manure and fuel. If the cyanogenic glycosides are detoxified it has scope to be used as a good cattle feed.

Kusum Timber

Timber is very useful. *Kusum* wood is hard to saw when dry. It is well known for it's toughness and hardness. Hence it is used in construction material and aricultural and household tools. Bark is rich in tannins and is useful in treating burns, wounds, inflammations pains etc. Juice is also used to treat malaria. The tree also yields a resin.

Kusum Leaves

Young shoots and leaves are lopped for fodder. Leaves are rich in sugars, fibre, protein and minerals. It is a good roughage for animals.

Crambe

Crambe abyssinica is an industrial oil crop. *Crambe* belongs to *cruciferae* family. It is native to mediterranean region extending from central Asia to Western Europe (Abyssinia). *Crambe* is a winter crop and can be grown in the *rabi* season (October to March) in India. Below –7^{0}C the crop gets killed. *Crambe* is an erect annual herb with many branches. The crop is grown at 60-90 cm spacing. It grows well on well drained soils of moderate fertility. It takes about 90 days for maturity. *Crambe* is a self pollinating crop. One flower bears one spherical seed which is present inside a papery thin pod. It normally yields 500 –800 kg/ha seeds. But under good management, seed yields upto 2000 kg/ha can be realised. The seeds can easily be separated from the mature pods. The seeds weigh 7.0 to 7.5g/1000 seeds. The hull content ranges from 20-40%.

The seeds contain 26-38% oil with a mean value of 32% (Earle and Peters,1966). The dehulled seeds contain 33-54% oil with a protein content of 30-50% (Earle, *et al.*, 1966). *Crambe* varieties grown in Kashmir region of India, have an average seed oil content of 18.4% and the deoiled meal, a protein content of 33.6% (Dhar *et al.*, 1988) Twelve *Crambe* genotypes have been analyzed for their seed oil, seed protein, and fatty acid composition (Nagaraj, 1998). The seed oil content is around 31.4% (range 30-34%), while protein content averages around 24.2% (range 23-26%).

The seeds are cleaned and the hull removed before oil is extracted using expellers. The residual oil in the cake can be removed through solvent extraction. The oil is refined, bleached and deodorised through normal procedures. The oil is light yellow in colour. It is a semidrying oil (I.V. =93). The oil has higher smoke point and lower level unsaponifiables. It has higher viscosity and hence serves as a good lubricant.

Table 134: Characteristics of *Crambe* oil.

Characteristic	Value
Specific gravity	0.908
Viscosity at 25^{0}C	85 CP
Refractive index 20^{0}C	1.472
Refractive index 40^{0}C	1.466
Melting point	6^{0}C
Iodine value	93
Smoke point	271^{0}C
Unsaponifiables	0.5%
F.F.A	0.2%

Source: Anonymous, 1962

The fatty acid composition of *crambe* grown in Hyderabad, India as post-rainy crop is palmitic =1.7, stearic 0.4, oleic=16.2%, linoleic=7.7, lenolenic=6.7 eicosenoic =3.9, and erucic acid =61.8%(Nagaraj, 1998). It is evident that *crambe* oil contains high levels of erucic acid. *Crambe* oil has more erucic and oleic acids along with some quantity of eicosenoic acid. The total monoenoic acid content is around 70-75%. This as well as the lower levels of lenoleic and lenolenic acids impart *crambe* oil a better shelf life than even mustard oil.

Table 135: Fatty acid composition.

Fatty acid	% composition
Erucic	55-60
Oleic	15
Linoleic	10
Linolenic	7
Eicosenoic	3
Palmitic	2
Behenic	2

Source: Anonymous, 1962

Utility of Crambe Oil

After Western countries have switched over to zero erucic (canola) rapeseed, *crambe* has become the major source of erucic acid. Erucic acid and its products are useful as slips agents, textile softeners, lubricants , emollients, pour point depressants, food emulsifiers and in hair care products. Erucic acid can be converted into erucamide which is an important chemical in the manufacture of nylon (13 and 1313). On oxidative ozonolysis erucic acid breaks down to pelargonic acid and brassylic acids which are useful in the manufacture of aviation lubricants. Hydrogenated derivatives of erucic acid are useful as waxes. These waxes are similar to sperm whale oil. Erucamide is useful as slip and antiblocking agent. Behenyl amine obtained from *crambe* oil is a corrosion inhibiter. Erucic acid also serves as a raw material for making adhesives (Neischlag and Wolff, 1971).

Crambe Seed Meal

Crambe seed meal contains 35-45% protein. The cake also contains around 35% carbohydrates and 8% minerals. The cake has a good potential for use as animal feed. The fibre content of the meal is very low. If the antinutrients like glucosinolates are removed, the cake can easily be put to better uses than as an organic manure.

Like rapeseed mustard, *crambe* contains glucosinolates which on hydrolysis by enzymes *in vivo* get converted to the thiocyanates. The major glucosinolate present is epi-progoitrin (Daxenbichler, *et al.*, 1965). The glucosinolates can be detoxified by heat or ammoniation (Kirk, *et al.*, 1966). Seed meal treated with soda ash, is free from epi- progoitrin and is more palatable to animals (Mustakas *et al.*, 1968, Mustakas *et al.*, 1976). Glucosinolates can also be deactivated by microwave radiation (Lesserman and Kirleic, 1979, Medeiros *et al.*, 1978).

Table 136: Composition of *crambe* seed and meal.

Constituent	Whole seed	Dehulled seed	Seed meal
Moisture	7.1	4.6	6.8
Crude fat	33.3	45.6	0.4
Protein (Nx6.25)	17.1	24.2	44.8
Crude fibre	14.0	3.1	4.6
Ash	5.3	4.2	7.9
N. free extract	23.2	18.3	35.5

Source: Mustakas, *et al.*, (1976).

The cake is also rich in amino acids like arginine, aspartic acid, glutamine, histidine, leucine, lysine, methionine, proline, serine, tyrosine and valine. In amino acid composition it is comparable to many traditional oilseeds. It is richer in glutamine, arginine, aspartic acid, leucine, proline and lysine. It is deficient in sulphur containing amino acids

Table 137: Amino acid composition of *Crambe* meal.

Amino acid	g/16 g	Amino acid	g/16 g N
Glutamine	15.5	Threonine	4.1
Arginine	6.4	Isoleucine	3.7
Aspartic acid	6.3	Phenylalanine	3.7
Leucine	6.1	Serine	3.6
Hydroxy proline	0.4	Tyrosine	2.7
Proline	6.1	Histidine	2.3
Lysine	5.3	Methionine	1.6
Valine	4.7	Alanine	4.1

Source: Mustakas, *et al.*, 1965

Crambe meal contains sinapine, the choline ester of 3,5 dimethoxy -4-hydroxy cinnamic acid at a concentration of 0.27-0.56% (Kirk *et al.*, 1966, Austin and Wolff,1968). Sinapine, a bitter constituent is also present in rapeseed mustard. It is bitter only to human beings. Sinapine during germination gets hydrolysed to sinapic acid and choline. Sinapic acid is useful in the biosynthesis of lignins, while choline is useful in the methylation cycle. Sinapine is not degraded by the endogenous enzymes. When it is fed to rats there is no effect on their growth, feed intake or feed efficiency (Austin and Wolff, 1968). There is thus a good scope for utilization of *crabme* meal when available in sufficient quantity.

Mango (*Mangifera Indica*)

Mango is the king of fruits and accounts for more than 50% of India's total fruit production. It is a wild or semi wild evergreen tree growing in tropical and subtropical climates. It seems to have originated in Indo-Burma (Myanmar) region. It belongs to the family *Anacardiaceae*. It is an evergreen large tree with a wide crown. It is grown in orchards, plantations and on roadsides. Mango trees are grown mainly for the delicious fresh fruits. It flowers in January–February and fruits in April-June period in India. Raw mango is used for pickles while the ripe mango is used as a fruit and liked by one and all, the world over.

The fruit has 85% of skin and pulp. The stone (seed) which is thrown away constitutes around 3-25% of total fruit (Lakshminarayana *et al.*, 1983).The kernel inside constitutes 54-85% of the stone which is kidney shaped and contains 9-14% oil. The average fruit weight among 18 cultivars varies from 155 to 1070 g. Air-dried stone content varies from 17 to 39% and kernel content varies from 61 to 83% (Ramalinga swamy, *et al.*,1980) of the stone. Sometimes the shell to kernel ratio in the dry stone may be 1:1. India has a potential for production of 35,000 tonnes of mango fat annually.

Mango Kernel

The kernel has medicinal value mostly as anthelminthic. Collection and decortication are being attempted, in view of the importance of mango kernel fat as tallow and cocoa butter substitutes. Mango stones are dried in the sun to a moisture content of 13%. They are decorticated by hand. A mechanical decorticator, if available is useful in handling large quantities. The kernels are edible and consumed in some parts of India, especially during scarcity times. It can easily be stored for a year. It is slightly astringent and eaten after roasting, boiling or leaching with water.

Mango kernel contains around 12% fat and 10% protein. The major constituent, however is starch at 70-75%. Minerals constitute around 4%. Soluble sugars and tannins are present in small quantities. The protein and starch are easily digestible. The ash portion consists of mainly Ca, Mg and P. Mango kernel is used in making pickles, fluor and starch. Mango kernel starch is used in textile, paper and jute industries. The fat content among 18 cultivars in the mango kernel varies from 5.8 to 11.3%(Ramalinga swamy, *et al.*,1980).

Table 138: Composition of mango kernel.

Constituent	%
Protein	9.5
Fat	11.0
Starch	72.8
Sugar	1.0
Tannins	0.1
Ash	3.6
Digestible proteins	70.0
Starch equivalent	67.0

Source: CSIR, (1962).

Processing Mango Stone for Fat

From the point of view of processing, mango is an important fruit providing large amount of waste material in the form of peels and stones. Mango seed kernel is comparable to cereals such as rice, wheat, maize and barley with respect to its content of carbohydrates, proteins, fat, calcium and phosphorus. It is a rich source of starch and fat. Because of their size and shape, and toughness of the shell wall, mango stones are difficult to decorticate for separating the inner kernel. Development of a mango stone decorticator by Narasimhachar *et al.*, (1975) is an important breakthrough in the utilization of mango kernels. The degree of decortication attained in this continuous-operation in the mechanical decorticator is almost 100%, and the yields are quantitative. Dry mango kernels can be stored without deterioration for a period of more than a year.

The air dried stones yield kernels which can be sun dried to a moisture content of 10%. The cooked kernels are passed through an expeller, and the resulting cake can be extracted with hexane. The fat from the kernels is extracted mostly by solvent extraction. The kernels are broken into small pieces, flaked by beating after steam treatment. Very thin flakes with 10-12% moisture are then extracted in continuous type extraction mills using hexane at 40^0-50^0C. Later the hexane is removed at 110^0C to obtain the fat. The deoiled meal also is desolventized. Mango kernel fat is obtained in 11% yield.

Mango Kernel Fat (MKF)

The fat obtained from fresh kernels is cream to pale yellow in colour. Because of higher melting point it is solid at room temperature. Blandness, plasticity and absence of toxic substances make it a potential edible fat.

The fat characteristics are as follows; acid value 2.1-6.2; iodine value 31.6-59.1; saponification value 189.9-194.7; unsaponifiable matter, 1.0-2.5% and melting point, 35.0-49.0°C. MKF contains 96.1% neutral lipids and 3.9% polar lipids with 2.9% glycolipids and 1% phospholipids (Hemavathy *et al.*, 1987). MKF contains 52% oleopalmitostearin and 19% oleodistearin.

Table 139: Characteristic of MKF.

Characteristic	Range
Acetyl Value	6-12
Iodine Value	32-60
Saponifiable Value	190-195
Refractive Index at 40^0C	1.4550-1.4570
Unsaponifiables %	0.75-2.4
Specific gravity	0.91
M.P	38- 51^0C
Slip point	34-43^0C
Bellier number	38.5

Table 140: Glyceride composition of MKF

SSS	1-4%
SSU	1-3%
SUS	38-64%
SUU	30-45%
UUU	4-13%

Generally fatty acid composition varies with location and the variety. MKF is rich in saturated fatty acids namely palmitic and stearic acids. The oleic acid content is around 45%. The fatty acid composition is as follows: 16:0=3.5-17.8; 17:0=traces-2.0; 18:0=24.0-56.9; 18:1=33.3-52.6; 18:2=1.3-13.4 and 20:0=0.4-2.6. In general MKF from India, Bangladesh, Malaysia and Egypt have similar oleic acid levels ranging around 35- 42%. Zaire samples have a little higher level upto 56% oleic acid. Indian and Malaysian samples are richer in stearic acid ranging from 42-48%. Zaire sample has the lowest level of 31% with Bangladesh and Egypt having levels ranging from 38-42% oleic acid. Saturated and unsaturated fatty acid levels must be related to the teperatures that prevail in the different countries at the time of fruit development and maturation.

Table 141: Fatty acid composition of MKF from different countries

F.A	% Composition				
	Zaire[1]	**India[2]**	**Bangladesh[3]**	**Malaysia[4]**	**Egypt[5]**
Oil %	6.8-12.6	9.1-12.3	7.1-10.0	7.5-8.8	9.7-10.8
12:0	0-0.2	0.3-0.4			
14:0	0.1-0.7	0.5			
16:0	7.0-9.1	5.1-8.0	7.0-10.0	6.9-7.3	7.6-8.4
18:0	31.0-43.0	42.3-47.6	38.2-40.2	44.3-44.4	39.7-40.4
18:1	43.5-56.3	34.9-42.0	41.1-43.8	38.9-42.1	41.2-43.3
18:2	3.5-7.8	4.2-8.3	6.0-7.6	4.5-7.4	5.7-6.3
18;3	0.2-0.9	0.2-0.4	0.6-1.0	Three-0.6	0.7-1.0
20:0	1.0-1.8	2.2-3.1	1.7-2.6	1.6-1.8	1.8-1.9

Source : [1] Van pee *et al.*, 1981, [2] Shah *et al.*, 1983, [3] Ali *et al.*, 1985, [4] Augustin and Ling 1987, [5] Abd El–Aal *et al.*, 1987

MKF as CBS

MKF and mango fat stearin are refined through phosphoric acid treatment to remove gums, and alkali treatment to reduce FFA. The oil is then dried under vacuum at 100^0 C and bleached using bleaching earth and carbon. MKF when utilized as CBS does not show any differences in taste, texture and odour of the ice cream, butter scotch and toffees (Muharram and Mustafa,1982). MKF with a M.P. of 36^0 C can easily be blended with cocoa butter without any change in the properties (Baliga and Shitole, 1981). MKF contains high levels of 18:1 and 18:2 in the sn-2 position and thus is similar and a better substitute of shea butter (Van Pee *et al.*, 1981). MKF is utilized mainly as CBS. As MKF is softer, its quality as CBS is improved by selective solvent fractionation and partial hydrogenation.

Hemavathy *et al.*, (1979) have carried out experiments on the fractionation and modification of mango kernel and sal (*shorea robusta, Dipterocarpaceae)* fat for use in foods. These two hard fats of tree origin have recently received attention in India for the purpose of preparing a cocoa butter substitute or extender. Fat fractionation is necessary to match cocoa butter characteristics. In general, solvent fractionation raises the stearic acid content of mango fat by about 5% with a proportionate decrease in the oleic acid content. Mango fat stearin is almost similar to cocoa fat. The cooling curves of mango kernel and sal fat are different from those of cocoa fat. Based on the cooling curves, mango stearin obtained by solvent fractionation appears more suitable as a cocoa butter substitute.

MKF is also utilized as an ointment base. It is better than paraffin base in releasing the drugs quickly. Mango fat when randomised and fractionated or co-randomised with coconut oil becomes a useful raw material for ointment base and its stearin fraction is a good suppositoty base (Biswas and Bhattacharya, 1989).

Utilization of Mango Kernel

Mango stone, like the pulp, finds many uses. The stony portion is used as a fuel in India, while the kernel is utilized for making such products as pickle, flour, starch, and cattle feed (Mittal and Bhagwan Singh, 1973). It is estimated that about 140,000 tonnes of starch can be obtained annually from kernels alone, together with the tannins and fats as by-products. This starch can be used in textile, jute and paper industries. The dried kernels with less than 10% moisture content can be used as animal feed, because of a balanced proportion of amino acids (Dutta and Bhattacharya,1959) present in them. Feeding experiments (Bose *et al.*,1952) conducted on poultry have shown that a mixture of maize meal and mango kernel flour to the extent of 20% have no detrimental effect on the development of chickens. Dried kernels can be fed as such to animals without any further treatment or processing. Mango kernel serves as a good manure for plants. Application of powdered mango kernel compost at half the dose of FYM has given better crop response and yields (Dum and Hilditch, 1947).

Defatted Meal

The proximate composition of defatted mango kernel meal is as follows: fat 0.41%, protein 6.2%, ash 3.6% and crude fibre 4.5%. The protein and ash contents in the defatted mango kernel meal among 18 cultivars varies from 5.9 to 8.1% and 1.4 to 2.4% respectively (Ramalinga swamy, *et al.*,1980). The defatted meal is rich in starch and thus is a source of industrial starch. It is edible and can be eaten after drying, though slightly astringent. Since it is poor in nitrogen it is not very useful as animal feed. However it is free from toxic constituents and hence it is normally utilized as cattle feed. Mango shell (outer portion or stone) can be used as fuel.

Moringa

Introduction

Drumstick is a popular vegetable in India. It is very delicious and is liked by one and all. The green unripe pods range in length from 50 to 90 cm. The pods are triangular in shape, with trigonal winged seeds inside. The tree is grown in the back yards in most of the households in India. Due to demand and popularity *moringa* plantations are gaining importance in recent years in India.

Moringa is a medium sized tree, growing upto a height of 10m. It is indigenous to North-West India and is most suited to alluvial soils and sandy beds of rivers and streams. The trunk and stem of the plant are thick, soft corky and deeply fissured. Leaves are usually tripinnate and leaflets elliptic. The flowers are white in colour with some fragrance.

Normally a good mature *moringa* tree can yield 500-1000 fruits. Since each seed weighs about 0.3g and there are around 10 seeds per mature pod, there can be about 3 kg of seeds per plant. Thus there can be a theoretical oil yield of 1 kg oil per plant. Since the bulk of the drumsticks gets utilised as vegetable, only a limited quantity of over mature pods will be available for oil expulsion. Inspite of the limited quantity of oil that may become available, it has a good potential because of its utility as edible, industrial and medicinal oil.

Seed Composition

The seed contains about 20% shell and 70% kernel. The kernel in turn is rich in proteins and oil with about 35% each of these constituents. *Moringa oleifera seeds of* Pakistan have 38-42% oil, 26-32% proteins, 6-9% fibre, and 5-8% ash (Farook and Bhanger, 2003). *Moringa stenopetala* seeds contain 35.7-45% oil (Lalas *et al.*, 2003). The indian seeds have a low oil content of around 21% (Nagaraj and Mukta, 2004) Both the proteins and oil are balanced with respect to their constituents namely amino acids and fatty acids. The protein is rich in leucine while the oil is rich in the mono-unsaturated fatty acid namely oleic acid. It is a non drying oil resembling olive oil in its characteristics. It has a pleasant odour and can be utilised as an edible oil and also as an ingredient of cosmetics.

Moringa Oil

The *Moringa* seed oil is non-yellowing, non-drying and has tremendous medicinal properties and finds use to cure rheumatism and goitre. The oil is particularly valuable for ointment as it does not undergo oxidation on exposure for long time. The oils from *M. oleifera* and *M. concanensis* constitute between 27 to 38 percent with varying chemical composition owing to regional variation and/or source of seed collection.

Table 142: Characteristics of *moringa* oil

Characteristic	Range
Density	0.904-0.908
Refractive index (40°C)	1.4590-1.4625
Iodine value	68-72
Unsaponifiables	0.6-1.1
Colour (1cm cell)	0.95-1.10R +20-35Y
Saponification value	177-184
Smoke point (°C)	198-202
FFA (% oleic)	0.5-0.74

Source: Farooq and Bhanger, 2003 and Farooq *et al.*, 2005

The characteristics of *moringa* oil are s similar to other vegetable oils. It's FFA, unsaponifiables and I.V. are low. It's S.V. is around 180 and smoke point is 200° C. It is a stable oil suitable for frying purposes.

The oil is oleic rich (76-83%). *Moringa oleifera* oil of Indian origin has the fatty acid composition of palmitic=7.3, stearic=6.0, oleic= 77.0, arachidic =4.0, behenic=3.0, palmitoleic= 2.0 and linoleic acid= 0.6% (Nagaraj and Mukta, 2004). Pakistan seed oil has 79% oleic acid followed by 7% palmitic, 7.5% stearic, 6% behenic and 4% arachidic acids(Farook and Bhanger, 2003). *Moringa stenopetala oil* contains 76.4% oleic acid with around 6% each of behenic and palmitic acids. The oil has very good frying stability (Lalas *et al.*, 2006). With respect to oleic acid content, moringa oil appears unique. In these days when scientists are trying to identify high oleic types in each and every oilseed *moringa* needs special attention.

The blending of *Moringa* oil with phulwara butter (*Madhuca butyracea*) can be exploited as potential source of value addition product of rheumatism as the oil of *M. butyracea* also has the potential to cure rheumatism and pain (Verma and Baneraji, 2003).

The blending of two oils in 1:1 and 1:2 ratio can be done to get a product having better property. The fat blends have SFA, MUFA and PUFA as 32.9, 56.9, 10.2 in 1:1 ratio while in 1:2 ratio it is 45.9, 43.8, 10.2.

The oil is rich in beta sitosterol (52% of total sterols), stigmasterol (17%) and campesterol (14%). Alpha, gamma and delta tocopherols at 98,45 and 82mg/kg are present (Lalas *et al.*, 2003). Somewhat different levels of tocopherols, namely, alpha, gamma, and delta at 114.50-140.42, 58.05-86.70 and 54.20-75.16 mg/kg oil respectively have been reported. The major sterol fractions of the oils are campesterol (14.13-17.00%), stigmasterol(16-19%), beta sitosterol (45-53%) and delta avenastero(9-11%). *Moringa oleifera* with good seed and oil composition appears to be a very good oleaginous crop (Farooq *et al.*, 2005).

Moringa Seed Cake

The oil cake remaining after extraction of oil is bitter in taste with a protein content of about 60%. It has a potential to be utilised as animal feed. It is however used as an organic manure. The trunk may not be useful as timber but can be used as fire wood. Thus *moringa* tree and its various parts have wide ranging medicinal, industrial and edible uses. *Moringa* oil cake has been recommended as a protein supplement to monogastric animals at a maximum of 13.5% of total feed intake (Kakengi, *et al.*, 2006).

Moringa Leaf Extract as a Plant Growth Hormone

Juice from fresh *moringa* leaves can be used to produce an effective plant growth hormone. It is useful in increasing yields by 25-30% of crops like onions, bell pepper, soya, maize, sorghum, coffee, tea, chilli and melon as a foliar spray. One of the active substances is zeatin: a plant hormone from the cytokinins group. This foliar spray has to be used in addition to other fertilizers, watering and sound agricultural practices.

Leaf juice is made in the following manner: a) The young *moringa* shoots (not more than 40 days old) are ground with a bit of water (about one liter per 10 kg fresh material). b) The solids are filtered out of the solution with a cloth followed by wringing out the liquid. The solid matter, with about 12-14% protein, can be used as livestock feed. c) the extract is diluted with water at a 1:32 ratio. This can be sprayed directly onto the plants (if the extract is not going to be used within five hours, it is best stored in freezer until needed).

About 25 ml is applied per plant at 10 days after the first shoot emerges, followed by another spray at 30 days prior to flowering. One more spray during seed maturation phase is useful. Yields upto 30% higher can be obtained with sprays of moringa leaf extracts.

Utility of Moringa

In addition to the pods which are valued as vegetable, flowers, leaves and seeds are also edible and utilised in many ways(CSIR,1962). The pods are cut into slices and utilized as a sole vegetable or as a constituent of culinary dish like *sambar*, a liquid spicy cuury, most popular in South India. It is also pickled and consumed. The dried and fried seeds taste like groundnuts. The seeds are useful in reducing fever. The fresh pods have the following composition: 85% moisture 2.5% protein, 0.1% fat, 3-7% sugars, 4-8% fibre and 2% minerals. It is rich in calcium, phosphorus, iron, copper and iodine. The pods also contain carotene (Vitamin A precursor), nicotinic acid and Vitamin C (Nagaraj,2000).

The leaves and flowers apart from their utilization as pot herb are cooked along with pulses and eaten. A little unpleasant taste present in the leaves can be removed if they are boiled and utilized. It is a special healthy and nutritious dish served in South Indian households. An ordinary tree can yield abut 150 kg leaves. The leaves are rich in calcium, phosphorus and vitamins A and C. They are helpful in preventing scurvy, beri beri and catarrh (discharge of fluid from mucous membrane). The leaves are also useful as emetic. The leaf paste helps in healing wounds. The flowers also serve as tonic, purgative and diuretic.

The seed oil is useful in treatment of gout and rheumatism. The seed powder is water purifier. The leaf juice in combination with honey and or lemon juice is a household remedy for night blindness, dizziness, acme, forgetfulness etc. The gum coming out of the tree is red brown in colour and is used in calico-printing.

The nutritional study of drumstick seed oil on rats has not shown any adverse effects. The rats fed with drumstick seed oil have exhibited satisfactory growth rates. So drumstick seed oil appears to be useful as an edible oil. (Raval and Toliwal, 1996).

Information on the composition of *moringa,* its various plant parts and products is somewhat limited. This is especially so with regard to the seed in relation to harvest of higher oil. More systematic studies are needed especially relating to it's various neutraceutical values.

Tumba

Tumba (*Citrullus colocynthis Schrad.*), is a xerophytic monoecious, perennial trailing herb. It can produce 1-2 quintals oil rich seeds per hectare after every soaking rain, in April or August. It has vast potential as an oil crop in arid lands. It can easily stablize the agricultural production and improve the economy of the arid region. It's numerous branches spread over 10-15 meters in all directions would provide the desired vegetative cover for binding the loose sandy soil beneath and improve the physical and chemical properties of the soil.

Prospects

Tumba plant is an efficient utiliser of the scant rain fall. As the problems of establishment and growth phenophases in conventional annual crops are absent in *Tumba,* its productivity is high and stable. Further the dense soil cover not only binds the loose sand and regulates the temperature. It appreciably reduces the impact of raindrop splash and thus is a boon to improve the fragile ecosystem of arid lands.

Tumba plant is a creeper and grows wildly in semi-arid tracts of Gujarat and Rajasthan in India. According to Joshi and Solanki (1988) *thumba* can yield on an average 1-2 quintals seed per ha. Khadi and Village Industries Commission of India (KVIC) has estimated the availability of seed to be 23,000 tonnes with about 2700 tonnes of oil.

Tumba seeds contain (20-30 per cent), apale yellow oil. The oil from decorticated *Tumba* seeds is very good nutritionally. Information on quality parameters from different institutes confirmed its balanced fatty acid composition.

The worth of *Tumba* as an oil crop with high and stable productivity has been established in the recent past and accepted by the farmers particularly of Barmer, Jodhpur and Jaisalmer districts of Rajasthan in India. At present nearly 4 lakh tonnes of *Tumba* seeds are crushed every year for oil while a similar quantity is fed as fruits to animals particularly goats (Joshi, 1987).

Though *Tumba* farming is a new proposition in India, it is being cultivated as a pure crop as well as mixed crop with maize in Nigeria in nearly 3.5 lakh hectares under the local name *Egusi*. It's seeds and oil are indispensable parts of human diet in that country. While decorticated seeds are the major ingredient in evening supper, *Egusi* oil is preferred over imported palm oil by the local people. There are a good number of sweet meats and other preparations made from *Tumba* seeds in Nigeria.

Tumba Oil

Analyses of 168 germplasm accessions (Mev Singh *et al.*,1991) have revealed the presence of considerable variability in oil content (11-23.2%) as well as fatty acid composition. Only 40 accessions have less than 20% oil content. The oil is rich in linoleic acid ranging from 51.9 to 63.0% followed by oleic acid (12 to 27.2%) and palmitic acid (9.4 to 14.1%). Among the accessions examined GP-3 has the highest linoleic acid and GP-132 has higher oleic acid while GP-133 has the least linoleic acid content. Iodine values range from 111 (GP-137) to 123 (GP-29). It is a semidrying oil. Refractive index values range between 1.4740 and 1.4770.

Table 143: Quality attributes of *tumba* oil

	NIN	TPI
Fatty acids (%)		
Palmitic acid	14.79	9.6
Palmitoleic acid	4.48	
Stearic acid	8.68	6.6
Oleic acid	30.10	21.1
Linoleic acid	41.82	59.2
Iodine value	129.5	124.6
Saponification value	185.1	197.4
Free fatty acid (%)	1.5	1.6

NIN: National Institute of Nutrition, Hyderabad

TPI: Tropical Product Institute, London (Davis *et al.*, 1983)

Tobacco Seed Oil

Tobacco is the most widely grown commerical, non-food crop, entering world trade entirely on the basis of its leaf. Out of 68 documented species of the genus *Nicotiana,* only *N.tabaccum* L. and *N. rustica* are grown commercially. Over 85% area is under *N. tabaccum*. Tobacco appears to have originated in South America. It is mainly propagated through seed. Nurseries are raised initially and the seedlings are transplanted in October-November. The leaf, the main product of economic importance, is harvested during February-April in India. If left untopped the seeds become available in April- May and can easily be collected.

Apart from its conventional usage as cigarette, *bidi,* chewing, *hookah, zarda,* snuff, *cheroot* etc., scientists all over the world are working on the alternate uses of tobacco as it is an excellent source of several phyto-chemicals of pharmaceutical and industrial importance. The most important and well known chemical is nicotine, an alkaloid. Solanesol-a tri-sesquiterpene alcohol, another tobacco chemical, has recently gained lot of importance as pharmaceautical compound. Besides these chemicals, a number of other possibilities also exist like oil from the seed, edible protein from green leaf, rutin from cured leaf, furfural from the stem, etc.

Tobacco Seed

Tobacco plant is a prolific producer of seed having 35 to 39% oil. Tobacco seeds are so tiny that around 10,000 seeds weigh only one gram. The protective coat of the mature seed has a prominent epidermis and below this there are parenchyma cells. The outer wall is composed of cellusose. The endosperm contains thin walled cells rich in aleurone and oil. Tobacco crop aimed at higher seed recovery should be closely planted and kept untopped. In such cases, seed can be a source of oil and leaf can be utilized for extraction of phyto-chemicals like nicotine, solanesol, proteins and organic acids.

The oil content in tobacco seed varies from 35-39% and is comparable to mustard (33%) and sunflower (35%) and higher than safflower (30%) and soybean (17%). The seed also contains 25% protein, 20% fibre and 2-3% minerals.

About 30% yield of oil from a seed of 37% oil content could be obtained in a single crushing. Hitherto, only seed left over after tobacco production used to get processed for recovery of oil, as good quality tobacco leaf production practices do not provide scope for seed collection. Thus, sufficient quantity of tobacco seed is not available for commercial purposes. It is estimated that about 1300-1500 tonnes of tobacco seed oil is expelled annually and exported to other countries for utilization in the paint industry(Rama Varma *et al.*, 1964).

Table 144: Composition of tobacco seeed (% dry weight)

Constitutent	Tobacco seed
Moisture	3.3-5.1
Raw protein	25.0-25.9
Lipids	47.2-48.0
Fibre	19.9-21.2
Ash	3.2-3.6

Source: Frega *et al.*, 1991.

Tobacco, appears to have a good potential as an oilseed crop (Chakraborty, 1998). Different types of tobacco have been assessed for their oil yielding capacity. Maximum seed yield has been recorded with the chewing tobacco with an oil content of 37% and an oil yield of 409 kg/ha (Patel *et al.*, 1998). This oil yield is higher compared to the average oil yield of 200-300 kg in groundnut in India. *Rustica* seed, however, has higher oil content, but the total oil yield is lower. This finding clearly focuses the importance of tobacco as an oilseed crop with very good oil yield per unit area when compared with other oilseed crops.

Table 145: Seed yield, oil content and oil yield of tobacco.

Character	Types of tobacco		
	Bidi	**Chewing**	***Rustica***
Seed yield (kg/ha)	818	1106	729
Oil content (%)	37	37	39
Production potential (kg/ha)	303	409	284

Source: Patel *et al.*, 1998.

Production of 1171 kg/ha seed with a recovery of 433 kg oil/ha from a chewing tobacco variety A 145 under modified agro-techniques at anand, Gujarat has firmly established the potential of tobacco as an oilseed crop (Deo Singh, 2003). Later studies have shown the possibilities of producing tobacco seed to the tune of 15 to 20 q/ha.

Tobacco Seed Oil

Trilinolein and palmitodilinolein are the main triglycerides in the oil, accounting for 90% of the composition of the oil. The chemical properties of tobacco seed oil compare well with other edible oils. Acid value is generally lower for the fresh oil. The iodine value is 135 indicating it's unsaturated nature. The S.V. of tobacco oil is similar to other oils. Its unsaponifiable content is lower at around 1.5%.

Linoleic acid is the major(66-76%) fatty acid in tobacco seed oil. It has a palmitic acid content of 15% followed by oleic acid at 13% and stearic acid at 5%. Linolenic acid content ranges from non-detectable to 1.4%. Quality of the oil is better if linolenic acid content is below 5% (Patterson, 1989). Ratio of mono unsaturated fatty acids (MUFA) to polyunsaturatted fatty acids (PUFA) governs the stability of the oil and in this respect, tobacco seed oil is comparable to safflower seed oil. Ratio of PUFA to saturated fatty acids (SFA) is an index of nutritional quality of the oil. General nutritional quality of tobacco seed oil, free from nicotine, is better than groundnut, mustard and cotton seed oil and on par with sunflower seed oil and less than that safflower seed oil.

Table 146: Fatty acid profile of tobacco seed oil

Fatty acid	Tobacco seed oil
Palmitic	15.2
Stearic	4.8
Oleic	13.2
Linoleic	66.7
Linolenic	traces
PUFA level	66.7

Source: Thakur *et al.*, 1998.

The lipid classes of *N. tabacum* seed oil have been isolated by thin layer and gas chromatographic techniques. The lipid classes identified are hydrocarbons (1.4%), wax esters (1.7%), sterol esters (2.4%), triacylglycerols (69.3%), 1,3-diacylglycerols (4.6%), 1,2-diacylglycerols (3.5%), free sterols (2.3%), 2-monoacylglycerols (2.1%), 1-monoacylglycerols (1.8%), phosphatidylethanolamines (1.7%), phosphatidylcholines (0.9%), lysophosphatidylethanolamines (1.5%) and phosphatidylinositols (0.6%). Palmitic, oleic and linoleic acids are the major components of all lipid classes studies (Amran *et al.*, 2001).

Utility of Tobacco Seed Oil

Refined tobacco seed oil is reported to be used as edible oil in Bulgaria, Turkey, Tunisia and Greece (Eshetu, 2000). Tobacco seed oil is also used in paint and soap industry due to its higher unsaturated fatty acid content. The oil can also be used for control of root parasite, broomrape or orobanche.

The tobacco seed oil due to its high EFA (18:2) is as good as the best edible oil available in the world and it is good for heart (Anonymous, 2002). Another positive feature of tobacco seed oil is its low peroxide value (4.86) and up to a period of eight months, lipid peroxidation of the oil remains within reasonable limits (86.9) and compares well with groundnut oil (110.1) and superior to cotton seed oil (224). Subsequently, deterioration sets in faster (236.5) as compared to groundnut oil (154.8) but less than that of cotton seed oil (319.1).

In order to establish the potential of tobacco seed oil as edible oil, the refined oil has been subjected to rat feeding studies (Thakur *et al.*, 1998). Tobacco seed oil has not caused any adverse effect on growth and physiology of the rats, with respect to the total lipid content of serum, heart, liver, brain and kidney and is comparable to groundnut oil. In case of total phospo-lipid content of different organs, tobacco seed oil is on par with groundnut oil and histopathological studies of the major organs has not shown any toxic effect.

Toabacco seed oil is classified as a semi-drying oil. Because of its valuable non-yellowing property, the oil is extensively used in paint industry. The high degree of unsaturation of the oil can render it susceptible to auto-oxidation and polymerisation, resulting in cross-linked and tough films upon exposure to air. Alone or blended with linseed oil, tobacco seed oil can be used in the syntheses of modified alkyd resins for air drying glossy paints (Eshetu, 2000).

Tobacco Seed Oil as Biofuel

Tobacco seed oil, being non-edible presently, has a great potential for it's utilization as biodiesel. The physical, chemical and fuel related properties of tobacco seed oil have been investigated. The propeties are comparable to those of other vegetable oils and to current European specifications for automotive diesel fuel. This oil is an appropriate substitute for diesel fuel. The enviromental advantages of tobacco seed oil as a fuel can be exploited for specific niche markets such as inner city vehicles or tractors. Tabacco seed oil as a fuel represents one possible hope for the future of tobacco cultivation (Giannelos *et al.*, 2002).

Tobacco Seed Meal

The composition of tobacco seed meal is similar to those from other oilseeds. The tobacco seed cake after extraction of oil is rich in nitrogen (6%) and could serve as an animal feed supplement.

Prospects

The fact that the tobacco seed oil is better or comparable to other seed oils from the quality point of view has enlarged the potential and future prospects of tobacco seed oil as an edible oil after refining. Future research efforts therefore need to be aimed at breeding varieties with higher seed yield and more oil content and developing production technologies suitable for different agro-ecosystems, and exploring the possibilities of its greater industrial uses. Anti-smoking compaigns being intensified world wide, should stimulate this crop to be utilized for recovery of phytochemicals and the seed oil.

Jojoba (*Simmondsia Chinensis*)

Jojoba botanically known as *Simmondsia Chinensis* belongs to the family *Bixaceae*. It grows into a big, spreading bush about 10-12 feet in height and width. Jojoba seems to have originated in the Mediterranean-type climate along the North American Pacific Coast (Gentry, 1958). It is a shrub native to the triangle of the Sonoran Desert of Mexico, California and Arizona. The plant is being grown in a big way in Australia and Japan and is also tried in many countries like Saudi Arabia, Egypt, Ghana, Iran and Israel(Eckey 1954). The major potential of jojoba as a modern industrial crop is embodied in the liquid wax which make up about 50% of the seed's dry weight.

Jojoba is usually restricted to well-drained, coarse, well-aerated desert soils that are neutral to alkaline (pH. 7.3 to 8.2). The soil should be rich in phosphorus. It grows best where the annual rainfall exceeds 30 cm and can exist under annual precipitation as low as 2 to 11 cm. It tolerates full sun and temperatures ranging from 0° to 47°C. Mature shrubs tolerate temperatures as low as –10°C, but seedlings are sensitive to frosts just below freezing. An average seed yield of 1.2-2.2 kg/ shrub is obtained after four years of planting (Benzioni and Forti,1989).

Jojoba Seeds

Jojoba is known by a number of names like bucknut, coffee nut, goat nut, pig nut and lemon leaf. The fruit is a green elongated oval capsule enclosing one seed. The seed resembles pecan nut in shape. It is 12-18 mm in length, 6-12 mm in cross section and weighs 0.40 to 0.89g. Cotyledons make up most of the jojoba seed (Rost and Paterson, 1978). The jojoba oil is obtained from its fruit nut. Jojoba seeds contain 14% protein and 44% oil, while the dehulled seeds contain 16% protein and 51% oil or liquid wax (Weiss, 1983). The liquid wax is unique in plant kingdom and resembles in composition and properties with sperm whale oil (Benzioni and Forti, 1989). The liquid wax is not a triglyceride fat. The oil is stable even at high temperatures and pressures.

Jojoba Oil

Jojoba oil is extracted by mechanical pressing. Solvent extraction can be followed to remove the remaining oil in the meal after mechanical pressing. The extracted oil is relatively pure, nontoxic, and biodegradable. It is stable and has a long shelf life, as it contains natural antioxidants. Repeated heating to temperatures above 285°C for four days as well as high pressures do not alter its properties (Benzioni an Forti, 1989).

Table 147: Characteristics of jojoba oil

Characteristic	Oil from expeller press	Industrial grade
Average molecular weight	600-606	600-606
Iodine value:	82-83	80-85
I.V. of alcohols	77	
I.V.of acids	76	
Saponification value	92	90-95
Unsaponifiable matter	51%	
Acid value	1-2	<1
Phosphorus (by extraction of the oil)	0.001-0.006%	

Source: Wisniak (1977)

The oil or liquid wax is golden or light yellow in colour. It has a pleasant, slightly nutty taste. It is stable to oxidation and rancidity. Jojoba oil is characterized by specific gravity 0.863-0.865, refractive index 1.465-1.467, solidifying point 6.7°C, melting point 7-12°C, smoke point 195°C, flash point 290-295°C, fire point 330-338°C. Iodine value 82-89 and saponification value 82-167 (Weiss 1983; Katoh and Kunimoto 1983; Iyengar *et al.*, 1986). Hydrogenation of jojoba oil produces a hard, crystalline wax, with a melting point of about 70°C, similar to carnauba or candelilla wax.

Table 148: Physcial properties of jojoba oil

Characteristic	Value
Melting point	6.8-7.0°C
Freezing point	10.6-7.0°C
Boiling point (at 757 mm Hg under N_2)	398°C
Smoke point	195°C
Flash point	295°C
Fire point	338°C
Refractive index (25°C)	1.4650
Specific gravity (25/25°C)	0.863
Viscosity (different methods, 25°C)	33-37 cp
Heat of fusion by DSC	21 cal/g
Dielectric constant (27°C)	2.680
Specific conductivity (27°C)	8.86×10^{-13} mho/cm

Jojoba wax can be considered to be a homogenous substance, since it is composed of a homologous series of a small number of similar esters. There are small differences in the rates of reaction of the various esters, depending on their total length, but since the reactions go to completion, the oil may safely be regarded as consisting of a single chemical. This simplifies the chemistry of the wax for small-and large-scale reactions, both in the research laboratory and for industrial production.

This liquid, usully referred to as an oil, is similar in composition and propeties to sperm whale oil. Unlike other vegetable oils, it is not a triglyceride fat, but a wax composed almost entirely of esters of straight long-chain unsaturated fatty acids and alcohols. More than 87% of the oil is composed of straight-chain C20 and C22 fatty acids and alcohols (Miwa, 1980; United States National Research Council, 1985). It contains straight-chain esters of 20:1, 22:1 and 24:1 fatty acids and fatty alcohols (Miwa 1980: NAS, 1985). Weiss (1983) has reported 33% eicosenoic acid, 14% docosenoic acid, 15% eicosenol and 34% docosenol in jojoba seed oil. Chilean seeds have higher C42 and lower C38 contents than other commercial sources. The simmondsin content with respect to other analogues is lower (Cappillino *et al.*, 2003).

Uses of Jojoba Oil

The liquid jojoba oil is used in lubrication of high-speed machines, cosmetics, and pharmaceuticals. It has been used as a hair restorer since ancient times. Efforts have been also made to extract hydrocarbons from jojoba oil.

The chemical modifications of jojoba liquid wax yield a variety of derivatives with potential uses in cosmetics, pharmaceuticals, lubricants, electrical insulators, disinfectants, surfactants, emulsifiers, resins, foam control agents, high-pressure oils, plasticizers, rubber products, linoleum, fire retardants and transformer oil (Wisniak 1977; Miwa 1980; Weiss 1983; Anderson 1983; NAS 1985; Iyengar *et al.,* 1986). It is, however, not suitable for edible purposes.

Jojoba oil, because of its higher stability as compared to other vegetable oils, can be a potential source for formulating lubrication oils. The scope of its applications can be widened if some of its properties are modified. Alcoholysis of jojoba oil using short chain alcohols to prepare esters with improved pour points have been reported (Bhatia, *et al.,* 1994).

The main use of jojoba liquid wax is, however in cosmetics (Katoh and Kunimoto 1983). The oil is also used as medicine and hair restorer (Vaughan 1970). The jojoba seed oil contains mono ethylenic fatty acids and alcohols, which have good skin penetration similar to whale oil. The extensive use of jojoba oil, either unmodified or as one of its soft derivatives, is in cosmetics. As many as 300 cosmetic products containing jojoba oil are available in the market. Tests have shown that the oil can be utilized as an antifoam agent during the production of antibiotics and also for treatemnt of skin disorders. This oil is claimed as one of the Nature's gift to human race or liquid gold from the desert.

The double bonds and ester moieties can be modified to yield new products. When the oil is isomerized at the double bonds, the liquid is transformed to a cream. Depending on the degree of isomerization, a series of waxes with different melting points can be produced. Hydrogenation of jojoba oil produces a crystalline wax whose hardness among waxes of plant origin is second only to that of carnauba wax. Incomplete hydrogenation yields a range of products of different degrees of hardness, suitable for use in polishes, coatings, impregnation, carbon paper, etc.

When sulphur or radicals containing both sulphur and phosphorus, chlorine or bromine are incorporated into the oil, its properties as lubricant additive are remarkably enhanced (Wisniak, 1977; Miwa, 1980). Jojoba oil and it's sulphurized derivatives are regarded as excellent substituties for sperm whale oil, even out-performing it as additives in high-quality, high-pressure lubricants (e.g.gear oils and greases), and in cutting and drawing oils. By blending jojoba oil with other waxes, high-quality candles are being produced. Jojoba oil has been proposed as a source of numerous other derivatives, such as unusual acids and alcohols, plasticizers, detergents, fire retardants, transformer oil, and use in the leather industry (United States National Research Council, 1985).

Table 149: Chemical composition of jojoba oil and meal

Chemical component	% of whole nut	% of deoiled meal
Liquid wax	50.2-53.8	
Meal, air dried	44-49	
Protein	14.9-15.1	31.5
Reducing sugars, as glucose	2.85-2.96	8.9
Nonreducing sugars, as sucrose	0.84-1.59	3.7
Other carbohydrates, as starch	17.4-20.3	16.0
Crude fiber	3.5-4.2	
Ash	1.4-1.6	3.08
Simmondsin and derivatives	2.63-3.09	

Wells (1955) and Verbiscar and Banigan (1978)

Utilization of Meal

The defatted jojoba seed meal contains 25-35% proteins, 9-11% fiber and 5-15% oil (Vaughan 1970; Weiss 1983). It is a nutritious animal feed. Jojoba meal proteins are well balanced, with all the amino acids except for methionine (Yermanos 1974). The seeds contain simmondsin, a cyanomethylene cyclohexyl glucoside and three other structurally related cyanoglucosides (Booth *et al.*, 1974). The presence of these toxic cyanoglucosides suppresses the appetite and makes the meal unacceptable for animal feeds. Methods have been developed for detoxification and palatability improvement of jojoba meal (Verbiscar *et al.*, 1980; Verbiscar and Banigan 1978). The detoxified meal can be used in animal feeds. Jojoba seeds may be roasted or dried, ground, and mixed with sugar and water to form a beverage (Vaughan 1970).

Jojoba Plant

The jojoba plant is used on a small scale in gardening and landscaping Although jojoba is tolerant to drought and salinity and cattle browse its canopy (Gentry, 1958), it's growth is very slow. Its palatability is too high to permit its successful establishement and utilization as an evergreen fodder shrub in pasture improvement projects in semi-arid areas.

Table 150: Potential applications of the jojoba plant and its components

Component	Application
Natural liquid wax (jojoba oil)	Lubrication, cutting oil, thermal oil, consmetics, ink, paints and coatings, papers, leather proudcts, skin treatment (burns, application in ointments and drugs), diet food, penicillin production.
Sulfurized jojoba oil	Lubrication, paint, ink, printing
Hydrogenated jojoba oil (solid wax)	Polish, coatings, cosmetics, candles
Trans (isomerized) jojoba oil	Cosmetics, coating candles, textiles
Oxidized jojoba oil	Cosmetics, disinfection
Acids and alcohols	Detergents, disinfection, lubrication, corrosion inhibition, plasticizers, cosmetics, paints, emulsifers.
Simmondsin	Appetite suppression
Meal	Food for animals
Whole plant	Ornamental, afforestation, groves and avenues

References

Abd-El-Aal, M.A., Gomaa, E.G. and Karan, H.S. 1987.Bitter almond, Plum, and Mango Kernels as Sources of lipids. *Fette Wissenchaft. Technology.* 89: 304-306.

Abigor, R.D. and Uadia, P.O. 2001. Lipid composition of *Jatropha curcas* L. Seed Oil. *Rivista Italiana delle Sostanze Grasse.* 78(3): 163-165.

Aderibigbe, A.O., Johnson, C., Makkar, H.P.S., Becker, K. and Foidl, N. 1997. Chemical Composition and Effect of Heat on Organic Matter- and Nitrogen-Degradability and Some Anti-nutritional Components of *Jatropha* Meal. Animal Feed Science and Technology, 67: 223-243.

Aguilera, J.M. and Lusas, E.W. 1986. Laboratory and Pilot Solvent Extraction of Extruded High-Oil Corn. Journal of the American Oil Chemists' Society, 63:239-243.

Aigbodion, A.I. and Bakare, I.O. 2005. Rubber Seed Oil Quality Assessment and Authentification. Journal of the American Oil Chemists' Society. 82(7):465-469.

Akiya, T. 1967. Edible Rice Bran Oil in Japan. Japan Agricultural Research Quality. 2(3):27-31.

Ali, M.A., Gafur, M.A., Rahma, M.S. and Ahmed G.M. 1985. Variation in Fat Content and Lipid Class Composition in Ten Different Mango Varieties. Journal of the American Oil Chemists' Society. 62: 520.

Allen, R.D. 1979. Ingredient Analysis Table: Feed Stuffs. 51:31.

Ambasta, S.P. 1986. The Useful Plants of India. Publications and Information Directorate, CSIR, New Delhi. Pp. 430-464.

Amran Waheed, Shahid Mahmud, J. and Muhammad Saleem, M.A. 2001. Studies on the Lipid classes of *Nicotiana tabacum* L. Seed Oil. Natural Product Sciences. 7(4):110-113.

Anderson, B. 1983. *Jojoba* In. Lubrication Industry. Proceedings of 5th International Conference on *Jojoba* and its Uses Through 1982., Tucson, AZ.

Anonymous, 1959. The Wealth of India, Rraw Materials, Council of Scientific and Industrial Research, New Delhi. 5: 293.

Anonymous, 1962. *Crambe*- A Potential New Crop for Industrial and Feed Uses. Agriculture Research Service, USDA ARS. Wasington DC. pp 34-42. India.

Anonymous, 2002. Procedings of the 11th Meeting of Tobacco Board, Guntur, India. Held on May 21, 2002.

Armour, R.P. 1959. Investigations on *Simarouba Glauca* in El Salvador. Economic Botany. 13: 41-66.

Augustin, M.A. and Ling, E.T. 1987. Compostion of Mango Seed Kernel. *Pertanika.* 10:53-59.

Austin, F.C. and Wolff, I.A. 1968. Sinapine and Related Esters in Seed Meal of *Crambe.* Journal of Agricultural and Food Chemistry. 16: 132-135.

Axtell, B.L.1992. Minor Oil Crops. FAO Agricultural Services Bulletin, 94. FAO of UN, Rome. pp 157-159.

Baliga, B.P. and Shitole, AD. 1981. CBS. From Mango Fat. Journal of the American Oil Chemists' Society. 58: 110-114.

Bangarayya, M., Sarma, C.B., Narasimhamurty, Y.Ch. and Prabhakara Babu. D. 1982. Suppression of Suckers with Edible Oils in FCV Tobacco. Tobaco Research 8(1):25-29.

Banu, C., Nedelcu, N., Musca, L., Enache, A. and Dinache, P. 1974. Utilization of Maize Germs. *Ind. Aliment Agricultural*, 91:681-87.

Banzon, J.A. and Velasco,J.R. 1982. Coconut Products and Utilization. Philippine Coconut Research and Development Foundation. Pasig, Metro Manila, Philippines, pp-171.

Basu, S.P., Veeresham, C.and Mahli, S.S. 1988.Fatty Acid Composition of Purified *Pongamia Pinnata* Seed Oil. Journal of the Oil Technologists' Association of India. 20(1):22-23.

Benzioni, A. and Forti, M. 1989. *Jojoba.* In Oil Crops of the World, Ed. G. Robbelen, R.K. Downey, and A. Ashri, New York: McGraw-Hill. pp. 448-61.

Bhat, S.G. 1988. Proceedings of Coconut Breeding and Management Symposium, Kerala Agricultural University, Trichur, India. pp. 275-77.

Bhatia, V.K., Kaul, S. and Kapoor, V.B., 1994. Synthesis of Shorter Chain Esters of Jojoba Oil. Journal of the Oil Technologists' Association of India, 26(1):23-25.

Bhattacharya, D.K., 1988. Composition and Utlization of Oil and Non-Oil Constituents of Sal Seed and Mowrah. In. Oilseeds and their Utilization, Ed. R.K.Suri and K.C.Mathur. International Book Distrbutors, Dehra Dun. India, pp 154-172

Bhattacharya, D.K. 2002. Lesser-Known Indian Plant Sources for Fats and Oils. INFORM, 13:151-157.

Bhattacharya, D.K., Dasgupta, J. and Banerjee, K. 1981. Proceedings of the Conference-Palm Oil Product Technology in the Eighties. Kuala Lumpur, Malaysia, June 22-24.

Biswas, S. and Bhattacharya, D.K. 1989. Utilization of Mango Fat as Ointment and Suppository Bases. Journal of the Oil Technologists' Association of India. 19(1):20-22

Blessin, C.W., Inglett, G.E., Garcia, W.J., and Deatherage, W.L. 1974. An Edible Defatted Germ Flour from a Commercial Dry-Milled Corn Fraction. Cereal Science Today. 19:224-55.

Booth, A.N., Elliger, C.A. and Weiss, A. C.E. 1974. Isolation of a Toxic Factor from *Jojoba* Meal. Life Science. 15:1115-20.

Bose, S., Thakral, B.M. and Narayan, S. 1952. Indian. Journal of Veterinary Science. 22:251.

Bringi, N.V. 1979. Proceedings of the Symposium on Production and Utilization of Forest Products. Regional Research Laboratory, Jammu Tawi, India. III.2.

Bringi, N.V.1987. Lesser Known Tree-Borne Oilseeds. In Non- Traditional Oilseeds of India. Ed. Bringi, N.V. Oxford & IBH Publishing Co. Pvt. Ltd. NewDelhi. 216-248.

Bringi, N.V. 1988a. Processing of Non-Traditional Oils. In Souvenir of Strategies for Making India Self Reliant in Vegetable Oils Ed. G. Nagaraj, ISOR, DOR Hyderabad. pp. 55-63.

Bringi, N.V., 1988b. Progress in the Chemistry and Technology of Non Traditional Oilseeds/Oils. Journal of the Oil Technologists' Association of India. 20(1):2-9.

Bringi, N.V. and Mehta, D.T.1987. *Garcinia Indica Choisy (Kokum)*, *Vateria Indica (Dhupa)* and *Bassia Butyracea (Phulwara)* Seed Fats. In. Non- Traditional Oilseeds of India. Ed. Bringi, N.V. Oxford &IBH Publishing Co. Pvt. Ltd. NewDelhi. pp. 97-117.

Bringi, N.V. and Mukherjee,S.K. 1987. *Karanj* seed (*Pongamia glabra)* Oil, In. Non-Traditional Oilseeds of India. Ed. Bringi, N.V. Oxford & IBH Publishing Co. Pvt. Ltd. NewDelhi. pp143-166.

Cappillino, P., Kleiman, R. and Botti, C. 2003. Composition of Chilean *Jojoba* Seeds. Industrial Crops and Products. 17(3):177-182.

Chakaraborty, A. and Bhattacharyya, D.K. 1988. Edible Quality Karanja Oil by Mixed Solvent Refining Procedure. Journal of the Oil Technologists Association' of India. 20(2):29-30.

Chakaraborty, M.K. 1998. Industrial Uses of Tobacco. Tobacco Symposium Souvenir. Indian Society of Tobacco Science, Rajahmundry, January 20-23, pp,74-76.

Chand, S., Srinivasulu, C. and Mahapatra, S.N. 1974. The Oils and Oilseeds Journal. 26:12.

Chatakanonda, P., Sriroth, K., Vaysse, L. and Liangprayoon, S. 2005. Fatty acid Composition and Properties of *Jatropha* Seed Oil and its Methyl Ester. Proceedings of 43rd Kasetsart, University Annual Conference. Thailand Feb. 1-4, Subject Animals. pp. 641-648.

Coleman, M.H. 1965. Journal American Oil Chemists' Society, 42:751.

Cornelius J.A. 1977. Palm Oil and Palm Kernel Oil. Progress in the Chemistry of Fats and Other Lipids. Pergamon Press. Great Britain. 15, pp. 5-27.

Cross, C.E. 1987. Oxygen Radicals and Human Disease, Annals of Internal Medicine 197: 526-545.

CSIR, 1956. Wealth of India, Raw Materials, Council of Scientific and Industrial Research (CSIR), Publications and Information Directorate, New Delhi. 4: 101.

CSIR, 1962. The Wealth of India: Raw Materials. CSIR, Publications and Information Directorate, New Delhi, (6): 426.

CSIR, 1969. The Wealth of India. CSIR, Publications and Information Directorate, New Delhi 8: 208.

Davis, J.B., Kay, D.E. and Clark. V. 1983. Tropical Products Institute Report G. 150, 1-150.

Daxenbichler, M.E., VanEtten, C.H. and Wolff, I.A. 1965. A New Thioglucoside. R-2-Hydroxy-3- Butenyl-Glucosinolate from *Crambe Abyssinica* Seed.Biochemistry. 4: 318-323.

Deo Singh, K. 2003. Tobacco Seed Oil- Potential and Future Prospects. National Seminar on Stress Management in Oilseeds for Attaining Self-Reliance in Vegetable Oils. January 28-30. ISOR, DOR, Hyderabad, pp.118-122.

Dev, S. 1978. Journal of Scientific and Industrial Research. 37: 391.

Dhar, A.K., Sapru, R. and Koul, S. 1988. Studies on the Growing of *Crambe Abyssinica* Hochst in Kashmir. Journal of Oilseeds Research. 5(1): 110-113.

Dhawan, S. 1976. Dr. A.L. Mudaliar Memorial Lecture. Indian Institute of Technology, Madras. India.

Dudley. J.W. 1974. Seventy Generations of Selection for Oil and Protein in Maize Madison. WI: Crop Science Society of America.

Dum, A. C., and Hilditch,T.P. 1947. Journal of the Society of Chemical Industry. London, 66:209

Dutta, J. and Bhattacharya, R.K. 1959. Proceedings 46th Session of Indian Science Congress Part III: 459.

Earle, F.E., Peters, J.E., Wolff, I.A. and White, G.A. 1966. Compsitional Differences among *Crambe* Samples. Journal of the American Oil Chemists' Soceity. 43: 303-33.

Eckey, E.W. 1954. Vegetable Fats and Oils, Reinhold Publishing Corp, p. 583.

Elson, C.E. and Qureshi, A.A. 1995. Coupling the Cholesterol and Tumor-Suppressive Action of Palm Oil to the Impact of its Minor Constutuents on 3-Hydroxy-3-Methylglutaryl Coenzyme A-Reductase Activity. *Prosta. Leuko. Ess. Fatty Acids.* 52:205-208

Eshetu, B. 2000. *Nicotiana Tabacum* L Seed Oil, *Seminarbeit im Rahmen Der vorlesung Olplanzen-Zuuchtung, Anbau, Verwerung im Wintersemester* 1999/2000, *Wen, 2 Febrar.*

FAO. 1984. Coconut-Tree of Life, Food and Agriculture Oraganisation of the United Nations. Rome. Italy.

Farooq, A and Bhanger M.I. 2003. Analytical Characterization of *Moringa oleifera* Seed Oil grown in temperate regions of Pakistan. Journal of Agricultural and Food Chemistry 51(22): 6558-6563.

Farooq, A. and Bhanger, M.I. 2005. Interprovenance variation in the Composition of *Moringa Oleifera* oilseeds from Pakistan. Journal of the American Oil Chemists' Society. 82(1):45-51.

Feuge, R.O. and Reddi, P.B.V. 1949. Rice Bran Oil. III-Utilisation as an Edible Oil. Journal of the American Oil Chemists' Society. 26: 349-353.

Frega, N., Bocci, F., Conte, L.S. and Testa, F. 1991. Chemical Composition of Tobacco Seeds (*Nicotiana Tabacum L.*), Journal of the American Oil Chemists' Society, 68:29-33.

Fuglie, 2005. New Uses of *Moringa.* Studied in Nicaraugua. Website- Internet. OriginallyAppeared. ECHO Development Notes. Issue, 68.

Gandhi, V.M. and Mulky, M.H., 1994. Neutralization of Castor Toxins by Sal Tannnins. Journal of the Oil Technolotists' Associationof India. 26(2): 39-42.

Gandhi, V.M., Cherian, K.M. and Mulky, M.J. 1990. Nutritional and Toxicological Evaluation of Rubber Seed Oil. Journal of the American Oil Chemists' Society 67(11):883-886.

Gascon, J. P., Noiret, J.M., and Meunier, J. 1989. Oil palm. In. Oil Crops of the World, Ed. G. Robbelen, R.K. Downey, and A.Ashri, McGraw-Hill. NewYork. pp 475-495.

Gentry, H.S. 1958. The Natural History of Jojoba (*Simmondsia Chinensis*) and its Cultural Aspects. Economic Botany. 12:261-95.

Godin, V.J. and Spensley, P.C. 1971. T.P.I. Crop product Digest No.1 Tropical Products Institue, London.

Goenka, O.P. 2007. Rice Bran Oil-Unique Gift of Nature. In. Changing Global Vegetable Oils Scenario: Issues and Challenges Before India. Ed. D.M.Hegde, I.S.O.R., Directorate of Oilseeds Research, Hyderabad. India. pp.361-366.

Gohl, B. 1975. Tropical Feed. Food and Agriculture Organization, Rome. pp.375-76

Gopala Krishan, A.G., Prashanth, P.A., Pragasam, A., Raghavendra, K.V., and Sakina Khatoon, 2005. Nutritional Components of Unsaponifiable Matter in RBO. Oils and Fats Today. 8(3):13-17.

Govindarajan, V.S. 1952. Journal of Scientific and Industrial Research. IIB: 477.

Gunstone , F.D. 2005. The Chemistry of Oils and Fats: Sources, Composition, Properties and Uses. Blackwell Publishing, Oxford, U.K. pp 1-35.

Gupta, H.O., Lodha, M.L. Sen, K., Singh, J. and Mehta, S.L. 1979 Journal of Food Science and Technology. 16(4):129-132.

Hass, W. and Mittelbach, M. 2001. Detoxification Experiments with the Seed Oil from *Jatropha curcas* L. Industrial Crops and Products. 12(2): 111-118.

Hartley, C.W.S. 1967. The Oil Palm. Longmans, London. pp. 1-70.

Hemavathy, J., Prabhakar J.V. and Sen, D.P. 1979. Abstracts. N0. 217, International Congress on Oilseeds. New Delhi.

Hemavathy, J. , Prabhakar, J.V. and Sen, D.P. 1987. Composition of Plant Lipids of *Alphonso* Mango (*Mangifera Indica*) Kernel. Journal of Food Science. 52: 833-834.

Hilditch T,P. and Williams P.N. 1964. The Chemical Constitution of Natural Fats, IV. Chapman & Hall, London, 314,

Hirai, S., Okamoto, K. and Morimatsu, M. 1982. Lipid Peroxide in the Aging Process. In. Lipid Peroxides in Biology and Medicine, Ed. K. Yagi, Academic Press, New York. pp.305-315

Hoffman. G. 1961. 3-Cis-Hexenal, the 'Green' Reversion Flavor of Soybean Oil. Journal of the American Oil Chemists' Society. 38: 1-3.

Hutagalung, R.I., Mahyaddin M.D., Vijchulata, P., Jalaludin, S. and Zainal, J. 1983. Nutrient Availability and Utilization of Feedstuffs for Farm Animals. In. Feed Information and Animal Production, Ed. G.G. Robards, and R.G. Packham Common Wealth Agricultural Bureaux, U.K.

Iyengar, E.R.R., Vasudeva Rao, P.H.V., Pandya, J.B. and Balakrishna, V. 1986. *Jojoba, Simmondsia Chinesis* (Link) Schneid-A Renewable Source of Liquid Wax. In Plantation Crops-Constraints and Opportunities, II, Ed H.C. Srivastava, B. Vatsya, and K.K. G. Menon. Oxford and IBH. New Delhi, pp. 141-45

Jayasree, H. and Sunanda, S. 2001. Preliminary Study on Seed Protein and Protein Quality of *Simarouba Glauca*. Karnataka Journal of Agricultural Sciences. 14(4):1065- 1066

Jayasree, B., Harishankar, N. and Rukmini, C. 1998. Chemical Composition and Biological Evaluation of *Mahua* Flowers. Journal of the Oil Technologists' Association of India. 30(4); 170-172.

Jena, K.S., Behera,S. and Rao,Y.R. 2003.Studies on Mechanical Expression of *Simarouba Glauca* Kernels . Journal of Oilseeds Research. 20(2):324-326.

Joshi P. 1987. Paper Presented At XXX All India Annual Kharif Oilseeds Workshop, Akola, 22-27 April, 1987. pp. 1-10.

Joshi, P. and Solanki, Z.S. 1988. Paper Presented at the National Seminar on Strategies for Making India Self-Reliant in Vegetable Oils Held at Hyderabad, Sept. 5-9.

Joshi, Shyamsunder and Joshi Shanta, 2002. Oil Tree-*Simarouba Glauca* DC. UAS Bangalore, and ICAR, New Delhi. pp 1-88.

Kakengi, A.M.V., Shem, M.N., Sarwatt, S.V., Hasegawa, K., Sasaki, Y., Hayashi, T. and Fujihara, T. 2006. The Safe Feeding Level of *Moringa oleifera.* Seedcake as an Alternative Protein Source to Livestock. Journal of Food Agriculture and Environment. 4(2): 116-119.

Kane, F.G., 1966. *Fette Seifen Anstrichmittel*, 68: 1.

Karoshi, V.R. and Hegde, G.V. 2002. Vegetative Propagation of *Pongamia Pinnata* (L) Pierre. Hitherto a Neglected Species, Indian Forester. 128(3). pp348-350.

Katosh, M., and Kunimoto, T. 1983. The Use of Jojoba Oil in Cosmetics in Japan. Proceedings the International Conference On Jojoba and its Uses Through 1982, Tucson, AZ.

Ketkar, C.M. 1976. Rubber Seed Cake. Report of the Modified Neem Cake Manurial Project, India. pp. 180.

Ketkar, C.M. 2004. Neem Tree- A Renewable Source as an Energy Storehouse with Remarkable Properties. Proceedings of the Workshop on Minor & Tree Borne Oilseeds and their Value Added Products.2nd July, 2004 Raipur, Chattisgarh, India. Pp. 39-41.

Kinsella, J.E. 1982. Relationship between Structure and Functional Properties of Food Proteins. In. Food Proteins, Ed. P.F. Fox, and J.J. Condon, New York: Applied Science. pp.51-65.

Kiritsakkis, A.K. 1984. Effect of Selected Storage Conditions and Packaging Materials on Olive Oil Quality, Journal of the American Oil Chemists' Society. 61(12): 1868-1870

Kirk, L.D. , Mustakas, G.C. and Griffin, E.L. 1966. *Crambe* Seed Processing for Improved Feed Meal by Ammonia, Journal of the American Oil Chemists' Society. 43: 550-555

Kohle, J.N., Anil Bhaskar, and Bringi, N.V. 1981. Lipids, 16: 775

Krinsky, N.I. 1993. Actions of Carotenoids in Biological Systems. Annual Reviews of Nutrition. 13:561-588.

Lakshmikantam, V. 1978. Tree Borne Oilseeds-A Resource for Gainful Employment. Directorate of Oils and Soap Industry, KVIC, Irla Road, Vile Parle, Bombay (Mumbai), India. 1-66.

Lakshmikantam, V. 1988. Chemistry and Industry of Tree Borne Oilseeds in Service to Citizen, In. Oilseeds and Their Utilization. Ed. R. K. Srui and K.C. Mathur, International Book Distributors, Dehradun, India. Pp 1-12.

Lakshminarayana, G. Rao, T.C. and Ramalingeswaran, P.A. 1983. Varietal Variation in Content, Characterstics and Composition of Mango Seeds and Fat. Journal of the American Oil Chemists' Society. 60: 88-89

Lalas, S., Gortzi, O. and Tsaknis, J. 2006. Frying stability of *Moringa stenopetala* Seed Oil. Plant Foods for Human Nutrition. 61(2):99-108.

Lalas. S., Tsaknis, J. and Sflomos, K. 2003. Characterisation of *moringa stenopetala* Seed Oil variety "Marigat" from island Kokwa. European Journal of Lipid Science and Technology. 105(1): 23-31.

Leley, V.K. and Agashe, S.D. 1971. Indian Oil and Soap Journal. 36(10): 283.

Lesserman, K.J. and Kirleic, A.W. 1979. Effect of Microwaves on Thioglucosidase Activity in Intact Seeds of *Crambe* (*Crambe Abyssinica.* Hochst). Crop Sciences. 19:189-191

Loncin, M., Jacobsberg, B. and Evrard, G. 1970. Palm Oil, a Major Tropical Product, Brussels, Tropical Products Sales, Unilever House,

Lucisano, M., Casiraghi, E.M. and Barbleri, R. 1984. Use of Defatted Corn Germ Flour in Pasta Products. Journal of Food Sciences. 49:482-88.

Lynn, L. 1969. Protein Enriched Food for World Needs. Ed. M. Milner,. St. Paul, Minnesota: American Association of Cereal Chemistry. p. 154.

Lynn, L., Steen, G.J. and Anderson, R.M. 1968. New Rice Oil, Food Technologist. 22: 1250-1252.

Mani, V.V.S. and Lakshminarayana, G. 1968. 23rd Annual Symposium and Convention of Oil Technologists' Association of India Hyderabad, India.

Mathur, N.B. and Das, L.M. 1985. Utilisation of Non-Edible Wild Oil and Diesel Engine Fuel. Proceedings of Bio-energy Society, 2nd Convention and Symposium. pp. 198-202.

McDonald, P., Edwards, R.A. and Greenhalgh, J.F.D. 1982. Animal Nutrition, 3rd Edn. Harlow, Essex: Longmans.

McGregor. W.G., Plessers, A.G. and Graig, B.M. 1961. Species Tials With Oil Plants I. *Crambe*. Canadian Journal of Plant Sciences. 41: 716-719.

Medeiros, H., Kirleic, A.W. and Vetter, R.J. 1978. Microwave Inactivation of Thioglucosidase in Intact *Crambe* Seeds. Journal of the American Oil Chemists' Society. 55: 679-682.

Mev Singh, Muralidharudu, Y. and Nagaraj, G. 1991. Oil Content and Fatty Acid Composition of Some Potential Sources of Vegetable Oil. Proceedings of the National Seminar on Strategies for Making India Self-Reliant in Vegetable Oils. Eds. V. Ranga rao, M.V.R. Prasad. ISOR, Hyderabad, India. pp 351-354.

Mittal, J.P. and Bhagwan Singh. 1973.Research And Industry (India). 18:26.

Miwa, T.K. 1980. *Jojoba*. Vol. I, *Jojoba* Plantation Products, Inc., Los Angeles, California, U.S.A.

Moharrum. Y.G. and Moustafa, A.M. 1982. Utilization of Mango Seed Kernel (*Mangifera Indica*) as a Source of Oil. Food Chemistry. 8: 267-276.

Morrison, F.B. 1949. Feeds and Feeding, 21st Edn. Ithaca: Morrison.

Morrison, W.R. 1978. Cereal Lipids. In. Advances in Cereal Science and Technology, Ed. Y. Pomeranz, St. Paul, M.N. American Association of Cereal Chemistry. Vol. II: pp. 221-311.

Mulky, M.J.1976. Journal of the Oil Technologists' Association of India. 8(3): 106.

Mustakas, G.C., Kopas, K. and Robinson, N. 1965. Preparation and Solvent Extraction of *Crambe*. First Conmercial Trial Run of New Seed Oil. Journal of the American Oil Chemists' Society. 42: 550A- 554A, 594A

Mustakas G.C., Kirk, L. D.,Griffin Jr. E.L. and Declaton. 1968. *Crambe* Seed Processing. Improveed Seed Meal by Soda Ash Treament. Journal of the American Oil Chemists' Society. 45: 1-5.

Mustakas, G.C., Kirk, L.D., Griffin Jr. E.L. and Broth,A.N. 1976. *Crambe* Seed Processing. Removal of Glucosinolates by Water Extraction. Journal of the American Oil Chemists' Society. 53: 12-16.

Nagaraj, G. 1998. Seed and oil quality of *crambe* genotypes-rich in erucic acid. Journal of the Oil Techologists' Association of India. 30(1): 17-18

Nagaraj, G., 2000. Drumsticks for Health . Science Express. New Indian Express August 15.

Nagaraj, G. 2007. Personal Communication- D.O.R. Data

Nagaraj, G. and Mukta, N. 2004. Seed Composition and Fatty Acid Profile of some Tree Borne Oilseeds. Journal of Oilseeds Research. 21(1): 117-120.

Narsimhachar, B.L., Thirumala Rao, S.D and Reddy, B.R. 1975. Research and Industry (India). 20:79

Narasimha Reddy, P., 2001. Evaluation of *Simarouba Glauca* Seed Fat-Characteristics, Composition and Edibility. Abstracts: International Seminar on "Oleochemicals, Surfactants and Specialty Products". I.I.CT., Hyderabad.

NAS. 1985. *Jojoba:* New Crop for Arid Lands, New Material for Industry, National Academy of Sciences (NAS). Washington.

Nasirullah T. and Nagaraja,K.V., 1987. Rice Bran Lipids: A Review. Journal of the Oil Technologists' Association of India. 19(1):2-6.

Nasirullah, T. Mallika, S. Rajalakshmi, Vibhakar, S., Krishanamurthy, M.N., Nagaraja, K.V., and Kapur, O.P., 1981. Characteristics of Sal Seed Fat. Journal of the Oil Technologists' Association. 13(3):120-122.

Nicholas, O.O. and Daniel, E.A. 2005. Use of Rubber Seed Oil as Seed Protectant Against *Callosobruchus Maculatus* F. Infestation on Stored Cowpea. Natural Rubber Research. 18(1): 25-30.

Nielsen, H.C., Inglett, G.E. Wall, J.S. and Donaldsen, G.L. 1973. Corn Germ Protein Isolate-Preliminary Studies on Preparation and Properties. Cereal Chemistry. 50:435-43.

Nieschlag, H.J. and Wolff, I.A. 1971. Industral Uses of High Erucic Oils. Journal of the American Oil Chemists' Society. 48:723-727.

Pandey, S.S. 1987. Studies on Composition of Cottonseed and its Oil. Ph. D. Thesis of the University of Mumbai (Bombay), India.

Patel, J.A., Patel, B.K. and Chakraborty, M.K. 1998. Production Potential and Quality Aspects of Tobaco Seed Oil. Tobacco Research, 24:44-49.

Patterson, H.B.W. 1989. Handling and Storage of Oilseeds, Oils, Fats and Meal. Elsevier Applied Science, London, pp.142.

Purseglove, J. W. 1975. Tropical Crops: Monocotyledons. English Language Book Society and Longmans, London. Pp.479-510

Quckenbush, F.W., Firch, J.C., Raburn, W.J., McQuistan, M., Petzold, E.N. and Karg, T.E. 1961. Analysis of Carotenoids in Corn Grain. Journal of Agricultural and Food Chemistry. 9:132-37.

Ramalingaswamy, P.A., Chandrasekhara Rao. T., and Lakshminarayana, G., 1980. Cultivar Variations in Characteristics and Composition of Mango Stones and Fat and Deterioration in Fat Quality on Storage of Air Dried Stones. Abstracts of Papers. Journal of the Oil Technologists' Association of India. 12(4):175.

Ram Varma, K., Reddy, B.R., Somayajulu, B.A.R., Allabaksh, M., Thirumala Rao, S.D. and Murti, K.S. 1964. Pilot Plant Studies on Processing of Tobacco Seed II. Processing Variables. Indian Oilseeds Journal. 8(4):369-375.

Ramajayam, 2000. Development of Value Added Processed Products from *Simarouba* (*Simarouba Glauca* DC) Fruits. M.Sc. Thesis, University of Agricultural Sciences, Bangalore. pp. 102.

Rankin, J.C. 1982. The Non-Food Uses of Corn. In Handbook of Processing and Utilization in Agriculture, Vol. II(1): Plant Products, Ed. I.A. Wolff. CRC Press. Boca Raton, FL, U.S.A. pp.63-79

Raval, D.A. and Toliwal, S.D., 1996. Nutritional Study on Drumstick Seed (*Moringa Oleifera*) Oil. Journal of the Oil Technologists' Association of India. 28(1):3-4.

Reiners, R.A. 1978. Corn Oil. In. Products of the Corn Refining Industry. Food Seminar Proceedings. Washington DC: Corn Refiners Association. U.S.A.

Reiners, R.A. and Gooding C.W. 1970. Corn. In. Corn: Culture, Processing. Products, Ed. G.E. Inglett. AVI Publishing Co. Westport, CT. pp.241-61.

Rost, T.L. and Patterson, H.E. 1978. Structural and Histochemical Characterization of the Cotyledon Storage Organelles of *Jojoba* (*Simmondia chinesis*). Protoplasma 95:1-10.

Rukmini, 1987. Chemical and Nutritional Evaluation of Neem Oil. Food Chemistry. 26:199-124.

Rukmini, C. 1987a. Toxicity Studies on Unconventional Oils. Proceedings of the Nutrition Society of India. 33:1-7

Salunkhe, D.K., Chavan, J.K., and Kadam, S.S. 1985. Post-Harvest Biotechnology of Cereals. CRC Press. Boca Raton, FL., U.S.A.

Sengupta, S., Chakrabarty, M.M. and Bhattacharya, D.K. 1981 *Fette Seifen Anstrichmittel,* 83: 230.

Sethi, H. and Maini, N.S.1970. Indian Oil and Soap Journal. 35: 189

Shah, S.G., Subrahmanyam V.V.R. and Rege, D.V. 1983. Composition and Characteristics of Non- Traditional Hard Fats. Journal of the Oil Technologists' Association of India. 15: 9-23.

Shroder, J.D. and Heiman, V. 1970. Feed Products from Corn Processing. In. Corn, Culture, Processing. Products, Ed. G.E. Inglett, AVI Publishing co. Westport, CT. U.S.A. pp.220-40

Sonntag, N.O.V. 1979. "Composition and Characteristics of Individual Fats and Oils", In. Bailey's Industrial Oil and Fat Products, Vol. 1, Ed. D. Swern, A Wiley-Interscience Publication, NewYork. U.S.A. pp. 311-398.

Srilakshmi, 1999. Development of Value Added Products from *Simarouba Glauca* DC. M.Sc. Thesis, University of Agricultural Sciences, Bangalore. pp. 101.

Srimannarayana, G. 1995. Insect Antifeedant and Medicinal Aspects of Neem Oil. Proceedings of the One Day Seminar Held at O.T.R.I., J.N.T.U., Anantapur, pp.20-24.

Srinivasulu, C. and Mahapatra, S.N. 1970. Indian Oil and Soap Journal. 35: 165.

Srivastava, A.K. and Pandey, G.K. 2001. Changes in Protein and Total Amoino Acid Contents in the Seeds of *Schleichera Oleosa* (*Kusum*) Due to Biodeterioration by Pathogenic Fungi During Storage. Journal of Mycopathological Research. 39(1):65-67.

Subba Reddy, D.V. 2001. Processing of Mahua Seed-Dehulling and Crushing. Abstracts: International Seminar on "Oleochemicals, Surfactants and Specialty Products". I.I.C.T., Hyderabad.

Sujatha, M. and Mukta, N. 1995. *Jatropha*. Information Bulletin. Directorate of Oilseeds Research, Rajendranagar, Hyderabad, 500 030, India.

Sutescu, P., Zamaria, I., Rujinski, A., and Gontea, I. 1972. Increasing the Nutritional Efficiency of Bread by Adding Maize Germ. *Igiena*. 21: 647-654.

Swift, C.E., Fore, S.P. and Dollear, F.G. 1950. Rice Bran Oil. V-The Stability and Processing Characteristics of Some Rice Bran Oils, Journal of the American Oil Chemists' Society. 27:14-16.

Takeshita, Y. and Oba, I. 1962. Hydrogenation of Rice Bran Oil. Yukagaku, 11: 168-174.

Thakur, L.S., Parekh, L.J. and Patel, B.K. 1998. Evaluation of Nutritional Quality of Tobacco Seed Oil. Tobacco Research. 24:65-69

Thampan, P.K. 1981. The Coconut Palm and its Products. Government of India, Green Villa, Cochin, Kerala. India. pp. 314.

Thippeswamy, H.T. and Raina, P.L. 1989. Lipid Composition of *Kokum (Garcinia Indica)* and *Dhupa (Vateria Indica)*. Journal of Food Scienc and Technolgy. 26: 322-326.

Thirumala Rao, S.D. and Murti, K.S. 1961. A Survey of the Current Position of Rice Bran Oil Industry in Various Countries. Indian Oilseds Journal. 5: 121-132.

Tsen, C. C., Mojibian, C.H. and Inglett, G.E. 1974. Defatted Corn Germ Flour, a Nutrient Fortifier Ingredient for Breadmaking. Cereal Chemistry. 51:262-71.

Uddin, K.M. 2005. Studies on Cottonseed and Cottonseed Meal Part (1). Bangladesh Journal of Scientific and Industrial Research. 40(1/2):107-114.

U.S.National Research Council, 1985. *Jojoba*: New Crop for Arid Lands-New Material for Industry. National Academy Press, Washington, D.C., U.S.A.

Van Pee, W., Bori, L.E., Foma, M.N. W. and Hendrix, A. 1981, Fatty Acid Composition and Characteristcs of the Kernel Fat of Different Mango (*Mangifera indica*) Varieties. Journal of the Science of Food and Agriculture. 32: 485-488.

Varghese, S., Patel, K.V., Vashi, R.G. and Patel, U.G. 2004. Variation in Oil, Gossypol and Free Fatty acid Content in different *Gossypium* species. Indian Journal of Agricultural Biochemistry. 17(1): 49-50.

Vasishtha, A.K., Madhu Bajpai and Chhaya Singh, 1991. Spectrum of Variability of Seeds and Oil Composition in Indian Maize. Journal of the Oil Technologists' Association of India. 23(2):24-28.

Vasishtha, A.K., Pathak, J.P. and Lehri, A. 1992. Studies on Solvent Extraction of Neem Seed (*Azadirachta Indica*). Journal of the Oil Technologists Association' of India. 24(3):75-80.

Vaughan, J.G. 1970. The Structure and Utilization of Oilseeds. Chapman and Hall. London.

Verbiscar, A. J.and Banigan, T.F. 1978. Journal of Agriculture and Food Chemistry. 26: 1456.

Verbiscar, A.J., Banigan, T.F., Weber, C.W., Reid, B.L., Trei, J.E., Nelson, E.A. Reffaut, R.F. and Kosersky, D. 1980. Detoxifaction of *Jojoba* Meal. Journal of Agricultural and Food Chemistry. 28:571-78.

Verma S.C. and Banerji, R., 2003. Moringa oil Blends for Possible Utilization in Rheumatism. 58th Annual Convention and International Conference and Technological Developments in the Field of Oil, Fats and Allied Products. November 21-23. Oil Technologists' Assocation of India.

Vigneshwaran, N., Balasubramanya, R.H. and Sheikh, A.J. 2007. Scientific Processing of Cottonseed and Value Addition to its By-products. CIRCOT, Mumbai, India. pp. 1-27.

Vijayramayya, T., Reddy, K.K., Sundara, Rao, V.S., Sastry, K.N.S., Ganesan, A. and Santappa, M. 1979. Proceedings of the Syposium on Production and Utilization of Forest Products. Regional Research Laboratory, Jammu Tawi, India. I: 230

Vimal, O.P. and Naphade, K.T. 1980. Recent Trends in Utilisation of Non-edible Oilseeds. Journal of Scientific and Industrial Research (India). 39, 197.

Vimal, O.P. and Naphade, K.T. 1988. Utilization of Non-Edible Oilseeds-Recent Trends in Oilseeds and their Utilization. Ed R.K.Suri and K.C.Mathur, International Book Distributors, Dehra Dun. India. pp.78-112.

Walton, J.R. and Packer, L. 1980. Free Radical Damage and Protection: Relationship to Cellular Aging and Cancer. In. Vitamin E, a Comprehensive Treatise, Ed. L.J. Machlin, Marcel Dekker, Inc. NewYork, pp. 495-517.

Watson, S.A. 1962. Yellow Carotenoid Pigments in Corn. Proceedings of 17th Annual Hybrid Corn Industry Research Conference. American Seed Trade Association. 17: pp.92.

Watson, S.A. 1982. Corn : Amazing Maize: General Properties. In Handbook of Processing and Utilization in Agriculture. Vol. II(I). Plant Products, Ed.I. A. Wolff, CRC Press, Boca Raton, FL., U.S.A. pp. 3-30

Weber, E.J.1978. Corn lipids. Cereal Chemistry. 55:572-76.

Weber, E.J. 1979. The Lipids of Corn Germ and Endosperm. Journal of the American Oil Chemists. 56:637-41.

Weber, E.J. and Alexander, D.E. 1975. Breeding for Lipid Composition in Corn. Journal of the American Oil Chemists' Society. 52:370-373.

Weiss, T.J. 1983. Commercial Oil Sourcess. In. Food Oils and Their Uses, AVI Publishing Company, Inc. Westport, CT: pp. 39-40.

Wells, F.B., Cereal Chemistry. 1955. 32: 157.

Wisniak, J. 1977. Jojoba Oil and Derivatives. Progress in the Chemistry of Fats Other Lipids. 15: 167-218.

Yermanos, D.M. 1974. Agronomic Survey of *Jojoba* in California. Economic Botany. 28:160-74.

Zaharia, T. and Sutescu, P. 1972. Bread with Addition of Maize Germs. *Ind. Aliment* 23:546-56.

Chapter 4

Processing

Post Harvest Technology

Introduction

The demand for edible oil has grown with the increase in world population. Improvement in living standards have made the population increase the consumption of oils and fats. It also has become more discriminating with respect to the oil quality. Oil producers, have to meet the increasing demand for edible oil, and improve product quality and variety as these have become major factors in the market-place.

Agriculture has been tuned through selective seed breeding and improved crop production practices to take care of the increasing demand. The emphasis has been to improve yield per hectare or tolerance to climatic variations. The ability to modify seed through traditional breeding and biotechnological approaches promises a revolution in agriculture. Using various approaches, not only quantitative improvement but also the qualitative changes especially fatty acid changes have been brought about. Customer support is needed for the genetically modified crops. A lot of controversy exists at present in accepting genetically modified crops. This aspect has to be studied in depth and needs to be resolved at the appropriate level as quickly as possible.

A prerequisite for the production of top quality oils and meals is a high quality raw material. Some attempts have been made to improve the physical characteristics of oils and meals through genetic, agronomic as well as through post production handling and processing technologies. The harvested quality grain should not be damaged during transportation, storage and processing. To produce a crude oil, the industry uses a number of processing steps, the application of which depends on the type of seed and the scale of the operation.

The purpose of Post Harvest Technology of oilseeds is to recover the maximum quantity of good quality oil and also make the deoiled cake useful as food or feed. Oil in oilseeds is located in spherosomes, which are the oil bodies (Salunkhe *et al*, 1992). When seeds are broken into small particles and the cells get ruptured the oil becomes available and is easily extracted. From the time of harvest, till the seeds are fed to the oil extracting machines, various processes are carried out to make good quality of oil available to the consumer apart from providing remunerative prices to the producer, namely, the farmer. The various steps involved in the oil extraction are proper harvest, drying, storage, seed cleaning, heat treatment and different types of machinery. The basic objective of every step in the post harvest technology is to minimize losses and to sustain or improve the quality of the product.

Harvesting and handling

Delay in harvest causes field losses depending on the crop and seed shattering quality of the crop. Generally crop harvest is carried out based on subjective judgements. People decide to harvest based on colour of the leaves, pods, and sometimes looking at the taste of the seeds. Seed moisture content at around 15% appears to be a more objective method for optimum harvest (Ali, 1986). Normally harvesting losses in oilseeds are around 5-15%. Groundnut pods mature in the soil and hence if digging for harvest of pods is carried out when the soil is dry, 20-30% of the pods are likely to remain in the soil. Similarly, shattering, in crops like mustard, sesame and soybean can cause seed losses up to 20%.

The physiological activity leading to dry matter production is essentially complete within 3-4 weeks after flowering at which time wet weight of the seed is maximum. High yield of better quality seed is possible if crop is harvested at the stage when grain is fully matured.

Farmers adjudge the maturity of crop either by visual observation of colour of seed and/or plant or through a biting test (taste of the seed) or by both. Harvesting is done when crop is sufficiently dried. At this stage the moisture content of seed approximates to 18-12% wet basis (Wb). This practice of harvesting at lower moisture content causes a significant reduction in yield (5-15%) due to losses caused by shattering, birds, insects, rodent, bad weather, etc.

Harvesting studies conducted for different crops have shown that harvesting at proper physiological maturity reduces field and handling losses and results in superior quality seed/grain. It also makes land available at least 7-10 days early for the next crop resulting in better land utilization.

Groundnut

Indeterminate nature of pod bearing in groundnut results in a crop with varying maturity (Nagaraj *et al.*, 1987). Bunch-type groundnuts usually mature 110-130 days after sowing (DAS), runner type 130-150 DAS, and once the majority of pods are mature, delay in harvesting results in substantial losses. Pods have 30-40% moisture wet basis and need drying upto 20-25% for threshing/stripping and 8-10% for storage. (Ali, *et al.*, 1986). The yield and oil content increase till correct stage of maturity followed by a decrease at over-mature stages (Nagaraj and Kailash Kumar, 1984; Pattee, 1969; Singh and Randhir Singh, 1974). Oil and protein contents in the seed are less by 1% at over maturity stages (Nagaraj, 2003).

Rape/Mustard

Rape/mustard matures when stems and pods become yellow, the seed very dark, rattle in the pod when shaken, and have a moisture content of about 15%. To minimize shattering losses, harvesting may be done in the early morning or late evening when atmospheric humidity is high. A potential yield of 10 t/ha from rape/mustard crop is possible. However, the present average yield ranges from 0.5t to 2.5 t/ha (Weiss, 1983).

Sesame

Capsules ripen irregularly on the sesame plant starting from the bottom, hence early or delayed harvesting may cause field losses upto 75% in shattering varieties and 9-20% in non-shattering varieties (Weiss, 1983). When the colour of leaves and stems turn to a yellowish/reddish tint the crop can be harvested. Delayed harvesting also causes considerable losses in handling the crop from field to threshing site. Threshing is done manually in most of the developing countries.

Soybean

It can be harvested when the majority of plants are mature. This stage comes when the majority of the leaves have fallen, the lowest pods are yellowish and dry, and the seed is wholly yellow. Seed should have a maximum moisture content of 14-15%. Seed moisture will usually be between 50-60% at physiological maturity, and 12-24 days are necessary for it to fall to 14%. Harvesting is done manually and mechanically. Soybean may be harvested when average bean moisture is about 15-18% wb.

Safflower

A mature safflower crop can be left-standing in the field for about a month with minimal loss. Birds flock to mature safflower field and cause considerable damage. Safflower is ready for harvest when the plant is quite dry but not brittle and the seed has a moisture content of 8% or less. Harvesting can take place some 35-40 days after maximum flowering. Seed maturity could be tested by squeezing several heads between the fingers. If seed separates easily from the head, it is ready for harvesting. Safflower crop should ideally be harvested when the moisture content of the seed is below 8% preferably 5%. Safflower seeds harvested at the correct stage of maturity has 2-3% higher oil content than those harvested either early or late (Nagaraj, 2003).

Sunflower

Seed is ready for harvest when the back of the flower head turns yellow or brown and seed moisture content is 10-12%. Delayed harvesting results in unfavourable changes in seed-oil composition. When the plants are physiologically mature, the seed yield and oil content are highest. Delay in harvest induces the FFA to increase (Robertson *et al,* 1978). Sunflower needs to be harvested at the correct stage of maturity, in which case seeds will have 2-3% higher oil content (Nagaraj, 2003).

Selective manual harvesting is possible only on small farms that too when labour is cheap and plentiful or, to provide employment when other work is scarce. Plant is cut, stacked in the field to dry and then threshed. The seed is cleaned and dried to less than 10% moisture. In hot weather, standing sacks with the necks open preferably under shade are used to store sunflower seeds.

Niger

Niger-normally matures 90-120 days after emergence. Since the seeds are not held firmly in the head as with safflower, those which ripen first begin to shed their seed before the later heads are matured. Shattering can substantially reduce the yield, upto 25%. Harvesting is done manually and plants are dried for 2-3days in the field before taking it to threshing site in bundles. The potential yield is about 1t/ha but the average yields are around 250-400kg/ha.

Handling the Harvested Produce

The harvested oilseed crops are bundled and left in the field, or transported to the threshing yard for further drying and or threshing. Higher moisture in crops in threshing floors result in dark colour of the seeds. Sometimes fungal infestation, loss of seed viability and increase in free fatty acids in the seed and oil occur. Damage to seeds or kernels induces the pro-oxidative changes and also reduces market value.

For threshing of oilseed crops, beating with sticks and or cattle treading are the common practices being followed in the developing countries. Some seed losses ranging from 2-5% occur in such a type of threshing. Mechanical harvesting and threshing are slowly replacing the traditional methods.

Cleaning and Grading of Oilseeds

Winnowing at farmers level after threshing is carried out to separate trash and other lighter fractions. It is mainly weather dependent. Stones and metal parts are picked up manually. Sometimes mechanical air blowing also is used to remove trash and related light fractions. Losses to the tune of 2-5% are reported during the various stages of cleaning.

Grading of oilseeds is done in organized sectors. Experienced persons use methods like cracking the seeds, having a feel of it, testing the taste of the seeds etc. Such crude methods are used at the rural level and at the small mills. Some tests related to moisture content, seed weight, foreign matter, splits, damage, breakage, and colour are used while buying the seeds. Objective tests related to oil content, protein percent, FFA, germination percent and hardness etc. are used by bigger mills or solvent extraction plants.

Drying of Oilseeds

Oilseed crops are harvested when the pod/seed moisture is around 20-40%. Hence, they should be dried to less than 8% moisture, which is a safe level for proper storage. The simplest and easiest method of reducing moisture in the produce is to delay the harvest itself. If proper sunshine is not available or shattering is a problem, delay in harvest increases seed losses. Such situations are normally avoided by the farmer. After harvest, the oilseeds are dried in the open sun in the field itself. Though this method is slow and labour intensive, it is cheaper and requires zero power and hence farmers in the developing world favour this method. Improper drying under such situations, sometimes, may favour growth of insects, moulds, fungi, discoloration of seeds etc.

The first drying is accomplished by delaying the harvesting of the crop. Secondly, the harvested crop is bundled and left in the field or transported to threshing site and stacked for further drying. When oilseed moisture reaches between 12-16%, threshing is done and after winnowing and cleaning, the seeds are finally dried in the sun or in mechanical dryers to 5-8% moisture level for milling or storage.

Oilseeds are generally dried in open sun in drying yards by the farmers as well as oil millers. Sun drying is slow, labour intensive and weather dependent. Hygienically it is not good as the oilseeds collect dust, moulds and micro-organisms which along with seed get crushed during oil extraction and become a part of the oil as an admixture. It deteriorates the oil quality. Mechanical drying avoids such problems especially in groundnut. When harvesting of groundnut and other oilseeds coincides with rain and cloudy weather, development of moulds and aflatoxins starts. Such a situation is also very conducive for FFA development leading to higher refining cost of the oil. Bulk as well as continuous drying of oilseeds are possible. Wet seeds, above 17% moisture, may be dried in two passes. More moisture removed in one pass may lead to shriveling and cracking of the seed as well as limiting the dryer performance. Drying to 12% moisture slowly and steadily in the first pass and then tempering may give desirable results.

Oil and seed quality deterioration are the likely results of inadequate open drying in the sun. The crops which come to maturity and harvest during summer do not face much problem in drying. For example soybean, safflower, castor etc. can easily be dried after harvest. This is because these crops come to harvest when sunshine is available in plenty. Crops like groundnut get caught in rains during harvest and problems like aflatoxins are hence associated with them. Such situations may or may not be regular in nature. Mechanical drying is the answer to avoid unusual weather related drying problems.

Though the safe moisture for storage of oilseeds is 5-8%, farmers try to dry only upto 15% moisture content for having higher yield and for getting more returns. Hence drying is essential for storage as well as oil extraction.

Drying of seeds to a proper moisture level is an important pre-requisite. Excess moisture in the seeds need to be removed as quickly as possible. The safe levels of moisture for storage are 6-8%. It would be advantageous if it's below 5%. Normally during drying the moisture present in the outer surface evaporates by reduction of moisture in the interior of the seed portion. It is sufficient to dry the seed to a level of equilibrium moisture content of the environment. Drying to lower moisture content or bone dry condition is not of much use, as the seeds return back to equilibrium moisture content, once brought back to such an environment.

Seed Dryers

Two types of driers are used for oilseed drying, namely, (1) vertical dryers and (2) horizontal rotary dryers. The former dryers are extensively used in the oilseed industry. Oilseeds flow downward and intimately come in contact with a gas stream. The hot air moves in the opposite direction taking out the moisture in the seeds. The air with moisture is mostly released to the atmosphere and a part of hot and dry air is re-circulated. The driers are operated at 70-95^{0}C and exhaust temperatures range from 38-44^{0}C.

Rotary cylinder type dryers are used to dry extremely wet material like copra and other oilseeds. The rotary dryer consists of a cylinder rotated upon a suitable trannion and thrust roll assembly. The cylinder is slightly inclined to the horizontal. Cylinder length is four to ten times its diameter. The wet seed is fed at one end of the cylinder and flows through it by action of rotation and slope effect. Dry seed is discharged at the other end due to the moisture evaporating action of the hot air moving along with it.

In hot air drying plants, the fumes that come out of the stacks are polluted by fines and odouriferous or toxic gases. The solid particles are separated by filtration along with chemical wash of the fumes to remove odours.

Seed Cleaning

Oilseeds should preferably be cleaned prior to storage to remove plant residues like stems, veins, foliage and other foreign material, which have a role in heating up the seeds, due to metabolic decomposition reactions. This is easily said than done due to the huge quantities involved.

The pre- cleaning should be carried out before the seeds enter the storage facility, either a silo or a flat store. This is usually performed using a coarse screen, to remove stalks, stones and other debris that could damage the mechanical-handling or processing equipment.

In many cases, the fine cleaning of seed for storage is unnecessary. However, if the seed contains a lot of silica, which would cause severe wear in the oil extraction plants, then fine cleaning should be used to remove it as a first processing step. In general, simple vibrating screens with variable aspiration are used for fine cleaning.

Storage

Proper storage is important to preserve the quality and, therefore, the value of the seed. As a natural product, seed deteriorates over time and this process is accelerated at higher temperatures. The critical factor which determines the storage life of seed material is it's moisture content. For example, soybeans are normally traded at 12% moisture content. At this moisture level the storage life is as short as 6-9 months. However soybeans that are dried down to 10% moisture can safely be stored for up to 4 years. Groundnuts and mustard can be safely stored at moisture content of 7-9% in impervious structures. Groundnut pods can be stored easily for one year while the kernel with more than 7% moisture cannot be stored for more than three months. At 35^{0}C and 7-9% moisture content the storage life of soybeans is around 4-8 months (Chouksey and Ali, 1983).

Seed moisture must be interpreted on whole seed basis, i.e. as it is sold on the market, which does not reflect the moisture available to develop acidification and other reactions in the silo. In fact, the dry non-lipid components absorb moisture, while the fatty constituents do not. Therefore, if we admit that the critical moisture of the non-fat part of a seed is 15%, a seed with 40% oil has an overall moisture of 9% and a seed with 47% oil has an overall moisture of 7.95%. These figures are very important because any excess moisture automatically activates a biological reaction that is the cause of exothermic chemical reactions. Once the temperature reaches higher levels, the seeds deteriorate in a matter of hours. This may lead to total destruction in a few days, after localized wet zones have been produced and colonized by fungi.

To avoid this, it is imperative to process the seeds as soon as they reach a temperature of 40^{0}C. Reliable temperature probes and adequate ventilation are an absolute necessity. Insects and rodents are also a problem, deserving particular attention. The use of non-toxic chemicals need to be undertaken. An important problem is that of oil acidification inside the seed during storage. This is generally due to seeds broken or damaged as a result of careless or inadequate handling, namely, poor equipment or hurried unloading. Thus it is clear that storage and its peripheral equipment have the most dangerous impact on the quality.

Storage Conditions

Oilseeds need to be stored safely following certain guidelines. The factors which influence adverse storage are (I) moisture (2) temperature and (3) molds and bacteria. Seeds being living organisms continue to have processes like respiration and related biochemical reactions, though at a reduced level. Storage compounds like sugars, starch, lipids etc. degrade and release CO_2 and moisture. Temperature, moisture and relative humidity enhance these reactions. To minimize these negative effects, seeds need to be stored under low moisture and temperature conditions. For example oilseeds stored with 8% moisture and 25⁰C develop only 10g CO_2 per tonne/day while it increases 100 fold if the moisture content is around 20% (Bernardini, 1985).

Moisture content should be checked regularly and kept below the safe limits in the storage bins. Temperature also should be around safe limits. If proper aeration is allowed, the temperature normally, does not rise. Due to processes relating to fermentation, temperatures can rise up to 60⁰C. Once this temperature is reached, the seed ceases to be alive. However, combustion process may continue and spontaneous fires may occur in oilseed godowns. Hence good aeration is an essential requirement.

Table 151: Safe storage moisture content of some of the oilseeds

S. No.	Oilseeds	Moisture content for safe storage upto one year % Wet basis
1	Groundnut pods	7-9
2	Groundnut kernel	4-6
3	Rape/Mustard	7-9
4	Soybean	8-10
5	Sesame	6-8
6	Safflower	4-6
7	Sunflower	7-9
8	Niger	6-8
9	Linseed	6-8
10	Castor	6-8

On an average oilseeds store well below 8% moisture content. In rural areas oilseeds are generally stored in storage facilities created using locally available materials. Oilseeds are sometimes just heaped in one corner of the house. Poor storage results in high moisture, relative humidity, temperature and fungal infestation. It also results in mould and/or insect damage and over-heating of seed consequently affecting oil quality in terms of colour, FFA, iodine and other values. The best is to rapidly transport the oilseeds from producers to processing plant and immediately extract the oil.

Metallic bins and or concrete silos are better than metal silos for oilseed storage. The former provide insulation against temperatures also. However, concrete silos are slightly costlier than metallic ones. A rise in oilseed temperature may be an indication of seed spoilage. Typical silos may have 1000- 8000m^3 volume. Shed storage in bulk on concrete floor is the cheapest but does not facilitate in easy ventilation and it is conducive to insect infestation. 4-5 m high bag stacking is also practiced. Stacking must have effective ventilation.

Soybean seeds can be stored successfully at farm level in improved indigenous storage facilities such as metal bins for a period of 4-8 months provided soybean is dried to 8-9% moisture, treated with thiram fungicide at the rate of 3g/kg of seed and placed in the coolest part of the house/building where grain temperatures do not exceed 35°C (Dakshinamurthy and Ali, 1986). The germination percentage of seed in such storage conditions can be maintained above 70%.

Various types of bins made of materials like straw, clay, stone, metal etc. are used for storage of oilseeds at household level. Millers use godowns/ warehouses. In tribal areas, oilseeds are stored in earthen pots and sacks which are prone to moisture and rodent attack. Depending upon the type and size of the oilseeds, safe storage methods need to be followed (Chouksey and Ali, 1983, Dakshinamurthy and Ali, 1986).

Raw materials that have deteriorated cannot be reconditioned and are the cause of very severe losses in capacity at the refining level, more specifically during degumming or neutralizing. They also create a need for expensive equipment and more reagents/ chemicals at the refining level. The crushing plant involved in pressing and/or extraction, has zero or negligible effect on raw materials either good or bad.

Seed Intake and Handling at the Oil Mills

Proper treatment of oilseeds prior to expelling or direct solvent extraction is essential for better oil yield and good quality oil. The important steps are (1) cleaning, (2) weighing, (3) dehulling (optional), (4) size reduction, (5) heat treatment (temperature & moisture control) and (6) pelletizing.

The seeds can reach the plant in batches ranging from a few bags to a few thousand tonne loads. The handling requirements for these extremes vary enormously. It is essential at this stage to control the weight and quality received. Weight control is easy. Control of seed quality is more sensitive but also vital, as it is an important factor that makes the difference between profit and loss. Weighing and sampling at seed intake protects the interests of both the buyer and the seller and provides the basis for any necessary price adjustments.

Impurities like dust, sand, iron and other extraneous material need to be segregated before the oilseeds go for pressing. Passage through shaking sieves effectively removes most of the foreign material. Cleaners can handle a variety of materials like trash, dirt, sand and other materials. Electro magnets, when fitted to cleaners, can remove iron, which damages the rolls when left in the seeds. Weighing is an essential prerequisite to monitor accurate oil yields and operating costs. Generally oil mills are equipped with pit type truck scales.

Dehulling or Decortication

Most seeds can be processed either complete with their outer shell or with a proportion of it removed. There are several reasons for removing the outer shell (or hulls). One important reason for removing the hulls is to increase the protein content of the cake or meal. This is the usual reason for decorticating sunflower seeds, cottonseeds and soybeans.

Whether to decorticate or not is usually dictated by the market rather than processing considerations, and the equipment required depends on the nature of the hulls. Thin or fragile hulls are removed by an impact huller in conjunction with multistage aspiration. Thicker hulls are subjected to a cutting action and often need a beating action to aid separation. Since the hulls frequently represent a waste disposal problem, the final hull content will be adjusted to maintain the desired protein content in the product while not removing excessive quantities of hulls.

Hulls of oilseeds are low in oil content but high in fibre. Dehulling is done to produce low fibre cake as also to increase the capacity of the oil mill. Dehulling is essential in the case of high hull (20-40%) oilseeds like sunflower, safflower, castor and cottonseed. Other seeds like soybean, groundnut kernel, mustard etc. can be directly used in oil extraction. Dehullers depend on sieving and air classification. Complete removal of hulls is not done. About 8-12% hulls facilitate efficient oil expelling.

Table 152: Hull and kernel content of oilseeds.

Oilseeds	Kernel %	Hull %
Groundnut pod.	75	25
Rapeseed- mustard	80	20
Soybean	93	7
Sunflower	65	35
Safflower	60	40
Castor	75	25
Cottonseed	62	38
Sesame	85	15
Linseed	57	43

Source: Shukla *et al.*, 1992

High capacity mills are equipped with dehullers. Two methods of dehulling are normally used. Front end dehulling, consists of cracking the seed and separating loose hulls and tail end system removes hull after oil expelling to obtain a high protein meal. It is easy to dehull soybean. One can dehull it up to 1% hulls (99% hull removal). Sunflower dehulling is slightly difficult. However 90% hulls can be removed at seed moisture level of 8%. Cottonseed decortication is also some what difficult, due to the presence of lint.

Breaking/ Size Reduction:

Size reduction is accomplished by hammer mills. Small oilseeds like rapeseed, sesame, linseed and niger are rolled directly before oil expulsion, while groundnuts are handled with or without rolling. For *ghani* expulsion this step is not essential, as size reduction takes place automatically. Some quantity of water is added to cement the broken pieces during expulsion. Water or moisture up to 12% in the seed enhances the oil expulsion.

Breaking or crushing, grinding and flaking are useful steps prior to oil expelling. Seeds that have not been broken during decortication must be broken to reduce the particle size before further processing. The only exceptions to this are very small seeds, such as rapeseed or canola, sesame and niger. Breaking is achieved either in hammer mills or on roll stands, with two pairs of fluted rolls mounted one above the other. The purpose of breaking is to reduce the particle size sufficiently to ensure uniform cooking without producing an excessive quantity of fine particles. It also reduces the power consumption in subsequent processing steps especially in the expeller or press.

Flaking

All oilseeds need to be flaked to aid efficient oil extraction. The flakes are produced by passing the seed between smooth rolls. Flakes of the size of 0.3-0.4 mm thickness are normally prepared. The rolls are usually hydraulically loaded and are typically 600-700 mm in diameter and 1000-1500 mm long. Flaking ruptures the cell structure and makes the oil more readily available to the press. Flaking can be carried out after cooking or conditioning of the oilseeds.

Sufficient care must be taken to avoid production of excessive levels of fines during the flaking process. The temperature and moisture of the seed prior to flaking needs to be monitored carefully. If the material is too cold or too dry prior to flaking, the seed will shatter rather than flake. The fines produced by the shattering of the seed cause many subsequent processing problems. Whenever ambient temperatures are low, it is normal to precondition the seeds prior to flaking. The seed temperature will be raised to around 60⁰C without significantly drying the seed. At this temperature, the seed becomes more pliable and better flakes are produced.

Higher oil yields can be achieved, when size reduction is carried out. Flakes of 0.3 mm thick need more extractions (4 times) as against one in the case of 0.15 mm thick flakes. Hence least possible thickness is the best for most efficient extraction. Small oilseeds like mustard, sesame, and linseed and medium sized seeds like cotton seeds are rolled before expeller processing while soybeans are rolled/flaked for solvent extraction. Oilseeds processed in *ghanies* or baby expellers are neither decorticated nor reduced in size. Seeds get automatically crushed and reduced in size.

Conditioning/ Heat Treatment

Heating the seed improves oil extraction. Heat denatures protein and breaks the oil-protein emulsion making oil easily available for extraction. Also heat makes the small oil droplets in the seed to coalesce to bigger ones, for better recovery during expelling or solvent extraction. Oilseeds with a moisture of 1-2% yield less oil than seeds with 10%. Oil cells become impermeable under low moisture conditions. At around 10% moisture the oil in the cells try to diffuse out of the cells, making it more efficiently extractable. In view of this oilseeds are normally cooked at 10-12% moisture at about 90⁰C for 15-20 min. The heat apart from improving oil extraction efficiency, destroys moulds and bacteria. Anti-nutrients if any also get detoxified. In effect the quality of oil as well as cakes improve. Heat treatment step need not be followed in *ghani* and small expeller extractions.

The cooking step enables the processor to compensate for variations in the moisture content of the raw material and has the following main objectives:

1. To decrease the viscosity of the oil, making it easier to remove,
2. To rupture oil cells by flashing-off intrinsic moisture as steam,
3. To coagulate the proteins in the seed,
4. To sterilize the seed by destroying enzyme activity and preventing the growth of moulds or bacteria and
5. To detoxify certain undesirable seed constituents, such as gossypol in cotton seed.

Multistage vertical stack cookers or large single-stage horizontal conditioners can be used. In either case, the heating medium is invariably steam at 6-10 bar gauge pressure. The seed is usually heated to 90-110°C and dried to 3-5% moisture content in order to ensure efficient operation of the screw press. The exact conditions will depend on the seed being processed and the pressing duty required. Seeds which are solvent extracted without prepressing, such as soybeans are conditioned to 60-65°C prior to flaking.

Cooking oilseeds at 10-12% moisture and 90°C for 15-20 min coagulates the protein and facilitates better extraction of oil. Crushing in uncooked condition in *ghanies* and small expellers, results in oil losses. The cooking process also imparts plasticity to the seed mass. It insolubilizes gums and related material helping in reducing refining costs. Also, the cooking kills the bacteria and other harmful micro flora. It also helps in reducing antinutrients.

Moisture content of 9-14% is necessary in oilseeds to obtain good results. Normal cooking ranging from 100-130°C has little effect on colour as well as nutritive value of oil. Over cooking, darkens the oil, which is difficult to refine.

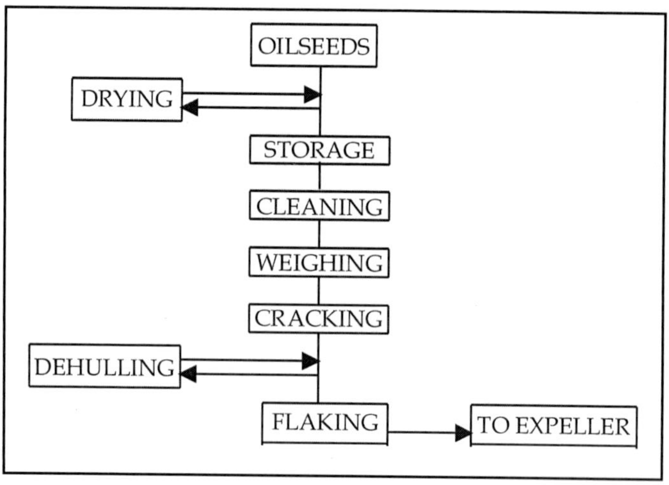

Basic operations prior to oil extraction

Oil Extraction

Oilseeds are processed mainly to obtain vegetable oil and good quality meal. Oils and oil cakes have various uses, both food and non-food. Different types of techniques have been developed to extract the oils and process them into final finished products. The objectives of all these processes are.

1. To obtain higher yield of oil
2. To minimize damage to the quality of oil
3. To obtain purer components, namely, oil and oil cake and
4. To produce oil cake with greater utility.

The operation of oil milling is essentially a mechanical process involving the use of pressure to squeeze out the oil from the oil bearing material. Before the oilseeds go for pressing, they need to be cleaned from extraneous material, followed by dehulling, size reduction, and heat treatment etc. Rupturing the oilseeds and pressing them to obtain oil is an age old process. Wood, iron and other heavy material have been used to remove oil. Even simple processes like boiling oilseeds have been practiced to obtain oil. In such cases the floating oil is scooped out, filtered and water present is removed by boiling.

Oil from oilseeds is mostly extracted with the help of traditional animal drawn *ghanies,* power *ghanies,* rotary oil mills, mechanical expellers and solvent extraction units. In the beginning of the twentieth century, hydraulic press superceeded all types of presses (Hurst, 1911). Wedge press has been used to extract oil from soybean in the olden days (Boatright, 1979). Mechanical screw press has been invented by Anderson in the year 1900. It has undergone various types of changes in quantity, quality and efficiency and has become the modern day expeller over the years. Expeller extraction is a major step in removing oil from various oil bearing material.

Seeds containing high oil (30 % or more) are pressed directly in a *ghani* or expeller to obtain a good quality cake with 12-16% oil and 30-40% protein. This is called pre-pressing during which process mild increase in temperatures occur. The cake is spongy and permeable and does not disintegrate during conveying. Application of higher pressures increase the temperatures, resulting in more oil and a cake rich in proteins.

Ghani Extraction

Oil extraction in India is mostly done by animal/power drawn *Ghanies* which are the predecessors of the present day expellers. Seed crushing for extraction of oil is generally done in conventional village *Ghanies*/rotaries which are normally in a small scale sector. The application of pressure effects the oil expulsion from the oilseeds in a *Ghani.*

The *Ghani* consists of a wooden mortar and a pestle. The mortar is fixed to the ground while, the pestle attached to one or a pair of animals (bullocks, buffaloes, camels etc.) is rotated in the mortar. The seeds get crushed by the generated friction or pressure. The oil moves out through an opening at the bottom and the cake is scooped out from the top in the mortar.

Ghanies are usually constructed of wood with the exception of a small iron band fixed around the top edge of the mortar. From design aspects, these *ghanies* have engineering concepts and are evolved keeping in view the local raw material availability, chemical characteristics of oil seeds, local set ups etc. Inspite of its divergent characters such as size, structure, area occupied etc., the traditional *ghanies* prevalent in different regions are constructed in the uniform basic design of mortar and pestle and exerting pressure on oilseeds on leverage principle. The pressing method of various oilseeds differ from place to place (Shukla *et al.*,1992).

The pestle is turned around at a speed of 5-7 rpm by means of a pair of bullocks fastened at the free end of the lever. While in operation, it makes up and down movement, crushing the material in the mortar. The expelled oil is collected from a small hole in the bottom. The traditional oil *ghani*, operated by bullocks has capacity of 6-10 kg/charge (daily crushing capacity 60-80 kg) of 90 minutes and requires a floor area of 36 m^2 (400 sq. ft). Normally, a '*Ghani* with a pair of bullocks, takes about 3 hours to crush one charge of 16 kg of mustard seed (Singh and Shukla 1991).

Power Ghani

The traditional *ghani* operated by animals is sometimes uneconomical for operation. An engine powered *Ghani* is now-a-days replacing the bullock powered *Ghani*. This mill provides improved output at relatively lower cost compared to traditional *ghani*. It is driven by a 2 HP/3 phase/1440 rpm electric motor which rotates the pestle at an average speed of 11 rpm through belt pulley arrangement. Other components remain almost the same as those in traditional *ghanies*. A portable model of power *Ghani* has also been developed.

The improved power *Ghani* has an efficiency equal to that of a small scale oil expeller. Power *ghani* increases the oil yield by about one percent and reduces the time for crushing by 20%. It also reduces the cost of motive power to the extent of 80% in terms of maintenance of animals. The improved overhead power *ghani* can crush 100 kg groundnut seed with an average capacity of 12-15 kg seed/charge of 60 minutes each. This improved *ghani* is mostly of wood. The *ghanies* can be fitted in concrete or wooden mortar or iron mortar. The overall floor area required for this improved *ghani* is 7.2 sq m

(80 sq ft). Such *ghanies* are getting changes depending on needs and performance.

The Operation of Ghanies

The oilseeds processed in the *ghanies* are normally neither reduced in size nor cooked prior to their crushing. Heat is developed during crushing. The rise in temperature of seed mass is appreciably very low which provides an insufficient heat treatment to oilseeds. About 4% water is added to oilseeds for the hydration of proteins which helps in releasing oil during crushing. The oilseed cakes obtained from *ghanies* have high percentage of oil (12-14%). These cakes are used in solvent extraction plant for further extraction of oil. However, such cakes are regularly used in rural areas as cattle feed, depending their acceptability.

Table 153: Performance of different types of ghanies used for expelling groundnut.

Test Results	Bullock Drawn Improved *Ghani*	Overhead power *ghani*	Portable power *ghani*
Floor space Required, sq. m	36.00	7.20	3.24
Capacity/charge, kg	6-10	12-15	12-15
Crushing time/ charge (min)	90	60	50
Crushing capacity kg/d (8h)	69-80	100	115

The oil obtained from *ghanies* is of premium quality. It has good flavour and also contain useful constituents like tocopherols etc. This is because of the low temperatures that prevail during *ghani* processing. At the village level in India, oils like groundnut, mustard, sesame, safflower, niger, linseed and castor are processed in *ghanies*.

Oil Expeller

Oil Extraction by Pressing

Efficient oil extraction by a screw press is highly dependent on having the correct pretreatment before pressing. The object of the pretreatment is to reduce the proportion of non-oil-bearing husks and break the cellular structure by a combination of mechanical action and cooking. The cooking will also control the moisture content of the feed. The screw presses are of two types. One is to obtain maximum oil yield from an oilseed. This is achieved by a single pressing through a screw press at high pressure. The other type is to press the high oilseed to cake of moderate oil content. The cake is then further processed, in a solvent extraction plant, to recover the remaining oil. The latter method is the most efficient oil extraction method and is being practiced widely now-a-days.

The extraction stage itself is carried out using a screw press, known as an expeller. The press will be fed by means of a variable speed conveyor within the feeder unit. The feeder regulates the flow of material into the press, and thereby controls the loading on the main motor of the press. Oil release along the length of the cage is allowed to drain into the base of the press, where it is collected. The solid material remaining within the press is finally discharged into conveyors to be removed for subsequent processing.

There are different types of manufactures around the world who manufacture and supply the machinery with variations in efficiency and durability. They can be grouped into three main categories: i) small local companies, which typically operate in one country or only in a part of the country; ii) bigger regional manufactures, which generally operate in several countries and iii) a few true international companies, which operate on a global scale.

The Mechanical Screw Press

Application of pressure has been the traditional process used for oil extraction since ages (Marie Fils,2004). A mechanical screw press extracts oil by applying pressure to the seed material by means of a decreasing volume of Archimedian screw contained within a drained barrel. An expeller or screw press is essentially a continuous device for gradually increasing the pressure on material fed to it as the latter progresses inside a closed barrel, with provisions for oil to drain out as it is squeezed from the feed stock (Ali, 1994).

The mechanical screw press has five essential elements: the main worm shaft and worms, the drainage barrel, the choke mechanism, the motor transmission and thrust bearings and the cooling system (Dunning, 1956). The main worms shaft and worms are designed to exert a pressure of 1000-1400 kg/cm^2 (Youngs, 1965) on the seed being processed and push out the seed via the pressure chamber. The choke mechanism permits a final adjustment of the pressure. The drainage barrel is made up of rectangular bars, which fit into a heavy bar frame. The individual bars are separated by spacing clips which provide spaces of about 0.25 mm. This permits the drainage of oil from the seed being pressed and keeps the cake within the barrel, which also acts as a coarse filter medium.

The screw press can be built up of a single horizontal shaft or a combination of a vertical and a horizontal shaft. A great deal of heat is produced during the pressing operation, especially when straight pressing is employed. In order to keep the temperature down, various cooling devices are required. The screw press may be equipped with water-cooled shafts and ribs in the bar cage or some of the expressed oil may be cooled and circulated over the exterior of the barrel.

The worm elements consist of a cylinder, bored to fit onto the shaft and driven by means of a key way. The flight is wrapped around the outside of the cylinder. The flight is an open-ended helix and has the effect of pushing material along the barrel as the worm is rotated. Between each worm element are distance pieces, which are similar to the worms but without flights. The inclusion of distance pieces allows 'knife bars' to be included within the cage. A knife bar is a stationary blade that is attached to the cage and projects into the barrel, in order to stop the rotation of seed material with the worm assembly; this helps to ensure that the seed material is pushed forward.

The worm assembly is arranged so that, as material is transported along the press, the bosses of the worm become larger and the pitch of the flight smaller. This has the effect of progressively reducing the volume swept by the worm flights. This progressive reduction in volume applies pressure to the fed material and causes the oil to be released. The release of oil and a greater degree of milling of the material within the press can be achieved by the inclusion of other devices, such as pressure pieces and mixing rings, into the worm assembly design.

The second major element of the press is the cage. The cage forms a drained barrel around the worm assembly that allows the oil to be expelled, while retaining the solid material within the barrel. It consists of a series of semi-circular 'barrel rings' linked together by longitudinal ribs, top and bottom. Into this half cage are placed 'linked bars', which are parallel bars that are laid in the cage and are separated by spacers. This has the effect of producing a series of parallel slots along the barrel, through which the oil can flow.

Compression Ratio

Compression ratio of an expeller is important and it significantly influences the performance of a press. It indicates the extent of pressure to which the oilseed material would be subjected. It is defined as volume of the material in the first section of the feed section of the screw to the volume of the last section in the plug section. A compression of 10:1 implies that there would be a reduction of 10 unit volume to one volume before the fed material would exit from the press. Normally compression ratios of 5:1 to 15:1 are used. The ratio of length and diameter (L:D) of the screw are also important for press performance.

Feed Moisture Content

This is one of the most important factors affecting the extent of oil recovery obtained during mechanical pressing. Extremely low moisture content of the feed causes choking/jamming of the screw while too high a moisture leads to plasticizing effect, and thereby causing peripheral rotation of the fed material. Generally a range of moisture content from 5 to 11% (wb) is optimum for most of the oilseeds.

Pressing Temperature

Another important factor is the pressing temperature for better oil recovery. Generally, the oil recovery increases with increase in pressing temperature. This is due to decrease in viscosity of oil with increasing temperature which enables more rapid flow of oil. Heating also inactivates some of the harmful enzymes like trypsin inhibitor in the case of soybean. However, too high a temperature for longer duration adversely affects the nutritional quality of oil and the deoiled cake. Hence, temperatures above 100°C are generally avoided during pressing.

Applied Pressure

Oil recovery generally increases with applied pressure. However, beyond a certain range it has very insignificant as well as some adverse effects on total oil recovery. Too high a pressure causes choking of the screw as well as loss of energy.

Pressing Time

Generally, the oil recovery increases with increasing pressing time. However, after some time an equilibrium is reached beyond which there is an insignificant increase in the oil recovery despite an increase in time of pressing. Hence, time of pressing needs to be optimized for each oilseed depending on other pressing conditions including applied pressure and temperature.

Oil Recovery

Depending upon the type of machine and oilseed, the oil content can be reduced to as low as 3%. When large quantities like 450 tonnes/day are handled and used for pre-pressing, the oil content gets reduced up to 10% only. Batch-presses have traditionally been used for the recovery of oil from the oilseeds. The cakes contain around 8-10% oil. It also needs the attention of a skilled labourer. Batch press or hydraulic press is not economical and it is not in wide use now-a-days. However, batch presses find use in castor oil expelling.

In the most efficient expellers, the oil left in the cake is around 8%. A power requirement of not less than 45 KWH per tonne of seed processed is necessary. In the expeller, the seed is subjected to high pressure and hence due to frictional resistance considerable heat up to 160°C may be generated. Such high temperatures adversely affects the quality of oil and cake or meal.

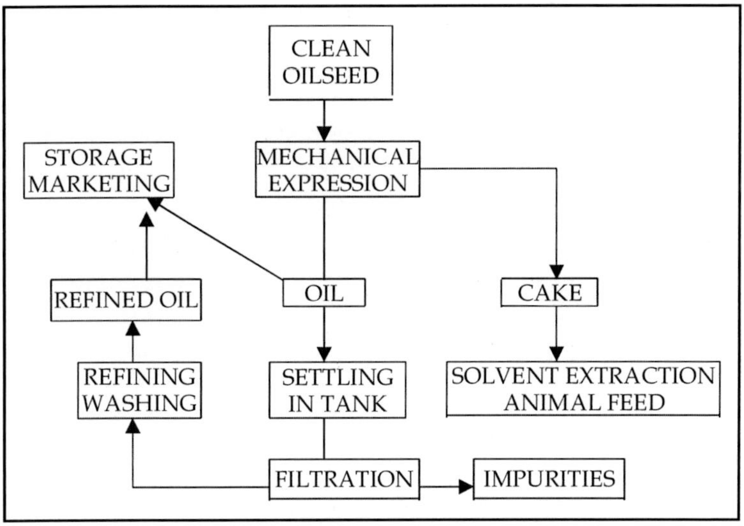

Steps in oil expeller

Cooked rapeseed flakes with about 3-4% moisture and 44-45% oil expels 75% or more of the oil reducing the cake oil to about 16% in an expeller. This mild pressing causes only a modest rise in temperature which unlike full pressing (3-5% oil in cake) does not threaten the stability and quality of the protein. Good cake is spongy and permeable and does not disintegrate during movement.

A screw press combining facilities of flaking, cooking, and pressing is also available. The machine repeatedly crushes and shears, compresses and then relaxes the flow of seed. The mechanical energy input provides adequate heating. Such machines are designed for only pre-pressing.

Principally there is no difference between presses used for straight pressing and those used for pre-pressing. Ordinary high-pressure screw presses may be operated at low pressure and at an increased capacity, but especially designed machines are now normally used for pre-pressing.

Disadvantages of Oil Expeller

Expeller presses are high power consuming plants. They also have high maintenance costs with lower capacities. In recent years, continuous - screw presses with capacities ranging from 200-400 tonnes per day have been developed. The main advantages of screw presses are that, (i) they have reduced capital cost, (ii) there is no danger of fire from combustible solvent as in solvent extraction plants and (iii) they can be operated by less number of skilled staff. The main disadvantage is that the oil recovery is not complete.

The process of straight pressing has the advantage over solvent extraction in being simple and of lower investment cost. Disadvantages are, however, the high power consumption and the wear and tear of the machinery, giving high working costs. Another major drawback is that it is only possible to press down to a residual oil content of 4-7 per cent, giving high losses in yield (Dunning, 1956). There is also the effect of the high temperature generated during straight pressing causing an oil and particularly a meal of lower quality. The cooking conditions employed are also more severe when straight pressing is used.

Oil Clarification

Certain amount of fine cake or meal called foots, come along with the expelled oil. To improve the quality of the oil, foots need to be removed. Generally screening tanks, filters or centrifuges are useful in removing the suspended fines.

Oil clarification is generally carried out in a two-stage process. Most of the solids can be removed simply by screening the oil over either a static or a vibratory screen. The screened oil is then finally clarified using a filter, usually a hermetically-sealed filter with stainless steel leaves, in which the material itself is used to generate a pre-coat through which the oil is filtered. On large pre-press plants, horizontal decanters are used in place of filters. The solids from the oil clarification processes are generally quite high in oil content. These are normally returned to the press feed stream for reprocessing.

Solvent Extraction

Extraction of the oil from the seed by dissolving it in a solvent (eg:Hexane) almost completely recovers the oil. The cake left behind is almost free from oil and is rich in protein and fibre. The quality of the oil obtained through solvent extraction is equivalent to that of expeller oil. A solvent extraction plant mainly consists of an extractor, in which oilseed and solvent come into contact, and the solvent removal parts from the cake as well as the solvent oil mixture. The oil is freed from the solvent in a series of stills and stripping columns with associated condensers. The oil is cooled and filtered for further storage.

As for back as 1850 carbon disulfide was used for chemical extraction of lipophilic constituents. Due to its bad odour, the solvent has been discontinued. But since 1870, light petroleum fractions have become available and real solvent extraction has found it's way. After world war II, it has become widespread.

'Solid–liquid extraction or leaching is the diffusive transfer of soluble components or solubles from an inert solid to a surrounding solvent. The extracted components, in this case oil, can be recovered without any change, by simple evaporation of the solvent. Solvent extraction is mainly used for recovering oil from soybean, oil cakes, minor oilseeds, rice bran and other low oil bearing materials. The rate of penetration of solvents into the particles of meal/oilseed is dependent on the thickness of the material. Hence the raw material is converted into flakes of 0.2 to 0.3 mm thickness. Better penetration of solvent through the flakes makes the extraction more efficient.

The most widely used solvent is hexane with a boiling point around 64^{0}C. Better recovery is possible due to the volatility of the solvent. The characteristics of hexane best satisfy the technical, economic and operational needs of all oil-millers. Several other solvents have been studied and tried but have not been found successful. Hexane, though flammable has many advantages like low volatility and viscosity.

The oil solution (hexane with oil) is filtered to remove the foots, etc. Distillation is then carried out to obtain the oil as well as the hexane solvent. The cake which is saturated with solvent when subjected to heat, gets freed from solvent. The de-oiled meal of good quality can thus be recovered. Processing of 100-300 tonnes seed/day is more efficient and economical than 30-50 tonnes/day which is the minimum that can be considered for solvent extraction. The loss of hexane is 5-15%.

Solvent Extractor

Solvent extraction consists of a sequence of four operations, namely, a) physical removal of oil from the seed in the extractor, called extraction, b) desolventizing-toasting of the de-oiled seed, i.e. meal, often combined with drying and cooling of this meal, c) distillation, to remove the solvent from the extracted oil and d)recovery of solvent, for reuse in the extractor. Different types of extractors are on the market. All of them basically work on the principle of systematic countercurrent washing of prepared seeds by a flow of solvent. Batch and continuous type solvent extractions are in practice.

Many phenomena are at work in solvent extraction. Some are natural, the others are linked to the process design. Some of the design parameters, like extraction speed, permeability, height of the bed and fines affect the process of solvent extraction. Extraction speed is the time required for the poor miscella (external to the particles) and the rich miscella (internal) to reach a state of equilibrium. The liquids must have the highest possible operating temperature, just below vaporization temperature, to optimize the effects of viscosity on all the phenomena described so far.

The prepared seed-flakes, collects (expandates) or cakes-are transported on a perforated belt through the horizontal body of the extractor. Fresh solvent is introduced at the discharge end of the extractor and is sprayed over the bed of product. The solvent percolates through the product and extracts oil to be collected in a hopper and pumped to a sprayer located next to the previous one, towards the extractor inlet. This process is repeated several times and, via a series of pumps, the solvent becomes richer in oil.

Maximum amount of oil is extracted during the first few minutes of extraction. About 75% is extracted in the first 25 min out of a total of 60-75 min residence time. After that, a prolonged time is required to complete the extraction, as the oil becomes less accessible to the solvent. Extraction speed also depends on percolation, or the liquids' flow through the seeds. Hexane and miscella spray distributors are designed to avoid channeling, contamination and dead zones. Oilseed preparation, generally, modifies the permeability of the feedstock. For example, the permeability of soybean flakes is 17 m^3/h and that of soybean pellets is >45m^3/h. A mixture of 60% pellets and 40% flakes gives a permeability of 30 m^3/h.

A deeper bed and longer contact time is a better option. There should be facility to modify the height of the bed material and the speed of extraction, for accommodating materials with different percolation parameters. Increase in plant capacity is limited by the drainage time. Expanders can always increase the capacity. Expanded pellets (of soybean) have a higher permeability than flakes. Both extraction and drainage become faster. Thus, it is possible to increase capacity by up to 20% without sending more solvent to the desolventizer-toaster. Extractors normally produce fines, concentrated in the last miscella. Solvent filters and hydrocyclones are used to remove fines.

Desolventization of the Meal

The spent meal leaving the extractor contains 25-35% solvent, which must be evaporated. In a solvent extraction plant, extracted meal desolventizing, drying and cooling have a marked influence on the quality of the finished meal.

Desolventizers are used to recover hexane. Initially steam is used to remove hexane, the moisture content of cake which goes up to 25% is removed by indirect heat and the flakes are dried to around 10% moisture. The heat apart from inactivating anti-nutrients, also denatures and reduces protein digestibility of the cake.

Vertical stack desolventizer-toaster-dryer-coolers are popularly being used now-a-days. Desolventization consists of different steps. They are predesolventizing, in which solvent is evaporated using only indirect heating steam; desolventizing-toasting, in which sprinkling steam and indirect heating steam are combined; stripping, in which live steam is used; and, finally, drying and cooling.

Distillation and Solvent Recovery

Distillation is carried out to obtain two products, namely, crude oil, and hexane for re-use in the extractor. The first-stage evaporator consists of a vertical evaporator, in which most of the solvent is distilled simply by recovery of the latent heat contained in the gases leaving the desolventizer-toaster. The gases leaving the desolventizer have a temperature of 72^{0}C and the miscella leaving the extractor, 55^{0}C. The vapours are a mixture of hexane and water. The water vapour condenses first, until the azeotropic concentration is reached. The pre-concentrated (65-75%) miscella which has a temperature of 50^{0}C, is preheated to a temperature of 75^{0}C at the inlet of the evaporator. The pressure of the steam is at 3-4 bar. Rich miscella concentration at the outlet of the evaporator is about 93% and its temperature around 80^{0}C.

To evaporate the last traces of hexane, it is necessary to increase the temperature or to reduce the pressure. As the crude oil may not tolerate temperatures above 105^{0}C to preserve quality, steam distillation is used to reach the required concentration. Maximum use is made of live steam injection to evaporate the hexane. The oil is first heated to 90-100^{0}C in a specialized pre-heater, in which steam is injected from the bottom to increase the heat transfer coefficient. Vapour liquid separation takes place in the upper part of the stripper. At the outlet, concentration of the oil reaches 99.2-99.5%. It is necessary to work at very low pressure to reduce the solvent content to less than 400 ppm.

Some air is passed into the extraction unit in the pores of the feedstock. This air which moves out carries with it a certain amount of hexane, depending on temperature. A dedicated absorption system recovers this hexane. To reduce the potential loss, the hexane-laden air is brought into contact with mineral or vegetable oil that absorbs the hexane gases. The absorption oil is heated to release the hexane, and cooled again to be returned to the absorber.

Extrusion Expeller

Extrusion is a high temperature short time process. Tissue disruption and heat treatment to plasticize the material are carried out. This process reduces microbial load, and at the same time retains the nutritional value of the oilseed. The process works on the principle of frictional heat, pressure and attrition in a well designed extrusion cooker. Expeller is later used to remove oil from the extruded material.

The extruder consists of a horizontal shaft rotating at a constant speed of 550 rpm on which is assembled the screw segments of suitable configuration (Wijeratne and Nelson, 2000). The screw is present in a barrel assembly. Steam locks of desired dimensions built in the screw assembly, and the axially ribbed internal surface of the barrel segments, create shear during operation. At the end of the barrel is a die through which the extrudate discharges into the atmosphere. Careful selection of screw configuration, the diameter of the die, and the back pressure on the die provide the desired temperature profiles in the barrel. The material remains for about 30 min and temperatures upto 135-145^{0}C are reached.

There are two types of extruders in the market, namely, single and twin screw extruders. The single screw extruder contains a smooth, deep-flighted Archemides screw within a grooved barrel. The twin screw extruders contain two co-rotating or counter rotating screws inside the barrel.

During the extrusion cooking process, the material is conveyed by the screw from the feed hopper to the die-section; the particles, cells and granules are ruptured, plasticized and restructured due to shear, heat and pressure. Gelatinization of starch, denaturation of proteins and degradation and reorientation of molecules take place during the process. In some cases, complex formation is noted between lipid, amylose, protein, sugars and amino acids. The plasticized matter is puffed, texturized, cooked or a combination thereof when it leaves the die nozzle. These changes however, can be controlled with processing conditions and chemical additives. Though molecular changes occur during extrusion-cooking, the basic raw material quality is still reflected in the product quality. Although extrusion cooking is highly exploited, the basic understanding of chemical changes that take place during the process is still not understood properly.

The process of extrusion results in disruption of cell tissues and release of oil within the extrudate. The extrudate discharges in a semi-fluid consistency due to presence of free oil with less viscosity at higher temperatures. Such a short time extrusion cooking not only reduces microbial load, but also destroys anti-nutrients, denatures proteins and reduces moisture content of the extrudate.

This extrudate is then passed through an expeller to obtain the oil. Adaptation of dry extrusion as a pre-treatment to soybean helps in expelling 70% of the oil present in the seeds (Nelson *et al*, 1987). The resulting cake is edible with 50% protein and 6% oil. The cake can be utilized as medium fat soy fluor. This technology is suitable for new soybean growing areas with no processing facilities (Singh and Ali 1992).

Developments over the past 15 years have led oil millers to use extruders or expanders more systematically. The basic advantage of expanding, used in combination with some other conventional preparatory equipment, is a much lower consumption of steam and hexane at the extraction level. Expanders are sometimes used in combination with expellers. The use of expanders varies considerably from country to country. It is a prerequisite in U.S. while it is considered an oddity in huge markets, such as China, Russia or India. In Europe, expanders are generally used to increase the capacity of an existing plant but are often excluded at the design stage of new plants.

Super Critical Fluid Extraction

Introduction

The solvent extraction process is favoured over mechanical expellers because it leaves less oil in the meal. Solvents have become very expensive and their possible escape from the system is a constant air pollution and explosion hazard. Economic and social factors have revived interest in searching for cheaper and safer solvents, like ethanol and isopropanol. However, it is super-critical fluid technology which appears to be a viable alternative to current extraction methods (Friedrich, 1983).

Supercritical fluid extraction (SCFE) is a developing technology, It is a way to strip soluble extractives from prepared plant materials without physical damage or chemical change. SCFE is the substitution of a fluid in its supercritical (SC) state for hexane in temperatures above which it cannot be liquefied. If external heat is applied to liquefied gas in a high pressure industrial cylinder, that liquid will go into supercritical phase when critical temperature is exceeded. There will be no change in volume and no addition of latent heat but the pressure will increase.

In engineering terms, the density and volume of gas and liquid phases are identical at the critical point. SC state is any condition of temperature and pressure above the critical point. Super critical fluids (SCF) are generally also called dense gases. They exist above critical temperature. SCF have solvency of liquid and the diffusivity of gas. SCF has the density of a liquid. SC pressures of interest to researchers are in the range of 20-2000kg/cm^2.

Supercritical Fluid Technology

Observations of SCFE have been reported in 1879 but the first practical extraction has been demonstrated in 1955. Present development has occurred after 1960. Ammonia, ethylene, toluene, and CO_2 all show promise for SCFE. Of these CO_2 offers unique advantages. It is abundant, non-reactive, non-toxic, and environment friendly. Minor leaks or accidental losses are of little consequence as CO_2 is non-combustible. All vegetable oils are soluble in SC-CO_2. Better performance has been observed in the temperature range of 35-70^{0}C. Successful use of SC-CO_2 could free 900m liters of hexane for other efficient energy input purposes. Supercritical carbondioxide (SC-CO_2) is easily removed from the extracted products.

Extraction efficiency in general terms is the weight of oil extracted per unit weight of CO_2 passed through the substrate. The rapid response of solubility to temperatures and/or pressures allows much faster extraction rates and provides more economical methods of oil recovery. As CO_2 is normally not compressible at high pressures, there are no recompression costs. Oil can easily be recovered by slightly reducing the pressure at constant temperature. Alternately, reducing the temperature at constant pressure also helps in good oil recovery.

The solubility of vegetable oil in SC-CO_2 varies with temperature and pressure; for example, at 50°C, the solubility increases with increasing pressure from less than 0.1% at 3000 psig to nearly 3.5% at 10,000 psig. However, above 8000 psig a decelerating rate of increase is observed. Although one might expect a series of parallel solubility isotherms, the 50°C and 60°C isotherms cross at 6000 psig. This cross-over of the solubility curves may be related to the densities of the SC-CO_2 between its critical pressure (1070 psig) and 6000 pisg. CO_2 is quite compressible. The density, which is related to solute holding power, changes rapidly in this range, whereas above 6000 psig the density does not change so rapidly.

The expected increase in solubility with increase in temperature can be observed at the higher pressures. At the lower pressures, the increased solubility effect due to temperature is overcome by the decrease in density and related decrease in solute holding power. Equilibrium solubility is readily established and maintained during extraction. So, an increase in pressure from 5000 to 8000 psig significantly improves extraction efficiency at constant temperature. Extraction efficiency should be maximized at the highest temperature that is consistent with product integrity and at the highest practical pressure above about 6000 psig.

Above 10,000 psig, unusual and unexpected solvent characteristics of SC-CO_2 are observed. At temperatures above 60°C, the solubility of seed oils increases dramatically with increasing pressure above 10,000 psig. The apparent maximum solubility (greater than 40%) is mass transfer controlled. It varies with seed type, configuration and morphology. Below this apparent maximum, all extracted seed triglycerides exhibit the same equilibrium solubilities under identical conditions of temperature and pressure.

Increasing the pressure increases the solubility of groundnut oil in SC-CO_2, and increasing the temperature increases the solubility above 350bar. Particle size is inversely proportional to oil extraction efficiency. Moisture has a positive influence while flow rates do not affect solubility.

The rapid response of solubility to temperature and/or pressure allows much faster extraction rates and provides more economical methods of oil recovery. As CO_2 is relatively incompressible at high pressure, recompression costs are greatly reduced. Thus oil can be recovered by slightly reducing the pressure at constant temperature. Reducing the temperature at constant pressure also affords good oil recovery.

Applications of SCFE

Full fat soyflakes, are readily extracted with SC-CO_2 at pressure of 200-700 kg/cm^2 at 50^0C. Soybean oil thus extracted is lighter in colour and contains less iron and about one-tenth the phosphorus of hexane-extracted crude oil from the same beans. Demonstrations have revealed that it is even possible to separate the waxes, gums, fatty acids and toxic contaminants from the meal upto a limit by careful control of the pressure and temperature during the depressurisation of the CO_2 miscella.

SCFE has been widely used for extraction of soybean oil. Better extraction has been observed at a pressure of 300 bar, a temperature of 40^0 C and a solvent flow rate of 1.8 litres /min at STP. Yields of oil upto 19.9% could be acieved (Dobarganes, *et al.*, 2002). Super critical fluid extraction is useful in obtaining higher recovery of tocopherols (Bruhl and Matthaus,1999). The SC-CO_2 soybean oil is light coloured, with low phosphatides. This is further evidenced by lower refining losses (low degumming). Oil yield through SC-CO_2 extraction is less (about 18.3%) than that of hexane extraction (19%). This is misleading as the latter contains 1.5% phospholipids. The cake also is of better quality.

SC-CO_2 extraction of soybeans and other oil seeds has been studied by several groups of workers(Stahl *et al.*, 1980 and Friedrich,1983). Previous studies suggest that for rapid and complete oil recovery, oilseeds must be ground or flaked to rupture the cell walls and to maximize CO_2 contact with the oil. As in the case of normal solvent extraction, decreasing the flake thickness increases the oil extraction efficiency by the SC-CO_2 method also.

The use of supercritical carbon dioxide (SC-CO_2) as a solvent to extract oil from oilseeds such as peanuts has potential advantages over currently used methods. Very high oil recoveries (about 100%) are readily attainable with CO_2 without the fire hazard and solvent removal operations associated with hexane extraction. The SC-CO_2 method also has the potential of being less susceptible than the screw press to mechanical damage if foreign objects such as stones are processed.

Commercial uses of SCFE include decaffeination of coffee, production of spice extracts, and a few costly food components and pharmaceuticals, from plant materials. Use of SCFE is in an infant stage. Versatile uses of the same are likely in future.

Palm Oil Production

Palm oil is obtained from the *Elaeis guineensis*, a tree native to Guinea, West Africa. Nowadays, the oil palm is grown on plantations of the equatorial tropics (latitudes 15° North-12° South) in South East Asia, Africa, Central and South America. Production of the oil has increased strongly in the last four decades and is now in excess of 17 million tonnes, more than ten times the volume of olive oil produced. Of this production Malaysia is responsible for 50% and Indonesia approximately 30%. The Ivory Coast, Nigeria, Columbia, Thailand and Papua New Guinea are the other main producers. Malaysia, Indonesia and Papua New Guinea are the main exporters of palm oil and its fractions.

The impressive figures for growth in production are partly due to the exceptionally high productivity of the palm tree, which yields up to 6 tonnes of palm oil, corresponding to 30 tonnes of fresh fruit bunches (FFB) per hectare per annum. At the same time, the crop produces significant quantities of palm kernel oil, equivalent to 10-12% of the palm oil output.

The Palm Tree

The early palm can grow upto a height of 25m. It is a source of oil for many inhabitants of equatorial zones. Harvesting is a difficult operation as the fruit grows in large bunches at the top of the tree. Damage to the fruit results from improper handling and the absence of mechanization of the milling process. The quality of the palm oil recovered is very poor, with the free fatty acid content rising to 20-25% due to the action of the lipolytic enzyme developing in the bruised fruit. In the last 40 years, constant research and development, have made it possible to develop short-stemmed, high-oil varieties. These new varieties grow to a height of 4-5m and the fruit contains 20% more oil than that obtained from older varieties.

The fruit bunch of a dwarf oil palm carries 800-2000 fruits (1200 on an average, each fruit containing one kernel) and weighs 10-50 kg, according to the age of the plant and the climatic conditions. The oil is contained in the mesocarp (palm oil). The procedure of oil recovery must be carried out with due care to maximize palm oil recovery with minimum breakage of nuts and kernels. Co-production of palm and palm kernel oil on an industrial scale to give a single oil production has recently been introduced in West Africa.

The most important objective of the processing of the fresh fruit bunches, from the moment of harvesting to the delivery of the oils produced into storage, are:

1. To ensure production of an oil of high quality and stability, meeting the standards set by the industry.
2. To maximize the separation of palm oil and palm kernel oil.
3. To maximize recovery of the palm and palm kernel oils.
4. To make the most economical use of the other components of the fresh fruit bunches.

Post Harvest Management of FFBs

Actual processing of palm oil does not start at the mill but in the plantation itself. The bunches of palm fruitlets need to be handled with great care, as they are prone to suffer rapid degradation. At the time of cutting, the level of FFA in ripe, unbruised fruit is low, generally 0.3-0.8%. Incorrect handling causes the acidity to rise rapidly in the outer part of the fruit, especially when it is damaged. This is due to a very active lipase, which splits the oil into fatty acids and glycerol once the cell structure of the fruit has been adversely affected. Storage of loose fruit after harvesting also causes severe microbiological damage within a few days resulting in oil with high FFA and impaired bleachability. The damaged fruit when processed gives poor quality oil with high FFA.

Correct and careful harvesting and fruit handling are therefore crucial for oil quality. For this reason, the techniques used for sterilizing and stripping have been modernized in order to cope with the larger flow of material from the plantations, changing from the traditional cages and mono-rail crane equipment to bigger capacity cages and belt coveying systems for the sterilized FFBs. These modern systems reduce and simplify the transport of the FFB and the operation of mill, and also reduce the investment in area and buildings. In addition, this leads to reduced maintenance costs.

Sterilization

Sterilization stops the development of lipolytic enzymes in fruit bunches. The process also eases mechanical stripping and prepares the kernels for further processing and preconditions the nuts to minimize kernel breakage during pressing and nut cracking. Immediately after arrival at the oil mill, the bunches are unloaded from the trucks through special transfer hoppers into cages which are then introduced into an autoclave (sterilizer). In small-scale palm oil mills, having a capacity of 0.3-5 tonnes/h FFB, the sterilizers are of the vertical type. In the larger modern mills, multiple horizontal sterilizers are used. These generally contain 3-10 cages each. Individual cages have a capacity of 1.5-3 tonnes FFB.

New space-effective handling systems are tested with larger cages having a capacity of 10 tonnes of FFB. This considerably reduces building dimensions, maintenance costs and fruits handling, and enables the mill to produce oil of higher quality. The bunches are processed for some 45-60 min, according to their size and ripeness. Injection of saturated steam is used to raise the pressure inside the autoclave to 2.5-3.0 bar. Each autoclave is able to carry out 12-14 sterilization cycles per day.

Sterilizing is normally a 3-step process in which different pressures and residence times are used. The cycle conditions are specific for each palm oil mill, according to the fruit processed. Automatic control equipment is used in the sterilizer of modern plants and can also be installed in existing mills. This guarantees that the required conditions of temperature (100-120^{0}C), pressure (2.5-3 bar) and time are applied.

Stripping

Stripping is the separation of the sterilized fruit and calyx leaves from the bunch stalks. The sterilized bunches are fed continuously into a beater arm stripper, suitable for the smaller palm oil mills, or a rotary drum stripper, used by the larger, more modern mills. These strip and separate the fruit from the bunch stalks. The rotary drum stripper comprises a drum made up of equally-spaced longitudinal U-bars, which revolve at low speed. Longitudinal flat bars are located radially inside the drum. The bunches are introduced at one end of the stripper and are raised by the action of centrifugal force and by the flat bars, eventually falling on the U-bars. The impact frees the fruits, which pass through the U-bars and are removed by a screw conveyor. The empty bunches are discharged at the other end of the drum and sent to the empty bunch hoppers for disposal or further processing (drying).

Digestion

From the stripper the fruits are conveyed into a digester. The purpose of digestion is to separate the fruit pulp from the nuts and to cause the oil bearing cells to burst, which facilitates the subsequent oil extraction by pressing of the oil.

This operation is of utmost importance as it affects the oil yield. Proper digestion ensures that most of the oil cells are ruptured, thus easing the task of the press, which will have to break a limited number of unruptured cells. The best digestion conditions are attained by mixing the fruits at 90-100^{0}C for about 20 min. The digester is a vertical cylindrical vessel, fitted with a central agitator and heated either by a double steam jacket or by steam injection. For high efficiency, it is extremely important that the vessel should be full throughout the digestion process.

Pressing

The mash consisting of digested pulp and palm kernel nuts is transferred to the continuous screw press, which separates the solids, including fibres and nuts, from the liquid phase, comprising water and oil. Continuous screw presses contain a perforated cage or barrel of horizontal cylindrical design, where the mass is subjected to increasing pressure by a variable pitch worm, thereby releasing the oil. With an adequate digesting time, residual oil in the fibres can be reduced to 5% and at the same time nut breakage rates can be kept below 15%. It is important to monitor wear on screw parts to avoid metal contamination of the oil, which would affect its refinability.

During the process of pressing, mixing of the palm oil with palm kernel oil can occur. Even when using screw presses of the low-pressure type, some nuts can get crushed at this stage. Five per cent admixture of palm kernel oil is accepted by international traders. In practice the admixture is negligible and unlikely to exceed 3%.

Press capacities range 1-18 tonnes FFB per hour. Modern plants generally favour double-screw presses. Two products are obtained at the press outlet; a mixture of water, oil and solid impurities (sand and vegetable residue); and a press cake containing fibre and nuts. The liquid stream from a screw press system typically contains 66% oil, 24% water and 10% non-oil solids, NOS (Maycock, 1990).

Crude Oil Clarification

In the first stage of the clarification treatment, significant proportions of the fibre, sand and shells present in the crude oil stream leaving the extraction stage are removed on single or double-deck vibratory screens. The crude oil is then transferred to continuous settling tanks, which use specific gravity differences to separate the palm oil from water and impurities.

The design of these tanks provides for a mean settling time of 2 h, which is based on the assumption that the sludge, which should not contain more than 15% oil, is given further downstream treatment in order to recover this oil. The water added to assist phase separation is determined by the amount of water added in the previous stages, the total amount normally comprising 75% of the tank contents. The operating temperature in the tank is in the range 85-87^{0}C.

These conditions achieve the best results in oil separation. The oil leaving the settling tank (top oil) contains about 0.5% moisture and a small amount of dirt. The solids discharged from the settling tank still contain some oil, and in large-scale mills this stream is recovered and sent for further settling. Hydrocyclones are used to remove sand and other heavy solids from the top oil stream. The sand content of the outgoing oil is reduced to 0.04-0.06%.

Use of sludge centrifuges and 3-phase decanters further improve the process. Three-phase decanters can be used on crude oil or on sludge, but the latter is more common. The decanters avoid the cleaning problems characteristic of centrifuges, and reduce the liquid effluent quantities as well as the maintenance.

Oil Drying

Palm oil contains a very small proportion of phosphatides, or gums. If the oil is not dried, these hydrate slowly over a period of time and become insoluble. They can precipitate in the storage tanks and are known as 'foots'. Two dehydration systems are used to remove residual moisture from the oil, these being the atmospheric pressure drier and the vacuum drier. The principal advantage of the vacuum drier is that it is operated at lower temperature (80°C) than the atmospheric pressure drier, which requires a temperature of 100-115°C. By removing residual moisture from the oil, the precipitation of foots in the storage tanks is avoided. The dry oil will also be less prone to hydrolysis, which would lead to an increase in free fatty acid content. Before transferring to storage, the oil should be cooled to a temperature of about 50°C in order to protect it from oxidative damage, which would affect its subsequent bleachability during refining.

Fibre Fruit Separation

The press cake is conveyed on a paddle conveyor to the nut-fibre separation unit. This conveyor, which may optionally be steam-jacketed, serves to reduce the moisture content from 40-45% to 30% and to break down the compacted cake in order to facilitate fibre-nuts separation. The fibre-nut separation unit (depericarper) consists of a pneumatic separation column, which draws off the fibres by suction, and a polishing drum, where the residual fibres and other waste material, e.g. stalks and stones, are removed from the nuts. After separation from the fibre, the fresh nuts are dried in silos from an initial moisture content of approximately 16% to a level of 10-12%, in order to separate the kernel from the shell. This is achieved using hot air at 85°C, which is blown through the soil.

Nut Cracking and Kernel Oil Extraction

The efficiency of cracking is a measure of the adequacy of the earlier sterilization of the FFB, and fine-tuning of the sterilizing section may be needed. A vibrating screen recycles the unbroken nuts. Large installations require two or more crackers, in front of which it is advisable to install a nut-grading screen to classify the nuts according to size.

The cracked mixture consists of free kernels, shells, unbroken nuts, partly cracked nuts and dust. The system ideally consists of two double stage winnowing plants, in which separation of kernels from remaining fibre, shell and dirt is carried out by drawing air through a vertical column. The clay bath separation system uses the difference in specific gravity between the kernels (1.06-1.16) and the shells (1.3-1.35) to achieve separation. Alternatively, hydrocyclones may be used for this separation.

After separation, the kernels contain about 20% moisture. Drying is therefore essential, as kernel moisture content of about 7% appears to be ideal to maintain stability and inhibit the formation of free fatty acids during storage. The kernels can then be conditioned prior to pressing or extraction. This process is similar to the steps discussed in expeller extraction. The composition of the palm kernels after drying is normally: dry matter 92%; protein 8.5%; oil 49%; fibre 5.8%; non-nitrogenous extracts 26.9%; and ash 1.8%. Expeller or other technology is used to extract oil from the kernels as in the case of coconut.

Refining Of Oil

Introduction

Vegetable oil refining is a necessary process to convert a crude oil into a desirable wholesome nutritious food oil for human consumption. The crude oil removed from an oilseed by mechanical pressing or solvent extraction has many undesirable components. Depending upon the seed, these components range from merely giving the oil a bad taste to being harmful for human consumption. More than 95% of the constituents of most crude oils and fats consist of mixtures of triacylglycerols. The oil delivered to the refinery should be free of fine meal particles and moisture, since both of these items increase refining losses.

The non-triacylglycerol fraction contains variable amounts of phosphatides, free fatty acids (FFA), oxidation products, unsaponifiable matter (tocopherols, sterols, hydrocarbons, etc.) and other impurities. The quantities of these minor components in crude oils vary with the oil source, extraction process, season and geographical origin. These components can have different effects on the nutritional, functional and organoleptic properties of the oil.

Components of Crude Oils

The general objective of refining is to remove the components that have an adverse effect on the overall oil quality with the least possible damage to the triacylglycerols. At the same time, minor components with a known positive impact have to be retained, as much as possible, in the oil. The primary non-oil components removed are free fatty acids (FFA), phospolipids, tocopherols, sterols, monoacyl glycerols and di-acylglycerols, colour imparting chemicals like carotenoids, metals, pollutants, if any, and oxidation products. The removal of these impurities improves the shelf life, taste and odour, as also gives an attractive golden yellow colour which are important to the consumers.

Some components, such as fatty acids, mono- and diacylglycerols and phosphatides, are of glyceride origin. Although phosphatides protect the crude oil from oxidation, they need to be removed because of their negative effect on the colour and taste of the oil. Other minor components can be isolated from the unsaponifiable fraction. The main components of the unsaponifiable fraction (0.3-5.0% of oil) are: sterols, tocopherols (vitamin E), hydrocarbons, pigments (gossypol, chlorophyll) and vitamins (carotene).

Table 154: Constituents of some crude vegetable oils (weight percent)

Oil	Triacylglycerols	Phosphatides	FFA	Unsaponifiables
Coconut	90	0.05	3	0.4
Corn	92-95	0.7-0.2	1-3	2.0
Cottonseed	92-98	1.0-2.55	0.5-2	0.6
Olive	90	0.01	1-3.5	0.8
Palm	90 0.	05-0.2	2-5	0.4
Rapeseed	90-92	1.5-3.5	1-2	0.9
Safflower	92-99	0.4-0.6	1-2	0.6
Soybean	90-93	2.0-4.0	0.5-1	1.2
Sunflower	93-99	0.5-1.3	0.5-1.5	0.7

The Refining Processes

The purpose of refining is to produce a bland-tasting, light coloured oil without odour or flavour. To obtain a fully refined oil from a crude oil, the following steps are necessary. (a) degumming, (b)neutralization, (c) bleaching, (d) winterization and (e) deodorization.

Degumming is the removal of non-oil constituents from crude oil and is the first step in the refining of an oil. Neutralizing means the removal of the free fatty acids contained in the crude oil. This is done by treating the oil with a caustic soda solution, which forms a soap with the fatty acids. The soap stock produced contains some oil, which is a refining loss.

Degumming, neutralization and sometimes, winterization (removal of solid fat by cooling from the oil) may be combined in one operation. Bleaching (removal of colour) is carried out by mixing the neutralized oil with activated carbon or bleaching earth. Winterizing may follow bleaching. Deodourizing is the final step which involves steaming of the oil under vacuum at high temperatures. This step removes volatile, odouriferous substances from the oil by distillation. The finished product (refined oil) is packed into various types of containers for supply to the consumers.

The refiner can only do an efficient job of refining when the crude oil received is of good quality. Many factors influence the quality of the oil extracted from the seed and they can be generally classified as pre-processing or processing-related conditions. Pre-processing conditions that relate to oil quality are seed variety, growing and harvest conditions and storage time and conditions.

The production of undesirable FFA occurs when seeds deteriorate and enzymatic action hydrolyzes the triglyceride molecule. The rate of deterioration depends on temperature, moisture, time and oilseed. The milling process at an oil mill also can cause damage to the oil. After the seed has been hulled or cracked, quick processing is essential. Prompt processing prevents damage from moisture and heat.

Degumming

This is a process followed to remove phospholipids or gums in the oil. It is mostly carried out by water or acid washing of the oil. Phospholipids, being powerful emulsifying agents, cause lot of refining losses, if not removed. They are also pro-oxidants. Crude oils with levels upto 1200 ppm of phosphorous should be degummed to a level of 20-50 ppm of phosphorus.

Phospholipids in the oils are present as hydratable or non-hydratable phospholipids (Hvolby, 1971). Phosphatidyl choline exists in the hydratable form, while the phosphatidyl inositol exists as a complex with K and Mg and is also fully hydratable due to the hydrophilic inositol group. The hydratable form can easily be hydrated and can be removed through water washing. It exists in the seed before any processing occurs. Enzymatic (phospholipase D) action can convert a hydratable phospholipid to a non hydratable form. This conversion makes a phospolipid unremovable by water washing. It is better to minimize the formation of non-hydratable phopholipids. The removal of phospholipids is essential to producing a quality finished oil.

The non-hydratable phosphatides exist as salts of Ca and Mg of phosphatidic acid and phosphatidyl ethanolamine. Non-hydratable phosphatide content increases mainly during processing of oilseeds. Controlling moisture content and inactivation of enzymes, can check the levels to below 10% in the flakes prior to oil extraction.

Hydratable gums can be easily removed through water washing. The non-hydratable gums can be removed by treatment with phosphoric acid (0.1 to 0.3% of 85% solution) or citric acid (0.1-1.0% of 30% solution). The non-hydratables are not recovered and are simply treated as refining loss. The hydratables are generally recovered, especially from soybean oil and are widely utilized as food emulsifiers.

Processors have the option of approaching the refining step in two ways. The phosphatides can either be recovered for their by-product value through degumming or treated as impurities that must be removed from the crude oil. Most of the vegetable oils processed are refined without degumming. The decision not to degum oil is based on energy conservation and capital savings. A separate combined degumming and refining can be carried out in one step. Vegetable oil degumming offers several potential advantages:

1. It is necessary for lecithin production. The hydrated gums are the raw materials for lecithin processing
2. It satisfies export oil requirements for a product free of impurities that settle out during shipment.
3. Gum removal prior to alkali refining often improves yield because the phosphatides can act as emulsifiers in a caustic solution to increase the neutral oil entrained in the soapstock.
4. It substantially reduces refinery waste load due to the lower neutral oil losses and the reduction of gums discharged.
5. Degummed oil is more suited to physical refining techniques due to the significant reduction in non-volatile impurities, such as phosphatides and metallic pro-oxidants.
6. The soapstock from alkali refining, after degumming is easier to acidulate due to a lower emulsifier content, and the acid water has less impact upon the wastewater treatment systems.

The hydrated gums are vacuum dried for crude lecithin processing. Approximately 2% water; by oil volume, is contacted with the crude oil by mechanical agitation in a mix tank. The proper amount of water is normally about 75% of the oil's phosphatide content. Too little water produces dark viscous gums and hazy oil, while too much water causes excess oil losses through hydrolysis. Complete hydration requires approximately 30 minutes of agitation at 60 to 71°C for batch processing.

Temperature is important because degumming is less complete at higher temperatures due to the increased solubility of the phosphatides in the oils, and at lower temperatures the increased oil viscosity makes separation of the phosphatides more difficult. For continuous systems, preheated oil (80°C) is treated with water and mixed in a holding tank for approximately 15 minutes or 1 minute maximum with continuous pipeline agitators. The water-oil mixture must be treated very gently to avoid developing an emulsion; high shear stress in the feed pump and at the centrifuge inlet must be avoided. After hydration, centrifuges separate the sludge and degummed oil phases. The degummed oil can be vacuum dried and pumped to degummed oil storage or can proceed directly to the refining process. Three types of refining namely, traditional caustic, physical and micella refining are in practice all over the world.

Caustic Refining

Vegetable oil refining started as batch or kettle refining which is still a popular small scale activity globally. The batch method is inexpensive and is easy to operate. The oil is put into a vessel and heated; alkali is added while the mixture is stirred. The resulting refined oil and soap-stock is allowed to settle and then the two layers are separated. The separation is carried out based on gravity settling. The batch method is slow and inefficient.

The introduction of centrifuges have speeded up the separation and have made the refining practice more acceptable. Continuous caustic refining is the most common method of refining vegetable oils today. The process involves mixing the crude oil with sodium hydroxide, and centrifuging to remove the soap stock formed from the reaction of the caustic soda with free fatty acids in the oil. Normally the oil is mixed with hot water and centrifuged a second time to remove the last traces of residual soap stock. This step is referred to as water washing (Carr,1976).

The amount of alkali added to react with the oil is a critical parameter. The treatment or process consists of two parts. One part is the caustic soda used to neutralize all free fatty acids and the second part is enough excess alkali to remove other non-triglyceride impurities, like colour. The percentage of caustic soda to neutralize the fatty acids (expressed as oleic acid) can be calculated by multiplying the % FFA by 0.142. The percent excess of alkali added depends on the condition and type of crude oil. Excess caustic soda required for common vegetable oils are 0.01-0.05% for degummed soybean and 0.15-0.40% for cottonseed. Any amount of caustic soda added in excess results in additional and unnecessary refining losses.

Miscella Refining

Miscella is defined as a mixture of solvent and oil. Miscella refining differs from alkali refining in that the oil is refined in the miscella before the solvent is removed (Cavanagh, 1956). Miscella refining is mostly used to refine cottonseed oil.

The process has evolved over the years to improve oil quality, oil recovery, and energy costs (Hendrix, 1984). The crude miscella is concentrated to about 65% oil in the solvent in the first stage evaporator. It is then heated to about 55^{0}C and mixed with caustic soda before being passed through retention vessels to give time to allow reaction and agglomeration of soapstock particles. The centrifuge separates the heavy dark soapstock from the lighter coloured refined miscella. A light tan soapstock usually indicates either a large amount of unreacted caustic or a high amount of saponified neutral oil. These conditions indicate that the refinery is incurring a higher than necessary refining loss.

The soapstock is normally added to the meal in the top ring of the desolventizer-toaster where it is mixed with meal and desolventized along with the meal. Adding the soapstock to the meal has several advantages, such as increasing fat content and reducing meal dustiness. Soapstock added to cottonseed meal is sold for the price of hulls because soapstock contains no protein. It is also possible to acidulate the soapstock and sell it as a separate by-product. The refined miscella is sometimes water washed to remove traces of soap or it can be sent straight to desolventization and vacuum drying. The miscella can be directly utilized for processes like bleaching and winterization.

Physical Refining

The physical refining of oils is the process of passing superheated steam through the oil under high temperature and vacuum (Young, 1989)). Temperatures of 180-260^{0}C and vacuum of 2-4 mbar are used in physical refining. The process is almost similar to deodorization. The steam serves as a vigorous agitator and a physical stripping media of the oil and allows the high vacuum to efficiently remove low to medium molecular weight compounds from the oil. Though steam is the most common agent used, nitrogen also has scope to be used. The compounds removed are free fatty acids (oil breakdown products), tocopherols, tocotrienols (antioxidants), and odouriferous compounds. These impurities may be useful as by-products. Many-a-time they are recovered and utilized.

Palm kernel, coconut, and palm oils are good candidates for physical refining. The process is successful with oils that are low in phosphatide content, low in metal content, and low in heat-sensitive pigments. Hence crude oils rich in gums etc. need pretreatment to reduce their phosphatides, to be good feed stocks for a physical refinery. A light degumming and minimal treatment with bleaching earth is a prerequisite. Cottonseed oil because of its moderate phospholipid levels of 0.7-0.9% is a good candidate for physical refining. But the gossypol present in crude cottonseed oil causes a dark color to be fixed in the oil when heated and gossypol is not removed by simple water washing.

A fully refined oil should have less than 0.05% FFA, 0.5% moisture, 0.5% peroxide, 0.1% metals and 25 ppb polynuclear aromatic hydrocarbons, (Gunstone, 2004). Due to low FFA, the smoke point will be above 220^{0}C for the refined oils.

Neutralization

This process is meant to neutralize the oils and fats. Oils treated with acid or those containing FFA need to be treated with alkali. However, loss during this step should be as low as possible. Alkali treatment or steam distillation (physical refining), aim at the same result. Physical refining is followed with oils of low phosphatide content. The resulting soap (alkali salt of fatty acids) can be hydrolysed to obtain fatty acids for their further use in soap manufacture or as animal feed. Soap stock can also be converted to biodiesel using chemical or enzymatic methods.

Depending on the FFA content, alkali at concentrations ranging from 8-24% are used. The FFA and the non-hydratable gums are converted to water soluble sodium salts. After treatment, the oils are washed with water once or twice to reduce the soap content to less than 50 ppm. Neutralized oil is dried under vacuum to remove moisture.

Winterization-Dewaxing

Winterization-dewaxing is applied to remove small quantities of solids from oils due to which cloudiness develops at low temperatures or when refrigerated. These substances are mostly saturated fatty acids or waxes. The latter compounds are slightly different from triacylglycerols. They are esters of long chain alcohols with fatty acids present in triacylglycerols. Waxes are present in quantities ranging from 150-2500 ppm in some oils. Sunflower oil is a typical example which can contain high levels of waxes. Kernel oil has 600-800 ppm while non-dehulled (whole seed) oil can contain upto 3500 ppm. Corn, conola and rice bran oils also contain waxes which need to be removed prior to their marketing.

The process of winterization is simple. The oil is cooled slowly and kept at a temperature ranging from 5-10°C for about six hours or more to allow the solids and waxes to crystallize. Later, the temperature is increased upto 15°C to decrease viscosity and help quicker filtration. Filter aid at about 0.1% per 100 ppm of wax is added to facilitate the filtration. Dewaxing through centrifugation can also be achieved. This method is not efficient and hence not practiced widely.

Bleaching

Bleaching process is the costliest of all the refining steps; This is because of the cost of bleaching earths and their consumption. The functions of the bleaching step are to remove or reduce the levels of pigments (colour), oxidative products, phosphatides, soaps and trace metals in neutralized oil. The colour is due to compounds like carotene and chlorophyll.

Water washed, vacuum dried oil from the refinery must be bleached to reduce the colour pigments to an acceptable level. A properly refined, water washed and vacuum dried oil may still contain small quantities of soap (10-100 ppm), which must be removed by the bleaching step. Poorly refined oil, with high soap content exceeding 100 ppm, will inhibit the bleaching activity of the clay and hence decrease the de-colourizing effect.

The oil is de-aerated and dried to a moisture of 0.1 %. De-aerated earth is then added to the oil. The oil is heated at 90-120°C, in the absence of air with bleaching earths like, activated charcoal, bentonite, silica or fullers earth (0.2-2% of the weight of oil). This is an essential step to be followed prior to physical refining. In addition to colour, traces of phospholipids, soaps, metals and oxidation products get removed during bleaching. Activated charcoal also removes the carcinogenic polynuclear aromatic hydrocarbons (PAH'S).

Continuous bleaching is usually employed because it protects the oil from oxidation during the period of elevated bleaching temperatures. The dried oil is heated to approximately 100°C as it enters the bleacher, where it comes in contact with a continuous addition of deaerated bleaching earth. The quantity of bleaching earth and the oil flow are ratio controlled, to ensure that a constant percentage of bleaching earth of approximately 0.5 to 1.5% activated clay is in contact with the oil. The flow rates are adjusted to provide a contact time of 15 to 45 minutes, depending upon the type of equipment and bleaching clay used.

The bleached oil is continuously pumped from the bottom of the bleacher into a filter to separate the spent clay from the oil. Open filter presses, which are manually emptied and cleaned, are still used for the filtration operation. However, new types of automatic or mechanical discharge presses are being used to protect the oil from exposure to air at the high filtration temperatures. A typical refined and bleached oil can have FFA upto 0.04% (as oleic); colour, 2.5 red and 25 yellow; peroxide value, 1 meq and acetone insolubles, 0.02%.

Deodourisation

This is a final step in oil refining. Deodourization involves removal of oxidation products responsible for off-flavaours. This is achieved by heating the oil between 170 and 250°C under reduced pressures, with sparge steam. Deodourizer distillate is a source of sterols and tocopherols. One needs to be careful in deodorizing an oil with higher 18:3 (around 10%). The fatty acids in such oils get transformed to trans acids. α- linolenic acid can get converted to the trans acid upto 20% depending on the conditions of deodorization. However, similar conversion of linoleic acid and oleic acids is either low or zero and 10 to 100 times less than the conversion of cis α-18:3 to trans α-18:3. This aspect needs proper attention while refining soybean and rapeseed oils rich in α -linolenic acid. The level of trans fatty acids in edible oils should be in the range of 0.5-1.0%. (Greyt and Kellens, 2004)

Depending upon the characteristics desired in the finished product, liquid oil, hydrogenated oil, or various blends of the two are prepared and evaluated prior to a steam distillation, or deodorization operation. The large difference in volatility between the low volatile triglycerides and their natural flavour and odour components makes steam deodourization feasible for stripping these relatively volatile substances. In addition, the process destroys peroxides in the oil, removes aldehydes and ketones, or other volatile products resulting from atmospheric oxidation and reduces the oil colour by the destruction of unstable carotenoid pigments.

In most cases, the deodourization operation is carried out in "continuous" or "semi-continuous" vertical stainless steel deodorizers fitted with approximately five to seven shallow trays. The oil is departed under vacuum and then charged to the top steam heating tray from where it falls to the second, or Dowtherm heating tray, in which it is heated to approximately

250°C. The oil then proceeds through two deodorization trays, counter current to the stripping steam flow. A three-stage steam ejector system maintains volatilization of the impurities and to protect the quality of the neutral oil, dwell time in each of the trays is approximately 20 minutes. Under normal operating conditions, the stripping steam requirement pressures are approximately 4.5 kg per 45 kg of oil. The deodourized oil then exits the second deodorizer tray and is cooled in a water chilled cooling tray, before being filtered and transferred to a deodourized oil storage tank. Typical deodourized oil characteristics would be: a) colour= 0.4 red, 4 yellow, b) FFA= 0.02%, c) PV= 0.0 meq and d) tocopherols=15mg/100 ml.

Bland, deodourized fats and oils from the deodourized oil storage tanks are now ready. They can be packaged as cooking oils, salad oils, shortenings, salad dressings and margarines.

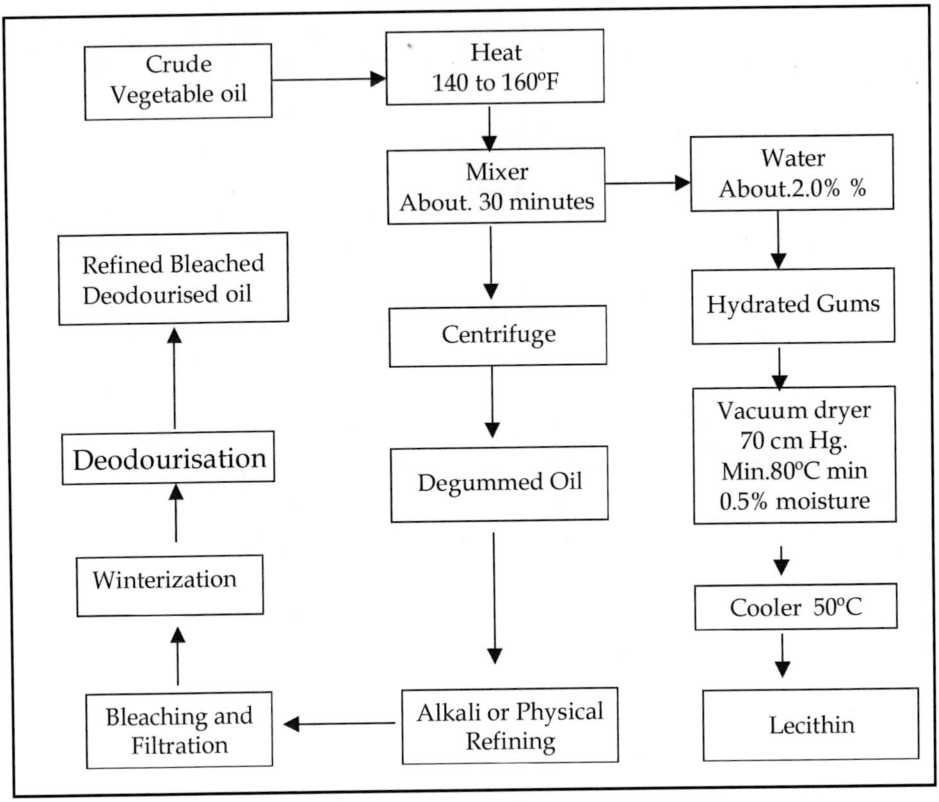

Fats and oils processing

References

Ali, N, 1986. Quality Aspects of Soybean Grains. Paper presented in the National Seminar-Cum-Workshop on Agricultural Machinery for Production of Oilseed Crops. CIAE, Bhopal. India. December, 30-31.

Ali, N. 1994. Present Status of Post Harvest Processing Technology of Oilseeds and the future Strategies to Augment Edible Oil and Protein Resources of India. In. Sustainability in Oilseeds- Proceedings of the National Seminar on Oilseeds Research and Development in India- Status and Strategies. ISOR, D.O.R. Hyderabad. India. pp 457-466

Ali, N., Alam, A. and Shukla, B.D., 1986. Research Digest 1973-1985. PHT Scheme (ICAR), CIAE, Bhopal.

Bernardini, E. 1985. Oilseeds, Oils and Fats. Volume I. Raw Materials. B.E.Oil Publishing House, Roma, Italy. pp 1-600.

Boatwright, J.N. 1979. Apropriate Technolology 6(2):24-25.

Bruhl, L. and Matthaus, B. 1999. Extraction of Oilseeds by SFE – A Comparison With Other Methods for the Determination of the Oil Content. Journal of Analytical Chemistry. 364(7): 631-634.

Carr, R.A. 1976. Degumming and Refining Practices in the U.S.. Journal of the American Oil Chemists' Society. 53: 347-352.

Cavanagh, G.C. 1956. New Integrated Process for Edible Oils. Journal of the American Oil Chemists' Society. 33:528-531.

Chouksey, R.C. and Ali, N. 1983. Optimum Conditions for Storage of B*engalgram,* Mustard and Groundnut. Proceedings of the International Warehousing Conference. CWC, Nov., 3-6. II-3. pp, 193

Dakshinamurthy, A. and Ali, N. 1986. Proceedings of the National Seminar on Soybean Processing and Utilization. Nov. 22-23, CIAE, Bhopal. India.

Dobarganes Nodar, M., Molero Gomez, A. and Martinez de la Ossa, E. 2002. Characterisation and Process development of Supercritical Fluid Extraction of Soybean. Food Science and Technology International. 8(6): 337-342.

Dunning, J.W. 1956 . Journal of the American Oil Chemists' Society.33:462.

Friedrich, J.P. 1983. Supercritical Fluid Extraction. Northern Regional Research Centre, Peoria, Illinois, U.S.A. Technology and Product-mix Forecast-oils and Fats in 2000 A. D. Dec., 11-12. pp 63.

Greyt, W. De. and Kellens, M. 2004. Refining Practice in Edible Oil Processing. Ed. by Wolf Hamm, and Hamilton, R.J., Blackwell Publishing, Shefield, U.K. pp, 110.

Gunstone, F.D. 2004. The Chemistry of Oils and Fats: Sources, Composition, Properties and Uses. Blackwell Publishing, Oxford, U.K. pp 1-288.

Hendrix, W.B. 1984. Current Practices in Continuous Cottonseed Micella Refining. Journal of the American Oil Chemists' Society 61:1369-1372.

Hurst, G.N.1911. Lubricating Oil, Fats and Greases: Their Origin, Preparation, Properties, Uses and Analysis. Scott Greenwood and Son, London.

Hvolby, A.J. 1971. Removal of Nonhydratable Phospholipids. Journal of the American Oil Chemists' Society. 48:503-509.

Marie Fils, J. 2004. The Production of Oils. In Edible Oil Processing. Ed. Wolf Hamn and R.J. Hamilton. Blackwell. Publishing, Shefield, U.K. pp, 47-78.

Nagaraj, G. 2003. Effect of Seed Maturity on Seed and Oil Quality in some Oil Seeds. Abstracts. 58th Annual Convention and International Conference. Oil Technologists' Association of India. November, 21-23. Kolkata. India.

Nagaraj, G. and Kaliash Kumar, 1984. Accumulation of Lipids, Proteins and Sugars in Developing Pods of Four *Arachis Hypogaea* L.Varieties. Journal of Plant Physiolology and Biochemistry. 11(2):40-46.

Nagaraj,G., Sheela Chauhan and Ravindra, V. 1987. Influence of Maturity on Oil, Protein Content and Yield in Three Spanish Bunch Groundnut Varieties. Seeds and Farms. 13(8): 53-54.

Nelson, A.I., Wijeratne, W.B., Yeh, S.W., Wei, T.M. and Wei, L.S. 1987. Dry Extrusion as an Aid to Mechanical Expelling of Oil from Soybean. Journal of the American Oil Chemists' Society. 64(9):1341-1347.

Pattee, H.E. 1969. Changes in Carotenoids and Oil Content During Maturation of Peanut Seeds. Journal of the American Oil Chemists' Society. 46: 629-31.

Robertson, J.A., Chapman, G.W. and Wilson, R.L., 1978. Sun News. 4(2): 14-17.

Salunkhe, D.K. Chavan, J.K., Adsule, R.N. and Kadam, S.S (Ed), 1992. World Oilseeds: Chemistry, Technology and Utilization. AVI Publication. New York. p.9

Shukla, B.D., Srivastava, P.K. and Gupta, R.K. 1992. Oilseeds Processing Technology. Central Institute of Agricultural Engineering (CIAE), Bhopal. pp 54-61.

Singh, D.P. and Randhir Singh, 1974. Effect of Growth Habits on Carbohydrate and Oil Content During Germination and Seed Formation of Peanut (*Arachis Hypogaea L.*). Indian Journal of Agricultural Research, 8:61-63.

Singh, J. and Shukla, B.D. 1991. Post Harvest Technology of Oilseeds. Central Institute of Agricultural Engineering, Bhopal

Singh, J. and Ali, N. 1992. Recent Developments in Edible Oil Extraction Technology. Journal of the Institutions of Engineers (India). 72: 23-27.

Weiss, E.A. 1983. Oilseed Crops. Longman, New York. pp. 540.

Wijeratne, W.B. and Nelson, A.I., 2000. Soybean Processing by Extrusion Cooking and Mechanical Expelling. In. Modern Technology in Oils and Fats Ed: S.C. Singhal and J.B.M. Rattray, Oil Technologists' Association of Iindia, Delhi. India. pp.31-35.

Young, F.V.K. 1989.Proceedings of the American Oil Chemists' Society World Conference, October 1-7, Maastricht, pp. 124-135.

Youngs, C.G.1965. Rapeseed Meal for Livestock and Poultry - Review. Canadian Department of Agriculture, Ottawa. pp.24.

Chapter 5

Oilseed Products

Nutritional Quality of Edible Oils

Fats improve the taste and flavour of foods. Fatty foods satiate a hungry person fast because fats remain in the stomach for a longer period than either protein or carbohydrate (sugar) foods. Fats also carry essential vitamins like A, D, E and K. They also provide the body with essential fatty acids. Thus, fats have a very important role to play in human nutrition.

Spectrum of Fats and Oils

It is not just the butter and *ghee* that provide us the necessary fats. Vegetable oils, meat and fish too add fats in our diet. There is an array of vegetable oils available to human beings. They are groundnut, rapeseed, sesame, sunflower, safflower, soybean, niger, cotton seed, coconut, palm and rice bran oils. *Ghee*(butter fat), however, is the main animal fat utilized widely. The quality and composition of these oils vary as also the preference of people.

What are Fats and Oils?

Fats and oils are chemically known as 'triglycerides' because every molecule of fat has one molecule of glycerol linked up with three molecules of fatty acids, either the same acid or three different fatty acids. The fatty acids are either saturated or unsaturated. Both kinds contain molecules made up of varying numbers of carbon, hydrogen and oxygen only. The saturated fatty acids (SFA) are so called because no more hydrogen atoms can be linked directly to any of the carbon atom in them. The unsaturated fatty acids however, have one (monounsaturated fatty acids-MUFA) or two or more (polyunsaturated fatty acids-PUFA) carbon atoms which can still link up with hydrogen.

Some of the polyunsaturated fatty acids like linoleic acid are not synthesized in our body and are supplied through the food. Hence, they are called essential fatty acids (EFA). Absence of these fatty acids in the diet results in hair fall, eczema like skin eruptions, slow healing of wounds and dry, scaly skin.

Functions of Dietary Lipids

Fat is an important component of human diet and fulfills several nutritional functions. It is a concentrated source of energy and helps to increase the calorie density of diets. Fat is more important in the diet of young children. It ensures adequate intake of energy without making the diet bulky. It is a source of essential fatty acids (EFAs), namely, linoleic and alpha linolenic acids. It is a carrier of fat soluble vitamins and facilitates their absorption and mobilization. It also contributes to flavour and palatability of the diet. In recent years, several other nutritional and health benefits of individual fatty acids have been elucidated. To understand the importance of fats, it is necessary to look into the process of digestion, absorption and metabolism of dietary lipids (Thompson, 1989; British Nutrition Foundation, 1992; 1993; FAO, 1994).

Digestion, Absorption and Metabolism of Lipids

Triglycerides with long chain (LC) fatty acids are mainly hydrolyzed by pancreatic lipase to 2-monoglycerides and free fatty acids from position 1 and 3 of the triglycerides. The 2-monoglycerides are readily absorbed in the intestine and after entering in the mucosal cells the fatty acids and 2-monoglycerides are re-esterified into triglycerides. These are then packaged with about 2% protein, 8% cholesterol and 7% phospolipids into chylomicrons and transported to blood. Chylomicrons supply triglycerides to the tissues particularly fat depot which take up fatty acids released from chylomicron and use them to synthesize triglycerides for storage. The short and medium chain triglycerides are completely hydrolyzed to glycerol and free fatty acids. The latter get bound to albumin and are directly transported to liver as free fatty acids which are easily oxidized. The rate of absorption depends on the type of fatty acids and the emulcification in the intestine.

Triglycerides can also be formed in the liver from carbohydrates, but these are transported in blood stream in a different lipoprotein package, the so called very low density lipoproteins (VLDL). The triglycerides of VLDL are hydrolyzed in capillaries of adipose tissue and muscle and a new lipoprotein complex called low density lipoproteins (LDL) are formed. LDL is the major carrier for the transport of cholesterol. Another lipoprotein, called high density lipoprotein (HDL), is an important scavenger of surplus cholesterol, transporting it from cell membranes to the liver where it is degraded or converted into bile acids (reverse cholesterol transport). The process of assimilation of triglycerides and cholesterol from the diet and production of these substances within the tissues of body are interrelated and regulated to keep the levels within certain limits. Disturbances in the metabolism of lipoproteins increase the risk of diet-related chronic diseases (metabolic syndrome, coronary heart disease, diabetes etc.).

SFA and MUFA are synthesized in the body in addition to being obtained from diet. Linoleic acid (LA) and alpha linolenic acid(ALNA) cannot be synthesized *de novo* in humans but fulfill crucial roles in the body. Therefore, they need to be consumed as part of the diet. All fatty acids generate energy. When in excess, they are stored in adipose tissue.

Fatty Acids and Nutrition

The fatty acids present in the fats determine their value as cooking media. Oils rich in saturated fatty acids and MUFA are stable and can be stored for longer periods (6 to 12 months). They can also withstand repeated deep frying. Oils rich in PUFA are more nutritious, due to their ability to supply EFA. A premium food fat should have equal proportions of saturated fatty acid, MUFA and PUFA.

Groundnut, rice bran, sesame and cottonseed oils have nearly equal amounts of the three categories of fatty acids and hence can be considered as good food fats. Safflower, sunflower, niger and soybean oil containing more of polyunsaturated oils are rich sources of EFA. Coconut and palm oils are saturated fats and are nutritionally poor. However, the essential fatty acids required for human beings may or may not be provided by any of these oils (visible fats). Food items like rice, wheat, pulses, vegetables and fruits in which fats are not so prominent, have significant quantities of EFA. They usually account for the required quantity of 2% EFA. Hence normal healthy persons need not be so selective about the visible fats for getting their EFAs. The main criteria to choose a cooking oil has always been their easy availability and the cost.

Cooking and Frying Medium

A good cooking oil should be bland without any taste. It should not be spoiled on frying for a long time. A golden brown colour is appetizingly attractive. The frying medium should not easily turn rancid or get smoked on heating. MUFA fats serve as good frying media with reasonably good keeping and nutritional qualities. Fats like safflower and sunflower cannot be stored over longer periods. Soybean oil looses its original flavour after one deep frying.

When the free fatty acid content of fats is above 2%, their quality is considered to be poor. Repeated deep frying and long periods of storage increase the fat acidity which in turn decreases the smoking point. Such a deterioration in quality is more in the case of PUFA oils. Fishy odour develops in these oils due to oxidation. However, vegetable oils contain natural antioxidants like Vitamin E which to some extent prevent the oxidation. Partial hydrogenation or addition of antioxidants are some of the techniques employed to avoid rancidity development in high PUFA oils. It is also the basis of *vanaspati* industry where unsaturation is reduced through the addition of hydrogen. Rancid food is unpalatable and can even be harmful. Therefore, in the case of children who depend mainly on one type of food, especially fats like *ghee,* care should be taken to feed only fresh fats instead of rancid ones.

Essential Fatty Acids and Prostaglandins

EFA deficiencey in humans results in abnormal skin conditions such as scaliness and dermatitis, increased water loss, reduced regeneration of tissues, increased susceptilitity to infection and increase in the ratio of 20:3 to 20:4 acids in the serum phospholipids. These effects are probably due to the lack of prostaglandins produced from specific C_{20} polyene acids.

EFA deficiency can usually be cured by linoleic acid because this acid which cannot be produced in animals and is entirely of plant origin can be metabolised to the required C_{20} and C_{22} polyene acids. Such acids are preferentially incorporated into cell membrane phospholipids where they have an important structural and functional role.

EFA status is best assessed biochemically by the ratio of 20:3 (n-9) acid produced from oleic to 20:4 (n-6) acid produced from linoleic acid in the serum phospholipids. Values ranging from 0.3 to 0.4 are generally regarded as acceptable. The recommended daily intake of EFA for adult humans is at least 3% of energy requirement or 2-10g per day of linoleic acid.

Natural prostaglandins (PG) are produced in many animal tissues from the 20:3, 20:4 and 20:5 polyene acids which furnish the PG1, and PG3 series of compounds respectively. The several members of each series vary in the degree of oxygenation. All prostaglandins are derivatives of the cyclic C_{20} acid, namely, prostanoic acid. These compounds are not stored in mammalian tissue, but are elaborated in response to various stimuli. Prostaglandins are used in child birth to induce parturition. They are used for treatment of gastric ulcers and bronchial asthma and they inhibit platelet aggregation.

Dietary Fat and Coronary Heart Disease (CHD)

The relationship between dietary fats and CHD has been extensively investigated with strong consistent associations emerging from animal experiments, and clinical, metabolic and epidemiological studies in diverse human population (Hu *et al.*, 2001, Hooper, *et al.*, 2001, WHO, 2003). The effects of dietary fat are mediated through the atherogenic effects of plasma lipids namely, high concentrations of LDL, modified LDL, lipoprotein and triglycerides and low concentrations of HDL. The effects of dietary fats on insulin resistance (IR), thrombosis, endothelial function as well as the pathways of inflammation have been more recently understood (Bray, 2002).

Earlier recommendations have placed emphasis on a low-fat diet. However, more recently it has become clear that higher fat intakes (upto 30-35% energy) are not necessarily bad, provided the energy balance is maintained and the dietary fat provides optimal balance of fatty acids. Further, the triglyceride structure (FAO, 1994; Kritchevsky, 1995; Jon Kabara, 2005) and minor constituents (FAO, 1994; Shahidi and Shukla, 1996; Przbylski and Michael Eskin, 2006) of dietary fat as well as other components of diet influence the nutritional and health effects of dietary fats.

Fats/Fatty Acids and Health

Dietary fat is related to the level of blood cholesterol, the main culprit linked to the cardiovascular diseases. Human beings with high blood cholesterol levels (260mg/ 100 ml) have double the risk of developing a heart disease than a person with normal cholesterol level (200 mg/100 ml). Cholesterol is carried in the blood by either low density lipoprotein (LDL) or high density lipoprotein (HDL). It is the LDL which is found to have links with the incidence of heart attacks. On the other hand the HDL removes cholesterol from arteries and reduces the risk. Saturated fats increase the level of LDL in the blood. But PUFAs increase HDL and thus are beneficial. Even MUFAs have been shown to decrease cholesterol and LDL. Therefore, lower intake of saturated fat reduces cholesterol and thus reduces the chances of developing arterial thrombosis, malfunctioning of heart, increase in blood pressure and onset of diabeties in adults.

The PUFAs are the precursors involved in the biosynthesis of different prostaglandins – a class of harmones. EFAs, being precursors of eicosanoids, are involved in regulation of cell function, blood flow, homeostatis and immune system. However, higher levels of EFA/PUFA in the diet have been reported to hasten ageing and are implicated in tumor production due to their ability to produce free radicals- very highly reactive oxygen or hydrogen atoms.

SFA

Acting through both atherogenic and thrombogenic mechanisms, SFA increase the risk of CHD. All SFAs are not similar in their ability to raise serum cholesterol levels (Khosla and Hayes, 1994). Fatty acids with chain length up to 10 carbons do not increase serum cholesterol since they are absorbed directly into the blood and rapidly metabolized in the liver. SFA with 12-16 carbon atoms (lauric, myristic and palmitic) increase serum total and LDL cholesterol (atherogenic). Like oleic acid (cis MUFA), the 18 carbon SFA (stearic) does not have cholesterol raising effect possibly because it is rapidly metabolized to oleic acid.

Myristic and lauric acids have greater effects than palmitic acid but the latter is more abundant in food supply. The cholesterolemic effect of myristic acid is greater than those of lauric and palmitic acids. When dietary cholesterol is increased, the cholesterolemic effects of these three SFAs are further increased. SFA is associated with thrombogenicity. An increased consumption of foods rich in SFA is associated with higher plasma glucose levels and decrease in insulin sensitivity. Studies in animals and humans suggest that dietary SFA may contribute to increase in insulin resistance (Lovejoy, 2001; Panico and Lannuzzi, 2004).

MUFA (Cis)

Oleic acid is the major cis MUFA in humna diets. In metabolic studies, substitution of oleic acid for SFA has been found to lower serum total and LDL cholesterol concentrations. In contrast to a diet low in fat and high in carbohydrate content, a diet containing high oleic acid does not lower HDL cholesterol (anti-atherogenic lipoprotein fraction) or elevate serum triglycerides. A shift from PUFA to oleic acid generates LDL which is less resistant to oxidation.

TFA (Trans Fatty Acid)

In human diets, small amounts of naturally occurring trans isomers with trans unsaturation positions between 6-10 of 18:1 (major TFA is trans vaccenic) are obtained from ruminant fats. The conjugated linoleic acid (CLA) unlike other TFA has double bonds with methylene interruptions. Elaidic acid (18:1) is the major TFA in hydrogenated oils. TFAs inhibit the metabolism of LA and ALNA to long chain- PUFA. TFA from hydrogenated oils behave like SFA in terms of their ability to increase serum LDL cholesterol fraction (Katan *et al.*, 1995; Zock and Mensink, 1996). Epidemiological studies have shown a strong positive association between TFA intake and CHD risk. Other observed negative effects of TFA are increase in serum triglycerides and lipoprotein (Ascherio, 1999). A positive association between TFA intake and type 2 diabetes has been observed (Panico and Lannuzzi, 2004; Andrew *et al.*, 2006).

n-6 PUFA

Vegetable oils rich in LA decrease total and LDL cholesterol when substituted for dietary SFA. Also, dietary intervention trials using high LA diets are more effective than those using low-fat, high-carbohydrate diets in lowering total serum cholesterol as well as rates of CHD. However, LA intake more than 10% energy also reduces the HDL cholesterol fraction (anti-atherogenic). A strong inverse association between dietary PUFA and CHD has been observed. PUFAs enhance peripheral glucose utilization, improve insulin action and reduced adiposity. Decreased concentrations of tissue PUFA (a biochemical marker of PUFA intake) are associated with decreased insulin sensitivity, measured as fasting plasma insulin concentrations. In humans prospective studies have suggested that PUFA intake is inversely associated with risk of developing diabetes (Salmeron *et al.*, 2001; Panico and Lannazzi, 2004).

n-3 PUFA

ALNA has total LDL cholesterol lowering effects. Increased dietary consumption of ALNA may decrease platelet aggregation and reduce formation of pro-aggregatory thromboxane A2 by inhibiting the conversion of LA to AA (Arachidonic Acid). An association between higher ALNA intake and risk of CHD has been observed (Petithory, 2001). Lot of evidence has been documented that LC n-3 PUFA (EPA-Eicoisa Pentenoic Acid, and DHA-Docosa Hexenoic Acid) reduce CHD risk. (Mori and Beilin, 2001). The protective effects of LC n-3PUFA are mediated through multiple mechanism. Replacement of AA in membrane phospholipids shifts eicosanoids balance towards the '3' series which in turn produce anti-inflammatory, anti-aggregatory and vasodilatory effects. There is also a decrease in plasma triglycerides and postprandial lipemia, improvement of endothelial dysfunction and prevention of ventriclar arrhythmias.

Specific Fatty Acids

Medium-chain fatty acids (MCFA) are digested and absorbed directly and are transported directly to liver and are readily oxidized to produce energy rather than be stored as body fat. They are, therefore, routinely used in treatment of many digestive and metabolic health conditions and in infant formulae. Medium chain triglycerides have a greater satiating effect as compared with LC triglycerides. However, they have been known to increase triglycerides which are undesirable. Fast rate of oxidation of MCFA leads to greater energy expenditure and therfore less body weight gain and decreased size of fat depots after several months of their consumption. (Jon Kabara, 2005).

Erucic acid (22:1) in rapeseed oil causes a rapid but transient accumulation of lipids in the heart. Concerns about possible deleterious effects of feeding high erucic rapeseed oil to humans has led to developing low/zero erucic rapeseed varieties (FAO, 1994). Consumption of large amounts of this oil has been discouraged due to its suspected involvement in heart disease. However, carefully conducted studies have revealed that at lower levels of consumption in the tropical countries, there is no need for any panic.

Ghee, Vanaspati and Animal Fats

Ghee is considered the supreme fat by Indians. Butter and *ghee,* derived from milk are often mistaken to be saturated fats. *Ghee* contains only 60% saturated fatty acids, the rest being MUFA. *Ghee* is unique in that it contains a good amount of (about 10%) short chain fatty acids (mainly butyric acid), which are easily digested and do not contribute to blood cholesterol. Hence the Indian view that it is a supreme edible fat is not unfounded. Butter comes next to it.

The vanaspati which has become popular as a cooking medium is nothing but a hydrogenated oil. It serves mainly as a butter substitute. During hydrogenation unsaturated fats are converted to saturated fats by the addition of hydrogen. Hence vanaspati can be stored for longer periods without deterioration. However, the hydrogenated fat gets enriched in trans-fatty acids, which are not always good for our health. Metabolically trans fats function as saturated fats. The higher melting point of vanaspati also makes it less digestible. Trans fatty acid intake increases LDL cholesterol and therefore increases the risk of coronary heart disease (CHD). They also accentuate EFA deficiency and are also implicated in tumor development. Hence it is better to reduce or avoid consumption of vanaspati.

Animal fats except fish oils are more saturated than vegetable oils. Hence they are not considered to be nutritionally good. Fish oils are rich in EFA and eicosapentenoic acid which are effective in lowering blood cholesterol.

Minor Fat Components

Fats and oils also contain endogenous components referred to as unsaponifiable matter (non glyceride fraction, NGF). The components of NGF have hypocholesterolemic and/ or antioxidant effects. The latter property is useful in imparting stability to the oil and serving as free radical scavengers (Shahidi and Shukla, 1996; Przbylski and Michael Eskin, 2006).

Table 155: Composition of minor compounds in processed vegetable oils.

Components	Range (%)
Diacylglycerols	0-8.0
Monoacylglycerols	0-0.2
Free fatty acids	0.2-9.5
Phospholipids	0.01-0.1
Sterols	0.4-2.0 (ppm)
Tocopherols	150-2000
Tocotrienols	0-1500
Phenolics	0-50
Chlorophylls and derivatives	0-2.0
Carotenoids	0-500

Phytosterols are present in vegetable oils in free and esterified forms. The chemical structure of phytosterols is similar to cholesterol, differing only in the side chain. Phytosterols mimic and compete with cholesterol at the site of absorption and reduce absorption of dietary cholesterol. Red palm oil provides carotenes. Carotenes are the precursors of vitamin A and have antioxiant effects (Sies and Stahl, 1995).

Tocopherols and Tocotrienols

All vegetable oils have varying amounts of different isomers of tocopherols. Palm oil and rice bran oil have tocotrienols in addition to tocopherols. The antioxidant properties of tocopherols and tocotrienols in biological systems have been well established. Different isomers of tocopherols like α, β, γ and δ exist in oils. The δ isomers exert the highest and α isomers exert the least antioxidant effects. Tocopherols and tocotrienols capture and destroy damaging free radicals that have been suggested to play a role in cellular aging, atherosclerosis and cancer. δ tocotrienol has been shown to be effective in ibhibiting breast and liver cancer cells (Gohl 1994). Tocotrienols have a role in molecular signalling (Sen, 2006). Being free radical scavengers, tocopherols, reduce the carcinogenic properties of PUFA. An adult human needs to consume 8 to 10 mg of tocopherols daily.

Rice bran oil contains oryzanols which have hypocholesterolemic and antioxidant effects (Rong *et al.*, 1997). Sesame oil has lignans (sesamin, sesamolin and sesamol) which are powerful antioxidants that make the polyunsatured oil remain fairly resistant to oxidation and thermal decomposition. Studies in rats have shown that sesame lignans inhibit endogenous synthesis of cholesterol and have antioxidant effects (Hirose, 1991). Further, combination of sesame lignans and tocopherols have higher hypocholesterolemic and antioxidant effects as compared to the effects of lignans (Yamashita *et al.*, 1995, Nakabayashi *et al.*, 1995). Cotton seed oil contains gossypol. Almost all vegetable oils contain phospholipids. Even squalene is present in some oils. All these constituents have anti-oxidant properties.

Soybean oil provides isoflavones which have several health benefits. Natural phenols, either simple phenolic acids and their esters or more complex molecules such as flavones, and flavonoids present in vegetable oils have antioxidant activity. Groundnut contains resveratrol at 200-250 μg/100g of seeds. This compound has a role in reducing heart disease risk. Resveratrol is present in redwine at a level of 800 μg/glass.

Recommendations for Dietary Fat

Initially, FAO had recommended equal levels of SFA, MUFA, and PUFA out of a total 30 % energy from fats and oils (FAO, 1977). The current recommendations for safe range of intake (WHO, 2003) of dietary fat for optimal health and prevention of diet-related chronic diseases are as follows (% energy): total fat 20-30, SFA<10, LA 6-8, ALNA 1-2.5, LA/ALNA 5-10, MUFA 10-12, cholesterol <200 mg/day and adequate antioxidants. Trans fatty acids have no health benefits and hence there is no requirement.

In an infant, the rate of synthesis of fatty acids such as arachidonic acid, docosahexaenoic acid (DHA) and eicosapentaenoic acid (EPA) does not meet the body's requirements and they need to be supplied separately. A developing foetal brain is again critically dependent on DHA and arachidonic acid. Hence women who intend to become pregnant, as well as those who are pregnant, should have adequate dietary intakes of essential fatty acids and their derivatives. Even lactating mothers should have adequate supplies of EFA and their derivatives in their diet to ensure that these fatty acids are available to the breast fed infant. Use of visible fats/oils from wide sources provides greater variety of minor components which may increase the hypocholesterolemic and antioxidant effects due to synergism in the biological effects of the minor components. Further, the consumer has the option to select oils according to availability, cost and culinary practice.

The ratio of PUFA/SFA should be 1 for an edible oil to be considered good. Groundnut, cottonseed, mustard, rice bran, sesame and soybean oils have values between 1and 3. Other oils like sunflower, safflower, linseed, palm and coconut are either very rich or poor sources of PUFA resulting in very high or very low PUFA/SFA(P/S) ratios. Blending of oils can be done to obtain a balance of fatty acids as per the recommendation. Use of two or more oils, depending on the purpose, to offset their imbalance in fatty acid composition is another way of obtaining balanced nutrition.

Conclusion

Saturated fats have a sound relationship to coronary heart disease. Hence a reduction in saturated fats can be suggested for certain individuals. It is difficult to cure coronary heart disease by altering the fat content of the diet. The principle "A little fat will do good" should be kept in mind and practiced.

Keeping Quality of Oils

The oils and fats are used as frying medium in both shallow and deep-frying modes. Fried foods are popular and appetizing components of our diet. Frying is usually carried out at a temperature of 165-185°C and is an efficient method of heat transfer that allows quick cooking and adds flavour to fried food. Some oil is absorbed by the fried food. In shallow pan frying, surplus oil is cleaned away at the end of the frying operation. In deep fat frying residual oil is re-used until such time as it has to be discarded because of its poor quality. This low-grade material may be added to animal feed as an energy source or it may be converted to methyl esters and used as bio-diesel.

Frying Operation

Frying is carried out on a domestic scale, in restaurants and fast food outlets on a batch scale using 4-20 kg of oil, and under industrial conditions in continuous mode with one tonne or more of oil to produce fried products for retail outlets. During frying, oil is transferred to the food so that fried foods contain additional fat at a level of 10-40 percent. Fat from the food is also transferred to the frying oil. The oil soon becomes contaminated, for example, with fish oil or with animal fat depending on the food being fried.

Under the conditions of frying, a number of changes occur in the oil. These include hydrolysis, producing free acids and partial glycerol esters, oxidation, producing flavour notes some of which are considered to be desirable and others not so desirable, and thermal changes leading to polymeric products and acyl groups with *trans* unsaturation and with five and six-membered ring systems. Volatile products are lost under the conditions of steam distillation which exist during frying. But compounds of higher molecular weight remain in the frying oil. With continued use, the oil begins to smoke, foam and become more viscous. Oil absorbed by the fried food has to be replaced by fresh oil and turnover and replacement of fat are important factors in a good quality frying operation.

Quality of Cooking Oil

A good frying oil should have high oxidative stability, a high smoke point (low level of free fatty acid) and show minimum colour darkening. The oil may be chosen because it gives a distinctive flavour to the fried food. A good frying oil should be low in saturated and in trans unsaturated acids for nutritional reasons and low in polyunsaturated fatty acids to increase oxidative stability. Consequently, it can be high in cis-monounsaturated acids.

A good cooking oil should be bland and neutral in taste and flavour. It should have a long shelf life. An appetizing golden brown colour would be an attraction. The frying medium should not develop off-flavours and no smoke should emanate. The oil should impart good taste and texture to the fried food and should not impart greasiness on the surface of the food items.

A good cooking oil also needs to have a long shelf-life ranging from at least 6-12 months. Development of rancidity (off- flavours) and quality deterioration can occur during storage over longer periods and during deep fat frying. These changes are dependent on various factors like type of oil, exposure to light, temperature, air, storage container, method and material used in frying.

Fatty Acids and Cooking Quality

Oils and fats derive their quality mainly from the fatty acids they contain. Three main types of fatty acids, namely, saturated fatty acids (SFAs), mono unsaturated fatty acids (MUFAs) and poly unsaturated fatty acids (PUFAs) are present in oils. The oils rich in SFAs are generally considered inferior from nutritional point of view because of their involvement in cholesterol production and the consequent cardiac ailments. But these SFA oils are highly stable and can be stored over long periods and have long fry-life.

PUFA oils are more nutritious because of their ability to supply the essential fatty acids like linoleic acid. Because of their higher levels of unsaturation, the PUFA oils get deteriorated quickly during storage and deep fat frying. MUFA oils are intermediate with respect to nutritional and keeping qualities. Hence there is no wonder these oils have been the main cooking medium over the ages and geographical locations.

The available cooking oils can be considered most stable from the point of view of storage and deep fat frying in the following order: coconut, palm kernel oil, palm, vanaspati, olive oil, groundnut, sesame, mustard, cottonseed, rice bran, niger, sunflower, corn, safflower, soybean and butter oil (*Ghee*). In addition to stability and nutrition, the consumer cares more for the cost of the oil while buying, in preference to taste and tradition. Palm oil, cottonseed, soybean oil or other oils, depending on the cheapness in that location, are being used irrespective of the nutritional, frying and stability considerations.

Factors Influencing Stability

Light and air have an important role in the keeping quality of an oil. Larger surface area means more exposure to these elements. Hence it is better to reduce the surface area. This is possible if the container is wide at the bottom and narrow at the top. Coloured glass or plastic offers protection to the oil from light. But the consumer prefers to have a look at the oil and satisfy himself about its clarity and attractive golden yellow colour before buying. To facilitate this the bottom of the container is kept transparent by some manufacturers.

Frying is simply a method of cooking. During the process, heat is transferred to the food through the cooking oil. Hence the oil itself needs to be resistant to temperature at which frying is carried out. There are two types of frying, namely, shallow or pan frying and deep fat frying. The former is used for preparing *chapati* or *dosa* (pancakes) and involves limited or no deterioration in oil quality. Other items (like *vada, murukku, samosa,* puree, *bhajji,* fish or mutton, etc.) need to be deep fat fried. Rancidity development in such cases is likely, depending upon the time and type of oil used. Commercial preparation in sweetmeat shops and catering establishments sometimes lead to production of rancid food products.

FFA and Oil Quality

Free fatty acid (FFA) content is an important constituent which determines the quality of an oil. In the case of edible oils, FFA content should not exceed 2%. Freshly extracted raw or refined oils generally have lower FFA values. However storage as well as cooking processes increase this value. At higher FFA levels, the smoke point (temperature at which smoke starts emanating) decreases. The smoke point of most of the cooking oils is around 230°C. Higher FFA and the consequent excessive smoking at lower temperatures renders the oil unsuitable for edible purposes. It is better to avoid stored or rancid oils for frying purposes.

Rancidity

A sequence of reactions known as peroxidation with release of fee radicals and peroxides occur during rancidity development. The products of lipid peroxidation can cause damage to body tissues. The damage can be minimized if the lipid or the organs contain sufficient amount of antioxidants like vitamin E. There is an inherent advantage in the oils and fats namely, they become inedible due to rancidity before the oxidation products reach toxic levels. Hence based on odour and taste itself, consumers avoid rancid foods. However, care is necessary while feeding infants and children.

Hydrogenation is an important process developed to offset deterioration of unsaturated oils during storage and frying operations. The product popularly known as vegetable *ghee* (Vanaspati) has a long shelf-life. But it contains the undesirable trans-fatty acids and saturated fatty acids and hence are not considered good for health. Trans fats are linked to CHD and even diabetes. Hence it is better to avoid use of trans-fats in food preparation.

The first effect of rancidity development is the deterioration in flavour and appearance. The other changes are related to the deterioration of unsaturated fatty acids, particularly the essential fatty acids. Thus the nutritional quality gets affected. Rancid oil also reacts with essential amino acids in the foods being cooked in them. Roasting and frying at high temperatures favour the production of carcinogenic constituents. In view of the above factors, it is better, if the oils and fats are handled and utilized with sufficient care. Apart from the economy that is likely, it would help maintain nutritional quality and product acceptability.

Tips for Better Storage

1. Add a pinch of jaggery to improve the shelf-life.

2. Store the oil after purchase in an opaque container made of tin or plastic with an air-tight lid and a tapering top.
3. Do not retain the oil in the frying pan itself after use as it takes a longer time for cooling in it.
4. After each deep fat frying, filter the oil to remove suspended particles. These particles aid in early rancidity development.
5. As far as possible avoid use of excess oil for each deep fat frying. Use of the oil for a second frying is not advisable. Any oil remaining after first fat frying could be used for purposes of seasoning, or as lamp oil etc.
6. When frothing occurs due to moisture, add a pinch of tamarind or a few drops of vinegar to the oil.

In general it can be mentioned that MUFA oils with moderate levels of PUFA are better for cooking/frying purposes while high PUFA oils can easily serve as salad or table oils.

Hydrogenated Fats

Introduction

Hydrogenation is a process in which liquid fats are converted to semi-solid and solid fats. The unsaturation in the oils is reduced through addition of hydrogen. Hydrogenation is a versatile means to stabilise oils, fats and fatty acids and to modify their melting characteristics. The hydrogenated fats have lower iodine value and very good keeping quality. Hydrogenated fats are prepared from refined deodourized oils. The refined, bleached liquid oils can be converted to semi-solid, plastic fats suitable for shortening or margarine manufacture.

Hydrogenation is carried out under pressure at 100°-120°C in the presence of catalysts prepared from nickel, platinum or palladium for a period of 1-3 hr, till the fat is hydrogenated to the desired level. During hydrogenation, the double bond present in unsatureated fatty acids take up hydrogen and one or more of the following reactions take place.

Oleic acid	+2H	→	Stearic acid
Linoleic acid	+2H	→	Oleic acid
Linoleic acid	+4H	→	Stearic acid
Lenolenic acid	+2H	→	Linoleic acid
Lenolenic acid	+4H	→	Oleic acid
Linolenic acid	+6H	→	Stearic acid

Hydrogenation is preferential in the sense that certain fatty acid radicals hydrogenate more rapidly than others. More unsaturated fatty acids get hydrogenated first. The first reaction is hydrogenation at the double bond of linolenic acid; second at that of linoleic; and finally at that of oleic. This preferential nature of hydrogenation fortunately permits control of the process to such an extent that the molecular structure of glycerides can be modified as desired.

For the hydrogenation reaction to take place, gaseous hydrogen, liquid oil and the catalyst are combined by mechanical agitation, at a specific temperature and pressure in a closed reaction vessel. The rate at which the hydrogenation reaction takes place depends upon the temperature, the nature of the oil being hydrogenated, the activity and concentration of the catalyst, and the rate at which hydrogen and unsaturated oil molecules are combined at the active points of the catalyst surface.

Hydrogenation equipment and nickel catalysts used in the process have come a long way since their first industrial application in 1906. Efficiency, reliability and process control have greatly been improved over the years. Hydrogenation has become highly specialized. Highly flavourful and appealing hydrogenated fats are being produced using the latest advanced technologies.

The Catalyst

The method of preparing the catalyst has a considerable influence on its activity and its resistance to external influences. For example, nickel which has been reduced at as low a temperature as possible is more effective than that reduced at a higher temperature. Such a catalyst is also more sensitive to the action of catalyst poisons. The activity of a catalyst has also some effect on the nature of the reaction, namely the formation of isomeric glycerides of oleic acid which always accompanies hydrogenation of fatty oils.

Catalyst poisons such as sulphur and phosphorus compounds can strongly influence selectivity and rate of the process. The rate of the hydrogenation process is strongly dependent on the mixing rate that can be achieved in the autoclave. Activity differences between catalysts can be masked by hydrogen transfer limitations. Fatty acids are hydrogenated in spite of the fact that they themselves are poisons to nickel catalysts. To avoid dissolution of much of the catalytic nickel into the fatty acid it is necessary to control the hydrogenation conditions and to use purposely designed catalysts.

Hydrogenation

During the heating up period, the vacuum de-aerates and dries the oil before it reaches the elevated reaction temperature. When the required hydrogenation temperature is reached, the vacuum line is closed, and hydrogen is pumped into the vessel until the pressure builds to the desired level. One of the main essentials for the successful hydrogenation of oils is a supply of pure hydrogen. The gas must be free from sulphur, chlorine, arsenic and phosphorus, and contain only negligible quantities of oxygen or oxides of nitrogen. Typical hydrogenation conditions are as follows:

a) Catalyst concentration	0.03% to 0.1% Nickel
b) Reaction temperature	150 to 185°C
c) Hydrogen pressure	3 to 5 atmospheres
d) Agitation speed	approximately 150 rpm
e) Reaction time	10 to 60 minutes

The reaction end point is controlled in the hydrogenation department by determining the refractive index, which relates relatively closely to the iodine value. The solids content and melting point evaluations are later performed in the laboratory. At the end of the reaction, hydrogen flow is stopped and the hydrogen in the head space, with accumulated impurities, is blown off to the atmosphere. Vacuum is again applied to the vessel and the hydrogenated oil is cooled to approximately 80 to 85°C before being discharged from the vessel.

The first step following hydrogenation is removal of the catalyst by means of filtration. It has been shown that to a great extent the required filtration properties of a catalyst pose a limit to the activity of the catalyst. The cooled, hydrogenated oil is then sent through a catalyst filtration unit and a post bleaching operation is carried out before being sent to the storage tank.

Shortenings

"Shortening and shortening oils" include the partially hydrogenated plastic fats used in baking as well as the more heavily hydrogenated products designed for heavy-duty frying. For baking breads, cakes and pastries, most institutions, as well as households, use a plastic fat prepared from blended, hydrogenated vegetable oils. Blending is done to assure the stability of the fine crystal structure that is necessary of plasticity. Traditionally, fat plasticity has been considered the key to its successful performance in cake baking. Institutions and food processors can vary in their choice of solid or liquid fat for frying. Because of their increased level of hydrogenation, solid fats exhibit more stability.

However, recent interest in nutrition suggests that liquid oil may be preferred as the form of fat for cakes of the future. Liquid oils provide the double benefit of reducing the fat content of baked products and reducing the need for hydrogenation of the fat used in them. However, liquid oils are easier to handle and approximately 5 to 10% less fat is absorbed by the food, because the oil drains more readily when the food is removed from frying. Most consumers pan-fry more frequently than they deep-fry. When they fry, they use more liquid oil than solid fat or table speards.

A good frying fat must be bland in flavour and heat stable; that is, it should resist breakdown under temperatures of 175 to 190°C. Frying stability is assessed by measuring how much certain properties change during extended use, such as: increase in free fatty acids, increase in viscosity, and the drop in smoke point.

Frying Shortenings

In deep fat frying, the shortening or oil functions as a heat-transfer agent. Frying is more efficient than dry oven heat and faster than boiling in water and unlike these cooking methods, the frying oil becomes part of the food. The high temperatures used cause rapid heat penetration and provide short cooking times, a prerequisite of the fast food industry. The frying shortening reacts with protein and carbohydrates in food to develop unique flavours and odours that can provide consumer appeal.

Types of Shortenings

1. Liquid frying shortenings provide longer frying stabilities than either refined or salad oils. These have hard fraction suspended in a liquid oil to produce an opaque pourable product. Products fried in this type have a semi-oily appearance.
2. All purpose shortenings have been the first plastic vegetable products used for frying. These are used for specialty products with high fat absorption. They require a dry fried appearance.
3. Vanaspati has maximum frying stability due to reduction of polyunsturation through hydrogenation. Such fats have longer frying life. Solid consistency and non-pourability are the drawbacks of vanaspati.

Methyl silicone which effectively retards polymerization and oxidation may be added to the frying shortening at 0.5-2 ppm. This increases frying stability by 3-10 times.

Margarines

Introduction

Table spreads are becoming popular due to changing food habits globally. Their flavour and mouth feel are their attractions. They can be stored easily at refrigerator temperatures, where they keep best. They are amenable for spreading at ambient as well as lower temperatures. From the nutritional standpoint, there is also an increasing interest in low-calorie products which contain only 40% fat, rather than 80% fat in butter.

Margarine consists of small fat crystals trapped in oil and water globules. The spreadability depends on the proportions of the liquid oil and crystalline fat. Soft margarines are spreadable at refrigerator temperatures since they contain up to 70% liquid oil. Butter and the stick margarines are brittle when cold and become spreadable at room temperature, when the solid crystals begin to melt. These differing consistencies are achieved by a combination of the hydrogenation process, plus the blending of varying oil types and solid contents.

Composition of Margarine

The aqueous phase in margarines, generally consists of reconstituted milk powder, brine and water. Milk is pasteurized by rapidly heating it to 75°C and holding it at the temperature for 30 minutes. It is cooled rapidly to 5°C and held at that temperature until required. Salt, at about 2%, is added to the milk to prevent the growth of micro-organisms and improve the flavour. Vitamins, colouring and flavouring agents and emulsifiers, such as monoglycerides and lecithin, are added to the oil phase.

Practically all the margarines contain partially hydrogenated oils and therefore, have appreciable levels of trans fatty acids. Attempts are being made to reduce the levels of such acids on nutritional grounds. In reality, it is not possible to make spreads without a proportion of solid triacylglycerols which contain saturated and/or trans acids. If the level of trans acids is decreased, the content of saturated acids has to be increased. Interesterification can be attempted to get proper distribution of fatty acids.

Shelf life

Margarines can have a shelf-life of about 12 weeks. With good ingredients and absence of pro-oxidants (copper), oxidative deterioration is unlikely to be a problem. Care must be taken to avoid microbiological contamination in the aqueous phase. This is avoided by hygienic practices during manufacture, the addition of some salt (8-10 per cent in the aqueous phase, corresponding to a little over 1 percent in the margarine) control of pH of any cultured milk that is used, and careful attention to droplet size in the emulsion.

Fat Crystals

Individual crystals are between one tenth and several micrometres (μm) in size and form clusters or aggregates of 10-30 μm. One gram of fat phase may contain up to 10^{12} individual crystals. The aqueous phase is present in droplets, generally 2-4 μm in diameter, stabilised by a coating of fat crystals.

Fats usually crystallise in two different forms known a s β′ and β. Of these the β form is thermodynamically more stable and is present in many fats and fat blends. But sometimes the fat remains in the β′ form. For margarines and other spreads the β′ form is preferred since the crystals are smaller, are able to entrap more liquid to give firm products with good texture and mouth feel and impart a high gloss to the product. β crystals, on the other hand, start small but end to agglomerate and can trap less liquid. It is therefore desirable to choose a blend of oils which crystallise in the β′ form.

Margarines and shortenings made from rapeseed/canola, sunflower and soybean oil after partial hydrogenation tend to develop β crystals. This can be inhibited or prevented by the incorporation of some cottonseed oil, hydrogenated palm oil, palmolein, etc., all of which stabilise crystals in the β′ form. The canola, sunflower and soybean oils all have high levels of C_{18} acids, while the remaining oils have appreciable levels of C_{16} acids (or other chain length) along with the C_{18} acid and thus, contain more triacylglycerols with acids of mixed chain length.

Solids and Speadability

It is desirable that margarine taken from the refrigerator at 4°C should spread easily. The proportion of solids should be 30-40 per cent at this temperature. For the sample to 'stand up' at room temperature (and not collapse to an oily liquid) it should still have 10-20 per cent solids at 10°C. Finally, for the margarine to melt completely in the mouth and not have a waxy mouth feel, the solid content should be less than three per cent at 35°C. These important parameters can be attained with many different fat blends.

Formulations have to be changed slightly to make the product suitable for use in hot climates.

Margarines is consumed for table use, baking and frying purposes at house hold level and by the professional bakers. For the table, the margarine with a pleasant flavour, which is released fully and quickly is what is desired. It must be soft, spread easily over a wide range of room temperatures and melt rapidly in the mouth without much sensation of fattiness. For baking, the chief qualities required by the homemaker are plasticity and creamability (the ability to take up air when beaten). For frying, the fat must remain stable at high temperatures and must not contain excessive free fatty acids and the problem of smoking should not arise.

For the professional baker, taste is not of much importance. Plasticity and creamability are the main qualities required by the baker. The manufacturers combine the properly designed margarine oil with an aqueous phase, so that the water globules are finely dispersed but combine loosely enough for the emulsion to break easily on melting.

Table Margarines

These are water/oil (W/O) emulsions which contain at least 80% fat, milk products, vitamins, etc. The prospect of development of a variety of products to suit the liking of the consumer exists for the table margarines. Softer varieties of table margarines rich in EFA are becoming popular because of the health benefits associated with them. Spreads with 40 percent or 20 percent fat emulsions are also value added products which could be developed with high EFA content. Such items have low calories and adequate EFA content.

Spreads with reduced levels of fat (down to 40 per cent or less) are popular with consumers (as an alternative to restricting the amount of normal spread consumed). As a consequence there is much less margarine with the statutory level of at least 80 percent of fat and the word 'spread' has largely replaced the word 'margarine'. Spreads contain more water than full fat spreads and require emulsifiers (monoacylgylcerol or polyglycerol esters). It is also usual to add thickeners such as gelatin, sodium alginate, pectin, and carrageenan to the aqueous phase.

Baker's Margarines

These are formulated for the preparation of cakes, icings, biscuits, bread and other yeast-raised baked products. These margarines are formulated for their plastic consistency with good creaming properties. These characteristics are obtained by the use of 10-20% hydrogenated cottonseed or palm oil for a â′ crystal structure. These crystals tightly knit to produce good plasticity and creaming properties.

Pan and Grill Products

These can be prepared from any edible oil, soybean oil being the preferred one. Colour, butter flavour and lecithin are added to all the products to provide antisticking properties for pan and grill use. Both liquid and solid products have been designed. The solid products contain soybean oil, coconut and cottonseed oils. These products are useful in pan frying and grilling operations, for preparing various foods like soups, sauces, gravies, seasonings, pizzas, etc.

Frozen Deserts

Taurine (2-aminoethane sulphonic acid) has beneficial effects on the heart and eye. It has been incorporated in baby foods for effecting benefit in protein and fat digestibility. It may be of considerable benefit in congestive heart failure, atherosclerosis and hypercholesterolemia and could be beneficially added to various shortenings to produce food products with health benefit. One could identify many other fat soluble ingredients with positive health attributes to enrich fat based products as not only providing nutrition but also contributing to improved health.

Oilseed Extractions

The increased production of oilseeds is adding a new dimension to our food and feed front by making available larger quantities of protein-rich oil cakes. These oilseed meals have traditionally been utilised as feed for cattle, fish and poultry and the inedible cakes as organic manure. After processing the oilseeds for the oil through expeller and the solvent extraction mills, the residue, namely oil meal or cake is treated as a by-product by the oil industry. The cakes are generally handled carelessly and under unhygienic conditions. Such products lose their commercial value and importance.

Oilcakes in Food Products

The oil cakes are by-products of the oil milling industry. Upto 60-70% of the seeds return back as oilcakes from oil industry. The cakes are richer than the oilseeds in proteins, carbohydrates, minerals and even vitamins. These cakes are rich in proteins (20-50%), carbohydrates (20-30%) and minerals

(5-8%). As long as the oil extraction units have been in the village level *ghanies* some oil cakes like groundnut and sesame have been utilised as components of dietary items.

The advent of modern vegetable oil processing mills have rendered the residue as by-product suitable only as livestock feed. Hygienically processed cakes can be utilised as protein rich fluors which can be incorporated as part of the household foods and snack food items. They are potentially used as raw material in many food industries. They can serve as milk and meat analogues.

Oil meals are mostly utlized as animal feed (upto 75%) with about 10-15% being used as organic manure. Another 5-10% is being utilized for food purposes. There is need for changing this pattern. The meals can be upgraded for human food consumption. Oil cakes can be utilized in the production of low cost protein fluors. Other products that may be considered are protein isolates or protein hydrolysates (amino acids). These can be used in commercial production of weak, convalescent or infant foods. Low cost protein flours can be of help in subsidized nutrition programs.

Fabricated Foods

Texturized vegetable proteins (meat analogues), through extrusion cooking may be prepared alongwith other cereals and pulses. Other foods such as simulated dairy analogues (milk, infant foods, coffee whiteners, icecreams, malted foods), bakery and confectionery products (biscuits and toffees), fermented foods like *Miso, Natto* etc., may also be prepared.

Simple Mixes

The oil cakes can be ground into flour and mixed with cereals and pulses to prepare various energy rich foods like *Balahar* (infant food)and *Incaparina.* In this case, oil cakes with different fat contents may be used to meet the nutritional requirement of different categories of people.

Protein isolates and protein concentrates can be prepared and used for various functional and nutritional properties in food formulations. They contain about 90% and 70% proteins, respectively. They are used in infant formulae, medical foods (neutraceuticals)for oral and external feeding, weight-loss products and body building supplements, as coffee creamers, protein fortified juices and other acid beverages, baked food and confections, dairy analogues and non diary sour creams and frozen desserts.

Quality of Oilseed Meals

The protein content of the oilseed meals range from 20-50%. Oilseed extractions can be placed in three categories based on their protein content. High protein meals include those of soybean, groundnut and sesame with 40-50% protein, meals of niger, rapeseed-mustard and linseed have 30-40% protein while the others have lower protein contents of 20-30%. Proteins are necessary for physical and mental growth as well as for the maintenance of good health and vigour. An adult human being needs to consume 50g of protein per day. Animal protein is costly and hence oilseed extractions can play a major role in reducing the protein malnutrition in the developing countries.

The protein of oilseeds are very good sources of essential amino acids required in our food. Sesame protein is the best followed by niger, groundnut and soybean. Sesame protein is the best followed by niger, groundnut and soybean. Soybean and rapeseed are rich in lysine. But soybean is deficient in sulphur containing amino acids namely cysteine and methionine apart from trytophan. Sesame protein is rich in methionine. Hence, sesame cake can be blended with other oilseed cakes to improve their nutritional quality. Oil meals are also rich in carbohydrates and minerals. Sugar content of the meals range between 26 and 30%. They are less harmful to diabetic patients. The mineral contents range from 4 to 10%. They are rich in calcium, phosphorus, magnesium and iron which are essential nutritional constituents for the human body.

Undesirable Factors and their Elimination

Oilseed meals contain some toxins and anti-nutritional factors like polyphenols, protease inhibitors, oxalates, phytates, glucosinolates and aflatoxin. These reduce the nutritional quality of the meal apart from causing illness and other health problems. Such low grade cakes can be improved with some treatments. The cakes can be treated with ammonia gas to improve their texture, detoxify aflatoxin and allergen and enrich them with nitrogen (protein). Dehulling is another simple process to upgrade the meal quality by reducing the crude fibre. Steaming is useful in removing toxins from rapeseed, castor and linseed meals. There are a variety of simple techniques to improve the cake quality from feed to food and inedible cake to animal feed. Such improved cakes and those processed under hygienic conditions apart form being useful as food and feed can earn more foreign exchange to the producing countries.

The undesirable factors mainly associated with husks or hulls may be eliminated through dehulling/dehusking prior to crushing. The off-flavours such as the musty odour of mustard meal, catty smell of cotton seed, nutty/beany flavour of soybean may be eliminated by certain pre-treatments like soaking in alkaline solutions or extractions with ethanol/isopropanol. These treatments may add to the cost of the whole system but yet yield acceptable edible grade oil cakes. Antinutritional factors are eliminated through thermal processing either dry heat/moist heat or supersaturated steaming. Biotechnology is also helping in breeding oilseed crops without these undesirable constituents. Mention may be made about cotton varieties without gossypol, groundnuts tolerant/ resistant to *A.flavus,* rapeseed varieties without glucosinolates and soybeans without trypsin inhibitors which are being developed and released in several countries.

Quality of Cake as Affected by the Method of Extraction

The method of extraction of oil affects the quality of the resultant oilcakes. There are different methods of oil extraction namely traditional *ghanies,* mechanical expeller, solvent extraction and supercritical CO_2 extraction systems. The oilcakes obtained from the mechanical expression contain about 8-10% oil. There is less thermal destruction of desirable nutrients. However, unhygienic conditions exist in these operations which most of the times yield non-edible oilseed cakes. But when the oilseeds are subjected to solvent extraction, less than 1% oil is left in the cake. Thus, by adopting different extraction methods oilcakes containing varied amount of fats and proteins are obtained which may meet the requirement of different groups of people.

Feed Uses of Oilseed Meals

Now-a-days a large number of feed formulations include oilseed cakes as concentrates. The good quality protein improves the egg-laying capacity in chickens, milk yield in cattle, quality of fish catches and meat production. The quality and utility of different oilseed meals as animal feed are discussed below.

Groundnut meal: Groundnut meal has been a major source of vegetable protein in animal diets all over the world.. Solvent extracted groundnut meal contains 43 to 48% crude protein. It is an excellent source of arginine but deficient in lysine, cysteine and methionine. Supplementation of synthetic lysine to groundnut meal can make it an ideal protein source in feeds. However, it has been seen that groundnut meal is associated with mycotoxins like aflatoxins. Groundnut seed coat also contains tannins which lower the quality of protein compared to soybean meal.

Soybean meal: Soybean meal contains 40 to 48% protein. The amino acid profile of soybean meal is well balanced and is rich in lysine but deficient in methionine content. Soybean seeds have certain anti-nutritional factors like protease inhibitors, goitrogens and estrogenic factors that can affect the production performance of chicken and other animals. But they are all thermo-labile and can be destroyed by roasting/heating or autoclaving. An adequate balance of amino acids by substituting methionine to soybean meal can successfully replace even fish meal.

Mustard meal: Mustard meal has an average crude protein value of 37% with more lysine and methionine contents than groundnut meal. It's usage in animal feeds is limited due to the presence of deleterious factors like glucosinolates, tannins, erucic acid, high crude fibre etc. Glucosinolates are potential giotrogens which cause enlargement of thyroid gland, consequently reducing the performance of layers and broilers. The toxin sinapin affects palatability of feeds and imparts a fishy flavour. The process of solvent extraction after initial pressing of the seed destroys the goitrogens and isothiocyanates and reduces the toxic effects of the cake. The tannins can be checked to some extent by addition of calcium hydroxide and sodium carbonate in dry form in the animal diets.

Sesame meal: Sesame meal contains 40% protein and 8% crude fiber. The protein is rich in arginine, leucine and methionine but low in lysine. Since the sesame meal is deficient in lysine, its combination with soybean meal is useful. Sesame meal has high calcium content but its availbility is questionable because of phytate content in the hulls of the seed. The phytic acid reduces the availability of calcium and zinc from the diet. Nevertheless the diet can be included in animal feeds in combination with other lysine rich oil cakes.

Sunflower meal: The sunflower meal contains 30-40% of protein with variable levels of crude fiber depending upon the level of dehulling of the seed. The net protein utilization of sunflower meal is higher than groundnut cake, probably due to higher methionine and cysteine content. Sunflower meal is generally not recommended beyond 20% in the diets due to its high crude fibre content and also due to the presence of polyphenolic compounds (mainly chlorogenic acid) which inhibit the activity of hydrolytic enzymes. To overcome this problem supplementation of additional methionine and choline in the animal diet is suggested.

Cottonseed meal: Cotton seed meal contains around 40% crude protein but protein quality of cotton seed meal is poor due to the presence of gossypol. Gossypol inhibits the activity of digestive enzymes and reduces the palatibility of the diet. Mechanical pressing of seed followed by solvent extraction reduces the gossypol content to 0.02-0.05% level. Dietary levels of gossypol upto 0.015% levels are safer in poultry. Detoxification of cotton seed meal with a solvent mixture containing hexane, acetone and water has been useful after initial cooking of the meal. Addition of ferrous sulphate at the rate of 4 parts to one part of gossypol prevents egg yolk discoloration due to gossypol. Cotton seed has a desirabe profile of amino acids except lysine and the digestibility of cotton seed meal is poor due to higher fibre.

Neem seed meal: Neem seed meal varies in its protein content from 20 to 32% depending on the hull portion left in the meal. Neem seed meal as such is not recommended even at 2.5% level in the poultry feed due to the presence of the oil bound toxic factors like azadirachtin, meliantriol and salanin which have growth depressing effects on chicks. Hexane extraction of oil from the meal greatly reduces the toxic factors. The meal is bitter due to the presence of nimbinin, nimbidin, nimbosterol etc., Complete extraction of oil using a solvent and saponification of the same with potassium hydroxide minimises the toxic principles in the meal.

Castor seed meal: The protein content of castor seed meal varies from 21 to 48% depending upon the extent of decortication. It has an ideal amino acid profile with moderately high cysteine, mithionine and isoleucine. But its anti-nutritional substances ricin, ricinine and the allergens restrict its use in animal rations even at low levels of inclusion. Processing of castor meal by heating at 80°C temperature in the presence of 6N ammonia or an acid or alkali reduces the toxins substantially. Detoxified meal needs to be effectively corrected for lysine. It's net protein utilization value is higher than other oilseed meals used in poultry.

Low-Fat, High Protein Products

Edible proteins from oilseeds, constitute a potential source of energy for humans and animals. In the processing of oilseeds for recovery of oils, proteins present in the seed material undergo certain changes. Very often, the proteins present in pristine conditions, get partially or fully denatured due to the drastic processing conditions. No attention is paid to preserve the quality of edible proteins present in the oilseeds during the processing of oilseeds. This results in a tremendous drain of high quality proteins as resultant oilcakes are generally used in the feed stuffs of livestock and not as food stuffs for human consumption.

The pristine quality of proteins present in the original seeds can be well preserved if suitable care is taken. The development of low fat-high protein oilseeds is one route through which the quality of oilseeds can be maintained for food uses. The outlines of the steps that need to be taken are: cleaning of seed, selection of pure seed kernels, hydraulic pressing of seed kernel to extract the oil from the seed material partially followed up by a process of reconstitution of the seed material to regain the original shape of the seed and finally drying the reconstituted seed material. After processing, the reconstituted seed material is used for table purposes as low-fat, high-protein edible products for human consumption.

The demand for food products may simply outstrip the supplies. The importance of maintaining a viable balance between demand and supply of foods and feeds cannot be overemphasized. In the field of edible proteins, a sea change is expected on the world scenario. Among the various food products, the protein-rich materials of vegetable origin, will play a significant and crucial role in human nutrition. Edible proteins from unconventional sources like oil seeds and oil-bearing materials is not a new concept. Proteins of vegetable origin are as important as any other. Protein deficiency in the diets of people, especially, from developing countries is one of the major nutritional problems. Meat protein, which is the natural choice of our teeming population is becoming more and more expensive. Shortages, violent fluctuations in prices and millions of depleting numbers in livestocks have brought many miseries to the common people. To surmount these problems, there is no other choice except to depend upon the vegetable proteins which can be tapped easily and economically from oilseeds. A wide variety of raw materials are used in the production of high-protein supplements. Products such as oilseed meal prepared from groundnut, soybean, coconut, cottonseed, sunflower, sesame, etc., are extensively used for this purpose. These vegetable oilseed proteins offer an excellent potential as a direct foodstuff for human consumption.

Groundnut : Partially deoiled groundnut kernels are made to regain their original shape by reconstitution. During pressing, a prime quality oil is recovered which can be used for edible purposes. The pressed and reconstituted seed is low in oil and rich in proteins. These kernels would serve as tasty snack food and would be suitable for crunching and munching during intervals(Ramachar et al., 1974). The groundnut proteins are highly digestible. The biological value of groundnut proteins is one among the highest of all the vegetable proteins and equal to that of casein. The groundnut proteins compare well in food value with more expensive foods of animal origin.

Sunflower: Extending the above technique, high-protein sunflower seed kernels can also be prepared. Sunflower kernels are hydraulic pressed to recover oil partially (Yield 25-26%). The pressed kernels are reconstituted to attain their original shape and the moist reconstituted kernels, are dried in the open sun. The oil content of the reconstituted and sun dried kernel are 23-24% with a protein content of 34.2-34.9 % as against 50.1% oil and 25.6% protein in the corresponding original seed kernels. Sunflower is a source of high quality protein. The characteristic grey-green colour of sunflower seed protein concentrates or meal is due to the presence of chlorogenic acid in seeds, which darkens to a green or brown colour upon oxidation. However, it does not detract its feed value and palatability.

Sesame seed: Both brown and white varieties, as well as decuticled seeds of brown variety can be utilized in the preparation of low-oil and high-protein products (Surendranth *et al.,* 1984). The low-fat high protein sesame seeds are made in a way similar to that of groundnuts. Sesame seed protein is rich in sulphur containing amino acids, calcium and phosphorous. However, it is low in lysine and also contains about 5 per cent phytic acid, which reduces the bio-availability of mineral nutrients such as calcium, magnesium, zinc and possibly Iron. Decuticling of sesame seed reduces the fibre content and bitter taste and increases palatability, digestibility and protein content.

An optimum heat treatment of 60-70°C of the raw materials i.e., oilseeds is a must. This increases the protein efficiency ratio, palatability and digestibility and reduces or completely destroys the heat labile toxic and anti-nutritional factors in the oilseeds.

Other Applications

Some of the undesirable factors that are isolated from the oil cakes such as glucosinolates from mustard and gossypol from cotton have commercial applications like insecticides and contraceptives, respectively. Incorporation of soy proteins in diet therapy protects us from breast cancer, colon cancer and heart attacks. Most oilseeds function as neutaceuticals. Ricinine and ricin from castor have insecticidal and medicinal values. Hence utilization of various oilseeds based on their composion needs careful attention.

Oilseed Based Food Products

Oilseeds mainly contain oil which is a concentrated source of energy. They also contain proteins, carbohydrates, minerals and useful vitamins. All these together make the oilseeds as good sources of energy for the human being and animals. Oilseeds provide 500-600 calories of energy per 100g. Groundnut, sesame, soybean, sunflower kernels, kopra, olive and niger are directly edible. Minimal processing like dehulling, roasting, cooking etc. makes them more acceptable and hence they are widely consumed as food items all over the world. Other oilseeds like safflower, linseed and rapeseed-mustard also are consumed to a limited extent as part of food items.

Among edible oilseeds, groundnut and sesame contain higher oil and proteins followed by niger, rapeseed and sunflower while soybean and safflower contain lower levels of oil. However, soybean serves as the richest source of protein. In general, all the oilseeds contain 10-25% sugars and 3-5% minerals. Oilseeds also contain sufficient levels of vitamins. The main vitamins that are present are vitamin B complex group, and vitamin E (tocopherols). Vitamins A, C and D are present at very low levels. Oilseeds serve not only as concentrated sources of energy, but also as balanced sources of most of the food nutrients.

Groundnut, sesame, soybean and niger seeds and their products are being utilised in the preparation of household food items. Precessing like dehulling, roasting or boiling etc. makes these oilseeds more flavourfull and tasty food items apart from increasing the nutritional value. All these oilseeds and their defatted fluors serve as good energy sources. Oilseeds and nuts are more popular as snack food items all over the world. Protein concentrates and protein isolates can also be manufactured from them. Milk and butter substitutes, *chikkis, laddus, chutneys*, dry powers, fermented and coagulated products, fortified foods like biscuits and breads, meat analogues and texturised products can be made either at home or by the industry which are presently in good demand. The oil rich foods are more satiating, help in delayed hunger and are more balanced with respect to most of the nutritional constituents.

Groundnut Based Foods

Groundnuts can be used as components of any food. The peanuts are unique in that they find a wide range of uses. The kernels are used after roasting, frying, salting, or boiling. They are used in a variety of dishes and in the preparation of confectionery products. Peanuts eaten with some jaggery form a cheap, tasty and nutritious food. They can be mixed with rice, wheat, sorghum, maize etc. and can be eaten as part of the foods prepared. Groundnuts can also be used in soups or stews. They are broken up first and boiled in the soup, together with meat, fish and or vegetables. They are consumed after first roasting and then pounding into a paste. This may be either a coarse or a fine paste, similar to the peanut butter. Groundnut stews are popular in many countries.

Roasted groundnuts: Roasting of groundnuts frequently with a little table salt is a very common practice all over the world. Groundnuts are roasted either by applying dry heat or in some vegetable oil. Dry roasted groundnuts are used in the preparation of butter, confectionery, or bakery products. Roasting reduces moisture content, develops a pleasant flavour, and makes the meal more acceptable for consumption. (Kabirullah *et al.*, 1977). The reduction in moisture content during roasting of groundnuts prevent molding and reduces staleness and rancidity.(Woodroof, 1945). Proteins are denatured during roasting. Excessive heating during roasting lowers the nutritional quality of the protein (Sharma *et al.*, 1983).

Amino acids such as glutamic acid, aspartic acid, phenylalanine, and histidine react with sugars to contribute to the flavour quality of roasted groundnuts(Cobb and Swaisgood, 1971). These amino acids are released from a large peptide, (Mason, *et al.*, 1969) while glucose and fructose are released from sucrose (Newell, *et al.*, 1967). Roasting decreases the methionine-rich proteins and aggregates arachin proteins (Basha and Young, 1985).

Cake and meal: Groundnut cake (defatted), when powdered, gives the meal. Full-fat meal is also obtained by direct powdering of groundnuts. The defatted peanut meal contains about 43 to 65% protein, 6 to 20% fat, (depending upon the method of oil extraction), and some vitamins of the B-complex. (Cherry, 1977). The defatted groundnut meal has higher PER, NPU, and digestibility than the full-fat groundnut meal (Kabirullah *et al.*, 1977). The chemical score of groundnut flours can be improved by blending with other plant proteins. These blends could be used in bakery products. (Abdul-Baki *et al.*, 1980; Ranhotra *et al.*, 1980).

Heat-processed groundnut meal is lower in lysine, methioine, and threonine (McOsker, 1962). Heating groundnut cake at 120°C or autoclaving at 105°C for 30 min does not improve or deteriorate the protein quality (Hete *et al.*, 1975). The feed efficiency of groundnut meal in chicken feeding is lower than that of raw meal (Saha, 1977). Autoclaving and oven heating of groundnut cake reduces the protein digestibility, PER and available lysine.

Protein isolates: Production of groundnut protein in the form of concentrates and isolates for human consumption has become a reality. They have lot of uses in the preparation of many types of food products. They impart beneficial characteristics to foods in which they are incorporated. Isolated protein (90-95%) from groundnut is a sophisticated product. It offers immense possibilities in the development of protein enriched foods and meat analoguess including milk-like products. The isolates can be textured and flavoured to meet a wide variety of consumer preferences. The biggest advantage is that it is bland and does not interfere with the natural flavours of the foods enriched with it.

Lawhon *et al.* (1980) have reported that groundnut protein isolates can replace about 80% of milk solids without texture change and about 60% without loss of colour, or overall acceptability. Protein products with high protein solubility, bland taste, and desirable light colour are obtained from groundnut by combining aqueous extraction process (Rhee *et al.*, 1977) and membrane isolation process (Lawhon *et al.*, 1981).

Presoaking of split groundnut kernels in 4% NaCl solution overnight followed by hot-water washing helps in washing out the water solubles and flavour components. The protein isolate thus obtained has high solubility, white colour and is free from a nutty flavour (Subramanian *et al.*, 1981). Groundnut protein isolates and oil have scope to be used in the manufacture of cheese analogues (Chen *et al.*, 1979). Groundnut proteins have been used to extend meat products without adverse effects (Cross and Nicholas, 1979; McWatters and Heaton, 1979).

Groundnut (peanut) butter: Manufacture of groundnut butter is gaining importance all over the world. It is prepared by grinding roasted and blanched kernels to which 1 to 4% common salt is added (Sparado, *et al.*, 1981). Groundnut butter flavor is a major factor in determining the acceptability of the product. Groundnut butter contains 25% protein, 50% fat, 23% carbohydrates and 4% minerals. It is rich in calcium, iron, phosphorus, sodium and potassium. It also has good levels of vitamin A, thiamine, riboflavin and niacin. The calorific value of groundnut butter is around 590 cal/100g. This is similar to an item called *chutney* made in India. The only difference is that chillies, tamarind(for sour taste), salt and spices are added to the *chutney*.

Infant and weanling foods: Milk, the ideal food for babies, infants and other vulnerable groups of population is an expensive item. A nutritionally sound infant food based on protein isolate and skim milk powder has been developed. This spray dried product with 26% protein is easily dispersible in water and can be fed to a baby from the bottle. Two thirds of its protein is derived from the groundnut protein isolate and one-third from milk. At 15% protein level in the diet, the PER of this product is about the same as that of milk.

Spray dried infant foods based on groundnut protein isolate, soya flour, and skim milk powder, have also been developed with added hydrogenated fat and hydrolysed starch. The PER of this food is 2.7 and when fortified with methionine is 3.0 or equivalent to milk.

Groundnut milk: Good-quality peanut kernels are cleaned of impurities and roasted lightly to facilitate removal of red skin. The testa or skin can be removed in a blanching machine. The germ is separated by sieving and the spoilt seeds by hand picking. The cleaned kernels are ground to a smooth paste. The paste is mixed with 7 times the weight of water in a blender. Calcium hydroxide solution is added till the pH of the milk is adjusted to 6.8. A mixture of disodium phosphate and acid potassium phosphate having pH 7.0 is added to stabilize the milk.

The milk is filtered through a fine cloth and fortified with vitamins A, D, riboflavin, folic acid, vitamin B12 and minerals (calcium and iron salts). Cane sugar is added at 7% level. The milk is homogenized, steamed, bottled and kept in a refrigerator till distributed.

A milk-like product or a milk analogue called "miltone" has been made in India. It consists of groundnut milk extended with buffalo milk. It has the same nutritive value as that of milk and can be used as its substitute, flavoured beverage or for making curds by lactic fermentation (Swaminathan and Parpia, 1967, Beuchat and Nail, 1978).

Fermented products: Groundnut cake or meal can be used for human consumption after partial hydrolysis of protein by the process of fermentation using certain molds. Such products are readily digestible, tasty, and nutritious. "*Oncom*" is a fermented groundnut press cake and is a popular product in Indonesia. Worthington and Beuchat (1974) have reported elimination of sucrose, raffinose, and stachyose during the first 24 hrs of fermentation of groundnuts with *Neurospora intermedia*. The fungus *Rhizopus oligosporus* is less effective in eliminating these sugars. There is no change in the fatty acid composition (Beuchat and Worthington, 1974) and practically no change in the protein content during fermentation of peanuts. (Van veen *et al.*, 1968). "*Oncom*" contains 3-9% oil, 20-30% protein and upto 22% carbohydrates.

During fermentation, large molecular weight globulins, such as arachin, are hydrolyzed to smaller components. (Beuchat *et al.*, 1975). As a results of protein breakdown, the nitrogen solubility profiles of groundnut protein at virous pH values change drastically (Quinn *et al.*, 1975). The increased digestibility of fermented groundnuts has been attributed to the hydrolytic activity of the molds (Worthington and Beuchat, 1974). Thiamin, riboflavin, and niacin levels are increased in fermented groundnut flour. (Quinn and Beuchat, 1975). There is degradation of phytic acid during the preparation of *oncom*. (Fardiaz and Markakis, 1981). Protein efficiency ratios (PER) of fermented groundnuts are however, not increased over the nonfermented materials. (Quinn and Beuchat, 1975; Van Veen and Steinkraus, 1970).

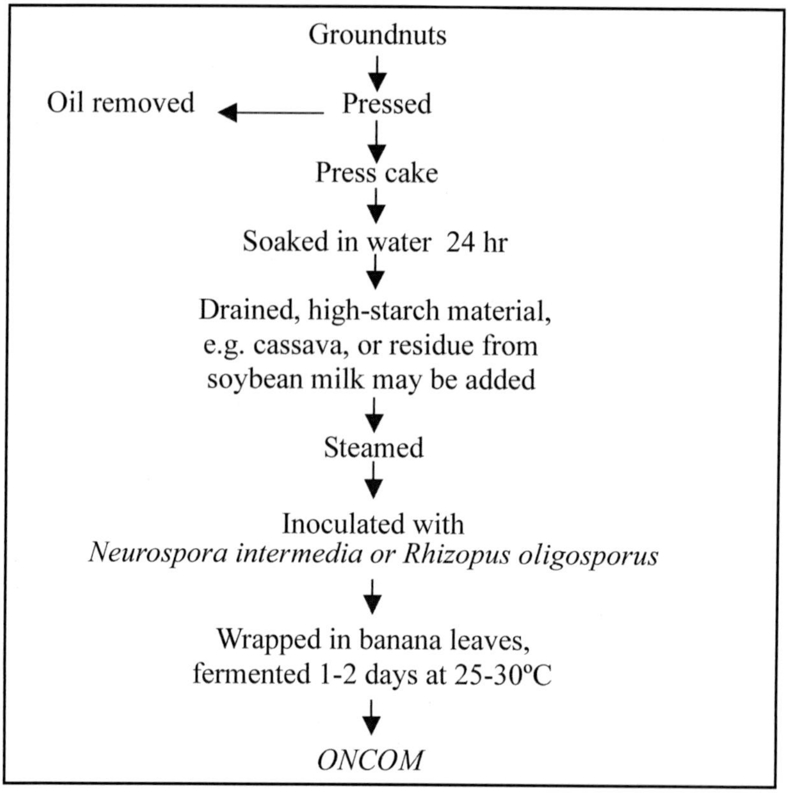

Preparation of *oncom* (Worthington, and Beuchat, 1974.)

Composite flours: Groundnut flour can be used for the preparation of composite flours. Such composite flours have a higher protein content and better nutritive value. Whole wheat and maize flours are supplemented with 10, 20, and 30% levels of a 1:1 mixture of groundnut(peanut) and chickpea flours(PCF). Supplementation increases the protein content of the wheat and maize blends by 20 to 61%. A significant increase in other proximate constituents as well as lysine also takes place. The chemical score of wheat flour increases from 53 to 72 and that of maize flour from 49 to 71 with 30% PCF. Biological evaluation of wheat breads at 10% protein level and maize breads at 8% protein level in the diet has shown a significant improvement in the protein quality of PCF-supplemented breads as judged by gain in body weight, protein efficiency ratio, net protein utilization, biological value and utilizable protein content. A supplementation level of 20% is adequate to achieve the desired nutritive benefits (Khalil and Chughtai, 1984).

Mysore flour: It is made of 75% of edible grade tapioca flour and 25% low fat edible grade groundnut flour in India. This product has 12-13% protein and could be used to prepare a number of traditional foods such as *chapati, parotha, puri* and several other savoury and sweet dishes. It helps to upgrade the nutritive value of tapioca.

Paustik atta (nutritious flour): Two formulations of this product are prepared. Formula-A has 75% whole wheat flour, 17% tapioca flour and 8% groundnut flour. This is more suitable for use in areas where indigenously grown tapioca flour is available at low price. Formula-B consists of 5-10% edible grade low fat groundnut flour added to whole wheat flour. Both the formulae have very good acceptance. A vast variety of enriched flours can be developed and used for several purposes.

Traditional products: Groundnuts are utilized for preparation of products like *laddu* and *chikki* in India. For preparation of *laddu,* peanut kernels are roasted and decorticated. The separated cotyledons are mixed with thick and hot jaggery syrup. Small portions of the mixture are pressed with the hand to obtain balls or *laddus* of about 3 to 5 cm diameter. *Laddus* may also be prepared from roasted and decorticated groundnut kernels after coarse grinding to obtain particles of about 1 to 5 mm diameter which are mixed with a small quanity (2:1 or 3:1) of crushed jaggery. Small portions of this mixture are pressed between the hands to obtain balls or spread on a non-sticky tray to have 0.5 to 1cm cakes.

Chikki is a very popular product being liked and utilized all over India. It is prepared by mixing roasted and decorticated peanut cotyledons with a hot slurry of jaggery. The mixture is spread on a tray or table or any surface with a uniform thickness of about 1.0 to 1.5 cm. After cooling, the material is cut into small pieces and packed.

Soy Food Items

Soybeans being very rich in protein have received lot of attention for preparation of many foods. A large number of traditional technologies have been developed to make fermented and textured products from soya. The problem with most soy based products is their strong flavour which is preferred in the far eastern cultures but not in South Asia. Lactic acid fermentation minimises this flavour.

Full fat soyflour: Full-fat soy flour comes from whole, dehulled soybeans that have been ground into a fine powder. It retains the fat naturally found in soybeans. However, defatted flour has the fat removed before grinding.

Although soy flour is not a common ingredient in home baking, it is used extensively by the food industry. Soy flour is approximately 50% protein on a dry weight basis.

One of the simplest ways of using soybean in human diet is in the form of full fat soyflour in combination with wheat, sorghum, maize or rice flour. 10% addition of soyflour is recommended with cereals to start with. Preparation and use of recipes from soy-cereal blended flour does not involve any change in the traditional food habits of the people. The process adapted/ developed for the production of fullfat soyflour consists of cleaning, dehulling, soaking/steeping of soy-splits in bean to water (containing 0.5-1% $NaHCO_3$, w/v) ratio of 1:3 followed by immersion cooking in water for 15-20 min, drying of cooked splits to about 8% moisture. The splits are then milled to obtain soyfluor.

The soyflour thus produced contains about 20% oil, 40% protein with PER value of about 2.0 with water absorption capacity of about 200%. Urease and trypsin inhibitor activities are almost nil and the soyflour can be stored safely upto 3-months. Organoleptic and sensory evaluation experiments on use of fullfat soyflour in various soybased food recipes have shown encouraging results.

Partially defatted soyflour: Cleaned and dehulled soybeans are conditioned with steam for 15-18 min at atmospheric pressure. They are flaked followed by oil expelling which is done in three passes. Soy flakes/cakes are powdered, cooled and packed. The partially defatted soyflour has about 5-6% oil. The cake obtained is of edible grade. Flour is creamy yellow. It offers potential for producing oil as well as food grade partially defatted soyflour.

Soy *paneer*: It is a coagulated and pressed soyprotein. Milk-*paneer* is very popular among high-income group urban people for making vegetarain-curry and *Paneer Pakoda.* However, the cost of milk-paneer is very high and hence, it is out of the economic reach of poorer sections of people. In this context, soypaneer is an appropriate alternative to milk-paneer as the cost of the former is only one-third the cost of milk paneer and gives all the nutritional advantages and psychological satisfaction of eating a *paneer*-dish.

The process developed for the production of soypaneer consists of cleaning and grading, dehulling, soaking of soysplits in water (containing 1% $NaHCO_3$, w/v basis) with bean to water ratio of 1:3 at room temperature for 4 hours or for one hour at 60°C, grinding of soaked and thoroughly washed splits using hot water (bean to water ratio = 1:8), boiling of soypuree for 15-20 min, and then removing the roughage through filtering, cooling the milk thus obtained to about 70°C, addition of 0.02 M $CaSO_4/MgSO_4$ and stirring the mixture 2 or 3 times, allowing it to settle for 30 min in order to get the protein coagulated. The proteinate complex is filtered through muslin cloth lined paneer forming boxes and pressing the soypaneer at 10 g/cm^2 for about an hour. The soypaneer blocks are put in cold water and after keeping for about 30 min. They are taken out and stored in water till use.

1.5-2 kg soy *paneer* can be obtained from one kg cleaned and dry soybean. Soypaneer at about 75% moisture content contains approximately 12% protein, 7% oil, 4% carbohydrate and 2% ash. In an organoleptic assessment of soypaneer, the mean score values for overall acceptability of soypaneer has been 7.5 as compared to 9 for milk-paneer. Paneer prepared out of soy-animal milk blends at 25, 50 and 75% levels has shown encouraging results in overcoming the initial hesitation of people in accepting soy paneer.

Soydal: Vegetarians generally consume cereals along with cooked pulse in a soup/broth form. Appropriate fortification of these pulses with soybean can provide a better protein with all essential amino-acids at low cost. Experimental results have shown that blending of soy*dal* with other pulses namely pigeon pea, green gram, black gram and lentil upto 40-50% and pressure cooking provides a well cooked, dispersed and organoleptically acceptable *dal* containing about 32% protein.

Soyflakes: It is another product which can be used along with rice, maize or sorghum flakes. The process developed for making soyflakes consists of cleaning of raw soybeans, to remove impurities, brokens and shrivelled grain, dehulling and splitting of cleaned soybeans, blanching of soy splits for 1 hour and then drying to 25-30% moisture, flaking of the semi-dried soysplits drying of flakes to 8% moisture, grading and packaging.

Soyflakes thus obtained contain approximately 44% protein and 19% oil with no trypsin inhibitors. Storage of soyflakes at 6-8% moisture in polyethylene bags or metallic containers is safe for atleast 6 months. Soyflakes/grits could be added to daily dietary preparations like *curries, dal,* etc., at 10-15% level. The broken soyflakes are good food ingredients in preparation of rice based fermented products like *Idli* and *Dosa.*

Soy-protein ingredient products: Soy-protein ingredient products are commercially available sources of soy protein that are used in the preparation of other foodstuffs. Soybeans are first dehulled, flaked, and defatted to make "white flakes". These white flakes are made into defatted grits or flours, soy concentrates, or have their flavour compounds and flatulence-producing sugars extracted to produce isolated soy protein.

Soy protein has become an ingredient of many common foods, including salad dressings, creams, soups, and candy coatings. In addition to the superior nutrition provided by the consumption of soy proteins, they have functional properites that make them important ingredients in many processed foods. For example, soy proteins absorb water and fat, act as emulsifier, aid in whipping (aeration), and improve texture in foods to which they are added.

Soy protein concentrates: Soy protein concentrates are made by removing a portion of the carbohydrates from defatted and dehulled soybeans. Alcohol extraction is the method most commonly used to manufacture soy protein concentrates. It results in the loss of isoflavones and saponins, which are alcohol soluble. Soy protein concentrates retain some of the fiber in the original soybean and contain about 70% protein on a moisture-free basis. They are most often used as functional or nutritional ingredients in a wide variety of food products.

Soy protein isolates: It is the most concentrated source of soy protein. It has at least 90% protein on a dry-weight basis. Most isolated soy protein is manufactured by water extraction of protein from defatted and dehulled soybeans. Since extensive water washing can remove isoflavones and other compounds, processing must be carefully controlled to retain these naturally occurring components. Isolated soy protein is heat treated during processing to ensure inactivation of trypsin inhibitors. It is used as a source of high-quality protein and serves as a functional ingredient in many familiar food products. The protein isolate from soybean is prepared using the following steps.

1. Solvent extraction of edible soybean to remove fats,
2. Extraction of proteins with dilute sodium hydroxide at pH 8,
3. Precipitation of proteins at pH 4.5 from the extract by adding hydrochloric acid,
4. Filtration of proteins and washing with water, and
5. Solubilizing the wet protein in water by adjustment of pH to 7.0 and spray drying.

Textured vegetable protein and meat extenders: Soy protein isolate has been used as a meat substitute. The meat extender may be prepared in several ways. The two main manufacturing processes consist of either obtaining an alkaline dispersion of the protein, which is then mixed with colour and flavour and steam extruded, or the dispersion is mixed with gum, colour and flavour and extruded through fine dyes of 0.18 mm diameter. The extruded material is stretched and layered on reciprocating tables, coagulated with live steam and thus given a fibrous, more meat-like structure This is in contrast to the rather amorphous material obtained with the first method. These meat substitutes are used extensively by mixing with commercial meat products in the manufacture of sausages and hamburgers. They can also be part of various vegetarian *curries.*

Fabricated products made with spun protein analogues, even at 50% level, are very similar to their meat counterparts and have high acceptability. The flavour and texture problems associated with many other meat extenders which restrict their use to 10-30% can be overcome by use of spun protein analogues. These spun protein analogues have varied applications. They actually improve product quality when used as part of other edible items.

Extruded products: Extrusion cooking, a process of high temperature, high pressure short-time cooking under controlled conditions of moisture, is particularly suited for high-protein low-fat materials like solvent extracted soy flour or groundnut flour.

Soy milk and infant foods: Milk prepared from soybean has long been used in China and other Asian countries for feeding infants. Improvements have been made recently in the processes by several workers (Swaminathan, 1987). Preparation of soybean milk without any beany odour can be carried out using the following steps: (i) cleanining and dehulling of soybean in a mill; (ii) soaking dehulled split beans in water for 5-10 hr; (iii) steaming the soaked beans for 30 min to destroy lipoxidase, trypsin inhibitors and other antinutrients; (iv) grinding the cooked beans in boiling water; (v) the ground mass is then filtered and sugar, minerals and vitamins are added to the filtrate; and (vi) the final product is then homogenized, steamed and bottled.

Soy milk can also be prepared from commercially available defatted flour. The defatted flour is suspended in hot water (80°C), blended and the pH of the suspension is adjusted to 10.0 to solubilize proteins. The suspension is centrifuged and the filtrate's pH is then adjusted to 7.0-7.4 and centrifuged again. The protein content in the filtrate is adjusted with water and the mixture is blended with sugar and fat, homogenized, bottled and sterilized. Flavoured milks can be prepared from soy milk using sweeteners and suitable colouring and flavouring agents. Dried soymilk powders have also been prepared. Protein concentrates and isolates can also be used to prepare soymilk.

Soysnack: Cereal based snack foods are quite popular all over the world. Addition of soybean improves the nutritional value. A process developed for making soy-based snack food consists of cleaning of cereals and treated soybean splits and grinding of the soy-cereal mixture into flour. The dough made from fluor is extruded (through a form of extruder) at about 100°C, semi-drying the extrudate, flaking, drying of flakes to 8% moisture and packaging. The flakes thus prepared from soy-wheat/sorghum/maize (30:70) have approximately 20% protein, 7% oil, 55% carbohydrates.

Soy-badi: It is a dehydrated and spherical shaped pulse based vegetable prepared in villages for house-hold consumption. It is generally made using green gram pulse along with other ingredients such as spices, salt etc. The process of soy-*badi* making consists of cleaning of soybean, dehulling and splitting, soaking of soysplits in water containing 1% $NaHCO_3$ for 4 hr at a bean-water ratio of 1:3, grinding of soaked soydal to make paste, addition of salt, spices, etc. as per taste, cooking for about 30 min, forming *Badis* and then drying to 6-8% moisture. The soy-*badi* thus made has about 43% protein, 19% oil and negligible trypsin inhibitor which is further eliminated during cooking of the *Badis*.

Coconut Food Products

When unripe, the nut contains coconut milk-a pale whitish liquid with a strong taste of coconut. This liquid is gradually absorbed as the nuts ripen. The coconut meat is either sun-dried or kiln-dried and is known as copra.

Dehydrated coconut (copra): It is prepared from ripe nuts and is available in two forms, viz. ball copra and cup copra. Ball copra is obtained from mature unhusked nuts stored in the shade for 8-12 months. During this period the coconut water is absorbed and kernel dries up. The husk and the shell are then carefully removed. Cup copra is made from fresh or stored nuts by cutting the kernel into two halves and drying them under the sun or special driers or kilns. Copra dried to 5% moisture content does not deteriorate quickly. When the moisture content exceeds 6%, mould and insect attack occurs resulting in detetioration.

Dessicated coconut: It is prepared from the white fleshy layer of the kernel commonly known as 'meat'. The white meat is shredded or disintegrated and dried in a hot air drier to less than 2% moisture. Dessicated coconut is used in the manufacture of cakes, pastries and chocolates. It is an important ingredient of a variety of sweets and savouries.

Coconut milk: The coconut milk and cream production is essentially a wet-milling process. The testa of the fresh clean endosperm is removed and the kernels are washed, sterilized and scraped into a fine meal. The meal is then squeezed in a special continuous process to press out the milk. The milk is filtered and spray-dried under high vaccum (Nambiar 1983). It is a white emulsion and does not have good stability (Monera and Rosario 1982). It is susceptible to oxidation and microbial attack. It can be kept for about 5 hr without any sign of spoilage at 2-5^{0}C (Som, 1980). The spray-dried product is more stable and can be later reconstituted with water.

Coconut cream is obtained by centrifuging the coconut milk. The ground endosperm is mixed with water and passed through an expeller once or twice. Coconut cream is a useful product for confectionery industry. Addition of coconut cream at a 9% level in soya milk produces an acceptable *tofu* with improved hardness, cohesiveness, gumminess and calorific value (Escueta *et al.*, 1985).

The separation of cream from whole coconut milk yields skim milk as a by-product. It can be dehydrated to a skim milk powder (Hagenmaier *et al.*, 1974; Hagenmaier, 1983). To improve its quality, partially hydrolyzed starch may be mixed with it before spray-drying. Skim milk powder with 25% starch contains 24% protein, 5% moisture and 6.0% oil. It is a white powder that may be mixed with water to make a beverage (Hagenmaier, 1979).

Krauss-Moffei process, has been developed in Philippines. In this process the coconuts are wet-rendered and screw pressed to give 'milk' which is centrifuged to recover fat. The 'skim-milk' which is a by-product is quite rich in protein and can be concentrated and reconstituted as a drink (Parpia, 1991). This material can also be used along with groundnut protein and hydrolysed starch to obtain a spray dried infant food. The residue from Krauss-Moffei process can be used for making '*chutney*' and other products of daily consumption instead of the whole fat coconut.

Protein products: Coconut protein isolates can be prepared by extracting the fresh endosperm in dilute acid, alkali, or 1 M sodium chloride solution at pH 7.0, followed by precipitation of proteins at pH 2 to 3.9. Over 50% of the total proteins can be recovered by this process (Samson *et al.*, 1971). On dry basis, it contains 59% protein and has a well-balanced amino acid composition. It can be added to several products such as biscuits, cakes, breads or mixed with milk (FAO 1984).

Copra meal and flour: The cake is obtained from hydraulic pressing of copra while the meal is obtained in expeller pressing. Removal of water during the drying of coconut into copra, and the subsequent extraction of oil, improves the protein content in cake or meal from 4-4.5% to about 20%. The protein content in the cake or meal is influenced by the method and efficiency of oil extraction. The cake or meal is rich in fiber (10-15%) and carbohydrates (35-45%).

Good-quality cakes and meals have almost a white colour, with a reddish brown tint due to the presence of testa particles. Overheating of copra during expeller or hydraulic pressing produces a dark-coloured product which is less appreciated in animal feed because of its low digestibility (Thieme, 1968). The cake and meal obtained during dry milling are useful as animal feeds.

Coconut flour is the name commonly applied to the food-grade product made from coconut endosperm by drying to make a good-quality desiccated coconut, followed by removal of the oil using pressing and/or solvent extraction to yield coconut meal. The meal is finally milled to flour. Desiccated nuts are pre-pressed, flaked and hexane-extracted to remove the oil. The white coconut flour contains 25% protein and 0.5% oil.

Coconut Toddy: Toddy is an alcoholic beverage obtained by the natural fermentation of coconut palm sap. The coconut inflorescence sap is traditionally collected in clay pots sterilized by inverting over a flame for 5min., and allowed to ferment in open pots for up to 2 days (Nathanael, 1955) during collection of sap. During this period, microorganisms from the atmosphere enter the clay pots and multiply in the palm sap which contain 15 to 18% sucrose, transforming the sugar into ethanol and other products. The resulting liquid containing about 7% ethanol is known as *toddy*, which is consumed as liquid beverage like soft drinks and beer, in coconut-producing countries.

Rapeseed and Mustard Products

There is a great potential for preparation of high quality oilseed protein concentrates from rapeseed- mustard. The amino acid profile of rapeseed meal is better than that of soybean. Glucosinolates and their breakdown products,namely, the iso-thiocyanates are not desirable in the protein meal. Mustard and rapeseed meal when consumed cause enlargement of thyroid, adrenal gland, kidney and liver haemorrage. Hence it is better to detoxify/ remove these antinutrients, prior to utilization of the meal. They can be removed by methods such as extraction by suitable solvents, enzymatic decomposition, treatment with chemicals etc. Such techniques when utilized will make available, a million tonnes of high quality protein for human consumption.

Rapeseed meal can be utilized in preparation of protein concentrates, isolates and textured products(Salunkhe *et al.*, 1991). Blaicher *et al.*, (1983) have prepared two protein isolates containing about 60% and 12% of meal protein, respectively. The colour of the protein isolate has been improved by using sodium bisulfite in the solvent for protein extraction.

The nutritional quality of glucosinolate free rapeseed protein is comparable to that of casein with a PER of 2.4 and NPU value of 65. The following foods can be prepared using mustard protein.

1. Using mustard protein concentrate (MPC), protein rich biscuits with 15% protein can be prepared. The biscuits have desirable texture and taste.
2. High protein candies incorporating MPC can also be prepared. This type of product finds ready acceptance with children and provide them protein supplement.
3. Weaning foods: MPC can also be used for preparing weaning foods containing mixtures of cereals and pulses. Germinated cereals and pulses can be used to prepare such foods. The *in vitro* digestibilities of protein of such products are 67 for ungerminated and 65 for germinated categories.

Sesame Based Foods

Sesame protein is a rich source of sulphur amino-acids in which most other vegetable proteins are deficient. It is an oilseed of choice with many uses. Nearly 20% of sesame seeds produced is consumed directly in confections and other foods.

Production and use of sesame protein presents several problems but they have now been solved. It's husk is hard to remove and has undigestible fibre. Besides, the oxalic acid content of the husk is also quite high. The protein concentrate made from dehusked sesame is an excellent supplement of sulphur amino-acids for incorporation with other oilseed concentrates and grain legumes to make foods with high PER, especially for consumption by children.

Sesame, the nourishing food: Sesame kernels serve as very good nourishing food. The seeds are digestive, rejuvenative and anti-aging. They are a rich source of quality oil (50%), protein (25%) and corbohydrates (15%). Seeds can provide 640 calories/100g and are rich in vitamins E, A, B complex and minerals like calcium, phosphorus, iron, copper, magnesium, zinc and potassium. Sesame seeds contain extraordinary quantities of methionine and tryptophan which are essential amino acids, for healthy liver and kidneys and utilization of B complex. This unique composition of sesame coupled with balanced proportion of saturated, mono-unsaturated and poly-unsaturated fatty acids makes sesame seed a nearly perfect food.

Sesame has very little possibilities of causing any allergy and therefore it is the best alternative source of food to the people with milk allergies. The husk is also very rich in minerals and vitamin B1, but these are unfortunately non-extractable by human digestive system due to the presence of anti-nutritional factors, oxalic and phytic acids. Oxalic acid binds calcium to form non-available calcium oxalate. Therefore, dehulling is preferred before direct consumption of sesame seeds.

Protein concentrates and isolates: Sesame is widely processed into several high-protein products such as flakes, flours, protein concentrates and isolates (Sastry *et al.*, 1969). The protein concentrate contains more protein (about 70%) than the flour while protein isolate contains about 90% or more protein. The defatted flour and isolates prepared from sesame do not contain any undesirable pigments, off-flavors or toxins (Toma *et al.*, 1979; Jonson *et al.*, 1979).

Sodium hydroxide solution (0.04M) is the most suitable solvent, which can extract about 90% of the protein. Rivas *et al.*, (1981) have reported a miximum extraction of proteins with 0.5 to 1.0M NaCl in the flour to solvent ratio of 1:40 and at pH 11.0 (alkaline medium). A low-phytate protein isolate has been prepared by Taha *et al.*, (1987) by dissolving the protein by counter current procedure and precipitating it at pH 5.4. The protein isolate contains 91.4% protein.

The chemical compositions of defatted sesame flour, protein concentrate and protein isolate are significantly different from each other (Rivas et al., 1981). The protein isolate contains very high levels of protein and is free from oil, ash, crude fiber and phytate phosphorus with very low levels of nitrogen-free extract. The chemical composition of protein concentrate is intermediate between that of defatted meal and protein isolate.

The essential amino acid composition of alkali isolate is comparable to that of sesame flour (Prakash and Nandi 1978, Rooney *et al.*, 1972). The salt isolate, however, contains more threonine and valine and less lysine and methionine than the other two products.

Functional properties : The functional properties such as water absorption, fat or oil absorption, emulsification, foaming and viscosity are important in determining the applications of food products in various food systems. Sesame flour and protein concentrate exhibits less water absorption than soy flour. Sesame products show high fat absorption than soy products. The emulsifying capacity and emulsion stability of sesame products are similar to that of soy products. The foam expansion and foam stability are better.

Lawhon *et al.,* (1972) have observed that the protein extractability and whipping potential of sesame flour extract to be lower in comparison to other oilseeds. As per Dench *et al.,* (1981) the functional properties of the sesame protein products show poor water absorption but better fat absorption, and lower bulk densities than the soy products. Emulsifying capacity and stability of sesame products are generally lower but the foam expansion and stability are comparable to or better than soybean products.

Sesame chikkies and laddus: Sesame seeds are used in the preparation of traditional confections such as *halva, laddu* and *chikki.* They are also eaten whole after roasting. A confection called *laddu* is prepared from roasted groundnuts and sesame seeds by pounding with jaggery in the proportion of 2:1:2. Small balls are prepared by hand. *Laddus* are also prepared from sesame seeds by mixing them with hot jaggery syrup. *Chikki* is another confection popular in India. It is prepared by pouring sesame seeds in boiling jaggery solution to obtain a thick slurry. The slurry is spread uniformly on a metallic sheet or table and cut into small rectangular pieces.

In India *laddus* (balls) of sesame and jaggery are prepared mostly at household level. These *laddus* keep people healthy, young and energetic and protect from cold and cold related ailments. Sesame seeds in combination with jaggery, cow's ghee and other ingredients complement and supplement each other to produce synergistic effect on human body to overcome debility and revitalize the system. Besides improving general health and stamina, this preparation acts as laxative for bowel movement, lubricates the joints to relieve rheumatic pain, induces resistance to cold and reduces other problems associated with old age. If taken twice daily in the morning and evening with tea/milk, the *laddus* reduce the process of aging and debility.

Ready-to-eat foods: Minor millet fluors like pearlmillet, sorghum and finger millet(*ragi*) are used to prepare ready to eat foods in Indian households. Roasted *bajra* (pearl millet) flour (60g) is mixed with roasted green gram *dhal* (15g), roasted groundnut (10g) and sesame seeds (5g). The mixture is pounded/ ground to obtain a fine powder. It is mixed with boiling water or milk to the desired thickness. Sugar or salt are added to taste. *Ragi*/ sorghum instant foods are prepared in the same way by replacing bajra flour.

Sesame seeds are utilized in the production of *tehineh* (sesame butter) and *halwah* (*halva*). *Tehineh* is made from a paste of dehulled roasted seeds similar to peanut butter. *Halwah* is a sweet made up of *tehineh,* sugar, citric acid and *saponaria officinalis* root extract. *Tehineh* is added to a variety of foods, bread and bakery products (Sawaya *et al.,* 1985).

Sesame seeds and kernels are used in commercial bakeries for the preparation of quick breads, rolls, crackers, coffee cakes, pies and pastry products (Weiss 1983; Farrell 1985). The seeds are lightly roasted and used in salads and salad dressings (Farrell 1985).

Cholam (sorghum) porridge is prepared by mixing roasted flours of *cholam* (50g), green gram *dhal* (15g), and sesame cake (25g) with hot water. The batter is poured in boiling jaggery solution while stirring and bioled for 10-15 min.

Sesame flour has been used as a methionine supplement in the preparation of fermented foods, namely, *vada* and *dosa* (Gulati *et al.*, 1979; Chopra *et al.*, 1982). The sesame flour is used to replace 5-20% of rice-blackgram flour (Chopra *et al.*, 1982). Sesame meal is used as a vegetable in preparation of curries in the Eastern part of India.

Sesame flour has been used in the preparation of bread and cookies (Hoojjat, 1982). When used as a part of wheat flour, sesame flour performs better than the sunflower flour. High-protein biscuits can be prepared by mixing wheat flour with roasted Bengal gram and roasted sesame flour to prepare the dough. Sesame fluor can be used in formulating high protein beverages. A nutritious beverage could be prepared using 70% soy protein and 30% protein and 30% sesame protein.

Sunflower Based Foods

The sunflower seeds and kernels are used for human consumption after roasting, similar to roasted peanuts (Robertson *et al.*, 1988). Roasted and salted sunflower seed, either dehulled or whole, is a popular snack food all over the world (Weiss 1983). The broken or whole kernels with or without roasting can be used in a variety of bakery products, salads, candies, and some food dishes1 (Robertson, 1975). The kernels, with or without roasting, are used for sprinkling on syrup or pancakes and waffles, blended with honey, butter and salt to make a spread, or used as a fondue dip (Millette 1974).

Confectionery sunflower: Sunflower seed is consumed as a snack food. Confectionery sunflower (generally bolder seeds) is consumed in-shell or dehulled. Kernel is used in a broad variety of snacks and food products such as breads or as part of nuts and dried fruit mixes.

Confectionery sunflower seeds are larger than oil varieties, normally black with white stripes and approximately five eighths of an inch long. The heavy hull accounts for 50% of the weight of the seed and is loosely fixed to the kernel inside. Confectionery sunflower seeds fall into three categories: in shell, kernel and birdseed. The seeds (in shell) are roasted, salted and packaged for human consumption and are classified as either large or jumbo. Medium-sized kernels are also packaged for human consumption. They are mostly utilized by the bakery industry. The black-shelled oilseed variety is richer in oil and are used for the production of sunflower oil.

In shell seeds are normally roasted and seasoned. They are eaten as a snack by cracking the shell with one's teeth, discarding the hull and eating the delicious morsel within. The shells are removed mechanically and the resulting kernels are roasted and utilized directly as snack food and or incorporated as components of food items to improve their taste and nutrition.

Sunflower, the neutraceutical: Sunflower seeds are a good source of vitamin E, mainly alpha-tocopherol, the most biologically active form. A one-ounce serving of sunflower seeds contains a whopping 76% of the Recommended Dietary Allowance for vitamin E. Sunflower seeds provide vitamin E, along with healthful, unsaturated fat. They are also good sources of the antioxidants copper and selenium, which work in conjunction with vitamin E. Sunflower seeds provide six grams of protein and two grams of fiber per serving, and are power-packed with folate, pantothenic acid, vitamin B6, manganese, iron and zinc.

Some of these nutrients have been shown to offer a variety of potential health benefits, including protection against heart disease and cancer, as well as a role in memory and cognitive functions. Sunflower seeds contain high levels of the phytochemicals tocopherols (vitamin E), choline, betaine, and phenolic acids, as compared to other nuts. In addition, the kernel is a good source of lignans and arginine. The high levels of these phytochemicals, makes the sunflower kernel a powerhouse of health benefits.

Sunflower meal fluor: A mixture of sunflower meal, corn, roasted chickpea flour and sesame in the ratio of 65:15:10:10 has been used in the school lunch programme in India (Devadas *et al.*, 1977). The mixture is effective in increasing the heights of growing children. Sunflower meal (defatted) and protein products can be utilized in the preparation of bread, cakes, cookies, muffins and pies as a partial substitute for wheat flour or nuts (Lawhon *et al.*, 1975, Robertson 1975, Sirko and Bukhantsov, 1973, Talley *et al.*, 1972). The concentration of sunflower above 15% results in poor loaf volume and dark color (Rooney *et al.*, 1972). Autoclaving of flour, roasting of kernels or the use of commercial dough conditioners help restore bread texture and volume (Lorenz, 1978; Khan *et al.*, 1980; Rooney *et al.*, 1972; Sosulki 1979).

Addition of 5-10% sunflower protein concentrate to wheat flour can increase the protein content and improve the amino acid balance of bread without changing the crumb colour (Sirko and Bukhantsov, 1973). Sunflower meal and protein concentrates have been extruded to form meat-like, chewy, texturized vegetable protein particles (Sosulski 1979b; Rossi, 1988). Sunflower protein isolate incorporated in batter/breading system can impart its antioxidant effect to the meat of the fried beef patties (Rhee, 1984).

Sunflower cakes of various grades and the protein isolates have been utilized in the preparation of some Indian snack food items like *pakodi*, biscuits and *chapatti*. Sunflower cake flour at 10% level enhances protein content by 3-4% and protein isolate at 5 % level increases it by 15-25% and at 10% it is 18-29% higher. Taste panel acceptability has been 50-90% for the snack foods as against 100% for the control food items. There is scope for utilization of sunflower products in food items (Praveena *et al.*, 1999, Nagaraj *et al.*, 2000).

Sunbutter-a peanut butter alternative: The American Academy of Allergy, Asthma and Immunology estimates that 3 million Americans are allergic to peanuts and/or tree nuts. Because children make up a growing number of those allergic, some schools have opted not to allow peanut butter in their cafeterias. Sunflower seeds are a popular alternative for preparing peanut butter. Sunbutter is prepared in a way similar to peanut butter using sunflower kernels. Sunbutter looks like peanut butter and has a mild and distinctive sunflower seed flavour.

Safflower Based Foods

Safflower is an important traditional oilseed in many parts of the world especially the developing countries. It is a rich source of poly-unsaturated fatty acids and proteins along with useful minerals and vitamins.

Safflower seeds are utilized as snack food. Fried seeds with hulls are eaten like groundnuts. The kernels are a better option. The seeds are generally disliked because they contain phenolics which impart bitter taste. However, those accumstomed easily relish them. They are especially useful as neutaceuticals because of the high PUFA, proteins, minerals and useful vitamins present in them. Roasted seeds, generally mixed with chickpeas, barley or wheat, are eaten as snack food. The ground kernels are mixed with sesame and eaten. In India the seeds are used to make a spicy item called *chutney*.

A paste of seeds is useful in hastening curd formation. In Ethiopia, finely pounded safflower kernels are mixed with water to prepare a drink called '*fitfit*', which is used on fast-days or mixed with '*teff*' bread and spices to form a porridge. Crushed dried seeds are also used to grease the baking plate for the '*teff*' flatbread.

Utilization of cake and flour: The meal obtained from completely decorticated seeds is light in colour, bitter in taste and contains about 60 to 67% proteins. The removal of bitter compounds by methanol extraction further increases the protein content upto 70% (Salazer-Zazueta and Price, 1989). Such meals have promise in human foods in the form of high-protein flours, protein isolates and concentrates, in a variety of bakery products and in weaning or traditional foods as a protein supplement.

The defatted meal can be further pulverized to flour or used to prepare protein isolates and concentrates. This facilitates the removal of hull or fiber. A high-protein (41 to 45%)product can be prepared by impacting the meal and classifying the material on the basis of particle size. A low-fiber, high-protein flour and edible concentrate have been prepared by screening and aspirating cracked safflower seeds, extracting the oil, milling and sieving (Goodban, 1967). The light coloured, bitter flour contains more fibre than the recommended 5% maximum for human consumption. A protein concentrate has been prepared by extracting the bitter and cathartic glucosides from the flour with 70 to 80% ethanol (Goodban, 1967; Kohler, 1969).

The edible safflower flour (70% proteins) can be incorporated into meat-like patties and bread at a level of 5% (Elmquist, 1965). Due to the presence of bitter compounds, the utilization of defatted flour in various processed foods has remained limited to the experimental stage. The defatted flour holds promise as a protein supplement in a variety of processed foods such as bakery products, weaning foods and traditional cereal-legume-based products.

Protein isolates: The functional properties of safflower protein isolates have been studied. The safflower protein isolates show inferior foaming capacity and stability, and whipping capacity as compared to other oilseed protein isolates. Heating treatments decrease whipping and water-holding capacities of protein isolates. Isolates prepared by the micellization procedure show superior functional properties than those obtained by isoelectric precipitation (Paredes-Lopez and Ordorica-Falomir, 1987). However, the recovery of proteins is lower in the former process.

Protein isolates prepared from debittered meal can be used to fortify bread, pasta and nutritional drinks. Among amino acids, lysine is limiting, while methionine and isoleucine are on the border line.

Niger Based Foods

Niger seeds are edible. Their seeds fried in ghee can accompany other foods to improve their palatability. They are parched and mixed with pulses as snack food. A spicy preparation called *chutney*, preparaed by mixing red chilli powder with roasted and pounded niger seeds is a common preparation in some parts of rural India (Nagaraj, 2002). Till date no antinutrients have been reported in niger seeds.

Niger seeds are parached and ground with water to make a drink. Other minor uses include mixing the parached seed with pulses as a snack food, and making a stew or "wot" with ground seed and spices (Seegler, 1983). Niger flour is also baked into a bread or sprinkled on bread prior to consumption (Riley and Belayneh, 1989). The seeds have good market and they are utilized as bird feed in some countries (Subramanian, 2003).

Niger cake in addition to being rich in proteins, is also rich in soluble sugars. Niger cake and its protein are balanced with respect to essential amino acids. Its quality is similar to that of sesame cake. Hence, attempts are needed for wider utilization of niger cake as a food ingredient (Nagaraj, 2002).

Some valuable work has been done on its protein. Protein from dehusked seed is quite rich in sulphur amino-acids, but deficient in lysine. When the deoiled meal is mixed with grain legumes or cereals, it gives a PER quite comparable to casein.

Oleo-Chemicals

Oils and fats have traditionally been utilized as food. However, they have more varied uses in the non-food sector. The oldest use has been as lamp oil and lubricant. In recent times they are greatly being used in the manufacture of paints, lacquers and soaps. Fats, fatty acids and their derivatives also find use in detergents, cosmetics, plastics and as rubber additive. Their wide use and adaptability is presently under check because of the availability of cheaper priced petroleum derivatives.

Vegetable oils have some advantages over petro-products. They are a renewable resource, they are biodegradable (environment friendly), they have excellent technical properties and they cause fewer medical problems (less carcinogenic) and allergies. The main disadvantages of vegetable oils are that they are high priced, they have limited oxidative stability, they have high viscosity and they are too reactive.

Non-food uses of vegetable oils and or fatty acids have received less attention because of predominant interest in human nutrition. Because of some of the advantages mentioned above and also in view of the increasing production of vegetable oils world over through traditional and non-traditional sources, there exists a greater potential for use of vegetable oils in the industry.

Oleochemicals

Oleochemicals are the chemical substances which are produced from oils and fats. In the past decade, this term has been used quite often, especially after the realization that oils and fats can replace petroleum feed stocks for a variety of chemicals. The petroleum resources are getting depleted gradually and the world is realizing the importance of oils and fats as alternative resources.

Oleochemicals find diverse uses. All of them have a common functionality namely the processes at the surface. The long hydrocarbon chain favours the oily or hydrophobic phase, while the carboxyl group favours the aqueous or hydrophilic phase. In most of the situations, it is the carboxyl group that is modified through oleo-chemical reactions to fulfill specific functions, keeping intact the hydrophobic chain.

The oleochemicals derived from oils and fats can be divided into two categories. The chemicals which can be obtained by carrying out unit operations on oils and fats are called 'basic oleochemicals'. Other chemicals which are produced by derivative operations on oils and fats and/or basic oleochemicals are known as 'oleochemical derivatives'.The fatty acid molecules present in oils and fats are the 'building blocks' of these oleochemicals.

There are many derivatives of basic oleochemicals. The major derivatives of basic oleochemicals, fatty acids, fatty alcohols, fatty acid methyl esters, fatty amines and glycerol are the major oleochemicals from which several derivatives are made. Some derivatives can be made from more than one starting oleochemical.

Fat Splitting

Oils and fats are triglycerides of fatty acids and glycerol. These triglcerides can be hydrolysed to get fatty acids and glycerol. This reaction, called fat splitting, is the basic reaction for obtaining the basic oleochemicals. The acid group and the double bond are the reactive sites in fatty acids and they make fatty acids important starting materials for a variety of oleochemicals. The three hydroxyl groups in the glycerol molecule make it very reactive, and hence a very useful intermediate for a number of derivatives.

Oleochemistry emphasizes alkene, ester, hydroxyl and carboxyl reactivities. Industrial value of a vegetable oil generally depends on its content of specific fatty acid and the ease with which it can be modified. The general chain length of fatty acids ranges from C:12 to C:22 with saturated, mono, di or tri unsaturated acids. The main exception is ricinoleic acid which has a hydroxyl group in its molecule.

Triglycerides (oils and fats) are split into fatty acids and glycerol by 'fat splitting'. The mixture of fatty acids is distilled to get colourless, odourless, distilled grade fatty acids. By fractionation, the fatty acid mixture can be separated into fatty acids of different chain lengths. The glycerol-water mixture is passed through a series of evaporators to get glycerol of various grades, including chemically pure (CP) glycerine. Alternatively, the oils and fats can be subjected to alcoholysis using excess of glycerol or methanol in the presence of a catalyst to give monoglycerides or fatty acid methyl esters, respectively. Fatty acids and fatty acid methyl esters can be converted into fatty alcohols.

Glycerol and Derivatives

Glycerol is the hydrolysis product of oils. It is an important by-product of soap industry. It has wide uses, in pharmaceuticals, food emulsifiers, resins, cellulose, polyols, polyurethane, tobacco products, explosives etc. Glycerol directly is used as a humectant in various foods and tobacco. Triacetyl glycerol (triacetin) finds use in the production of cigarette filters. It is also used in the manufacture of alkyd resins (esterification of dibasic acids with glycerol). Partial fatty acid esters of glycerol are used as emulsifiers, since the free hydroxyl group confers hydrophilicity, for example glycerol mono stearate. Ethoxylated monoacyl glycerols are used in shampoos and foam bath preparations, transparent creams, and in suntan lotions. Glycerol monoesters (ex: monolaurin) find use in cosmetics, and as food preservative. At present glycerol is in abundant supply. But with increasing oil supplies, more industrial uses have to be found out for glycerol in future.

The consumption of glycerine is maximum (31.5%) in the manufacture of drugs and personal care products followed by tobacco/triacetin, alkyd resins and polyether polyols. More than 90% of all triacetin is used in cigarette papers and filters. Miscellaneous applications also consume significantly large quantities of glycerine amounting to 16 % share of the total market. Because natural glycerine is preferred, the production of synthetic glycerine is expected to be squeezed out by expanding natural glycerine supply.

Fatty Acids and Derivatives

Saponification of storage lipids produces fatty acids. Hydrolysis based on counter current splitting at high pressure is the method widely used by the industry. The resulting fatty acids are fractionated to C8/10, C12/14, C16/18 etc. or of any other defined chain length. Saturated fatty acids are separated through crystallization. For example stearin fraction crystallises first which can easily be separated through centrifugation from the liquid olein fraction.

The carboxyl group is the reactive portion in a fatty acid. Compounds like methyl esters, alcohols, amides, amines and quaternary ammonium compounds can be manufactured. Soaps are nothing but alkali metal salts of fatty acids and are utilized as such. Various grades of soaps are possible depending upon the proportion of lauric and stearic acids. Fatty acids find use as polishes, buffing compounds, household and industrial cleaning agents. They also find use in the preparation of high grade candles and rubbers for tyres. Metallic stearates are used in the preparation of moulds in the manufacture of plastics. They serve as lubricants for coated papers. In food and pharmaceutical industry, metallic stearates are used to coat tablets or spices to increase their flow and prevent sticking. Other applications of fatty acids are in the field of cosmetics, greases and varnishes. The applications of various fatty acid groups are presented below.

Caprylic-capric acid group: Coconut oil is the chief commercial source of caprylic-capric and lauric acids. These acids of intermediate chain length are highly important and useful for production of surface active agents, plasticisers, physiologically active compounds, perfumes, flavours etc. Low molecular weight fatty acids and their derivatives have wide uses in the cosmetic industry.

Lauric acid group: Coconut oil and palm kernel oil are the chief sources of commercial lauric acid. The fats of this group are distinguished not only by the presence of high content of lauric acid but also by the least unsaturation and yet relatively low melting points. Their sodium soaps are hard and stable towards oxidation and also soluble and freely lathering and hence they are particularly prized for soap making. They are also the sources of lower molecular weight fatty acids such as caproic, caprylic and capric which in-turn find many industrial applications.

Oleic-linoleic acid group: The oleic-linoleic acid group is the largest group of vegetable fats and is one of the most varied in composition and characteristics of the individual oils and fats. The saturated components in most of these fats is less than 20%. The fats from this group are the most useful and most adjustable of all the fats and oils. They form the largest group of edible oils and commercial fats which include cotton seed, olive, palm, peanut, safflower, sunflower, sesame, soybean, rice bran, tree oilseeds etc.

The oils rich in oleic acid are more stable and free from any serious tendency towards flavour reversion. For the same reason they find much use in soap making and produce firm stable soaps of desired properties. The vegetable fats from this group also find extensive applications in cosmetic formultations and are good ingredients imparting emolliency and good skin compatibility. Cocoa butter is a unique naturally occurring fat used in the manufacture of chocolates and cosmetics. The shea butter, illipe butter and mango kernel fat have been found as valuable sources of cocoa butter substitute. Sesame oil, apart from being a cooking and salad oil, also finds use as a soap fat, in pharmaceuticals and as a synergist for insecticides.

Linolenic acid group: The oils from the group are distinguished by the presence of large proportion of linolenic acid, in addition to oleic and linoleic acids. The important members from this group include linseed, perilla, soybeen and hemp seed, which grow well in a relatively cool climate. The oils from this group have drying properties and find applications in paint, varnish and allied products. Because of their high tendency for auto-oxidation and deterioration they do not find much acceptance as edible oils or as sources for the manufacture of soaps.

Erucic and behenic acids: Oleocheicals based on the long chain C22 fatty acids erucic and behenic acids have special properties and find increasing applications. The *Cruciferae* family of plants, (rape, mustard, *crambe*) provide the most economical and readily available sources of erucic acid. Behenic acid is produced today by the hydrogenation of high erucic acid rapeseed (HEAR) oil and occasionally, of fish oils followed by fractionation. The major limitation of these oils as sources of erucic and behenic acids is that only about half of the fatty acids are useful. The disposal of the by-product C18 acids assumes economic importance. A major high value application of erucic acid can be in the manufacture of Nylon 13 and Nylon 13, 13.

Other applications of erucic acid/benenic acid based oleochemicals would be in surfactants and detergents, plastics and plastic additives, photographic and recording materials, food and food additives, cosmetics and personal care products, ink additives, paper, textiles, lubricants and fuel oil additives.

Conjugated and acetylenic acid group: The fats from this group include tung oil containing eleostearic acid and oticia oil containing licanic acid, both containing conjugated double bonds. Eleostearic acid is octadeca-9, 11, 13-trienoic acid and licanic acid is, 4-keto eleostearic acid. The presence of conjugated double bonds in their fatty acids, favours oxidation and polymerization thereby making them unsuitable for edible purpose and soap making. But they find much use for the manufacture of varnish, enamels and protective coatings. The fats containing acetylenic acid 'tariric acid' (octadeca-11-en-9-yonic acid) in sandalwood seed fat serve similar industrial applications.

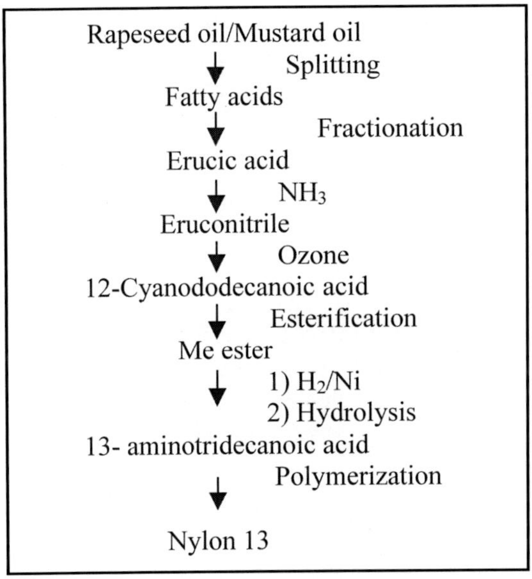

Manufacture of Nylon 13

Similarly Nylon 13,13 can be prepared according to the following scheme:

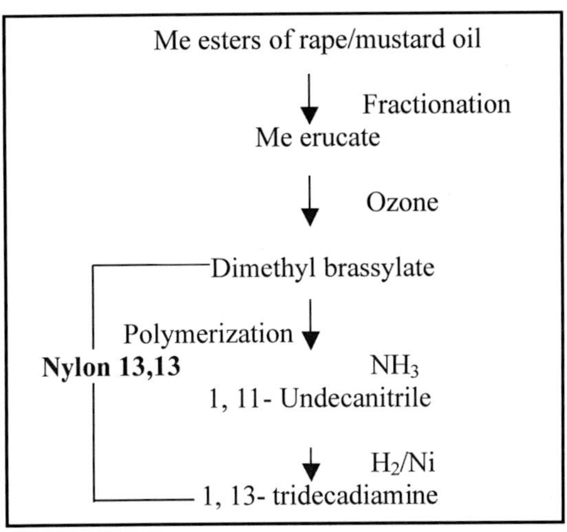

Preparation of Nylon -13, 13

Hydroxy acid group: Castor oil is the most unusual product, among all the vegetable oils. It is, by far more versatile (Boner, 1983) due to presence of ricinoleic acid, which constitutes about 90% of the oil. Mostly oils have one or two applications, such as for edible purposes (cotton seed, soybean, corn, peanut, rapeseed, sunflower etc.). Some have other uses as in coatings, inks, lubricants, detergents and soaps (Naughton, 1974). In contrast, castor has many more uses, with a wide diversity of commercial applications, all directly related to the unique hydroxy fatty acid, namely, ricinoleic acid. Castor oil is non-toxic, renewable and biodegradable. It's oil is used in medicines as purgative and also as hair oil (Irwin, 1990).

Castor oil containing a high content of nearly 90% of a hydroxy fatty acid, 'ricinoleic acid' (12-Hydroxy octadec-9-enoic) is unique in itself and differs from the other vegetable fats. It is not edible and is also not much used in soap making. Castor oil, a viscous liquid finds direct application in lubricating greases, polishes and plastics. Fully hydrogenated castor oil is a substitute for 'Carnauba Wax'. By catalytic dehydration process it can be converted to conjugated acid oil closely resembling tung oil and this dehydrated castor oil is extensively used in protective coating industry.

Castor oil has relatively high purity (high for naturally occurring material). Hence it can be subjected efficiently to a number of chemical processes to yield high purity chemical derivatives. The chemical bases for such reactions are the three points of functionality existing in ricinoleic acid. a) the carboxyl or ester groups, b) the single point of unsaturation (double bond) and c) the hydroxyl group. These functions are utilized in a wide range of esterifications.

Hydroxyl group can be acetylated or alkoxylated. The unsaturation can be altered by hydroxylation or expoxidation (Forbes, *et al.*, 1940). The hydroxyl group can be removed by dehydration to increase the unsaturation of the compound, which can further be made to yield drying compounds and polymers. The hydroxyl position is so reactive that the molecule can be split at that point by high temperature pyrolyis and by caustic fusion to yield useful products of short chain length (Achaya, *et al.*, 1964). Castor oil is used in the manufacture of cleavage products like sebacic acid, undecylenic acid and heptaldehyde and nylon through dimerization, polymerization etc., (Naughton 1974).

Castor oil on sulfonation yields sulfated castor oil, known as 'Turkey Red Oil' which is used in cotton dyeing and printing industry to give lusture and brightness. On alkaline fusion, castor oil yields, sebacic acid and 2-octanol (caprylic alcohol). Sebacic acid is used as plasticiser and it's alcohol is used a as defoaming agent and as a solvent in plastic and lacquer industry.

Pyrolytic decomposition of castor oil at temperatures above 250^{0}C yield, undecylenic acid and heptaldehyde. The undecylenic acid finds wide use as a fungicide and bactericide. Heptadehyde is used as raw material for a host of perfumery chemicals like alpha amyl cinnamic aldehyde, methyl heptine carbonate etc.

A fatty acid very similar to ricinoleic acid, but with a change in the position of hydroxyl group and double bond, known as 'strophanthus acid' occurs in some of the seed fats of *Apocynaceae* family. 12-Hydroxy-octadec-9-enoic acid is ricinoleic acid, whereas 9-hydroxy octadec-12-enoic acid is strophanthus acid. Another closely resembling fatty acid, Dimorphecolic (9-hydroxy octadeca-10, 12-dienoic acid) occurs in some of the *compositae* seed species. The industrial application of these fats on similar lines of castor oil deserves attention.

There are an astonishing number of industrial applications for castor oil and its derivatives, and new ones are continually being discovered (Naughton, 1974). When dehyrated, castor oil is converted into quick-drying oil. It is used extensively in paints and varnishes (Achaya, 1972). It's water resistant qualities make it ideal for coating fabrics and for protecting coverings, insulation, food containers and gums (Fru, 1969 and Kirk, 1923). Sebacic acid, the castor oil product, is the basic ingredient in the production of nylon and other synthetic resins and fibres (Belanger, *et al.*, 1966). The superior oiliness of castor oil and its ability to cling to very hot moving parts make it an outstanding lubricant. Castrol-R, racing motor oil, for high-speed automobile and motor cycle engines is made from castor oil.

Strange and exceptional acid group: The naturally occuring fatty acids are in general, normal monobasic carboxylic acids consisting of a single carboxylic group attached to the end of a straight hydrocarbon chain. Quite an exception to this, there occur in certain vegetabel fats, very strange and exceptional type of fatty acids containing cyclopropenoid, cyclopentenoid, furanoid and fluoro fatty acids.

The behaviour of *Sterculia foetida,* an oil that polymerises rapidly with explosive violence on heating is attributed to the presence of a cyclopropenoid fatty acid known as sterculic acid (9, 10-methylene octadec-9-enoic). The cyclopropeniod fatty acids occur widely in the fats of *sterculiaceae, malvaceae, bobbacea and theaceae, sapindaceae* etc., plant families. Cyclopropenoid fatty acids cause undesibrable biological effects when they are included in animal diets and also possess carcinogenic properties.

The cyclopentenoid fatty acids occur in *flacourtiaceae* family and six optically similar fatty acids with cyclopentyl rings at the end of the chain have been isolated. They are chaulmoogric, hydnocarpic, gorilic, hormelic, oncobic and manoaic acids. The chaulmoogra and other oils from this family are used extensively in the treatment of leprosy and other cutaneous diseases. Due to this property the oils also find application as raw material for *Ayurvedic* medicated soap.

Fatty Acid Esters

Monobasic esters of short chain alcohols namely n-propyl, isopropyl, n-butyl, isobutyl or hexyl alcohols are very useful industrial chemicals. They serve as internal and external lubricants, in plastic industry, in metal treatment like drilling, cutting or rolling, in textile manufacture. They are useful as emollients in cosmetics and personal care products. Synthetic waxes are made from esterification with long chain alcohols or alcoholysis of triglycerides. These have properties similar to natural waxes and are hydrophobic. They are useful as lubricants, plasticizers, cosmetics, polishes and buffing compounds.

Fatty acid esters can also be made from polyols like ethylene glycol, propylene glycol, polyethylene glycol, penta erythritol, sorbitol and sucrose. Such compounds are highly useful as emulsifiers, protective coatings, synthetic lubricants, detergents, auxiliaries in textile and plastic industries, agrochemical and anti-foaming compounds. Since these compounds have mild or zero levels of toxicity they find use in food, pharmaceuticals and cosmetics. Synthetic lubricants have thermal and oxidative stability. They are useful as jet engine lubricants, crank case oils for automobiles, metal working lubricants, gear lubricants, long life engine oils etc.

Fatty Acid Methyl Esters

Fatty Acid Methyl Esters (FAME) are derived from the reaction of fatty acids with methanol or direct methanolysis of triglycerides. The advantages of methyl esters are that they can easily be transported, they can be distilled since they have lower boiling points, and they have lower corrosive activity than free fatty acids. FAME serve as intermediates in the manufacture of alkanolamides, sucrose esters and methyl ester sulfonates. Some of these compounds perform well as detergents which are cheaper and bio-degradable. In recent years FAME have acquired importance as biodiesel. These are mostly prepared from non edible oils to sustitute the declining stocks of the non renewable fossil fuels, namely, petroleum products. In recent years, soybean and rapeseed oils have been used to manufacture bio-diesel. Jatropha and karanj are also being explored for this purpose.

Fatty acid methyl esters do not have many direct applications. They are used as intermediates to produce a number of olecochemicals. Fatty alcohols are the major derivatives of fatty acid methyl esters.

Fatty Alcohols and Derivatives

Hydrogenation under high pressure of fatty acids yield fatty alcohols. More than 60% of fatty alcohols are derived from petro-chemicals while the rest are from fatty acids. Fatty alcohols are environment friendly and are utilized as plasticizers, lubricants, textile auxiliaries and cosmetics. Detergents like alcoholic ethoxylates, fatty alcohol ether sulfates and alkyl sulfates are made from fatty alcohols. These detergents are easily biodegradable and are efficient at low temperatures, have good water solubility and better hard water performance. Alkyl polyglycosides (derived from reaction with starch etc.) have excellent lather properties and are used in hair shampoos and food technology.

Fatty Amines and Derivatives

Fatty amines are the most important nitrogen containing oleochemicals. Fatty amines are prepared by converting the fatty acids to their nitrites and their hydrogenation to the amines. They can also be prepared by special hydrogenation in the presence of NH_3 of FAME. Fatty amines can be produced from either fatty acids, methyl esters, fatty alcohols or olefins. The most common process is from fatty acids via the nitriles. Processing of fatty alcohols to amines is a less favoured route.

Different types of amines, namely, primary, secondary, tertiary and quaternary amines are possible. These compounds are useful as bacteriostats, textile and paper softeners, laundry detergents, emulsifiers, organophilic clays, viscosifiers in paints and varnishes, additives to lubricants, corrosion inhibitors etc. Oleoamines have manifold uses, in petroleum exploration industry, as well as drilling auxiliaries, and as anticorrosion agents. Tertiary and quaternary ammonium compounds function as flotation agents in metallurgy. Fatty amine oxides (eg. lauramine oxides) are shampoo ingredients, which clean the hair and give body and sheen and to make the hair more manageable.

The term 'fatty amine' is used to refer not only to different types of amines but also their derivatives, because they contain a nitrogen atom in their molecules. These derivatives include imidazole, quaternary amines, ethoxylated amines, amine oxides, betains and others. The end-uses of fatty amines and their major derivatives are quite varied. Amines and their derivatives can be used as intermediates in chemical processes. The major appplications are in laundry and household products.

Fatty Amides

Fatty amides are obtained through reaction of fatty acids with anhydrous NH_3. Primary amides ($RCONH_2$) like erucamides are useful as slip or anti-blocking agents in the production of polyolefin films and as internal and external lubrication additives. Monosubstituted amides (R'CONHR) or secondary amides are useful internal lubricants. They are used to reduce cohesive forces of polymer chain and improve flow of polymers. Alkanolamides are surface active compounds and are incorporated into shampoos and light duty detergents. Sulfosuccinate monoesters of fatty amides are useful as hair shampoos, carpet cleaners and rust prevention agents.

General Applications of Vegetable Oils

The main products that are derived from oils an fats and their utilization in various walks of our life are listed hereunder :

a) *Soaps:* Saponified fatty acids have been the simplest surface active agents since ancient times. Various proportions of C_{12}-C_{14} and C_{16}-C_{18} fatty acids are used to obtain a soap with good lather, water solubility and detergency. Though stearin from palm oil is available at reasonable rates, it does not produce white soap and hence is utilized in low quality laundry soaps.

b) *Cosmetics and pharmaceuticals:* Oils like coconut, seasame, groundnut and castor oils have medicinal uses. They are also utilized in making specialty soaps like face and shaving creams and hair lotions. Vegetable oils function as emollients in creams and lotions. They serve as carriers of vitamins. Lecithins are excellent emulsifiers.

c) *Plastics and polymers:* Esterification of dibasic acids with polyhydroxy alcohols produces alkyd resins which dissolve in common solvents and represent tough and flexible polymers suitable for protective coatings. Urethane resins are made by esterifying di-isocyanates with dibasic acids. Epoxy resins are made from poly-epoxide and diphenolic compounds.

d) *Surface coatings and inks:* Drying and semi-drying oils like linseed and soybean oils as well as heat polymerized and bodied oils are used in the manufacture of paints, enamels, varnishes and other protective and decorative coatings. Vegetable oils or fatty acids sometimes function as plasticizers rather than coatings. They are useful as educts for alkyd resins and reactive diluents.

Recently vegetable oil based printing inks have gained more importance. Because of different printing types and surface properties of paper, plastic films or metallic foils, a great diversity of inks is required to fill the needs of the printing trade. Inks used in offset or letter printing are pastes consisting of pigments, pigment carrier, or binding agent (alkyd resins). Heat bodied or blown soybean or linseed oils with varying degrees of viscosity are substituted for petro-products in the manufacture of inks.

Table 156: Major applications of oleochemicals.

Oleochemicals	Applications
Fatty Acids & Derivatives	Plasticizers, lubricants, surfactants, washing & cleaning agents, soaps, Alkyd resins, dyes, textiles, leather, paper, rubber, paints & varnishes Food additive, surface coatings, printing inks, oilfield chemicals.
Fatty Alcohols & Derivatives	Washing & cleaning agents, surfactants, cosmetics, textiles, leather, paper, mineral oil additives.
Fatty Acid Methyl Esters	Cosmetics, surfactants, washing & cleaning agents, Bio-diesel
Fatty Amines & Derivatives	Fabric conditioners, mining, biocides, road making, textiles, fibres, mineral oil additives.
Glycerol & Derivatives	Cosmetics, toothpastes, pharmaceuticals, foodstuffs, lacquers, plastics, synthetic resins, tobacco, explosives, cellulose processing, dyes.
Drying Oils	Lacquers, varnishes, paints, Linoleum
Neutral oils	Soaps, cosmetics, lubricating oils.

e) *Lubricants and hydraulic oils:* The basic function of a lubricant is to prevent moving surfaces from coming into direct contact with each other. Vegetable oils exhibit better adhesion properties than mineral oils. Undesirable frictional wear and temperature rise are also lower with plant oils. They have higher viscosity indices. However they are less stable than mineral oils. Sometimes hydrolysis products like fatty acids may cause corrosion.

Rapeseed and castor oils are widely used as lubrication oils. The popular 'castrol' is a special blend of lubricating oil containing castor oil. Some of the applications of vegetable oils include heavy duty lubricants, hydraulic oils and automobile brake fluids. At present the main disadvantage of vegetable oils is their higher cost.

f) *Oiled cloth, linoleum and rubber additive:* Oiled cloth is manufactured by impregnating cotton cloth with drying oils, paints, filler material and varnish or wax. The popular floor covering linoleum is prepared from drying oils, filler compounds (rosin, alkyds), pigments and ground cork or wood. Factice, the vulcanized oil is a rubber substitute. Along with rubber, it improves the strength of rubber.

g) *Metal treatment:* Core oils are used as binders for sand to make a mold in which castings are made. The cores are baked at 200-230^{0}C, at which temperature, the oils polymerize and get hardened. Oils also find application in quenching of metals, tinning, and rust proofing of metals etc.

h) *Textile and leather industry:* Surface active compounds prepared by treatment of oils with sulfuric acid have been very important in imparting desirable softness or stiffness to fabrics. Turkey red oil is an important oil utilized in textile dyeing. Vegetable oils are used in fat liquoring of leather after chrome tanning treatment.

References

Abdul-Baki, M.S., El-Samahi, S.K., Selcha, H. and Morad, M.M.1980. Cake Mix Supplemenation with Soybean, Sweet Potato or Peanut Flours, Baker's Digest. 54: 28.

Achaya, K.T. 1972. Chemical Derivatives of Castor Oil. Journal of the American Oil Chemists' Society. 48:758.

Achaya, K.T., Craig, T. and Youngs, S. 1964. The Component Fatty Acids and Glycerides of Castor Oil. Journal of the American Oil Chemists' Society. 41-54.

Andrew, O., Odegaard, B.A. and Mark, A. Pereira. 2006. Trans Fatty Acids and *Bsykub Resustabce* and Type 2 Diabetes. Nature. 64(8): 364-372.

Ascherio, A., Katan, M.B., Zock, P.L., Stampfer, M.J. and Willet, W.C. 1999. Trans Fatty Acids and Coronary Heart Disease. North England Medicine. 340: 1994-1998.

Basha, S.M. and Young, C.T. 1985.Changes in Polypeptide Composition of Peanut (*ArachisHypogaea L.*) Seed During Oil Roasting. Journal of Agricultural and Food Chemistry. 33: 350.

Belanger, R., Devoe, D. and Reynolds, T. 1966. Coating Compositions Comprising Epoxy Resin. U.S. Patent. 3294, p.721.

Beuchat, L.R. and Worthington, R.E. 1974. Changes in the Lipid Content of Fermented Peanuts, Journal of Agricultural and Food Chemistry. 22: 509.

Beuchat, L.R., Young, C.T. and Cherry, J.P. 1975. Electrophoretic Patterns and Free AminoAcid Composition of Peanut Meal Fermented with Fungi. Canadian Institute of Food Science and Technology Journal. 8: 40.

Beuchat, L.R. and Nail, B.J, 1978. Fermentation of Peanut Milk *with Lactobacillus Bulgaricus* and *Lactobacillus Acidophilus.* Journal of Food Science. 43: 1109.

Blacher, F.M., Elstner, F., Stein, W. and Mukherjee, K.D. 1983. Rapeseed Protein Isolates: Effect of Processing on Yield and Composition of Protein. Journal of Agricultural and Food Chemistry. 31:358.

Bray, G., Alovejoy, J.C. and Smith, S.R. 2002. The Influence of Different Fats and Fatty Acids on Obesity, Insulin Resistance and Inflammation. Journal of Nutrition, 132: 2488-2491.

British Nutrition Foundation. 1992. Unsaturated Fatty Acids, Nutritional and Physiological Significance. *Report of BNF Task force*, London. 6:208.

British Nutrition foundation (BNF). 1993. Nutrition Aspects of Fish. British Nutrition Foundation Briefing. pp,10.

Chen, S., Wan, P.J., Lusas, E.W. and Rhee, K.C. 1979. Utilization of Peanut Protein and Oil in Cheese Analogues. Food Technology. 33: 88.

Cherry, J. P. 1977. Potential Sources of Peanut Seed Proteins and Oil in the Genus *Arachis*, Journal of Agricultural and Food Chemistry. 25: 186.

Chopra, G. Man, S.K. Kawaira, B.L., Wahal, C.K. and Bajaj, S. 1982. Effects of Supplementation of Sesame Seed on the Protein Quality of *Dosa*. Journal of Research, Punjab Agricultural University. 19:256-62.

Cobb, W.Y. and Swaisgood, H.E. 1971. Roasted Peanut Flavour and its Relation to Growth Environment. Journal of Food Science. 36: 538.

Cross, H.R. and Nicholas, J.E. 1979. Palatability of Groundnut Beef Patties Containing Peanut Meal, Textured Soy Protein and Mechanically Processed Beef, Peanut Science. 6:115.

Dench, J.E., Rivas, N. and Caygill, J.C. 1981. Selected Functional Properties of Sesame Flour and Two Protein Isolates. Journal of the Science of Food and Agriculture. 32:557-64.

Devdas, R.P., Chandrasekhara, U., Vasanthamani, G. and Gayathri, V. 1977. Evaluation of a Mixture Based on Sunflower Meal, Bengal Gram Flour and Sesame on School Children. Indian Journal of Nutrition and Dietetics. 14:291-93.

Elmquist, L.F. 1965. U.S. Patent. 3:175.909.

Escueta, E.E., Bourne, M.C. and Hood, L.F. 1985. Effect of Coconut Cream Addition to Soy Milk on the Composition, Texture, and Sensory Properties of *Tofu*. Journal of Food Science. 50:887-90.

FAO/WHO, 1977. Dietary Fats and Oils in Human Nutrition; FAO Food and Nutrition. Rome, FAO. pp, 3.

FAO, 1984. Coconut: Tree of Life. Rome: Food and Agriculture Organization of the United Nations. FAO, Rome.

FAO, 1994. Fats and Oils in Human Nutrition. Report of a Joint FAO/WHO Consultation. Rome, FAO of United Nation. pp, 57.

Fardiaz, D. and Markakis, P. 1981. Degradation of Phytic Acid in *Oncom* (Fermented Peanut Press Cake), Journal of Food Science. 45: 523.

Farrell, K.T. 1985. Spices, Condiments and Seasonings, Westport, CT: AVI. Publishers.

Forbes, R. 1940. Catayltic Methods for Increasing the Unsaturation of Long Chain Fatty Compounds. Industrial Engineering Chemistry. 32 (4): 555.

Fru, K. 1969. Union Carbide Corporation. Kansas City, M.O. Castor Oil Modified Acrylic Coating and Composition. US. Patent, 3454, 509.

Gohl, S.H., Hew, N.F., Yadav, M. and Aorhanom, A.W. 1994. Inhibition of Tumour Promotion by Various Palm Oil Tocotrienols. International Journal of Cancer. 57:529-31.

Goodban, A.E. 1967. First Research Conference on the Utilization of Safflower, 25-26 May, 1967, Albany, California. U.S.A.

Gulati, T., Chopra, A.K. and Bhat, C.M. 1979. Effects of Supplementation of Sesame and Skim Milk Powder on the Nutritional Quality of *Vadas*. Journal of Research. Punjab Agricultural University. 16:349-52.

Hagenmaier, R.D. 1979. Aqueous Processing of Coconuts. Paper Presented at the Fifth Session, FAO Technical Working Party, Coconut Production, Protection and Processing, Manila.

Hagenmaier, R.D. 1983. Dried Coconut Milk and Other New Foods from Wet Processing. Community. APCC/GS/45/83:36-43.

Hagenmaier, R.D., Mattil, K.F. and Cater, C.M.1974. Dehydrated Coconut Skim Milk as a Food Product: Composition and Functionality. Journal of Food Science. 39:196-99.

Hete, G.L., Roy, D. and Netke, S.P. 1975. Further Studies on Wheat Bran and Groundnut Cake Diet for Chicks, Indian Journal of Animal Science. 45; 377.

Hirose, N., Inode, T. and Nishihara Ketal. 1991. Inhibition of Cholesterol Absorption and Synthesis in Rats by Sesamin . Journal of Lipid research, 32:629-638.

Hoojjat, P. 1982. Protein Quality and Functionality of Navy Bean and Sesame Flour in Baked Products. Dissertation Abstracts International. B43:187-88.

Hooper, L., Summer bell, C.D. and Higgins, J.P.T. 2001. Dietary Fat Intake and Prevention of Cardiovascular Disease: Systematic Review. British Medical Journal. 322: 757-63.

Hu, F.B., Manson, J.E., and Willet, W.C. 2001. Types of Dietary Fat and Risk of Coronary Heart Disease: A Critical Review. Journal of American Coll. Nutrition. 20(5):1913.

Irwin, T. 1990. Technical Report on the Medicinal and Toxic Studies of Castor Oil. F344/ Rats and B6 C3 F1 Mice (Feed Studies). National Toxicology Program, North Carolina. U.S.A.

Jon Kabara, 2005. Pharmacological Effects of Coconut Oil Vs Monoglycerdies. INFORM. 16(6):386-387.

Jonson, L.A., Suleiman, T. M. and Lusas, E.W. 1979. Sesame Protein: A Review and Prospects. Journal of the American Oil Chemists' Society. 56:463-68.

Kabirullah, M., Khan, M.A. and Faruque, O. 1977. Studies on the Nutrive Value and Properties of Oilseed Meals, Bangladesh Journal of Science and Industrial Research. 12: 192.

Kappas, G.A. and Kneeland, J.A. 1966. U.S. Patent 3,271,160, 6 September 1966.

Katan, M.B. Zock, P.L. and Mensink, R.P. 1995. Trans Fatty Acids and their Effects on Lipoproteins in Humans. *Annual Review of Nutrition,* 15:473-93.

Khalil, J.K. and Chughtai, M.I.D. 1984. Nutritional Evaluation of Wheat and Maize Breads Supplemented with a Mixture of Peanut-Chickpea Flour. Quality Planta. Plant Foods in Human Nutrition. 34: 285.

Khan, M.N., Wan, P., Rooney, L.W. and Lusas, E.W. 1980. Sunflower Flour: A Potential Bread Ingredient. Cereal Foods World. 25:402-4.

Khosla, P. and Hayes, K.C. 1994. Cholesterolaemic Effects of the Saturated Fatty Acids of Palm Oil. Food and Nutritonal Bulletin. 15: 119-125.

Kirk, P. 1923. Chemical Engineering in the Production of Coated Fabrics. Chemical Engineer. 28(23): 1107.

Kohler, G.O. 1969. Third Safflower Research Conference, 7-8 May 1969, Davis, California.

Kritchevsky, D. 1995. Influence of Triglyceride Structure on Lipid Metabolism. Malaysian Oil Science and Technology. 4(1): 23-7.

Lawhon, J.T., Cater, C.M., and Matil, K.F. 1972. A Comparative Study of the Whipping Potential of an Extract from Several Oilseed Flours. Cereal Science Today. 17:240-94.

Lawhon, J.T., Cater, C.M. and Matil, K.F. 1975. Sensory, Analytical Evaluation of Cake, Doughnuts Fortified with Protein from Oilseed Flours. Food Products Development. 9: 110-114.

Lawhon, J.T., Golightly, N.H. and Lusas, E.W. 1980.Utilization of Membrane Produced Oilseed Isolates in Soft-Serve Frozen Desserts. Journal of the American Oil Chemists' Society. 57: 302.

Lawhon, J.T., Rhee, K.C. and Lusas, E.W. 1981. Production of Oil and Protein Food Products from Raw Peanuts by Aqueous Extraction and Ultra filtration. Journal of Food Science. 46: 391.

Lorenz, K. 1978. Non-Oilseed Sunflower Products in Bakery Foods. Bakers Digest. 52: 28-30.

Lovejoy, J.C. 2001. Dietary Fatty Acids and Insulin Resistance. Current Atherosclerosis Reporter. 1: 2115-220.

Mason, M.E., Newell, J.A., Johnson, B.R., Koehler, P.E. and Waller, G.R.1977. Nonvolatile Flavor Components of Peanuts, Journal of Agricultural and Food Chemistry. 17: 728,

McOsker, D.E. 1962. Limiting Amino Acid Sequence in Raw and Roasted Peanut Protein, Journal of Agricultural and Food Chemistry. 10: 453.

Millette, R.A. 1974. Seeds from Sunflower. North Dakota State University, Fargo, Circular HE-120, p.3.

Monera, O.O. and Rosario, E.J. Del. 1982. Physico-Chemical Evaluation of the Natural Stability of Coconut Milk Emulsion. Annals of Tropical Research. 4:47-54.

Mori, A. and Beilin, L.J. 2001. Long Chain Omega 3 Fatty Acids, Blood Lipids and Cardiovascular Risk Reduction. Current Opinions in Lipidology. 12:1-7.

Nagaraj, G. 2002. Value Addition for Improving Utility and Profitability of Sesame and Niger. In Integrated Crop Management of Sesame and Niger. Eds. S.S. Duhoon, A.K. Tripathi and H.K. Jharia, P.C. Unit(S&N), JNKVV, Jabalpur, India. pp 26-31.

Nagaraj, G., Srinivas, C.V.S. and Praveena, B. 2000. Nutritional Quality of Sunflower Genotypes and Utilization of Deoiled Cake. (Abstracts), National Seminar on Oilseeds and Oils – Research and Development Needs in the Millennium, Feb. 2-4, 2000, D.O.R., Hyderabad-30, India. pp. 336-337.

Nakabayashi. A., Kitagawa, Y., Suwa, Y., Akinoto, K., Asami, S., Shimizu, S., Hirose, N., Sugano, M. and Yamada, H. 1995. α-Tocopherol Enhances the Hypocholesterolemic Action of Sesamin in Rats. International Journal of Nutritional Research, 65:162-168.

Nathanael, W.R.N. 1955. Toddy Yields from Coconut Palms in Ceylon, Ceylon Coconut Quarterly. 6:8-16.

Naughton, F.C. 1974. Production, Chemistry and Commercial Applications of Various Chemicals from Castor Oil. Journal of the American Oil Chemists' Society. 51: 65.

Newell, J. A., Mason, M.E. and Matlock, R.S. 1967. Precursors of Typical and Atypical Roasted Peanut Flavor, Journal of Agricultural and Food Chemistry.15: 767.

Panico, S. and Lannuzzi, 2004. Dietary Fat Composition and the Metabolic Syndrome. European Journal of Lipid Science and Technology, 106: 61-67.

Paredes-Lopez, O. and Ordorica-Falomir, C. 1987. Production of Safflower Protein Isolates: Composition, Yield and Protein Quality. Journal of the Science of Food and Agriculture. 37:1097-1103.

Parpia, H.A.B. 1991. Oilseed Proteins for Human Consumption. In. Proceedings of the National Seminar on Strategies for Making India Self Reliant in Vegetable Oils ISOR, D.O.R., Hyderabad, India. pp.550-557.

Petithory, D.L. 2001. α- Linolenic Acid and Cardiovascular Disease. Journal of Nutritional Health and Ageing. 5:79-183.

Prakash, V. and Nandi, P.K. 1978. Isolation and Characterization of α- Globulin of Sesame Seeds. Journal of Agricultural and Food Chemistry. 26:320-23.

Praveena, B. Srinivas, C.V.S. and Nagaraj, G. 1999 Utilization of Sunflower Cake and it's Protein Isolate in Some Indian Snack Foods. Journal of Oilseeds Research 16(2): 303-305

Przbylski, R. and Michael Eskin, N.A. 2006. Minor Components and the Stability of Vegetable Oils. INFORM. 17(3): 187-189.

Quinn, M.R., Beuchat, L.R., Miller, J., Young, C.T. and Worthington, R.E, 1975. Fungal Fermentation of Peanut Flour, Effects on Chemical Composition and Nutritive Value, Journal of Food Science. 40: 470.

Quinn, M.R. and Beuchat, L.R. 1975. Functional Property Changes Resulting from Fungal Fermentation of Peanut Flour. Journal of Food Science. 40: 475,

Ramachar, Thirumala Rao, S.D. and Reddy, B.R. 1974. Laboratory Prepared High Protein Low Oil Groundnuts. Indian Food Packer. 28(1): 79-82.

Ranhotra, G.S., Lee, C. and Gelroth, J.A. 1980. Nutritional Characteristics of High Protein Cookies. Journal of the Agricultural and Food Chemistry. 28: 507.

Rhee, K.S. 1984. Use of glandless cotton seed and sunflower seed products in batter/ breading for fried beef patties. Journal of Food Science. 49:1224-25.

Rhee, K.C., Natrajan, K.R., Carter, C.M. and Mattil, K.F.1977. Processing Edible Peanut Protein Concentrates and Isolates to Inactivate Aflatoxins, Journal of the American Oil Chemists' Society. 54: 245A.

Riley, K.W. and Belayneh, H. 1989. Niger .In. Oil Crops of the World Eds. G. Robbbelen, R.K. Downey and A. Ashri. McGraw Hill, New York, U.S.A. pp. 394-403.

Rivas, N.R., Dench, J.E. and Caygill, J.C. 1981. Nitrogen Extractability of Sesame (*sesumum indicum* L.) seed and the Preparation of two Protein Isolates. Journal of the Science of Food and Agriculture. 32:565-71.

Robertson, J.A. 1975. Use of Sunflower seed in Food Products. Critical Reviews of Food Science and Nutrition. 6:201-40.

Robertson, J.A., Lyon, B. G., Morrison, W.H. and Miller, J.F. 1988. Sensory and Chemical Evaluation of Stored Oil-roasted High Oleic Non-oil Sunflower Kernels. Journal of the American Oil Chemists' Society. 55:266-69.

Rong, N., Ausman, L.N. and Nicolosi, R.J. 1997. Oryzanol decreases Cholesterol Absorption and Aortic Fatty Streaks in Hamsters. Lipids. 32:303-309.

Roony, L.W., Gustafason, C.B., Clark, S.P. and Cater, C.M. 1972. Comparison of the Baking Properties of several Oilseed Flours. Journal of Food Science. 37:14-18.

Rossi, M. 1988. Textured Sunflower Protein for use as a Meat Extender. *Lebensm. Weiss. U-Technol.* 21:267-70.

Saha, G.1977. First limiting Amino acid Deficiency in Ration containing 33% Groundnut cake for Chicks. World Poultry Science Journal. 33: 100.

Salazar-Zazueta, A.J. and Price, R.L. 1989. Solubility and Electrophoretic Properties of Processed Safflower Seed (*Carthamus tinctorius* L.) Proteins. Journal of Agricultural and Food Chemistry. 37:308-12.

Salmeron, J., Hu, B.F. and Manson, J.E. 2001. Dietary Fat intake and Risk of type 2 Diabetes in Women. American Journal of Clinical Nutrition, 73:1019-26.

Salunkhe, D.K., Chavan J.K., Adsule, R.N. and Kadam, S.S. 1991. World Oilseeds-Chemistry, Technology and Utilization. Van Nostrand, Reinhold, New York. P 85.

Samson, A.S., Khaund, R.N., Cater, C.M., and Mattil, K.F. 1971. Extractability of Coconut Proteins. Journal of Food Science. 36:725-28.

Sastry, M.C.S., Subramanian, N. and Rajagopalan, R. 1969. Studies on the Wet Dehulling of Sesame Seeds to Obtain Superior Grade Protein Concentrate. Journal of the American Oil Chemists' Society. 46:592-96.

Sawaya, W.N., Ayaz, M., Khalil, J.K. and Al-Shalhat, A.F. 1985. Chemical Composition and Nutritional Quality of *Tehineh* (Sesame Butter). Food Chemistry. 18:35-45.

Seegler, C.J.P. 1983. Oil plants in Ethiopia. Their Taxonomy and Agricultural Significance. Center for Agricultural Publication and Documentation, PUDOC, Wageningen.

Sen, C.K., Khanna, S. and Roy, S. 2006. Tocotrienols: Vitamin E: Beyond Tocopherols. Life Sciences, 78:2088-2098.

Shahidi, F. and Shukla, V.K.S. 1996. Non-triacylgycerol Constituents of Fats and Oils. INFORM. 7:1227-1232.

Sharma, Y.K., Urs, M.K. and Murthy, K.S. 1983. Changes Occurring during Roasting of Groundnut Seeds (*Arachis hypogaea*), 3rd Indian Convention of Food Scientists and Technologists, Central Food Technological Research Institute, Mysore, India, June 2 to 4, (Abstract, 83).

Sies, H. and Stahl, W. 1995. Beta Carotene and other Carotenoids as Antioxidants. American Journal of Clinical Nutrition. 35:1315S-1321S.

Sirko, V.N. and Bukhantsov, V.A. 1973. Use of Sunflower Oil Proteins in Bread Baking *Izv. Vyssh. Uchebn. Zaved. Pishch. Tekhnol.* 129-30.

Som. M.N.M. 1980. Processing of Canned Coconut Milk and Coconut Butter. In. Proceedings of the International Conference on Cocoa and Coconut, Kuala Lumpur, Malaysia.

Sosulaki, F. 1979. Food uses of Sunflower Proteins. Journal of the American Oil Chemists' Society. 56:438-40.

Sparado, J., Sparado, I. and Pearce, H.1981. Direct Extraction Process for the Production of a White Defatted Food Grade Peanut Butter, U.S. Patent 187, 382, March 13, 1981.

Subramanian, A.N. 2003. Export of Oilseeds-Potential, Costraints and Remedial Measures. In. Souvenir of the National Seminar on Stress Management in Oilseeds for Attaining Self Reliance in Vegetable Oils Eds. Harvir Singh and D.M. Hegde, Indian Society of Oilseeds Research, Hyderabad, India. pp. 128-138.

Subramanian, N., Ramamani, S., Sastry M.C.S. and Shreedhara, N. 1981. Quality Characteristics of Groundnut Protein Isolate as Influenced by Processing Conditions, II. Indian Convention of Food Scientists and Technologists, Central Food Technological Research Institute, Mysore, India, February, 19 - 20, (Abstract. B3).

Surendranath, M.R., Gangi Reddy, N.C., Azeemoddin, D. and Thirumala Rao, S.D. 1984. Preparation of Low Fat High Protein Sesame Seed. Journal of Food Science and Technologists' of India. 21:425-426.

Swaminathan, M. 1987. Food Science, Chemistry and Experimental Foods, Bangalore Printing and Publishing Co. Ltd., Bangalore. pp. 161-67.

Swaminathan, M. and Parpia, H.A.B. 1967. Milk substitutes based on oilseeds and nuts, World Rev. Nutritional Dietary. 8: 114,

Talley, L.J., Brummet, B.J. and Burns, E.C. 1972. Sunflower Food Proudcts. Texas Agricultural Experiment Station, U.S.A.1026, p.10

Thieme, J.G. 1968. Coconut Oil Processing. FAO Agricultural Development Paper No. 80. Food and Agriculture Organization. Rome.

Thompson, A.B.R. 1989. Intestinal aspects of lipids absorption, Nutrition Today. 24(4): 16-20.

Toma, R.B., Tebekhia, M.M. and Williams, J.D. 1979. Phytate and Oxalate Contents in Sesame Seed (*Sesamum indicum* L.). Nutrition Reports International. 20:25-31.

Van Veen, A.G., Graham, D.C.W. and Steinkraus, K.H. 1968. Fermented Peanut Press Cake. Cereal Science Today, 13: 96.

Van Veen, A.G. and Steinkraus, K.H. 1970. Nutritive Value and Wholesomeness of Fermented Foods. Journal of Agricultural and Food Chemistry. 18: 576,

Weiss, E.A., 1983. Oilseed Crops. Longmans. NewYork. pp. 597-639.

Woodroof, J.G. 1945.Why moisture in peanut should be controlled/ Food Industrialist. 17: 1302.

World Health Organization (WHO). 2003. Diet, Nutrition and Prevention of Chronic Diseases. Report of a joint WHO/FAO Expert consultation Technical Report Series No. 916. WHO, Geneva, Switzerland.

Worthington, R.E. and Beuchat, L.R.1974. Alpha galactosidase Activity of Fungi on Intestinal Gas-forming Peanut Oligosaccharides, Journal of Agricultural and Food Chemistry. 22: 1063.

Yamashita, K., Lizuka, Y., Imai, T. and Namiki, M. 1995. Sesame Seed and It's Lignans Produce Marked Enhancement on Vitamin E Activity in Rats Fed a Low Tocopherol Diet. Lipids. 30: 1019-1028.

Zock, P.L. and Mensink, R.P. 1996. Dietary Trans Fatty Acids and Serum Lipoproteins in Humans. Current Opinions in Lipidology. 7:34-37.

Chapter 6

Antinutrients

Antinutrients

Safety is the most important aspect among all the quality attributes of food. Food can be considered safe, only if it does not contain any factor which is hazardous to human or even animal health. The various factors which affect the safety of the foods may be broadly classified into three categories. The first category includes the toxic factors occurring naturally in foods. The second category comprises a group of incidental contaminants which find their way at various stages of processing, manufacture, preservation and storage of foods and the third category includes harmful intentional additives or adulterants.

Various categories of anti-nutritional factors are present in the traditional and non-traditional oilseeds, oils and oilcakes. Some of them are heat-labile and do not pose any harmful effects in the body, while the heat-stable factors may cause deleterious effects in human beings and animals. They belong to different groups of compounds. It is necessary to know their behaviour, significance and toxicities. Their avoidance or elimination is important to protect human and animal health.

Antinutritional Factors

An unusually large number of thermostable and heat labile substances, capable of eliciting diverse nutritional, biological and physiological responses in humans and animals are present in oilseeds. Most commonly present toxicants in oilseeds are trypsin inhibitors, haemagglutinnins, goitrogens, saponins, phenolic compounds, oxalates, phytates, cyanogens, allergenic factors, flatus factors, sterols and triterpene alcohols, off-flavours and cyanogenic glycosides.

Antinutritional factors are broadly catagorised into three types namely, Type A- substances which interfere with the digestion of proteins or the absorption and utilization of amino acids, Type B- substances interfering with the absorption or metabolic utilization of mineral elements and Type C- substances which inactivate or destory vitamins or otherwise increase their metabolic requirements in the organisms.

Examples of type A antinutritional factors are protease inhibitors, heamgglutinins and saponins. The type B antinutrients are phytic acid, gossypol and glucosinolates. Fanally the type C group includes anti-vitamins cyanogens, favic agents, estrogenic factors, anti-enzymes and flatulence flactors

If the antinutritional factors are not minimised or eliminated, they may create certain physiological syndromes/disorders in humans such as pancreatic hypertrophy, enlargement of thyroid gland, inhibit growth, inhibit proteolysis, reduce protein digestibility, increase the sulfur containing amino acids requirements and reduce metabolisable energy. The importance, specific role and detoxification/reduction of some antinutrients in oilseeds are discussed below.

Trypsin inhibitors

Trypsin inhibitors (TI) are widely distributed protease inhibitors and exert their action by forming complexes with trypsin and chymotrypsin. These are resistant to proteolysis. The presence of TI reduces the nutritive value of oilseeds. Raw seeds if consumed are not digested and cause growth depression and loss of weight. Moist heat treatments decrease their toxicity. The inhibitors suppress the release of amino acids and thus inhibit the normal growth of animals.

Legumes are the major sources of protese inhibitors. Trypsin inhibitors are present in soybean and peanuts. There are two types of TI. They are Kunitz inhibitors with molecular weight of around 20,000 Daltons and Bowmark inhibitors with molecular weight of around 800 Daltons. Enzyme activity and assays do not distinguish between the two inhibitors.

The antinutritional factors can easily be inactivated through processing (Liener, 1979). At 100°C, about 15 min of steaming is required to inactivate the trypsin inhibitor and to achieve maximum PER. It can be readily destroyed by atmospheric steaming for only 20 min provided the beans are tempered to about 25% moisture content before steaming. When moisture of whole beans is raised to 60% or more by over night soaking, boiling for only 5 minutes is sufficient to inactivate the inhibitor. 5-10% of the TI is retained even after the heat treatment.

Haemagglutinins

Hemagglutinnins are proteinacious in nature and are also called phytohaemagglutinnins or lectins. They are glycoproteins containing mannosol (5.5%) and glucosamine (1%). They have the ability to clump or agglutinate the RBC in a manner similar to antibodies. Some haemagglutinnins are heat labile. In experimental animals, diets containing raw beans cause growth depression, loss of weight and death. In human beings insufficiently cooked legumes combine with the cells lining the intestinal wall similar to RBC agglutination, causing an impaired absorption of aminoacids. They also cause anemia.

Hemagglutinnins are present in castor, soybeans, groundnuts, kidney beans, blackbeans, and jackbeans. At 5% of the diet rats die in two weeks. They can induce mitosis in lymphocytes. Ricin in castor beans is the most toxic of all lectins known. Its oral toxic dose in man is 150-200mg. The intravenous dose is 20 mg. More details of this toxin are discussed seprately in this chapter. The lectins in soybeans appear to be less toxic orally. One percent in the diet retards growth in rats to approximately 75% of normal weight. But this amount is not lethal as no death is reported.

The toxic action of heamgglutinin may be related to their ability to bind to specific cell reception. With raw soybeans in the diet there is a significantly diminshed absorption of amino acid and thyroxine, the impairment of absorption of thyroxine may explain the goiterogenic effect of soybean, due to the interference in absorption of iodine.

The detoxification of lectins is usually achieved by traditional cooking methods. Dry heat is less effective in removing lectins from foods.

Germination is known to decrease the lectin activity in varieties of peas and species of beans. In fact soybean can lose as much as 92% of its activity even during first day of germination. These factors can possibly be eliminated by heat treatment.

Goitrogens

Goitrogens are natural plant products which cause hypothyroidism with an enlargement of thyroid gland when consumed by man and animals. All natural thioglucosides (non-goitrogen) are associated with enzymes (thioglucosidase) that hydrolyse them to give glucose and bisulfate when wet seed material is crushed. The organic aglycon portion of the molecule undergoes intermolecular rearrangement following the enzyme hydrolysis to give isothiocynate which is a goitrogen. Sinigrin (allyl thioglucoside), a nongoitrogenic compound has been identified in rapeseed-mustard.

The thyroid inhibiting effect of goitrin is due to the inhibition of organic binding of iodine in thyroid gland to inhibit the iodine absorption. This can be easily removed or destroyed during processing. Soybeans and peanuts produce goitrogenic effects in animals. The goitrogenic principle identified in groundnut is a phenolic glycoside which resides in the skin. It is suggested that the phenolic metabolites formed from this glycoside preferentially get iodinated and deprive the thyriod of available iodine. Thus, the goitrogenic effect of groundnut can effetively be countered by iodine supplementation.

Glucosinolates are naturally occurring compounds such as goiterin, thiocyanates and nitriles etc. responsible for pungent flavor in condiments, horseradish, mustard, turnip, cabbage and other vegetables. Among oilseeds, rapeseed- mustard group of oilseeds are rich in glucosinolates.

Saponins

Saponins are glycosides present in a variety of plant materials. Not all saponins are poisonous and the poisonous ones are termed as sapotoxins. There are two types of saponins namely, steroids and triterpenoids. The toxicity is related to lowering of surface tension in body fluids. Saponin content/ effect can be lowered by dry-heat treatment.

Phenolic Compounds

Some phenolic and polyphenolic compounds present in plant materials consumed as food are highly toxic in nature. Gossypol is a phenolic compound present in the cotton seed, meal and oil. It is present in pigment glands which form approximately 2.5% of the weight of the kernel. Gossypol is highly toxic to monogastric animals whereas it is not toxic to ruminants, because of prolonged time of mastication, water contact, and prolonged time of stay in the rumen which results in binding of gossypol to protein.

Gossypol is less toxic when present in small concentrations in oil. Symptoms of toxicity are loss of appetite, weight loss, hypoprothrombinemia, diarrhea, hair discolouration, lowering of haemaglobin, RBC and serum protein. Fluid accumulates in body cavities, lungs and heart. Degenerative changes in liver and spleen, haemorrhages in liver, small intestine and stomach occur. In poultry, discolouration of yolk and decreased hatchability of eggs are seen. The gossypol in the meal/cake can be removed by treating with chemical agents like sodium hypochlorite and anthranilic acid. By isolating the protein, the meal can be made free from toxicity.

Gossypols form an insoluble chelate with many essential metals specially with Fe^{++} and also bind with amino acids in proteins specially lysine. Thus amino acids become unavailable and secondary lysine deficiency occurs. Iron binds in 1:1 molar ratio leading to iron deficency anaemia as one of the manifestation of gossypol toxicity.

Though gossypol poisoning is widely established and studied in non-ruminants, very limited studies are available on its effect on humans. Bread containing moderate amounts of cotton seed meal has been shown to produce no ill effect in humans. Similarly no harmful effect has been observed when 60g cottonseed cake containing 0.11-0.20% free gossypol has been fed daily for 4-5 months. Nevertheless, the toxic potential of gossypols is well evident in animals, so the limit of gossypol content in food has been fixed as 0.05% and 1.2% free and total gossypol respectively.

The formation of an insoluble Fe-gossypol complex in presence of Calcium-hydroxide has been the basis for reducing the free gosspol content of cotton seed meal to a very low level.

Tannins

Tannins are polyphenolic substances of high molecular weight and have the property of protein binding. There are two types of tannins, hydrolyzable and condensed. Condensed tannins are carcinogenic. Dehulling/dehusking the seed eliminates most of the tannins.

Flatus factors

It is mainly due to the presence of glacto-oligosaccharides such as raffinose, stachyose and verbacose. They are not easily digestible. The chemical composition of these sugars is mainly glucose, fructose, and galactose molecules linked together. The glucose and fructose yield sucrose. The human system has invertase to hydrolyse the sucrose. But when galactose is added to sucrose it is converted into raffinose and stachyose. A-galactosidase is required to hydrolyse these sugars and it is absent in humans. Therefore, these sugars escape into the lower intestinal tract where they are acted upon by certain microflora. Colonic bacteria free them to produce acid and gas resulting into flatulence, belching, bloated feeling and gastric discomforts. As a result, gases like carbondioxide, hydrogen etc. are released.

Flatulence associated with the consumption of beans specially soybean and groundnuts is well known. This is caused by the presence of galacto-oligosaccharides viz. raffinose, stachyose and verbascose (as high as 5% in these oilseeds). These are composed of galactose, glucose and fructose linked by alpha glycosidic bonds. These sugars are given below :

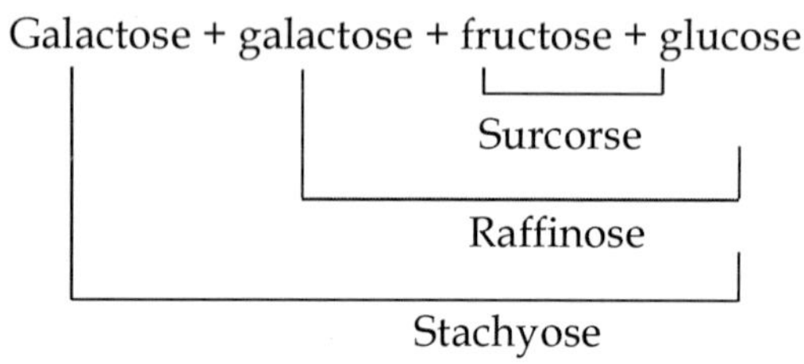

Since, flatulent oligosaccharides are resistant to moist as well as dry heat treatment, they survive all cooking treatments. However, germination and fermentation may reduce them to some extent.

Antiminerals

Substances interfering with the utilization of essential minerals are widely distributed in vegetables, fruits and grain. Although they are toxic *per-se,* the amount present in foods seldom cause acute intoxication under normal food consumption. However, they may harm the organism in a more subtle manner by interfering in absorption or metabolic utilization of essential minerals. Therefore, these substances acquire significance under suboptimal nutrition.

Oxalates: Oxalates are present in many foods of plant origin like sesame, green leafy vegetables, tea, wood apple etc. Ingestion of excessive oxalic acid by human beings can be fatal, associated with gastroenteritis, shock and convulsive symptoms. Ingestion of oxalate also interferes with calcium metabolism leading to renal damage. Sodium, potassium and calcium oxalates can cause acute poisoning. Oxaltes accumulate in the body and cause kidney stones. Soaking, washing and dehusking removes the oxalates present in seed material.

Phytates: Phytate also known as myoinositol (1,2,3,4,5,6-Hexa-cis dihydrogen phosphate) is widely distributed in plant foods including nuts, legumes, bran and germ portion. Phytate is usually associated with protein. It decreases absorption of minerals especially trace minerals like iron, zinc, magnesium, manganese, copper and calcium. Destruction of phytate during cooking depends on the nature and extent of extraction of flour, fermentation time, etc.

Phytic acid is a naturally occuring strong acid which binds to many types of bivalent and trivalent heavy metal ions, forming insoluble salts, thus making these essential trace elements/micronutrients unavailable. The phytate is widely distributed among foodstuffs, especially, cereals and legumes. Soybean and fababeans have as high as 400-500 mg phytate per 100g.

In many foodstuffs the level of phytate is reduced by phytase, an enzyme endogenous to plants. The phytase in soybeans is weak despite large amount of phytic acid. The increased activity of enzyme explains the decrease in phytate during the germination of soybeans. Application of polyacrylamide entrapped wheat phytase has shown some success in reducing phytate level in soyfoods.

Cyanogens and Cyanogenic Glycosides

There are various foods in nature that contain compounds which yield hydrogen cyanide (HCN) upon treatment with acid or appropriate hydrolytic enzymes. Legumes predominate in terms of their cyanide producing potential. HCN is released when tissues of the plant material are crushed or disrupted. The cyanogenic glycosides cause toxicity due to the production of HCN, a potent respiratory inhibitor (Liener, 1969). The site of inhibition is the enzyme cytochrome oxidase, the terminal respiratory catalyst of aerobic organisms. General cooking methods reduce the level of cynogens.

Though widely distributed in plants in form of plant cyanogenic glucoside, limabean is one of the rich source among legumes having HCN ranging from 20-30 mg/100g. Their natural form in plant is amygdalin, oburrin and linamarin. HCN is realsed in the presence of glucosidase and water. It is lost in urine as thiocynate after metabolizing in the body. However, it causes human poisoning leading to mental confusion, generalized muscle paralysis and respiratory distress. Hence, precaution must be taken to ensure that the HCN formed is thoroughly removed from food in the process of cooking and ensure that cooking water is discarded. Among oilseeds linseed contains linamarin the cyanogenic glucoside.

Allergenic Factors

Soybean allergenin is quite thermostable. Soymilk is hypoallergenic substitute for infants allergic to human and cows milk. Groundnuts also contain some allergens while castor allergens are also well known.

Unusual Components in Oils/Fats

There are certain components in oils which have some negative effects on the health of human beings and animals. Some of them are erucic acid , cycopropene fatty acids, hydroxy fatty acids, epoxy acids and long chain fatty acids.

Erucic Acid

Erucic acid is a long chain fatty acid with 22 carbons and one double bond (22:1, n-9) and is present in rapeseed/mustard in large amounts. Erucic acid causes myocarditis and other histopathological lesions due to depressed ability to oxidized substrate which results in decreased ATP synthesis. The pathological effects of rapeseed oil are more pronounced in young growing animals than old. Dietary erucic acid complexes with plasma albumin (transport protein) and due to decreased rate of β-oxidation, deposition occurs in the blood.

Cyclopropene Fatty Acid (CPFA)

Oils from plants of the order *Malvales* contain unsaturated fatty acids called cyclopropene acids, sterculic ($C_{19}H_{34}O_2$) and malvalic acids ($C_{19}H_{34}O_2$). The structure of sterculic acid is similar to that of oleic acid except in the arrangement of molecular structure. Both sterculic and malvalic acids are found in small concentration in cotton, okra and mesta seed oils. These CPFA induce delayed sexual maturity, cessation of normal egg laying, pink discolouration of yolk, pasty yolks and enlargement of liver and gall-bladder. CPFA can be destroyed by heating the oil as well as by hydrogenation. Generally frefind oils are devoid of CPFA.

Hydroxy Fatty Acids

Hydroxy fatty acids include hydroxy sterculic acid, dihydroxy stearic acid and ricinoleic acid (present in castor oil). Hydroxy fatty acids stimulate intenstinal fluid secretion and this in turn has laxative property. This property makes castor oil non-edible. Dihydroxy stearic acid present in sal seed fat is also known to interfere with synthesis of vitamin K in the intestine in addition to causing diarrhoea. Upto 2-3% of this fatty acid in oil/fat is considered to be safe.

Epoxy Acids

These are also known as oxygenated or oxide acids characterized by having oxirane ring in the fatty acid chain. Vernolic acid is a well known epoxy acid present in sal fat and soybean oil. Mesta oil also contains epoxy oleic acid and has been proved that this fatty acid survives digestion. Epoxy oleic and epoxy stearic acids are not toxic when present in low concentrations.

Undesirable Components

During solvent extraction polycyclic aromatic hydrocarbons and during hydrogenation trans fatty acids are formed which are toxic in nature. Trans fatty acids behave like SFA and induce high cholesterol level. The atherogenicity of trans fatty acids can be reduced by the addition of linoleic acid. Certain undesirable components like low molecular weight volatile fatty acids, highly isomerised and polymerized products and cyclic compounds are formed during cooking and frying.

Toxic Adulterants in Oils

High priced edible fats and oils are adulterated with a wide variety of non-edible oils like crude castor oil, argemone seed oil, used motor oil, mineral oil etc. Cheap edible oils such as linseed, cotton seed, water melon seed, rubber seed, tea seed are also used to adulterate oils. Orthotricresyl phosphate (TOP) an oily liquid used in the plastic industry has also been used to adulterate mustard and other oils. From health point of view, adulteration of edible oils and fats is a serious problem, wherein gastrointestinal and neurological damages are seen upon consumption.

Argemone Oil Toxins

High toxic principles present in the argemone oil are an alkaloid sanguinarine and a long chain fatty acid argemonic acid. Argemone seeds resemble mustard. Since argemone oil is a cheap source of oil, it is used for adulterating mustard or groundnut oils. Sanguinarine levels in the oil samples range from 0.16 to 2.2%. Other toxic factors like keto, hydroxy and epoxy acids are also present in argemone oil. Toxic manifestations of argemone oil poisoning include vomiting, diarrhoea, irregular fever, hyper pigmentation, erythema, anorexia and bilateral oedema of legs and whole body, covered under the term epidemic dropsy. Higher amount of ingestion apart from gastrointestinal disturbances, leads to death due to cardiac arrest. Argemone oil poisoning due to consumption of adulterated oil, even at lower levels is dangerous to health.

Mycotoxins in Oilseeds, Oils and Cakes

Mould growth on foods and feeds is a common problem which may cause organoleptic deterioration of products, economic losses and potential health hazards through the production of toxic mould metabolites known as mycotoxins. More details of mycotoxins are presented separately in this chapter.

Aflatoxin: Aflatoxin is a mycotoxin produced by *Aspergills flavus* in improperly harvested and stored groundnuts and other grains which have high moisture and protein levels. There are many types of aflatoxins (B_1, B_2, G_1, G_2, M_1, M_2, B_2a, G_2a etc.) with varying toxic potencies. Of all the types, aflatoxin B_1 is a strong hepatocarcinogen. Aflatoxin G_1 is more active than B_1 in inducing kidney tumours. Aflatoxin in small concentrations if continuously consumed over a period leads to diarrhoea, vomiting and finally liver damage.

General Methods of Detoxification

Both dry and wet methods like roasting, soaking, boiling and pressure cooking are generally considered good for eliminating some of the heat labile factors present in nuts and oils. The heat-stable factors present in the oilseed cakes/meals are to be processed differently by extracting and isolating the proteins. Toxin free protein isolates containing about 85 to 90% of the proteins from nuts and oilseeds such as soybean and peanut meals are available. The steps in the processing of protein isolates include solvent extraction of edible meals/cakes, extraction of protein with dilute sodium hydroxide at pH 8.0, precipitation of proteins at pH 4.5 from the extract by the addition of hydrochloric acid, filtration and washing with water and finally solubilising wet protein in water by adjusting to the pH 7.0 and spray drying.

The toxic components present in the saponifiable and unsaponifiable fractions of the expeller processed crude oils and solvent extracted oils can be removed by refining which includes alkali neutralization, bleaching with fuller's earth/activated carbon and finally deodourisation employing steam and vacuum. The alkaloids, pigments, lecithins/gums and free fatty acids present in high percentage, odouriferous volatile compounds are removed during refining. During hydrogenation process the long chain fatty acids and some of the keto and hydroxy fatty acid levels are reduced.

Correct application of heat in cooking can eliminate most toxic factors without impairment of nutritional value. Cooking also contributes to digestibility. Heat causes denaturation of the proteins responsible for trypsin inhibition, haemagglutination and the enzyme responsible for the hydrolysis of cyanogenic glycosides. The mode of application of heat is important.

Autoclaving and soaking followed by heating are effective. Another way of eliminating toxic factors is fermentation, which yields products more digestible and of higher nutritive value than the raw pulses (Shakunthala and Shadaksharaswamy, 1987). Germinated seeds are more nutritious and free from toxic constituents.

Ricin- The Toxic Protein

Introduction

In 1888, Stillmark discovered that the beans of the castor plant contain ricin, a toxic endosperm protein. He pulverized the seeds and extracted with 10% NaCl solution and he obtained a brownish yellow solid called ricin, which belongs to unformed enzymes. Stillmark's toxic studies revealed that the agglutination properties of ricin varied with respect to the blood corpuscles in different animal species. He also observed that ricin causes agglutination of erythrocytes and precipitation of serum proteins. It has been observed that the toxicity of ricin is because of its catalytic activity on eukaryotic ribosomes.

Kobert, (1891) has made ricin commercially available. He has shown that ricin can kill at a dose around 0.03mg/kg body weight by injecting ricin through intravenous route into the body of rabbit. He has compared this toxin with other existing toxins belonging to the same *euphorbiaceae* family namely abrin and has declared that ricin is a uniform substance.

Ricin has been used in the highly publicized assassination of the Bulgarian defector, George Markov. In 1978, a ricin-loaded platinum pellet has been fired into his thigh through an umbrella, in London. In recent years, there have been many cases of ricin toxicity. Paul Ehrlich, the Father of Chemotherapy, has induced immunity by feeding mice and rabbits small amounts of castor bean seeds. When specific serum proteins are introduced, they can precipitate and neutralize the toxic antigens. Also Paul Ehrlich has proved that during pregnancy, specific antitoxin activity is transferred from mother to the offspring through blood. The same protective components are transferred through milk during lactation. This pioneering work has become the foundation for the discipline of immunology.

In 1951, native ricin has been shown to have anti-tumour activity. This is the first example of any toxic protein to be used as a component of antitumour compounds called immunotoxins, or more specifically, chimeric toxins. No concerted effort has been made to design specific vaccines or to undertake prophylactic measures until the early 1990s, when ricin has been shown to be a powerful bio-warfare agent. Wide and easy availability, relative ease of extraction and purification and media popularization of ricin, all have contributed to its widespread use and abuse.

Ricin and RCA

Ricin is an extremely toxic glycoprotein produced by the plant *Ricinus communis* (castor bean). Ricin and Ricinus communis aggutinin (RCA) are the two highly toxic endosperm proteins of castor bean. These belong to the plant lectin family. Lectins are glycoproteins, which specifically bind carbohydrates and frequently agglutinate cells. Ricin is a polypeptide consisting of a 32 Kda A chain and a 34 Kda B chain, linked by a single disulfide bond. The A chain is an inhibitor of protein synthesis, while B is the carbohydrate binding chain. Ricin is one of the most toxic compounds produced in nature. It belongs to type II ribosome inactivating protein (RIP) family.

RCA is a four-chained polypeptide comprising of two A chains and two B chains, slightly less toxic than ricin and causes agglutination of RBCs in mammals. The two heterodimers in RCA are held together by non-covalent forces. Ricin is a potent cytotoxin but a weak haemagglutinin, whereas RCA is a weak cytotoxin and a powerful haemagglutinin. Endocytosis of bound ricin followed by disulfied reduction releases the A chain, which gains access to cytosol and there catalytically inactivates eucaryotic large ribosomal subunits. A single molecule of ricin A chain in the cytosol can inactivate 50,000 ribosomes and kill a eucaryote cell.

Ricin A chain specifically attacks a single highly conserved adenosine and removes the adenine base, leaving rRNA sugar-phosphate backbone intact. Removal of this single base inactivates a 60S large subunit. The A chain (ricin) then inhibits protein synthesis by enzymatically inactivating the EF2 binding portion of 60S ribosomal subunit. It has bean shown that in this immuntoxin, the B chain can be replaced by any binding moiety such as a harmone, a growth factor or an antibody, so that the binding specificity is changed, but the toxicity effect is retained.

Properties of Ricin

Ricin is a globular protein, and could be in the form of a dry powder, a mist (aerosol) or a pellet, or it can be dissolved in water or weak acid (liquid slurry). It is a stable substance which does not get affected much by hot or cold temperatures or by exposure to U.V. It can remain active in either an acid or a base or even in water. However, ricin with an isoelectric point of 5.9 is a multivalent anion and can form an insoluble trivalent polymer, in solution. It cannot be absorbed by the intact skin.

Ricin could be 99.4 per cent inactivated with 100-ppm of chlorine in 20 minutes, but not by iodine. If water is contaminated with ricin, then ricin could be successfully removed up to 99.8 per cent by reverse osmosis. Also water treatment systems employing charcoal effectively remove ricin or by treatment with 100 ppm free available chlorine in 20 minutes. But, coagulation and flocculation are ineffective in decontaminating the water polluted with ricin.

Estimation of Ricin

Due to the toxic nature of ricin, it is very important to detect ricin in the castor seed. Different methods have been devised to detect ricin from the castor seed pressed cake. A serological method for the detection of ricin has been developed This method can detect ricin when present at somewhat higher levels. Serological method is ineffective at very low levels of ricin.

Pinkerton *et al.*, (1999) have adopted a radial immunodiffusion (RID) assay to allow screening of castor lines for reduced levels of ricin and RCA concentration. The RID assay is based on incorporating a specific antibody against *Ricinus communis* lectin into a thin layer of agar. Seed extracts containing mixtures of antigens are placed into wells cut into the agar layer. As the antigens in the wells diffuse outwardly, they react with the antibody forming a precipitation halo around the well with the diameter of the halo providing an estimate of the ricin + RCA concentration. The RID assay has a minimum sensitivity of 1.0 mg of total ricin + RCA/g of seed. The genetic regulation of ricin + RCA concentration in castor is more complex (Pinkerton *et al.*, 1999).

An enzyme linked immunosorbent assay (ELISA) with a sensitivity of 0.01 mg/g is being developed to measure the levels of ricin, RCA and ricin + RCA in castor lines with very low concentrations of toxins (Poli *et al.*, 1994). This assay is more expensive to perform than RID assay, but its sensitivity and the ability to perform many assays on a single microtitre plate make it ideal for measuring the concentrations of ricin + RCA in castor lines with extremely low levels of toxins. Rabbit antisera against ricin will be coated onto the plastic wells and seed extracts added. Ricin will bind to the specific antibody on the walls of the microtitre wells and anti-ricin horse radish peroxidase conjugate will be used to quantify bound ricin + RCA. Toxin concentration will then be estimated spectrophotometrically/ colorimetrically.

Structure of Ricin

The structure of ricin and its subunits have been worked out by Funatsu, *et al.*, (1971). They have obtained ultra centrifugally and electrophoretically pure and homogenous ricin which has been designated as ricin D. Ricin is a two-polypeptide chain linked with a single disulfide bond. The A chain is a toxic enzyme while the B chain is a lectin that binds to a carbohydrate moiety (galactose). The molecular weight of the two subunits is around 30,000 daltons each (Rutenber, 1991).

Ishiguro, *et al.*, (1971) and Funatsu *et al.*, (1977) have determined the amino acid composition of ricin D and also studied the subunits of castor bean hemagglutinin, their amino acid composition and their cytoagglutinating activity. The physico-chemical properties of RCA have further been reported by Thomas, *et al.*, (1980). Villafranca and Robertus (1981) have reported the structure of the B chain of ricin using X-ray diffraction at 4A°.

The X-ray crystallographic studies of ricin at 2.8 A° by William, *et al.*, (1987). have shown that A chain of ricin is a globular protein and B chain is a lectin, which binds to two molecules of lactose. The two-polypeptide units of ricin are symmetrical having similar molecular weights i.e. 32,000 each. Milsna, *et al.*, (1993) and William, *et al.*, (1987) using X-ray crystallography at 2.3A° have confirmed that ricin A chain is a globular protein. They have proposed that there are 267 amino acid residues in RTA (ricin toxin A chain) while RTB (ricin toxin B chain) has 262 amino acid residues with an elongated dumbbell containing galactose moieties at both the ends of the B chain.

The active A chain is approximately 30 percent helical and contain 7 alpha helices. It also has about 15 percent beta structure, which consists of a free stranded beta sheet (Montford *et al.*, 1987). The B-chain is composed of 262 amino acid residues (Katzin, 1991). Srinivas (2003) has examined the stucture and the molecular weight of RCA and ricin A and B subunits. The molecular weight of RCA is between 95,000-97,000. The molecular weights of ricin and its polypeptides are 57,200,29,000 and 28,500 respectively for (A and B chains). MALDI(matrix assisted laser desorption ionisation) studies of ricin have shown a molecular mass of 57,257.32. The total helical content of ricin is higher at 53.6%. Amino acid studies revealed higher number of glycine molecules followed by alanine and glutamic acids.

Molecular Weight of Ricin

Kabat, *et al.*, (1947) and Takahashi, *et al.*, (1962a, 1962b) determined the molecular weight of ricin through electrophoretic fractionations. Both the workers reported similar values of 98,000. Later Olsnes and Pihl, (1972) Nicholas et al., (1974), Funatsu, *et al.*, (1974) and Cawley, *et al.*, (1978) confirmed using more advanced electrophoretic techniques and concluded that ricin has a molecular weight between 64,000-65,000 respectively while the molecular with of RCA protein range between 1,10,000-1,20,000. Later Roberts, *et al.*, (1985), Houston and Dooley, (1982), Ramos, *et al.*, (1998) have reported the molecular weight of ricin to be around 65,000.

Occurrence of Ricin

Ricin is produced in the endosperm tissue during the post-testa maturation of the castor seed (Roberts and Lord, 1981). Ricin, it's isoforms, storage albumins and crystalloid proteins are usually targeted to an organelle called the protein body for storage in the mature seed (Tulley & Beevers, 1976; Youle & Huang, 1976). The protein body is analogous to any other vacuolar compartment in the cell. Protein bodies are also present in the cotyledons, where again ricin and RCA, are part of the albumin fraction. Cotyledons contain much less lectins than the endosperm. Within these protein bodies, ricin accumulates to around 5 percent of total particulate protein and is usually degraded within a few days after germination of the seeds (Frigerio & Roberts, 1998).

Ricin content varies greatly with respect to location , genotype and maturity of seed etc. Osborne, *et al.*, (1905) have reported ricin at around 1.5% in the oil free meal. Ishiguro *et al.*, (1964) have reported about 4.0% of ricin in *Ricinus Sanguinensis* while Weiss, (1983) have reported 1.4% of ricin. Wei, (1973) has obtained almost similar yields of ricin. The ricin content varies (15-200 mg/ 100g cake) among genotypes, maturity levels and raceme positions (Nagaraj,1999). A genotyps, LRES-17, has 15mg while immature seeds have only 5.2 mg and the first pick or main stem seeds have 15-55 mg/ 100 g seeds.

Robertus, (1991) has shown that there are several isoforms of ricin including ricin D, ricin E, and the closely related lectin RCA. Together they account for more than 5% of the total protein present in mature *ricinus seeds*. Lord, *et al.*, (1994) point out that ricin cannot be the same every where and that there can be possible variations with respect to the seed, variety, temperature, which affect the content of ricin, its properties etc.

Isolation and Purification of Ricin

Several workers have isolated and purified ricin through different procedures in order to obtain pure ricin. A hemagglutinin from castor seed has been extracted by Mitsutaro, (1924) which also consisted of other impurities. Kabat, *et al.*, (1947), has obtained a higher toxic hemagglutinin preparation of ricin by fractional precipitation with sodium sulphate at different concentrations. This method is slightly a better one to obtain pure ricin. Kunitz and Donald, (1948) have isolated toxic crystalline protein by a method similar to the Kabat's method using salt precipitation. Later they have obtained the toxin using ultra centrigugation and electrophoretic separation technique. Their studies have shown some homogeneity in obtaining pure toxin.

George and Negi, (1958) have shown a single step scheme for isolation of ricin, ricinine and allergens from the same lot of castor bean, but their method included several steps in order to obtain the toxin. A more advanced procedure has been developed by Takahashi, *et al.*, (1962a). He has obtained castor bean hemagglutinin or RCA by fractional precipitation with ammonium sulphate followed by ion exchange chromatography on a column. This method is better and more simplified than that of the earlier ones which involve salt precipitations.

Ishiguro, *et al.*, (1964) have for the first time fractionated ricin in a crystalline form to toxic and non-toxic moieties. The toxic component does not show hemagglutination activity while the non-toxic component shows hemagglutinating activity. He has successfully separated the toxic protein from the non-toxic portion by performing hemagglutination tests. Funatsu, *et al.*, (1971) have obtained ricin by sodium chloride precipitation and again reprecipitated by using saturated ammonium sulphate solution. Nicholson and Blaustein, (1973) and Olsnes and Pihl, (1973) have separated ricin using 60% ammonium sulphate salt fractionation.

Purification of ricin has been carried out using different column chromatographic techniques. Funatsu, *et al.*, (1977) have separated ricin using affinity chromatography on a guar gum (alpha-galacto-beta mannopyranoside). Cawley, *et al.*, (1978) obtained ricinus agglutinin (RCA) which is 60 times more potent than ricin. Simmons, *et al.*, (1985) have demonstrated a single step to get ricin by affinity chromatography on a Galactose-Sepharose bed. Alderton, (1994) have recently reported the purification of ricin with immobilized triazine dyes. However, these methods are not universally applicable because ricin behaves differently depending on the material used for purification.

Mechanism of Ricin Toxicity

Ricin is classified as a cytotoxin, which inhibits protein synthesis by irreversibly inactivating eukaryotic ribosmes (Olsnes, 1973). These RIPs(ribosome inactivating proteins) are typically N-glycosylated. There are two types of RIPs which differ structurally. Type 1 RIP is a monomer without a carbohydrate moiety while Type 2 RIPs are made with two monomers one being a protein and the other a lectin (ricin). These monomers are joined by a disulfide bond to form the toxic heterodimers. Ricin is a heterodimeric type 2 RIP. The A chain of ricin inactivates ribosomes.

RTA (Ricin Toxin A) specifically targets sequences in ribosomal RNA (ribosome nucleic acid) which completely inactivates ribosomes. These ribosomes are the machinery for producing proteins in the cell. RTB(Ricin Toxin B) sticks to the outside of cells and can internalize into the cell.

Ricin as Immunotoxin

The 'ricin' has actually been used for a study of its effect on mouse tumor cell (B cell tumors). There can be a variety of these tumor cells each having its own characteristic ideotype (specific immuno-globulin or surface antigen), so that an anti-idiotypic antibody can be raised and conjugated with the toxin (ricin or ricin A, which is a chemically synthesized toxin resembling natural ricin) to give an immunotoxin. When tumor cells are incubated with an immunotoxin, protein synthesis decreases by 70-80%.

Experiments have also been performed on bone marrow cells (treated with immunotoxin) used for reinfusion into a patient treated with high doses of irradiation or chemotherapy leading to complete killing of the patients own bone marrow. Ricin has been used to target specific cells such a s cancer cells, by conjugating the RT-A subunit to antibodies or growth factors that preferentially bind the unwanted cells. Vidal, *et al.*, (1985) have used ricin as immunotoxin and successfully transplanted into bone marrow to destroy T lymphocytes, which he collected from histocompatible donors. He has found that the malignant cells or the T cells are destroyed completely without effecting the other normal cells. All these experiments have been carried out in *in vitro* conditions.

Another report by Houston, (1981) used ricin A chain for transportation in mouse leukaemia cells with positive results. Ricin A chain is being used in immunotherapy of human cancers and auto-immune disorders (Vitetta *et al.*, 1993 and Ghetie and Vitetta, 1994). Recently, Lord, *et al.*, (1994) and Jones, *et al.*, (1998) reported *in vivo* treatment of solid tumours and have also shown clinical use in cancer and AIDS treatment utilizing DNA recombinant technology.

Ricin and its Detoxification

Because of the presence of toxic constituents, castor cake is being used as organic manure (Kulkarni and Ramanamurthy, 1977). Castor cake has a good composition of proteins, carbohydrates, minerals etc. There is scope for it to be utilized for food, feed and industrial purposes when it is free from toxic constituents. The major toxic constituents are ricin, allergens, ricinine (alkaloid) and phenolic compounds (Coulson and Stevens, 1950).

Detoxification of castor meal, especially, with respect to ricin has been a long-standing major objective. Isolation of ricin, ricinine and the allergens from castor pomace has been described by Waller and Negi, (1958). They measured the nutritive value of the detoxified meal (pomace). They extracted with HCl at pH 3.8 to remove the toxins. Commercially processed pomace has 10 times lower toxins than laboratory processed pomace. They considered the detoxified meal as potential protein rich animal feed. Later several workers adopted different physico-chemical treatments in removing ricin. Corwin (1962) has felt that determination of toxicity as a means of assay is essential for categorising the pomace.

Jenkins (1963) has discussed all aspects ricin and its toxicity in "Allergenic and Toxic Component of Castor Bean Meal". Heating castor bean meal for one hour in steam at 15 psi can destroy ricin and inactivate the allergen. Cooper and Scheel, (1964) have used steam treatment to remove solvent after extraction of oil. The treated cake is no longer positive to hemagglutination test and significantly lower in the allergen content.

Ambedkar and Dole (1957), have detoxified the castor meal by washing with water, chemical treatment , steaming, cooking and boiling. They have reported that solvent extraction has no effect on the ricin. Steam cooking after pressing the re-pressed cake might have neutralized ricin. Petrosyan and Ponomorev, (1937) have detoxified the meal by boiling the oil cake for 1-2 hr. Steaming and autoclaving have also been used by several workers. Jakie, (1941) treated the castor cake with steam and has removed the excess moisture under vacuum. Another study has revealed, that autoclaving the castor meal for 15 min, at 125°C, almost completely destroys ricin with minimum of physical changes in the oil cake (Kodras, *et al.,* 1949).

Reports are also available to show that ricin could be detoxified by treating with sodium ricinoleate, potassium permanganate, hydrogen peroxide and halogens. Cake treated with water, alcohol and aqueous HCl gives the least toxic product (Eliane and Gregory, 1957).

Ethanol has been used for extraction of oil cakes by some workers. Repeated extraction with 95% ethanol has shown complete detoxification of the oil cake. Extraction with ethanol or autoclaving and steaming or boiling with water has yielded detoxified cake. Reymond, (1962) has treated the pressed cake with physiological serum and dilute formaldehyde to remove ricin. Ultra violet rays in the wavelength range of 225-250 nm have been found effective in removing ricin.

Extraction of the material with water, acids, and other solvents has been employed by several workers. Hydrolysis also destroys the toxicity (Gardner, *et al.*, 1960). Treatments like alkalies (0.5% $Ca(OH)_2$, 0.5% $NaHCO_3$) and autoclaving, reduces the ricin to non-detectable limits (Srinivas and Nagaraj, 1999). Gandhi and Mulky, (1994) have neutralized the castor toxins using sal tannins and the detoxified cake could be used as an animal feed.

Several attempts have been made to detoxify the castor seed meal by destorying the ricin and autoclaving the oil meal, washing the cake with water, alcohol and aqueous HCl and treatment with formaldehyde and lime (Gandhi *et al.*, 1994). But, these methods are possibly not commercially feasible because these reactions, which are drastic/harsh while destroying the ricin and RCA, also destroy the useful proteins. In another approach, it has been suggested that the ricin and RCA protein could be inactivated by use of Sal seed meal, which contains very high levels of tannins. However, the treated meal shows reduced protein content and also shows the presence of substantial levels of tannins (Gandhi *et al.*, 1994). Tannins, which non-selectively bind to the proteins and destroy their enzymatic activities also reduce the net protein availability of the treated castor meal. This meal impairs the digestibility of the proteins in animals, as the digestive enzymes are also complexed by tannins. Apart from these disadvantages, all the detoxifying methods need additional knowledge and inputs from the end user and also the extent of detoxification might vary subject to the method adopted and followed. There is thus always a fair amount of risk on the farmer's side.

Biochetchnological Approach

Reduction of unwanted proteins to acceptable levels could be achieved effectively by gene silencing technologies. Post transcriptional gene silencing (PTGS), which occurs in plants transformed with foreign or endogenous DNA results in the reduced accumulation of target RNA molecules with sequence similarity to the introduced nucleic acid. It has also been induced either through the use of DNA construct encoded double-stranded RNA or co-suppression constructs producing small interfering RNA.

The synthesis of ricin and RCA proteins could be blocked using PTGS strategies like antisense RNA, intron hairpin RNA (ihp RNA) and SHUTR (silencing by heterologous 3′ untranslated region). As ricin and RCA genes share very high levels of homology in their nucleotide sequences, by choosing appropriate gene segments conserved between their genes, it is possible to target the transcripts of both for degradation. Coupled with an efficient transformation protocol, it is surmised that transgenic castor plants with reduced/acceptable levels of ricin and RCA could be developed.

From the foregorin it is evident that various options are available to detoxify the castor meal. Until such time that ricin free genotypes, developed through biotechnology are available, simple chemical methods for detoxification need to be utilized. Low cost technologies like early harvesting to obtain low ricin produce also need to be explored. Castor cakes from such background can be used in composite feeds. Also industrial utilization of castor cake proteins can be explored as a better option.

Mycotoxins

Introduction

Certain fungi commonly known as moulds produce a wide range of metabolites. Many of them have biological activity. It may be expressed by their inhibitory or lethal effects on micro-organisms (antibiotics), on plants (phytotoxins) or on man and his domesticated animals (mycotoxins). It is the ability of moulds to grow on foods, feeds or other raw materials which makes them a nuisance in the spoilage of these commodities. It makes them potentially dangerous to health if they are also able to produce mycotoxins.

The Significance of Mycotoxins

Mycotoxins produce a toxicity syndrome in animals which is commonly known as mycotoxicoses. Mycotoxins have been of great concern since 1961 when aflatoxin, the first mycotoxin has been shown to be responsible for the death of one lakh turkey poults in England(Blount,1961, Sargeant *et al.*, 1961). Moulds and mycotoxin production can occur on food and fodder starting from crop growth stage through grain storage to processing. The toxins can be present and remain stable long after the moulds responsible have died. Mycotoxins are important environmental pollutants. They get into the living organism through ingestion, inhalation or contact.

Three kinds of fungi namely field fungi which grow on the living plant, decay fungi growing on decaying plant material and storage fungi can produce the mycotoxins. Field fungi and decay fungi need high moisture contents ranging from 20-25% while storage fungi can grow at lower moisture levels of 10-18%. Temperature is another important factor for the growth of moulds and for production of aflatoxins, 30-32^{0}C is the most favorable range.

Aspergillus; Pencillium and *Fusarium* are the most important agents of mycotoxicoses. *Aspergillus* and *Penicillium* are storage fungi (Cerutti and Vecchio,1989). Species of *Fusarium* are plant pathogens, but can continue to grow on moist stored products. Most of the food and food products serve as good subtrates for the growth of these organisms. Mycotoxin contamination has been observed on items like wheat, rice, sorghum, rye, barley, oilseeds and oilseed meals (particularly groundnut and cottonseed). Small grains appear less susceptible to mould growth than large grains. Even fruits and fruit juices occasionally get contaminated.

It is visually difficult to demarcate whether a commodity is contaminated with mycotoxins or not. However, the chances of shriveled, discoloured or broken kernels and chaffy or unfilled seed being associated with moulds and toxins are more likely. Hence such seeds may be rejected for eating. The significance of mycotoxins stems from the following factors (1) they can impair the health of animals and man and sometimes kill them, (2) because of their contamination in grains and food products, farmers, merchants and consumers can suffer heavy losses and (3) national economies can suffer due to a reduction in export of commodities and the necessity to import quality grains and foods to offset shortfalls.

The Mycotoxins

More than 300 mycotoxins formed by 350 species are known as on today. The important ones among them are aflatoxins, sterigmatocystin, ochratoxin, zearalenone, citrinin, luteoskyrin, patulin, rubratoxin, T2 toxin and vomitoxin. Aflatoxins (B_1, B_2, G_1 and G_2) are the most widely occurring mycotoxins. The causal organisms and their effects have been summarized in Tables 157 and 158. The mycotoxins have a wide range of chemical structures. They can be broadly classified into three main groups, namely (1) macrocyclic lactones (eg: zearalenone), (2) oxygen containing heterocyclics (eg: aflatoxins and patulin) and (3) alicyclic compounds (eg: T2 toxin).

Although approximately 300 mycotoxins have been identified, only a small proportion of these have been shown to occur naturally. Mycotoxins occur in a wide variety of animal feeds and human foods (Scott,1973). The consumption of contaminated feed by a domestic animal may result in the contamination of foods such as meat, milk and other dairy products. Some mycotoxins are carcinogenic in animals, while others interfere with the immune response system. Aflatoxin B1 is the most potent naturally occurring carcinogen known. Chronic toxicity may also result in severe economic losses caused by the decrease in productivity of domestic animals.

Table 157: Compounds involved in mycotoxicoses

Toxin	Fungi	Susceptible	Biological species Effects
1. Aflatoxin .	*A.parasiticus* *A. flavus*	Mammals, fish, bird, Man?	Hepatotoxin, Carcinogen
2. Citrinin	*Penicillium Viridicatum*	Swine	Nephrotoxin
3.Ochratoxin	P. *viridicatum* *A.ochraceaus*	Swine, cattle	Nephrotoxin
4. Trichothe-cines (T2 toxin)	*Fusarium tricinctums*	Cattle, sheep Swine, Poultry, man	Dermal haemorrhage Necrosis
5. Vomitoxin	*Fusarium graminareum*	swine	Vomiting
6.Zearalenone	*F.graminareum* *F. roseum*	swine abortion	vulvovaginitis,

Table 158: Compounds suspected to be involved in mycotoxicoses

Toxin	Fungi	Susceptible	Biological species Effects
1. Luteoskyrin.	*Penicillium islandicum*	Man	Hepatotoxin
2. Patulin	*P. patulum* P. *urticae*	Cattle	Death
3. Rubratoxin	P.*rubrum*	Cattle	Hepatotoxin
4. Sterigmato-cystin	A. *species*	Mammals	Carcinogenic

The Aflatoxins

The aflatoxins (B1, B2, G1, G2 etc.) are produced by two moulds, *Aspergillus flavus* and *A.parasiticus*. These moulds grow on food and feed materials especially the oilseeds and their products, both at pre-and post-harvest conditions. Aflatoxin M1 is a toxic metabolite, derived from aflatoxin B1, which occurs in milk when aflatoxin-contaminated feed is consumed by dairy cattle. Aflatoxins are the most potent toxic, mutagenic, teratogenic and carcinogenic metabolites.

The aflatoxins occur in edible nuts, oilseeds, cereals, pulses, cocoa, spices and dairy products, and have been implicated in human fatalities when badly contaminated cereals have been consumed under famine conditions. Human exposure to these toxins can arise from direct consumption of contaminated commodities such as corn, peanuts, coconuts, or farm animal products (egg, milk and meat) previously exposed to aflatoxins in feed (Park and Pohland, 1986; Park, 1993).

Ochratoxins

Ochratoxin A is produced by both *Aspergillus* (e.g. *A. ochraceus*) and *Penicillium* (eg. *P.viridicatum*) moulds, and occurs in maize, barley, wheat, oats, pulses and green coffee beans. It is a nephrotoxin and has been associated with porcine nephropathy in several countries. Balkan nephropathy is a human kidney disease of unknown etiology encountered in some areas of the Balkan Peninsula. Renal damage, associated with this disease, is similar to that observed in porcine nephropathy.

Citrinin

Citrinin is nephrotoxic and occurs in cereals as well as in apple juice and groundnuts. This toxin occurrs along with ochratoxin A in cereals associated with outbreaks of porcine nephropathy. Citrinin is produced by several *Penicillium* (e.g.*P.citrinum, P. viridicatum*) and a few *Aspergillus* (e.g.*A. terreus*) moulds.

Patulin

Patulin is a neurotoxin produced by numerous *Penicillium* species (e.g. *P. patulum*) and a small number of *Aspergillus* species (e.g.*A.clavatus*). It occurs in fruit and fruit juices.

The Trichothecenes

The trichothecenes are sesquiterpenes, and may be divided into three groups. The trichothecenes are produced by a selection of *Fusarium* moulds (*e.g.F.tricinctum, F. solani*), and occur in maize and other cereals. The ingestion of these toxins produces vomiting, diarrhea, anorexia, ataxia, haematuria and leukocytosis followed by severe leukopenia and damage to the gastrointestinal tract, the central nervous system, cardiac muscles and testes. T-2 toxin and deoxynivalenol (vomitoxin), are implicated in animal mycotoxicosis.

Zearalenone

Zearalenone induces oestrogenic and anabolic activity. It is produced by *Fusarium roseum* and occurs in cereals and animal feeds. Zearalenone occurs along with the trichothecene mycotoxins.

Occurrence in Oil Seeds

A flavus grows readily on groundnuts when they are stored traditionally in the open, under hot humid conditions. Levels ranging from 700-10,000 ppb have been suspected to be present in the feed responsible for the turkey catastrophe(Scott,1973). Groundnut oil extracted from *A. flavus* contaminated seeds have been reported to have levels of 1000 ppb aflatoxins and the refined oil one ppb. Groundnut cakes generally harbour the *A.flavus* mould and the occurrence of aflatoxins in the commodity is more common. A survey into the occurrence of aflatoxins has revealed that the groundnut kernels are less contaminated and the groundnut cakes are generally more contaminated with aflatoxins.

Cottonseed and soybean also support the growth *of A. flavus* and the production of aflatoxins. Only stray references to the occurrence of aflatoxins in linseed, safflower meal and sunflower seeds are available. Groundnut also appears to be associated with ochratoxin A to a level of 4900 ppb. Other mycotoxins associated with oilseeds are sterigmatocystin, citrinin, patulin, cyclopiazonic acid, luteoskyrin, trichothecines and cyclochlorotine. Information on the occurrence of aflatoxins in oilseeds other than groundnut is either limited or lacking.

Toxicology

Mycotoxins elicit both acute and chronic toxicities. Acute effects result in rapid, readily noticeable and sometimes fatal diseases. While chronic effects occur over a longer period and are less easily defined, Some mycotoxins are acutely toxic to liver (aflatoxins, patulin) while others attack the kidney (citrinin, ochratoxin) or the circulatory system (trichothecines) etc. The LD 50 value (dose at which 50% of the population will be killed) for aflatoxin is 18.2 parts per kg body weight. Aflatoxin group of mycotoxins are the most dangerous apart from being hapatocarcinogenic in nature.

Human diseases that have been identified as possible or probable mycotoxicoses date back to middle ages and include ergotism (St.Anthony's fire), alimentary toxic aleukia, cardiac beriberi (yellowed rice disease), Ray's syndrome, Balkan nephropathy and hapatocarcinoma. Aflatoxin B_1 when ingested by lactating farm animals gets converted into aflatoxin M_1 and gets its way into the milk of these animals. Through the milk the toxin gets an entry into the human food chain. M_1 has similar toxicity as that of B_1. Babies who depend upon milk alone are most vulnerable to the toxic effects. Countries the world over have prescribed the maximum limits up to which the food products can be contaminated with mycotoxins. In the case of aflatoxin this limit ranges between 5-20ppb in the foods or 20-50ppb in the feeds in most countries (Jelinek *et al.*, 1989). For zealenone it is 3-200 ppb and for patulin it is 50ppb.

Table 159: Maximum permissible Aflatoxin levels for human consumption

Country	Aflatoxin limit (μgkg^{-1})	Country	Aflatoxin limit limit (μgkg^{-1})
Belgium	5	Luxembourg	5
Colombia	10	Malaysia	0
Cuba	0	The Netherlands	0
Czechoslovakia	5	Norway	5 to 20
Denmark	10	Poland	0
Dominican Republic	0	Portugal	0
Finland	5	Russia	5
France	1 to 5	Singapore	10
Germany	2	Surinam	5
Hong Kong	20	Sweden	5
Ireland	5	Switzerland	1 to 5
Italy	5	United kingdom	4
Japan	0	Yugoslovia	10
Jordan	15		

Table 160: Aflatoxin Limits in Groundnut Cake/Animal feed in various countries

Country	Groundnut cake/animal feed	Aflatoxin limit (μgkg^{-1})
Belgium	Animal feed	20
Brazil	Groundnut cake	50
France	Groundnut cake	20
Germany	Animal feed	5
Italy	Animal feed	5
Japan	Groundnut cake for animal feed mixes	1000
The Netherlands	Animal feed	20
Norway	Oilseed cake	600
Poland	Animal feed	5
Sweden	Animal feed	5
UK	Animal feed for dairy cattle	5
USA	Groundnut cake for immature animal and dairy	20
	Groundnut cake for beef cattle, swine, poultry	100-300
European Union	Animal feed supplements for dairy animals	5
	Complete feed for pigs, sheep, cattle and poultry	20-50
Zimbabwe	Animal feed	50-400

Prevention and Detoxification

It is evident that the occurrence of toxigenic fungi in the food chain presents a potential hazard to human and animal health. Prevention or inhibition of mould growth is usually accomplished by modification of interseed environment. Moisture reduction to less than 8% is very useful. Hermetic storage at high CO_2 and low O_2 levels in airtight containers is a sure way of preventing mycotoxin production. Addition of chemicals like propionic acid help in preventing the growth of fungi on food grains. Insecticides like Diazinon, Dichlorovos, Laudrin and Sevin are highly effective in inhibiting *A. parasiticus* growth and also production of aflatoxins. β-ionone a plant metabolite has been shown to inhibit aflatoxin production by *A. flavus* and *A. parasiticus*.

In the case of groundnut it has been observed that varieties with less vitamin E harbour less growth of *A. flavus*. Similarly varieties with lower levels of protein content and higher levels of phenols also have lower levels of aflatoxins in groundnuts. Sugar content in the seeds is not related to *A. flavus* colonization and aflatoxin content (Ghewande *et al.*, 1993). Aflatoxin contamination in groundnut could be prevented during storage and transit through use of salts , plant products and aqueous leaf extracts of plant species(Ghewande and Nagaraj, 1987).

The possibility exists that cultivars of susceptible crops could be bred for resistance not just for infection of *A. flavus* and *A. parasiticus* but also for the inhibition of mycotoxin formation in the event of mould infection. Change from crops susceptible to mycotoxin contamination such as groundnuts and maize to alternative crops such as soybeans also is another mode of preventing the production of mycotoxins. If the preventive methods are not practicable, then detoxification of the contaminated food material has to be attempted.

Detoxification of Aflatoxins

From the view point of health and economics, low levels of aflatoxins such as 0.5-30 ppb and 20-300 ppb are prescribed in foods and feeds respectively by several countries (Park and Liang, 1993). However, achieving such low levels is an uphill task especially in tropical and third world countries, where the conditions are conducive for the mould to infect and elaborate the toxin on foodstuffs even prior to harvest. Although drying of foodstuffs to safe moisture levels (<10.0%) can reduce the post-harvest development of aflatoxins, the practice is beset with certain problems like inclement weather conditions, small farm holdings and lack of facilities for drying. Hence, approaches involving physical, chemical and biological methods have been made to detoxify aflatoxins in food and feed stuffs in the recent past.

Strategies for Detoxification

Examination of physico-chemical and biochemical characteristics of aflatoxin B1 molecule reveals two important sites for toxicological activity (Heathcote and Hibbert, 1978). The first site is the double bond in position C-8,9 of furfuran ring. The aflatoxin, DNA and protein interactions which occur at this site alter the normal biochemical functions of these macromolecule leading to deleterious effects at the cellular level. The second reactive group is the lactone-ring in the coumarin moiety. The lactone ring is easily hydrolysed and therefore a vulnerable site for degradation. The degradation treatments should be aimed at removing the double bond of the terminal furan ring or in opening the lactone ring. Once the lactone ring is opened, further reactions could occur to alter the binding properties of the terminal furan ring to DNA and proteins.

Among various aflatoxins that occur in food and feed, aflatoxin B1 is the most toxic. It has been found that any change in this molecule either by physical, chemical or biological means would reduce the toxicity. The relative toxicity and mutagenicity of many of the derivatives of aflatoxin B1 has been reviewed by Samarajeewa *et al.*, (1990). For example, a simple introduction of a hydroxyl group at furfuran ring of aflatoxin B1 will lead to a derivative aflatoxin M1 which takes place in cow and excreted in milk. This toxin is 3 times less carcinogenic to rain-bow trout and has the same order of toxicity to ducklings as aflatoxin B1 (Wogan, 1971) and 30 times less mutagenic to *Salmonella typhimurium* (Wong and Hsieh, 1976).

Hydration of double bond in furan ring leading to aflatoxin B2a takes place in microsomes of ducklings as well as under acid conditions. This is 200 times less toxic to ducklings (Lillehoj and Ciegler, 1969) and 1000 times less mutagenic (Wong and Hsieh, 1976). Similarly, physical (irradiation) and other chemical reactions would lead to varied derivatives that would have less toxicity.

Various physical and chemical agents have been used to decontaminate/detoxify food and feed materials. Several reaction products are obtained depending upon the treatment process and some strategies and specific criteria have been established for evaluating the acceptance of a given decontamination procedure (Park, 1993). The process must: a) inactivate, destroy or remove the toxin, (b) not produce or leave toxic residues in food/feed, (c) retain the nutritive value and feed acceptability of the product; (d) not alter significantly the technological properties of the product; and (e) if possible, destroy fungal spores.

Physical Methods

Physical removal of infected grains (coarse), by handpicking or electronic colour sorting, density segregation by floatation, removal by milking cereal grains, adsoprtion-cum-filtration technique for decontamination of groundnut oil, removal by suitable solvents and inactivation by heat and irradiation are some of the methods that have been found promising in reducing aflatoxin levels in contaminated products.

An effective method has been reported for reducing aflatoxin in corn (Huff and Hagler, 1985) and peanuts (Cole, 1989) by floatation and density segregation of toxic kernels. A systematic electronic and hand sorting of contaminated peanuts have been used extensively by peanut industry to reduce aflatoxin level in peanut products. By following a set of strategies like belt separation, shelling, colour and gravity sorting, blanching and colour and re-colour sorting, it has been possible to reduce aflatoxin in peanut kennels from 217 ppb to a level of 1.6 ppb amounting to 99.3% removal (Park and Stoloff, 1989).

When peanuts are processed in the expeller mill, the cake contains 85-90% of the aflatoxin orginally present in kernels and the oil contains only 10-15% (Basappa and Sreenivasa murthy, 1974). On further processing of cake to isolate protein, the protein fraction retains 60-70% of the toxin and rest in the whey and residue (Basappa *et al.*, 1972). Mixing of 2% Fuller's earth with oil and centrifuging results in adsorption of toxin to the earth material, thereby the clear oil contains only 10-15% of what has been originally present (Basappa and Sreenivasa Murthy, 1977). Based on this data, filter pads to suit plate-and frame filter press in pilot plant/oil mills have been prepared and aflatoxin could be removed successfully to the extent of 85% on filtration (Basappa and Sreenivasa murthy, 1979; Basappa, 1983).

The other chemsorbents such as clay, activated carbon and zeolite minerals can bind and remove aflatoxins from liquid media (Park and Liang, 1993). Machen *et al.*, (1991) have reported that a phyllosillicate clay effectively removes aflatoxin from contaminated peanut oil and prevents its mutagenicity and toxicity *in virto*. A phyllosilicate clay has been used in animal feeds to reduce levels of bioavailable aflatoxins by selective chemosorption (Park and Liang, 1993). The physical removal methods are not only cost effective but also conform to the specific criteria laid down for decontamination procedures.

Heat Inactivation

In general aflatoxins are resistant to thermal inactivation. Thermal processing of foods can only result in modest reductions in aflatoxin levels following treatments like pasteurization (about 64% of M1 in milk, Purchase *et al.*, 1972), roasting of peanuts (60-90%, Escher *et al.*, 1973), frying in peanut oil (50%, Dwarakanath *et al.*, 1969), microwave roasting of peanuts (95%, Luter *et al.*, 1982) and spraydrying of milk (87%, Purchase *et al.*, 1972). These procedures are of practical importance and are relevant in epidemiological surveys in order to assign actual consumption of aflatoxin through processed foods.

Irradiation

Irradiation has been shown to result in a marked reduction in the concentration of aflatoxin in contaminated peanut oil exposed to short wave and longwave UV light (Shantha and Sreenivasa Murthy, 1977). A 14 hour exposure to sunlight destroys 77-90% of aflatoxin B1 added to peanut flakes, whereas only 50% of the toxin is destroyed in naturally contaminated product (Shantha *et al.*, 1986). Aflatoxin M1 level in contaminated milk containing 0.05% peroxides after exposure to UV light at 25°C for 20 minutes decreases by 89.1% compared with a reduction of 60.7% in peroxide-free milk (Yousef and Marth, 1986) Oxygen appears to enhance UV mediated free radical degradation of aflatoxin. However, some of the UV degraded compounds of aflatoxin are toxic to the chick embryo. (Samarajeewa *et al.*, 1990). This indicates the limited prospects of UV radiation to degrade aflatoxins in foods.

The accumulated evidence from solar and UV irradiation studies indicates that the aflatoxin B1 is the likely site of photodegradation. Because this is also the site of aflatoxin binding to nucleic acids and proteins, aflatoxins bound to food through the C-8,9-positions are not susceptible to solar degradation (Samarajeewa *et al.*, 1990).

Removal by Solvents

Various solvent systems can be used for the removal of aflatoxin to the extent of 85-90% for different commodities with minimal effects on protein content or nutritional quality (Rayner *et al.*, 1977). These include 95% ethanol, 80% isopropanol, 96% methoxymethane and hexane-acetone-water etc.(Goldblatt, 1965, Rayner and Dollear, 1968, Rayner *et al.*, 1970). However, due to flavour alterations, nutritional quality changes, solvent residues, disposal of toxic extracts and cost factors, these procedures are impracticable (Park and Liang, 1993).

Chemical Methods

Structural degradation or inactivation of aflatoxins has been found to be possible by the use of chemicals such as chlorinating agents (sodium hypochlorite, chlorine dioxide and gaseous chlorine), oxidizing agents (H_2O_2, ozone, sodium bisulfite) and hydrolytic agents, acids (organic and inorganic)and alkalies (sodium hydroxide, ammonium hydroxide, potassium hydroxide, etc). Some of these chemicals are already in use in food industry and are less prone to consumer resistance (Samarajeewa *et al.*, 1990). However, according to Park and Liang (1993), most of these chemicals are impractical and are particularly unsafe because they form toxic residues or damage nutrient content, flavour, odour, colour, texture or functional properties of the products, although the involved processes can destroy aflatoxin efficiently in the range of 85-100% in meals as well as whole grains.

Ammoniation

Among various processes, ammoniation process has been extensively worked out and largely accepted inspite of certain resulting nutritional losses. Liquor ammonia, gaseous ammonia, *in situ* liberation of ammonia by reaction of urea and urease as well as reacting ethylamine with lime, have been used. It involves treatment of the contaminated food material (12-15% moisture) with ammonia gas at 2-3 atmospheres for 15-30 min. at 90°C.

The noteworthy losses affecting the nutritional quality of ammoniated animal feeds include 10% reduction of protein quality (Thiesen, 1977), a significant drop in lysine and methionine content (Conkerton *et al.*, 1980; Piva, *et al.*, 1981) and an irreversible reduction in the degree of unsaturation in lipids (Black *et al.*, 1978). The presence of residual toxicity arising from hydrolysed products (Schroeder *et al.*, 1985) along with the potential for covalent binding of aflatoxin B1 to proteins (Brekke *et al.*, 1979) and the losses in nutritional quality, make the ammoniation treatment processes less acceptable. However, under defined conditions the treatment of grains (peanuts, cotton seed, corn) and their meals with ammonia appears to be a commercially viable approach to detoxification of aflatoxins to the extent of 99% (Park and Liang, 1993) particularly for feed purposes.

Ammonia degradation proceeds through hydrolysis of lactone ring, and is followed by decarboxylation to produce non-toxic compounds. Due to severity of aflatoxin contamination in selected agricultural commodities, specific decontamination processes have been approved and put into use (Park, 1993; Park *et al.*, 1988). Treated peanut meal is widely used in animal feeds in Europe and elsewhere. Ammonia treatment processes for feed mill level and at farm level have been worked out intensively by Park *et al.*, (1988). Low ammonia concentration (0.2 to 2.0%), a high pressure (35-50psi) and high temperature (80-120°C) would require less time (20-60 minutes) as compared to high ammonia concentration (1-5%) at atmospheric pressure. Ambient conditions need more time (14-21 days) for treating feed materials containing 12-16% moisture (Park *et al.*, 1988) at feed mill level and farm level, respectively.

Toxicological studies have revealed that the product is safe and is being utilized as animal feed. Pathological and histopathological examinations with experimental and farm animals which have been fed with ammoniated meals does not show any signs of aflatoxicosis. Also, no differences in egg production and immunological responses in poultry have been observed. Excessively contaminated food material should not be used for edible purpose. Such material for example, maize can profitably be utilized in the production of ethyl alcohol.

Process of Sodium Bisulfite

Sodium bisulfite can react with aflatoxins B1 G1, M1 and aflatoxicol at various temperatures and concentrations at various times to form water soluble products (Hagler *et al.*, 1982). Addition of 0.04g of potassium bisulfite per 10 ml milk results in a 45% reduction in the level of aflatoxin M1 after 5 minutes (Doyle *et al.*, 1982). Potassium bisulfite is a common food preservative and does not pose any consumer resistance problem.

Degradation by Acids

Conversion of aflatoxin B1 to aflatoxin B2a has been reported in the metabolic system using rat liver microsomal enzymes (Patterson and Roberts, 1970). Because the treatment with strong acids may affect food quality, and aflatoxin B2a still possesses toxicity, the use of strong acids to degrade aflatoxins in foods is discouraged. However, complete degradation of aflatoxins without the production of toxic derivatives during manufacture of hydrolysed peanut meal with 3N HCI at elevated temperatures and pressure for 12 hours (Dutton and Williams, 1988) encourage further exploration of acid treatment for decontamination of aflatoxin in specific foods.

Degradation by Oxidizing Agents

Hydrogen peroxide and ozone have high potential as aflatoxin degrading agents. Hydrogen peroxide at 0.5% totally degrades aflatoxin B1, at pH 4 in peanut protein isolates (Rhee *et al.*, 1976), and at 6% at pH 9.5 for 30 minutes in peanut meal (Sreenivasa murthy *et al.*, 1967).

Ozone effectively degrades aflatoxin B1 and aflatoxin G1 in 4% dimethyl sulfoxide at room temperature within a few minutes. The treated products are non-toxic. Aflatoxin B1 levels get reduced by 91% in cottonseed meal with 22% moisture after treatment at 100°C for 2h. With peanut meal (30% moisture) the reduction is only 78% after exposure to ozone for 1h (Dollear *et al.*, 1968).

Microbiological Methods

Microbiological inactivation and fermentation have also been studied as means of degrading and removing aflatoxins (Marth and Hadpoyle, 1979). *Flavobacterium aurantiacum* (Ciegler *et al.*, 1966; Lillehoj *et al.*, 1971) and selected acid producing moulds have been successfully used to remove aflatoxins from liquid media. Transformation of aflatoxin B1 to less toxic B2a has been observed in case of fermented foods by *Lactobacillus delbrueckii* (Maing *et al.*, 1973) and yoghurt cultures (Megalla and Hafez, 1982; Resic *et al.*, 1991). Transformation of aflatoxin B1 to aflatoxicol (18 times less toxic than aflatoxin B1) is observed in cases of *corynebacterium rubrum* (Mann and Rehm, 1976) and several species of moulds belonging to the genera *Aspergillus, Mucor, Helminthosporium, Rhizopus, Absidia, Trichoderma and Doctylum* (Detroy and Hesseltine, 1969; Rabertson *et al.*, 1970; Mitsuo *et al.*, 1990). An enzyme responsible for conversion (reversible) of aflatoxin B1 to aflatoxicol has been isolated, partially purified and characterized (Mitsuo *et al.*, 1991c).

The use of microorganism(s) for degradation of aflatoxin in food or feed may have certain disadvantages. The organism would not only utilise the food for its growth but may also elaborate undesirable compounds. However, the use of selected lactic acid bacteria in food fermentaion like yoghurt and fermented milk may be of advantage.

Ther are many possibilities for detoxifying or reducing aflatoxins/ mycotoxins in foods and feeds. In view of the extreme toxicities of mycotoxins, especially, aflatoxins this step has assumed great importance. However, the negative effects of the detoxicants like nutritional degradation need to be kept in mind. The best would be to avoid growth of the causal organisms or develop genotypes resistant to their growth.

Glucosinolates

Introduction

Rapeseed meal contains 34-38% protein (at 8% moisture level). The aminoacid balance of rapeseed protein is quite favourable. The use of rapeseed meal in animal rations is limited due to it's glucosinolate content. The antinutritional and goitrogenic cleavage products from these glucosinolates have a pungent taste that decreases the palatability of the feed.

Glucosinolates are a group of plant thioglucosides found principally among members of family *Cruciferae*. Their presence in agricultural crop plants such as rapeseed-mustard and *brassica* forages is important because of the harmful effects of their breakdown products. Pungency of these crops is due to the hydrolytic products. Almost one hundred glucosinolates have been reported in plants primarily amongst the members of *Cruciferae*, but only 15-20 have been found in the genus *Brassica*.

General strucure of glucosinolates is as under:

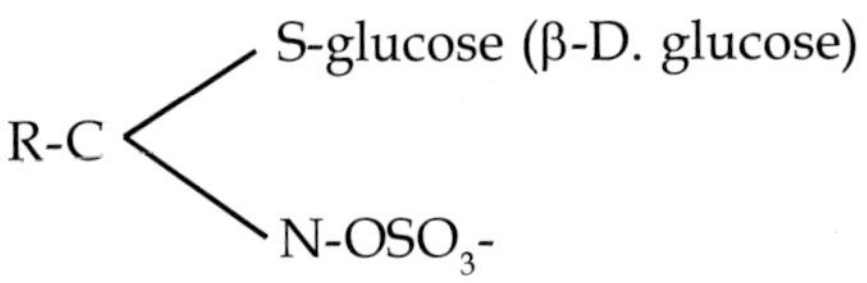

It consists of a thioglucose group linked via sulphonated oxime group to the carbon of a R-group. There are two major categories of glucosinolates, an aliphatic, having aliphatic side chain with terminal unsaturated bond, and aromatic, having commonly an indole group. Further, there may be present methyl, thio and hydroxyl groups which add to the diversity of glucosinolates. R-chain of important glucosinolates detected in *cruciferous* seed (Sang *et al.*, 1984) are given in Table 161.

Under some conditions cleavage products from glucosinolates containing sulfur dissolve in the oil. The sulfur containing compounds act as catalyst poisons and make hydrogenations difficult. A simple technique involves heating of the seed at an early stage in processing to inactivate the enzyme, myrosinase, to minimize the breakdown of the glucosinolates during crushing and oil extraction. This technique greatly facilitates utilization of oil and meal.

Breakdown Products of Glucosinolates

Enzyme thio-glucoside glucohydrolase (EC 3.2.3.1) commonly known as myrosinase acts on glucosinolates and hydrolyses them following cellular disruption. It causes cleavage of the thioglucoside bond resulting in release of free glucose along with aglucone intermediate (Fenwick *et al.*, 1989). Aglucone intermediate further breaks down into a number of toxic metabolites such as isothiocyanates, nitriles, oxazolidine, 5-viny 1-2 thione and sulphate.

Depending upon the conditions of the hydrolysis medium and the parent molecule, a number of products may be formed. Sinigrin breaks down to allylisothiocyanates, which are the major components responsible for pungency in mustard.

Table 161: Glcosinolates that occur in *cruciferae*

	IUPAC	Trivial Name	Chemical formula (R)
1.	2-hydroxy but 3-enyl	Progoitrin	CH_2=CH-CH-CH_2-
2.	2-propenyl or allyl	Sinigrin	CH_2=CH-CH_2-
3.	p-hydroxybenzyl	Glucosinalbin	p-OH-benzyl-
4.	4-methyl sulphinyl but-3-ethyl	Glucoraphenin	CH_3SO-CH=CH-CH_2-CH-
5.	4-methyl sulphinyl butyl	Glucoraphanin	$CH_3SO(CH_2)_4$-
6.	2-hydroxy pent-4-enyl	Napoleiferin	CH_2=CH-CH_2-CH-CH_2-
7.	But-3-enyl	Gluconapin	CH_2=CH-CH_2-CH_2-
8.	4-hydroxy-3-indolylmethyl	4-hydroxy glucobrassicin	-----
9.	3-methyl thiopropyl	Glucoibervirin	CH_3-S-CH_2-CH_2-CH_2-
10.	4-methyl thiobutyl	Glucoerucin	CH_3-S-CH_2-CH_2-CH_2-CH_2-
11.	pent-4-enyl	Glucobrassicanapin	CH-CH_2=CH_2-CH_2-CH_2-
12.	4-methyl thiobut-3-enyl	Glucoraphasatin	CH_3-S-CH=CH-CH-CH_2-CH_2-
13.	3-indolylmethyl	Glucobrassicin	--------
14.	2-phenyethyl	Gluconasturtin	C_6H_5-CH_2-CH_2-
15.	4-methoxy-3-Indolylmethyl	4-methoxy glucobrassicin	-----

Anti-Nutritional Aspects of Glucosinolates

Intact glucosinolates as well as their hydrolytic products can cause anti-nutritional and toxic effects. Hydrolytic products are more toxic than intact glucosinolates. It has been established that thiocyanate, isothiocyanate and oxazolidinethiones are goitrogenic (Bell *et al.*, 1972). Sinigrin is also goitrogenic. It is well known that thiocyanate ion competes with iodine and thus uptake of iodine and consequently the synthesis of thyroid hormone is decreased which results in goitre. The effect is evident only in iodine deficient diet. Inclusion of high glucosinolate meal in feeds of dairy cattle lowers the feed intake and reduces milk yield. Reproductive performance in animals is also affected. Continued intake of progoitrin affects liver and kidney weight and can cause liver haemorrhage (Vermorel *et al.*, 1986). Allylisothiocyanates may affect the key enzymes like liver succinic dehydrogenase, kidney xanthine oxidase etc.

Biosynthesis

Amino acids are the precursors for glucosinolate biosynthesis. Aliphatic glucosinolates are synthesized from methionine while indole glucosinolates arise from phenylalanine, tyrosine and tryptophan. All aliphatic glucosinolates follow the same biosynthetic pathway.

Different types of glucosinolates differ in the nature of R-groups and this diversity in R-groups is generated by two processes in the biosynthesis, namely, 1) modification of amino acid precursors before initial hydroxylation e.g. chain lengthening of valine to form leucine and methionine to form homomethionine and 2) modification of side chain after the last stage of glucosinolate biosynthesis.

In *B. Napus* seeds, 2-OH-but-3-enyl is in highest concentration, hence chain elongating enzymes have very low activity. During seed development a shift from 2-OH-pent-4-enyl to 2-OH-but-3-enyl occurs. This occurs due to the cessation of chain elongation step which decreases the synthesis of pent-3-enyl.

Pod is the main site for glucosinolate synthesis. Other parts of the plant also synthesize some of the glucosinolates or alternatively glucosinolate precursors may be transported from other parts of the plant for conversion into glucosinolates in the pod or seed. Decreasing glucosinolate contents in the pod tissue accompanying increasing seed glucosinolate content may represent turnover or translocation. Hence, glucosinolates or their precursors must also be supplied by plant parts other than pod tissues. The lower glucosinolate content in mature seed of low glucosinolate lines indicate that this must be due to a metabolic block in the biosynthesis of the main glucosinolate. This is located after the formation of 2-amino-6 (methyl thio) caproic acid preventing the synthesis of but-3-enyl glucosinolate.

Analytical Techniques

Two types of analyses are available, namely, analysis of total glucosinolate content and analysis of individual glucosinolates or their hydrolysis products. A very simple and commonly followed method is the estimation of glucose in the crushed seeds by using a glucose test paper impregnated with the enzyme system glucose oxidase, peroxidase and a chromogen. It is a very quick simple method and can be employed for screening of large number of samples. About-200-300 samples can be evaluated per day (Lein, 1970)

A number of colorimetric or titrimetric methods based on estimation of split products such as glucose, bisulphate ion, isothiocyanates and oxazolidinethione are also available. Glucose can be estimated with thymol (Tholen *et al.*, 1989), palladium can form coloured complex with glucosinolates, sulphate released can also be estimated by precipitating as barium sulphate (Mc Ghee *et al.*, 1965).

Analysis of total sulphur content can be carried out using X-ray fluorescence spectroscopy. This technique is sensitive and reproducible (Schung and Haneklans, 1988). Individual glucosinolates, can be estimated very quickly and efficiently using gas-liquid chromatography (Thies, 1976). Glucosinolates are first extracted using 70% methanol. They are purified and desulfated on microion-exchange column. These desulfated glucosinolates are derivatized to form volatile trimethylsiyl ethers and separated on a glass column packed with 2% OV-7. Separation of desulfated glucosinolates by HPLC has also been standardized, (Spinks *et al.*, 1984). Eluate from micro ion-exchange column which contains purified desulfated glucosinoates can directly be injeccted into the HPLC column without any derivatization and separated using a solvent system consisting of acetonitrile and water in varying gradients.

Glucosinolates in Brassicas

In seeds, the concentration of glucosinolates is higher (upto 5%) as compared to that present in vegetative tissue. In cultivars of *B.juncea, B. campestris* and *B.napus,* it varies from 2-3%. Range of glucosinolate content in seed meal of some *cruciferae* oil crops is as under (Josefsson and Appelqvist, 1968).

Brassical napus 3.4 to 7.8%

Brassica campestris 3.1 to 4.3%

Brassica juncea 3.6 to 6.5%

Sinapis alba 7.9 to 10.1%

Different *Brassica* species show variation in having major glucisinolates. Sinigrin is absent in *B. napus and B. campestris.* Progoitrin is completely absent in *B.nigra and B.juncea.* In *B.juncea,* major glucosinolates are gluconapin (but-3-enyl) and sinigrin (Allyl) while *B.campestris* and *B.napus* have gluconapin (but-3-enyl) and progoitrin (2-OH-but-3-enyl) as the major glucosinolates (Table 162).

Table 162: Glucosinolates in different Brassica species

Brassica Species	mg/g oil free meal All	But	Pent	Ozt	Mtb
Indian					
B. campestris (toria)	0.00	13.50	0.10	0.00	0.30
B. juncea	1.50	12.40	0.10	0.00	0.00
Canadian					
B. campestris	0.00	2.50	2.00	3.80	0.10
B. napus	0.00	3.70	1.00	15.90	0.00
B. juncea	7.80	0.00	0.70	0.00	0.00
Swedish					
B. campestris	0.00	2.20	0.00	1.20	0.00
B.napus	0.00	1.30	0.00	3.70	0.00

All= Allyisothiocyanate; But=3-butenyl isothiocyanate, Pent=Pentenyl isothiocyanate, Ozt=5-vinyloxazolidine-2-thione, Mtb=4-methl thiobutyl isothiocyanate

Source: Anand (1974)

In *B. campestris*, gluconapin is the major glucosinolate. In *B. juncea*, gluconapin and sinigrin and in *B. napus* progoitrin and gluconapin are major glucosinolates. *B.napus* has the highest content of total glucosinolates along with highest content of progoitrin, the most toxic glucosinolate (Rani, 1994).

Removal of Glucosinolates from Rapeseed Meal

Various methods are available for reduction/ removal of glucosinolates. Chemical treatments are always beset with nutritional degaradation. The best is to have genotypse with zero levels of these antinutrients. However for utilization of the seeds/cakes containing glucosinlates, some of the following methods of detosification can be utilized.

Enzyme Inactivation

If rapeseed (with 8% moisture) is heated at 90°C for 15 minutes, myrosinase is completely inactivated (Appleqvist and Josefsson, 1967). Meal produced from the rapeseed will not be toxic. It is, however, possible that animal may get myrosinase from other sources such as fresh parts of *cruciferae* plants. Myrosinase activity is also present in a variety of bacteria particularly *paracolobactrum* which is common in the intestinal tract of man.

Autolysis and Distillation

Water is added to meal of *Brassicas* along with crushed seeds as a source of myrosinase. Glucosinolates are split during this treatment and steam volatile isothiocyanates are removed by distillation.

Autoclaving and Steam Stripping

Most of the glucosinolates are destroyed in a meal that has been autoclaved for 60 minutes at 1.2 kg per cm^2 pressure. When steam is passed for one hour through rapeseed meal maintanined at 110°C, most of the glucosinolates get destroyed. These treatments deteriorate the protein quality of the meal.

Treatment with Alkalies and Heat

Treatment with ammonia and heat destroys the glucosinolates effectively. But this is associated with palatability problems. Glucosinolates are destroyed by cooking the crushed seeds or defatted rapeseed meal with sodium carbonate under pressure (Mustakas *et al.*, 1968). Feeding tests with cows demontrate that palatability of the soda treated meal is better than that of untreated meal. Soda treatment is inexpensive and may be of practical value.

Extraction of the Toxic Factors

Rapeseed meal has been extracted with cold or warm water or with buffers. For this type of extraction, it is advantageous to use myrosinase-free meal since cleavage products are less water soluble than intact glucosinolates. Myrosinase free meals of rapeseed can be made free of 80% of the glucosinolates by extraction with H_2O (Belzile and Bell, 1966) or 80% ethanol. 75% to 80% acetone is also effective in removing intact glucosinolates. Acetone removes hydrolysed products. Solvent extraction methods are expensive since cost of solvent is more and evaporation of solvent also involves higher cost.

References

Alderton, W.K., Low, C.R. and Thatcher, D.A.1994. Purification of Recombinant Ricin-A Chain with Immobilised Triazine dyes. Journal of Chromatography, 2:289.

Ambedkar, V.R. and Dole, K.K.1957. Detoxification of Castor Cake. Indian Journal of Dairy Science. 10: 107.

Anand, K.J. 1974. Mustard Oil Glucosides of the Indian *Brassicae*, Plant Biochemistry Journal. 1:26-31.

Appleqvist, L.A. and Josefsson, E. 1967. Methods for Quantitative Determination of Isothiocyanates and Oxazolidine Thiones in Digests of Seed Meals of Rape and Turnip Rape. Journal of the Science of Food and Agriculture. 18:510-519.

Basappa, S.C. 1983. In: Mycotoxins in Food and Feed. Eds. K.S.Bilgrami, T.Prasad and K.K.Sinha.The Allied Press, Bhagalpur. pp. 251-273.

Basappa, S.C. and Sreenivasa Murthy, V.1974. Journal of Food Science and Technology. 11:137

Basappa, S.C. and Sreenivasa Murthy, V. 1977. Journal of Food Science And Technology. 14: 57.

Basappa, S.C. and Sreenivasa Murthy, V. 1979. Indian Journal of Technology. 17:440

Basappa, S.C., Shankar Murti, A., Sreenivasa Murthy, V. and Parpia, H.AB. 1972. Indian Journal of Technology. 10:395.

Bell, J.M., Benjamin, B.R. and Giovannetti, P.M. 1972. Histopathology of Thyroids and Livers of Rats and Mice Fed on Diets Containing *Brassica* Glucosinolates. Canadian Journal of Animal Science. 52:395-406.

Belzie, R.J. and Bell, J.M. 1966. Growth Depressing Factors in Rapeseed Oil Meal. VII. Effects of Myrosinase Activity on Toxicity Following Treatment with Buffered Solutions. Canadian Journal of Animal Science. 46: 165-169

Black, L.T., Spencer, G.F. and Brekke, O.L. 1978. Journal of the American Oil Chemists' Society. 55:526.

Blount, W.P. 1961. Turkey "X" Disease. Journal of the British Turkey Federation. 9(2): 52-77.

Brekke, O.L., Peplinski, A.J., Nofsinger, G.W., Conway, H.F., Stringfellow, A.C., Montgomery, R.R., Silman, R.W., Sohns, V.E. and Begle, E.B. 1979. Transactions of the American Society of Agricultural Engineer. 22:425.

Cawley, D.B., Hedblom, M.L. and Houston, U. 1978. Homology Between Ricin and *Ricinus Communis* Agglutinin: Amino Terminal Sequence Analysis and Protein Synthesis Inhibition Studies. Archives of Biochemistry and Biophysics. 190(2): 744.

Cerutti, G. and Vecchio, A. 1989. New Indices of Quality: Mycotoxins in Food. In. Nutritional And Toxicological Aspects of Food Processing, Eds. R. Walker and E. Quathrucci, Proceedings International Symposium Held at the *Institute Superiore Di samita,* Rome, Italy. 14-16, April, 1989. pp. 315-320.

Ciegler, A., Lillehoj, E.B., Patterson, R.E. and Hall, H.H. 1966. Applied Microbiology. 14: 934.

Cole, R.J. 1989. In : Mycotoxins and Phycotoxins. Eds.Natoria, S, Hashimato, K. and Ueno, Y., Elsevier Publishing House. pp.177-184.

Cooper, C. and Sheel, K. 1964. Occupational Hazards from Castor Bean Pomace. Industrial Hygiene Journal. 25: 431.

Corwin, A.H. 1962. Toxic Constituents of the Castor Bean. Journal of Medicine and Pharmaceutical Chemistry. 4:431.

Conkerton, E.J., Chapital. D.C., Lee, L.S. and Ory, R.L. 1980. Journal of Food Science. 45: 564.

Coulson, S. and Stevens, L. 1950. The Allergen Content of Castor Bean and Castor Pomace. Journal of the American Oil Chemists' Society. 48: pp, 616.

Detroy, R.W. and Hesseltine, C.W. 1969. CAN. Journal of Biochemistry. 48:830.

Dollear, G.B., Mann, G.E., Codifer, L.P.Jr., Gardner, H.K., Koltun, S.P. and Vix H.L.E. 1968. Journal of the American Oil Chemists' Society. 45:862.

Doyle, M.P., Applebaum, R.S., Brackett, R.E. and Marth, E.H. 1982. Journal of Food Protection. 45: 964.

Dutton, M.G. and Williams, K. 1988. Food Reviews. 15:70.

Dwarakanath, C.T., Sreenivasa Murth, V. and Parpia, H.A.B. 1969. Journal of Food Science and Technoogy. l(6): 1

Eliane, B. and Gregory, P. 1957. French Patent. 1,001, 461, Chemical Abstracts. 51: 4597.

Escher, F.E., Koehler, P.E. and Ayers, J.C. 1973. Journal of Food Science. 38:889.

Fenwick, G.R., Heaney, R.K. and Mawson, R. 1989. Glucosinolates. In. Toxicants of Plant Origin. Vol. II, Ed. Cheeke, P.R. CRC Press, Florida. pp. 1-41.

Friedrich, J.P., List, G.R. and Christianson, D.D. 1982. Petroleum Free Extraction of Oil from Soybeans with Supercritical CO_2. Journal of the American Oil Chemists' Society. 59(7): 288-292.

Frigerio, L. and Roberts, L.M. 1998. The Enemy within: Ricin and Plant Cells. Journal of Experimental Botany. 49: 1473-1480.

Funatsu, G. and Funatsu, M. 1977. Biochemical studies of ricin. Agriculture Biological Chemistry. 41(7): 1211.

Funatsu, G., Ishiguro, M. and Funatsu, M. 1974. Purification Studies on Ricin. Agriculture & Biological Chemistry. 38(1): 65.

Funatsu, M., Funatsu, G., Ishiguro, M., Nanno, S. and Hara, K. 1971. Structure and Toxic Function of Ricin D. Proceedings of Japan Academy. 47: 715-786

Gandhi, V.M. and Mulky, M.H. 1994. Neutralization of Castor Toxins by Sal Tannins Journal of the Oil Technologists' Associationof India. 26(2): 39-42.

Gandhi, V.M., Cherian K.M. and Mulky M.J. 1994. Detoxification of Castor Seed Meal by Interaction with Sal Seed Meal. Jounral of the American Oil Chemists' Society. 71:827-831.

Gardner, W. 1960. Detoxification and Deallergination of Castor Bean. Journal of the American Oil Chemists' Society. 37(3): 142.

George, R.W. and Negi, S.S. 1958. Isolation of Ricin, Ricinine, and the Allergenic Fractions from Castor Seed Pomace from the Different Sources. Journal of the American Oil Chemists' Society. 35: 409.

Ghetie, V. and Vitetta, E. 1994. Immunotoxins in the Therapy of Cancer from Bench to Clinic. Pharmaceutical Therapy. 63: 209.

Ghewande, M.P. and Nagaraj, G. 1987. Prevention of Aflatoxin Contamination through some Commercial Chemical Products and Plant Extracts in Groundnut. Mycotoxin Research, 3: 19-24.

Ghewande, M.P., Nagaraj, G., Desai S. and Narayan, P. 1993. Screening of Groundnut Bold-Seeded Genotypes for Resistance to *Aspergillus Flavus* Seed Colonisation and Less Aflatoxin. Seed Science and Technology. 21:45-51

Goldblatt, L.A. 1965. In: Mycotoxins in Foodstuffs, Ed. G.N. Wogan, The MIT Press, Cambridge, Massachusts. U.S.A. pp.261.

Goodrum, J.W. and Kilgo, M.B. 1987. Peanut Oil Extraction with SC-CO_2 Soubility and Kinetic Functions. Transactions of the American Society of Agricultural Engineers. 30(6): 1865-1868.

Hagler, W.M., Jr. Hutchins, J.E. and Hamilton, P.B. 1982. Journal of Food Protection. 45:1287.

Heathcote, J.G. and Hibbert, J.R. 1978. Aflatoxins: Chemical and Biological Aspects Elsevier Scientific Publisheing Co. Amsterdam. pp. 112-130.

Houston, L.L. 1981. Transport of Ricin-A Chain After Prior Treatment of Mouse Leukemia Cells with Ricin B Chain. Journal of Biological Chemistry. 257(3): 1532

Houston, L.L. and Dooley. 1982. Binding of two Molecules of 4-Methyl Umbelliferyl Galactose or 4-Methyl Umbelliferyl N-Acetylgalactososamine to the B Chain of Ricin and *Ricinus Communis* Agglutinin and to Pufified Ricin B Chain. Journal of Biological Chemistry. 257(8) :4147.

Huff, W.F. and Hagler, W.M., Jr. 1985. Food Protection. 48:416.

Ishiguro, M., Funatsu, G. and Funatsu, M., 1971. Biochemical Studies on Ricin. Agricultural Biology and Chemistry. 35, 5, 729.

Ishiguro, M., Takahashi, T., Funatsu, G., Hayashi, K. and Funatsu, M. 1964. Biochemical Studies on Ricin. Journal of Biochemistry. 55(6):587.

Jakie, M.H. 1941. Patent, 124, 975, Chemical Abstracts. 35:2354.

Jenkins, F.P. 1963. Allergenic and Toxic Components of Castor Bean Meal. Journal of the Science of Food and Agriculture. 14: 773.

Jones, D. and Marascow, H. 1998. Antibodies for Gene Targetting and Intracellular Expression. Advances in Drug Delivery Reviews. 31(6):153.

Josefsson, E. and Appleqvist, L.A. 1968. Glucosinolates in Seeds of Rape and Turnip Rape as Affected by Variety and Environment. Journal of the Science of Food and Agriculture. 19:564-570.

Kabat, E.A., Heidelberger, M. and Bezer, A.E. 1947. A Study on the Purification and Properties of Ricin. Journal of Biological Chemistry. 168: 629.

Katzin B.J., Collins E.J. and Robertus, J.D. 1991. The Structure of Ricin A Chain at 2.5A°. Proteins. 10; 260.

Kobert, R. 1891. *Beitragezvr Kenntnis Deg Vegetabilischen Hamagglutine. Versuchsstationen.* 79:97.

Kodras, R., Whitehair, C.K. and MacVicar, R. 1949. Detoxification of Castor Meal. Jouranl of the American Oil Chemists' Society. 26: 641.

Kulkarni, L.G. and Ramanamurthy, G.V. 1977. Castor, ICAR, New Delhi.

Kunitz, M. and McDonald, M.R. 1948. Isolation Studies of Ricin. Journal of Gen. Physiology. 32: 25.

Lillehoj, E.B. and Ciegler, A. 1969. Applied Microbiology. 17:516.

Lillehoj, E.B., Stubblefield, R.E., Shanon, G.M. and Shotwell, O.L. 1971. Mycopathol. Mycol. Appl., 45:259.

Liener, I.E. 1969. Toxic Constituents of Plant Food Stuffs, Academic Press, New York. 2: 283

Liener, I.E. 1979. Significance for Humans of Biologically Active Factors in Soybeans and other Food Legumes. Journal of the American Oil Chemists' Society. 56:121-128.

Lein, K.A. 1970. Methods of Quantitative Determination of Seed Glucosinolates of Brassica Species and their Application in Plant Breeding of Rape Low in Glucosinolate Content Z. *Pflanzenzuchtung.* 63:137-154.

Lord, M.J., Roberts, L.M. and Robertus, J.D. 1994. Ricin: structure, Mode of Action and some Current Applications. FASEB. 8:202.

Luter, L., Wysolonzit, W. and Kashyap. S.C. 1982. Canadian Institute of Food Science and Technologists' Journal. 15:236.

Machen, M.D ., Mayura, K., Clements, B.A. and Phillips, T.D. 1991 Toxicologist. 11:280.

Maing, H.Y., Ayers, J.C. and Koethler, P.E. 1973. Applied Microbiology. 25:1015.

Mann, R. and Rehm. H. J. 1976. European Journal of Applied Microbiology. 2:297.

Marth, E. and Hadpoyle, M.P. 1979. Food Technology. 39:81

McGhee, J.E., Kirk, L.D. and Mustakas, G.C. 1965. Methods for Determining Thioglucosides in *Crambe Abyssinica.* Journal of the American oil Chemists' Society. 42: 889-891.

Megalla, S.E. and Hafez, A.H. 1982. Mycopathologia. 77: 89.

Mehan, V.K. 1999. Supplying Aflatoxin Free Groundnuts and Groundnut Products: A Real Challenge. CIPHET, Ludhiana, India.

Milsna, D., Monzingo, A.F. and Robertus, J.D. 1993. The Structure of Recombinant of Ricin Chain at 2.3 A°. Protein Science. 2: 429.

Mitsutaro, T. 1924. Chemical Abstracts. 18: 995.

Mitsuo, N., Satoshi, M., Kuzuo, S., Kenji, F., Taichiro, N. and Nubahiko, K. 1990. Applied Environmental Microbiology. 56: 1465.

Mitsuo, N., Satoshi, M., Kuzuo, S., Kenji, F., Taichiro, N. and Nabuhiko, K. 1991. *Applied Eise Kagaku*, 37:288.

Montford, W ., Villafranca, J.E., Monzingo, A.F., Ernst, S.R., Katzin, B., Rutenber, E., Nuyhen, H.X., Hamlin R. and Robertus, J.D. 1987. The Three-Dimensional Structure of Ricin at 3.8 A°. Journal of Biological Chemistry. 262: 5398-5403.

Mustakas, G.C. Kirk, L.D. Griffin jr., E.L. and Clanton, D.C. 1968. *Crambe* Seed Processing. Improved Feed Meal by Soda Ash Treatment. Journal of the American Oil Chemists' Society. 45:53-57.

Nagaraj, G. 1999. Antinutritional Constituents in Oilseeds and Oil Cakes. Final Report Of the ICAR Cess Fund Scheme. pp. 50.

Nicolson, G.L., Blaustein, J. and Marilyn, E. 1974. Characterization of Two-Plant Lectins from *Ricinus Communis* and their Quantative Interaction with a Murine Lymphoma. Biochemistry. 13: 196.

Olsnes, S. and Pihl, A. 1972. FEBS. 20: 327.

Olsnes, S. and Pihl, A. 1973. Different Biological Properties of the Two Constituent Peptide Chains of Ricin-A Toxic Protein Inhibiting Protein Synthesis. Biochemistry. 12: 3121.

Osborne, T.B., Mendel, L.B. and Harris, J.L. 1905. American Journal of Physiology. 14: 259.

Park, D.L. and Stofoff, L. 1989. Regulation Toxicology and Pharmacology, 9:109.

Park, D.L. 1993. Food Additives Contamination. 10:49.

Park, D.L. and Liang, B. 1993. Trends in Food Science and Technology. 4:334.

Park, D.L., Lee, L.S., Price, R.L. and Pohland, A.E. 1988. Journal of the Association of Official Analytical Chemists. 71:685.

Patterson, D.S.P. and Roberts, B.A. 1970. Food Cosmet. Toxicology. 8:527.

Petrosyan, E. and Ponomorev, V. 1937. Chemical Abstracts. 31: 7554.

Pinkerton S.D., Rolfe, R., Auld, D.L., Ghetie,V. and Lauterbach, B.F. 1999. Selection of Castor for Divergent Concentrations of Ricin and *Ricinus communis* Agglutinin. Crop Science. 39: 353-357.

Piva. G.E., Santi, E., Pietri, A. and Fiorentini, L. 1981. *Riv. Ital. Sostanze Grasse,* 58:289.

Poli, M.A., Rivera, V.R., Hewetson, J.F. and Merrill, G.A. 1994. Detection of Ricin by Colorimetric and Chemiluminescence ELISA. *Toxicology.* 32:1371-1377.

Purchase, I.F.H., Steyn, M., Rinsama, R. and Tustin, R.C. 1972. Food Cosmet. Toxicology. 10:383.

Ramos, M.V., Mota, D.M.and Teixeira, C.R., 1998. Isolation and Partial Characterization of Highly Toxic Lectins from *Abrus Pulchellus* Seeds. Toxicology. 36(3): 477.

Rani, B. 1994. Comparative Study of Glucosinolates Accumulation at Various Stages of Seed Development in the *Brassicas,* M.Sc. Thesis, PAU, Ludhiana.

Rayner, E.T., Dollear, F.G., 1968, Journal of the American Oil Chemists' Society. 45:622.

Rayner, E.T., Dollear, F.G. and Cofider, L.P. Jr. 1970. Journal of the American Oil Chemists' Society. 47:26.

Resic, J.L., Skrinjar, M. and Marker, S. 1991. *Mycopathologia.* 113:117.

Rhee, K.C., Natarajan, K.R., Cater, C.M. and Mattil, K.F. 1976. Journal of the American Oil Chemists' Society. 53:467.

Reymond, F., 1962. French Patent, 1,283,300, Chemical Abstracts. 57: 12976.

Robert, L., Lamb, F.I., Pappin, D. and Lord, J.M. 1985. The Primary Sequence of *Ricinus Communis* Agglutinin. Journal of Biological Chemistry. 260: 156-82.

Robertson, J.A., Teunisson, D.J. and Bourdeaux, G.J. 1970. Journal of Agricultural and Food Chemistry.18:1090.

Robertus L.M. and Lord, J.M. 1981. Protein Biosynthetic Capacity in the Endosperm Tissue of Ripening Castor Bean Seeds. Plantation. 152:420-427.

Rutenber, E. and Robertus, J.D. 1991. The Structure of Ricin B-Chain at 2.5 A° Resolution. *Proteins.* 10:260-269.

Samarajeewa, U., Sen., A.C., Cohen, M.D. and Wei. C.I. 1990. Journal of Food Protection. 53:4890.

Sang, J.P., Minchinton, I.R., Johnstone, P.K. and Truscott, R.J.W. 1984. Glucosinolate Profiles in the Seed, Root and Leaf Tissue of Cabbage, Mustard, Rapeseed, Radish and Swede. Canadian Journal of Plant Science. 64:77-93.

Sargeant, K., Sheridan, A., I'Kelly, J. and Carnaghan, R.B.A. 1961. Toxicity Associated with Certain Samples of Groundnuts. Nature. London. U.K. 192:1096-7.

Schnug, E. and Haneklaus, S. 1988. Theoretical Principles for the Indirect Determination of the Total Glucosinolate Content in Rapeseed and Meal Qualifying the Sulphur Concentration Via X-Ray Fluorescence (X-RF Method), Journal of the Science of Food and Agriculture. 45: 253-254.

Scott, P.M. 1973. Mycotoxins in Stored Grains, Feeds and other Ceral Products. In: Grain Storage-Part of a System, Eds. R.V. Sinha and W.E.Muir. Avi Publishing, USA. pp, 343-365.

Schroeder, T., Zerifer, U., Sagelsdorff, P., Friederich, U., Luthy, J. and Schlater. C. 1985. Journal of Agriculture and Food Chemistry. 33:311.

Shakunthala, M.N. and Shadaksharaswamy, M. 1987. Foods: Facts and Priniciples. New Age International Pvt. Ltd, New Delhi.

Shantha, T. and Sreenivasa Murthy, V. 1977. Indian Journal of Technology., 15:453.

Shantha, T., Sreenivasa Murthy, Y., Rati, E.R. and Prema, V. 1986. Jouranl of Food Safety. 7:225.

Simmons, B.M., Russel, S.U. 1985. A Single Affinity Column Step Method for the Purification of Ricin Toxin from Castor Bean. Analytical Biochemistry. 146: 206.

Spinks, E.A., Sones, K. and Fenwick, G.R. 1984. The Quantitative Analysis of Glucosinolates in Cruciferous Vegetables, Oilseeds and Forage Crops Using High Performance Liquid Chromatography. *Fette Seifen, Anstrichmittel.* 186:228-231.

Sreenivasa Murthy, V., Parpia, H.A.B., Srikanta, S. and Murti, A.S. 1987. Journal of Association of Official Analytical Chemists. 50:350.

Srinivas, C.V.S. 2003. Chemical Composition of Castor Seed and Studies on Isolation, Purification and Characterization of Ricin-the Toxic Protein. Thesis Submitted to the Faculty of Science, Osmania University for the Degree of Doctor of Philosophy in Chemistry. Pp.1-134.

Srinivas, C.V.S. and Nagaraj, G. 1999. Factors Influencing Ricin, the Toxic Protein in Castor Cake and its Detoxification. Journal of the Oil Technologists' Association of India, 32(1): 21-23.

Stahl, E. E., Sechutz H.K. and Mangold, 1980. Extraction of Seed Oils with Liquid and Supercritical Carbon Dioxide. Journal of Agricultural and Food Chemistry. 28:1153-1157.

Stefanis, U.S. and Ponte, J.G. 1968. Separation of Sugars by Thin Layer Chromatography. Journal of Chromatography 34:116-120.

Takahashi, T., Funatsu, G. and Funatsu, M. 1962a. Biochemical Studies on Castor Bean Hemagglutinin. Journal of Biochemisty. 51: 288.

Takahashi, T., Funatsu, G. and Funatsu, M. 1962b. Biochemical Studies on Castor Bean Hemagglutinin. Journal of Biochemistry. 52: 50.

Thies,W. 1976. Quantitative Gas Liquid Chromotography of Glucosinolates on the Micro-Liter Scale, *Fette Seifen Anstrichmittel.* 78:231-234.

Thiesen, J. 1977. Animal Food Science and Technology. 2:67.

Tholen, J. T., Shifeng, S., Truscott, R.J.W. and Roger, J.W. 1989. The Thymol Method for Glucosinolate Determination. Journal of the Science of Food and Agriculture. 49: 157-165.

Thomas, T., Linn, S. and Steven, S. 1980. Purification and Physicochemical Properties of Ricin and Aggutinin from *Ricinus Communis.* Euopian Journal of Biochemistry. 105: 453.

Tulley R.E. and Beevers, H. 1976. Protein Bodies of Castor Bean Endosperm: Isolation, Fractionation and Characterization of Protein Components. Plant physiology. 58: 710-716.

Vermorel, M., Heaney, R.K. and Fenwick, G.R. 1986. Nutritive Glucosinolates. Journal of the Science of Food and Agriculture. 37:1197-1202.

Vidal, P., Casellas, P. and Jansen, F.K. 1985. Studies on Immunotoxin: Studies on Purification of Ricin and its Subunits and Influence of Unreacted Antibodies. International Journal of Cancer. 36: 705.

Vitetta, E.S., Thorpe, P.E. and Uhr, J.W. 1993. Immunotoxins: Magic Bullets for Misguided Missiles? Immunology Today. 14: 252.

Waller, G.R. and Negi, S.S. 1958. Isolation of Ricin, Ricinine, and the Allergenic Fraction from Castor Seed and Pomace from Different Sources. Journal of the American Oil Chemists' Society. 35: 409.

Wei, C.H. 1973. Journal of Biological Chemistry. 248:3745.

Weiss, E.A. 1983. Oilseed Crops. Longmans, London.

William, M., Vilafranca, J.E., Monzingo, A.F., Ernst, S.R., Katazin, B., Earl, R., NuyHen, H., Ron, H. and Robertus, J.D. 1987. The Three Dimensional Structure of Ricin at 2.8 A°. Journal of Biological Chemistry. 262: 5398.

Wogan, G.N., Edwards, G.S. and Newborne, P.M. 1971. Cancer Research. 31:1936.

Wong, J.J. and Hsieh, D.P.H. 1976. Proceedings of the National Academy. U.S.A. 7:2241.

Youle, R. and Huang, A. 1976. Protein Bodies from the Endosperm of Castor Beans, Sub Fractionation: Protein Components, Lectins and Changes during germination. Plant Physiology. 58:703.

Yousef, A.E. and Marth, E.H. 1986. *Dairy Scencei.* 69: 2243.

Chapter 7

Procedures

Methods of Analysis

Organic material especially, seeds, grains, fruits and nuts contain mainly three groups of compounds. They are carbohydrates, proteins and fats/oils, mostly in that order. They also contain 3-5% minerals and smaller quantities of vitamins and other minor constituents. Cereal grains contain more of carbohydrates while oilseeds contain more oil. Tree nuts and even oilseeds may contain 30-50% protein while cereal grains contain 5-8% protein. The starch and sugar contents of oilseeds may be less than 10% while the oil content of cereal grains ranges around 5%. The actual content and the composition of these groups of compounds really account for the quality as well as their commercial value.

The composition and quality of grains and other plant products vary with respect to their genetic background, location or the soils under which they are grown, season or the environment, the maturity of the grain/product or its harvest stage and the post harvest processes including storage etc. Crop scientists have in recent years been bringing out made to order or designer seeds, whose quality and composition need to be examined to satisfy the needs of the quality conrol and consumer requirements. Keeping this aspect in view this chapter has been designed.

As this book deals with oilseeds, emphasis has been placed on methods of evaluation relating to thier composition and quality. This chapter deals with methods of oil, protein and sugar analysis. Oil being the major constituent of oilseeds, importance has been given to methods of oil quality evaluation including fatty acid profile. Antinutrients especially aflatoxins, glucosinolates, ricin, oxalate, phytic acid, oligosaccharides, etc. also find a place. Analytical methods of some vitamins and minerals have also been included. Only a few important and most essential procedures relating to oilseed quality have been presented. The analyst can find more details and methods of evaluation of other constituents in standard laboratory analytical manuals.

Fats, Oils and Lipids

Lipids are a diverse group of chemical molecules. They are generally insoluble in water but soluble in organic solvents. Lipids include long-chain hydrocarbons, alcohols, aldehydes, fatty acids, and their dervatives especially, glycerides, phospholipids, glycolipids, and sulfolipids. Sterols, fatty acid esters of sterols, vitamins A, D, E, and K, carotenoids, and other poly-isoprenoids also are part of the lipids. Oils and fats are mainly triglycerides and neutral molecules. Those in the liquid state at ambient temperatures are called oils while those in the solid state are called fats. Lipids and lipid components have varied uses in the food and non-food sectors. Vegetable and animal matter contain a variety of these constituents in different proportions. It is neccessry to evaluate the composition and quality of these constituents from the point of view of their utility.

Extraction of Crude Total Lipids

The samples are first ground in a hand operated/ electric grinder or wiley mill. Take 1 gm of ground sample immediately after grinding in 250 ml flask. Add 50 ml of petroleum ether to it. The mixture is stirred in an electric shaker for one hour and kept overnight. Then the mixture is filtered through sintered funnel. Two more washings are given to the residue with 5-10 ml petroleum ether each time. The filtrate collected and petroleum ether is removed completely on a boiling water bath. The weight of the residue indicates the content of lipids.

Bligh and Dyer (1959) procedure is less time-consuming and requires less sample manipulation than other methods. The organic materials are mixed with chloroform and methanol in amounts that yield a monophasic solution when combined with water in the tissue (Folch, 1957). Upon dilution with water or chloroform, a biphasic solution results. The lipids are retained in the chloroform phase, and non-lipid materials are retained in the methanol-water phase. The chloroform layer is isolated, and the lipids can be purified and saparated into classes as desired. The recovery of total crude lipids after separation from the nonlipid contaminants should exceed 95%.

Extraction and Purification of Lipids

Reagents and Materials

1. Absolute methanol and chloroform (Analytical grade, freshly distilled reagents).
2. Filter Paper: Whatman no. 1
3. Porcelain Buchner funnel: 43mm
4. Suction flask: Glass 125 ml

Procedure

The following modified Bligh and Dyer procedure may be used, for extaction of lipids: 100g of plant material is cut into small pieces or finely powdered and immediately added to 300 ml of methanol-chloroform (2:1 v/v) and 50 ml H_2O. The mixture is blended for 2 min in a mixer or waring blender. The homogenate is filtered with suction, and the residue is reblended with a mixture of 300 ml of methanol-chloroform (2:1 v/v) and 50 ml H_2O, washed with 150 ml of methanol: chloroform (2:1 v/v). The homogenate is again filtered with suction, and the residue is washed with 150 ml of methanol-chloroform (2:1 v/v). To the combined filtrates in a separatory funnel are added 250 ml of $CHCl_3$ and 290 ml H_2O and the phases are allowed to separate. The $CHCl_3$ layer is withdrawn, diuted with benzene and concentrated in vacuum. The residual lipids are immediately dissolved in 250 ml $CHCl_3$ - MeOH (1:1) and the solution cleared by centrifugation, if neccessary and stored at -10^{0}C.

Extraction of Seeds

Powder 2- 5g seeds in a Wiley mill or a mortar and pestle. Transfer the powdered sample to 50 ml centrifuge tube and extract with 20ml boiling isopropanol with vortexing for several min. Centrifuge and remove the supernatant. Repeat with 20 ml isopropanol. Combine the extracts. Evaporate on a rotary evaporator. Re-extract seed with 38ml $CHCl_3$:MeOH:H_2O (1: 2: 0.8 v/v). Add extract to isopropanol residue. Add 20 ml each of $CHCl_3$ and H_2O mixture and contrifuge. Remove the lower $CHCl_3$ phase. Dilute with benzene. Evaporate and dissolve the residue in $CHCl_3$: MeOH (1:1), centrifuge if neccessary. Store at -10^{0}C.

Quantification

If neccessry, take an aliquot of the extract into a tared glass or porcelain container, and evaporate to dryness under a stream of nitrogen on a water bath at 35 to 40^{0}C to give a dry crude total lipid preparation. Work in a fume hood. Weigh the amount of crude total lipid on an analytical balance.

Storage of Lipid Extracts

Due to general lability of lipids towards peroxidation and hydrolysis they should never be stored for extended periods of time (i.e. more than 1-2 years). Remove water by azeotropic distillation with benzene by dissolving in $CHCl_3$ containing 10-50% methanol and storing at –10 to -15^{0}C. For storage periods of upto 2 years dissolve in benzene containing 10-15% methanol or 99% ethanol(N_2 flushed). An antioxidant such as tocopherol or BHT (2,6 di-tert-butyl-4 methyl phenol) at 0.05% should be added. Store at -40^{0}C or – 80^{0}C.

Purification (Removal of Water Soluble Contaminants)

To the lipids in 10 ml of $CHCl_3$: MeOH (2:1), add 4.5ml of H_2O, mix and centrifuge. Remove upper phase and wash again with MeOH: H_2O (10:9). Dilute with benzene, bring to dryness under N_2. Dissolve in $CHCl_3$ and use the lipid extract for further analyses wherever required.

Chromatographic Purification of Lipids

Reagents and Materials

1. Solvents: Mix redistilled chloroform (200 parts) and methanol (100 parts) with 75 parts by volume of water, shake vigorously in a separatory funnel, and allow the mixture to separate into an upper phase (solvent I) and a lower phase(solvent II). Collect the phases separately and store them at room temperature.
2. Sephadex G-25, Fine bead form.
3. Chromatography tube: 1 cm by 15cm with sintered-glass bottom and teflon stopcock.

Procedure

Soak 25 g of Sephadex G-25 beads overnight in 100 ml of solvent I. Allow the beads to settle, decant the solvent, and rinse the beads four times with solvent I.

Prepare a thick slurry of beads (50-ml total volume) in solvent I. De-gas the slurry by placing it in a vacuum flask, and draw a vacuum with an aspirator for 10 to 15 min. Swirl occasionally. Pour the slurry of beads into a column containing a few ml of solvent I. Continue adding resin slurry until a resin bed, 10 cm high is obtained. Apply pressure of 1 to 2 lb/in from a nitrogen tank to settle the resin bed, and place a glass-fiber filter disk on top of the resin bed. Rinse the column with 10ml of solvent 1 and 10 ml of solvent II.

Dissolve 150 to 200 mg of the dry crude total lipids in 2 to 5ml of solvent II. Filter the dissolved lipid fraction through a medium-porosity sintered-glass funnel directly onto the Sephadex column. Rinse the funnel with 2 ml of solvent II. Apply pressure of 2 to3 lb/in^2 with a nitrogen tank to achieve a flow rate of 0.5 to 1.0 ml/min. Add solvent II and elute the column until 25ml of solvent is collected.

Concentrate the eluate under reduced pressure at room temperature until a cloudy aqueous emulsion forms. Lyophilize this emulsion to dryness. Dissolve the dried lipid in chloroform:methanol (2:1 v/v). Transfer it to tared glass stoppered vials, and evaporate the solvents under a stream of nitrogen at 35 to 40^{0}C. Complete the drying in an evacuated vacuum desiccater containing paraffin shavings. After drying, weigh the vial containing the dried lipid extract.

Recovery of lipids can be checked by using known amounts of purified total lipids in the extraction and purification steps.

Oil Analysis By Soxhlet Extraction

Principle

Lipids are extracted into non-polar organic sovents like hexane or petroleum ether and the quantity determined gravimetrically. Waxes, resins organic acids, colouring matter, and other lipophilic constituents along with fat get extracted.

Apparatus and Chemicals

1. Soxhlet extraction apparatus with flasks, condensers and heatere etc.
2. Thimbles
3. Analytical balance
4. Water bath
5. Oven, preferably vacuum oven

6. Separating funnels
7. Petroleum ether (40-60^{0}C)
8. Methanol.

Procedure

Part I: Total lipids

1. 10-15 g of oilseed sample is weighed into a thimble and the same is put inside the Soxhlet condenser.
2. Add solvent either petroleum ether (40-60^{0}C) or hexane approximately 150 ml. Run the Soxhlet for 4-6 hr.
3. The extraction is complete when a drop of the solvent taken from the drippings of the extractor leaves no greasy stain on the filter paper.
4. Remove the thimble and heat so that the extractor is filled up to about 2/3rd with petroleum ether and only a small quantity is left in boiling flask, disconnect the flask, filter the residual pet roleum ether (containing fat in solution), through sintered funnel and collect the filtrate in a weighed 100 ml beaker or a glass dish.
5. Wash the funnel and flask, four times with redistilled petroleum ether till a drop of the filtrate leaves no greasy spot. Evaporate petroleum ether in the dish and find the weight of the residue. Keep in an oven for 3 to 4 hrs for evaporation of the solvent. If necessary use vaccum to remove all solvent.
6. Dry the dish till the consecutive weights are constant. Weight of the dish with the total lipid is recorded.

Calculation

$$\text{Per cent total lipids} = \frac{\text{Weight of the dish with extract} - \text{Empty weight of dish} \times 100}{\text{Weight of the sample} \times (100\text{-M})}$$

M = Moisture percent in the sample.

Part II: Neutral lipids

1. After weighing, the total lipids are dissolved in 10 ml petroleum ether.
2. Pour the same into a separating funnel.
3. The container should be washed two more times (5ml each time) with petroleum ether and the washings transferred to the separating funnel.
4. Add 10 ml methanol (95% aqueous).

5. Shake and separate the lower layer.
6. Again shake-two times with 10 ml portions of methanol (95% aqueous).
7. Separate, pool the methanolic lower layers which contain polar lipids.
8. The upper layer contains non-polar lipids, mainly triglycerides/ oil.
9. Collect upper layer into an appropriate flask.
10. Keep it in an oven for 3 or 4 hr and weigh to obtain the neutral lipid content.

Calculation

Percent neutral lipids = % total lipids – % polar lipids

Percent polar lipids = % total lipids – % neutral lipids

Oil Content (Cold Percolation Method)

Principle

Oil / fat is extracted in the cold by repeated washing with petroleum spirit and the same estimated after removal of the solvent. This method is very simple and is useful when a Soxhlet extraction apparatus is not available.

Apparatus and Chemicals

1. Anhydrous sodium sulphate
2. Percolation tube or a funnel of suitable size
3. Oven or water bath or a hot plate
4. Porcelain cups

Procedure

1. Weigh the oilseed sample (about 10 g).
2. Crush and powder the sample with one spoonful of anhydrous sodium sulphate.
3. Transfer into a percolation tube with cotton or a funnel closed with cotton plug.

4. Give 3 washings at one hr intervals with 15ml petroleum spirit each time.
5. Collect the solvent which gets decanted to a cup whose empty weight is known.
6. Evaporate the solvent in the cup and dry in an oven for 5 to 10 min to remove all the solvent.
7. Weigh the cup with the oil

Calculation

$$\text{Per cent lipids} = \frac{\text{Weight of the cup with oil} - \text{Empty weight of cup} \times 100}{\text{Weight the sample.}}$$

Note

1. Check for any fine particles/sediment in the extract and filter or centrifuge before evaporation of the sovent
2. Completion of extraction can be checked by the absence of greasiness on a filter papers at the time of thrd washing.

Oil Content (NMR Method)

(Jambunathan, *et al.*, 1985)

Principle

The nuclei of certain materials when placed in a magnetic field absorb at a specific radio- frequency. This phenomenon is known as nuclear magnetic resonance (NMR). In the 630 gauss field provided by the permanent magnet of the analyzer, hydrogen nuclear resonance occurs at 2.7 MHz. By means of an electronic gate the narrow resonance from hydrogen in mobile liquid like moisture or oil is distinguished from the broad resonance from hydrogen in the solids.

Equipment

1. Balance,
2. Oven,
3. MK III A New Port analyzer. Standard settings: R.F. level, 100A; A.F. gain, 3000; Intergration, 32 s; Gate width, 5 gauss (Instruction manual of the instrument should be used to set up the equipment).
4. NMR tubes

Procedure

1. Fill known quantity of pure oil up to the mark in the NMR tube.
2. Place the NMR tube in the sample compartment of the Newport analyzer that is set to the standardized instrument conditions.
3. Record the NMR signal of reference oil.
4. Now fill the seed sample in another NMR tube up to the etched mark.
5. Note the weight of the sample filled.
6. Dry the sample in an oven at 105^{0}C for 16 h.
7. After drying, stopper and cool the NMR tube.
8. Note the weight of dried sample and calculate the moisture content.
9. Reset and record the signal of dry seed sample, in the NMR insturment.

Calculations

NMR oil equivalent (OE) is obtained by the equation (In dry seeds % O.E. = % Oil):

$$OE\% = \frac{\text{Weight of oil}}{\text{NMR reading of oil}} \times \frac{\text{NMR reading of sample}}{\text{Wet weight of the sample}} \times 100$$

$$\text{Oil \% (NMR) at 5\% moisture level} = \frac{(100-5) \times \text{NMR oil \%}}{(100 - \text{sample moisture \%})}$$

Precautions

1. Warm the instrument for at least 30 min before use.
2. Do not keep any iron objects near the instrument lest it would permanently damage the magnet.

Oil and Moisture Contents by Pulsed NMR (ISO 8292)

Principle

When a test sample is inserted into the magnetic field of a pulsed NMR spectrometer, an alternating electromagnetic field in the form of an intense

90° radio frequency (RF) pulse excites all the hydrogen nuclei. The free induction decay (FID) following the 90 ° pulse signal is proportional to the total number of protons from the water and oil phases of the sample. Application of a second RF pulse, a so-called 180 ° pulse, produces a spin-echo signal when only the signal from the oil phase contributes to the FID. The maximum amplitude of this echo signal is proportional to the oil content. The difference between the two amplitudes is proportional to the moisture content. Automatic conversion of the measured signals, after suitable calibration of the apparatus is made to obtain the percentages of oil or moisture.

Samples for Moisture/ Oil Content Calibration

To obtain a reliable calibration curve, the moisture contents of the calibration samples should be 6-9% for oilseeds, in general and 6 -12% for soya beans. Samples shall be of the same species (same variety or hybrid etc.) as the test samples and of similar fatty acid compositions (especially for the analysis of rapeseeds which are rich in erucic acid, or sunflower seeds etc. which are rich in oleic acid or linoleic acid or castor seeds which are rich in ricinoleic acid). Oil content has to be determined using a reference method. In general samples for analysis should be repesentative, homogeneous and free from impurities, especially metallic objects.

Pulsed Low-Resolution NMR Spectrometer

Suitable for measurement of the oil and moisture contents of oilseeds with the following characteristics.

a) Proton resonance frequency: about 10 MHz,

b) Production of 90 ° and 180 ° RF pulses for the spin-echo method,

c) Magnetization time or pulse repetition time:1 s,

d) Excitation time (e) or pulse time: *e*= 5μs approximately for the 90° pulse, *e*= 10μs aproximately for the 180 ° pulse,

e) Dead time (t) = 20μs,

f) Total dead time (Tt), comprising the excitation time (*e*) and the dead time (*t*), where: T*t*=25 μs for 90° pulses, T*t*= 30 μs for 180° pulses, *t*=20μs.

g) NMR signal sampling: 1)50μs after the 90 ° pulse on the FID, 2)7000 μs after the 90° pulse on the spin-echo.

h) Interval between the 90° and the 180° pulse: 3500 μs. Interval between the 180° pulse and the echo maximum: 3500 μs (the echo maximum corresponds to the oil signal only).

i) Number of NMR measurements per sample: 16 to 25 (i.e. the total time of measurement is between 16s and 25s).

j) Facility for storage of calibration curves (straight lines, least-squares method) of moisture and oil contents.

k) Homogeneity of magnetic field for a desirable sample volume of 35 ml

l) Start system, which automatically starts the measurement after the introduction of the sample tube.

m) Temperature-regulated magnet, in order to maintain all the NMR parameters stable.

Other Apparatus

1. Sample tubes, made of glass, suitable for use with the NMR spectrometer.
2. Analytical balance, electronic, with an accuracy of $\pm$ 0.01g, preferably linked to the NMR spectrometer so that the sample mass is recorded directly by the NMR.
3. Drying oven, capable of being maintained at 103°C$\pm$2°C. Dishes, made of glass or metal, of diameter 7cm to 10cm, and provided with lids.
4. Desiccator, containing an efficient desiccant.

Procedure

Use the set-up parameters of the NMR spectrometer recommended by the manufacturer and optimize them by preliminary tests. For all calibration and measurement operations, follow the user's manual. Ensure that all operations during calibration and measurement are carried out under the same conditions and, in particular, at the same temperature ($\pm$ 2 ° C).

A minimum of three calibration samples is necessary, although six to eight are desirable. Enter the parameters for the measurements (pulse sequence, attenuation, etc) of the oil or moisture content into the NMR spectrometer and specify a number under which the calibration curve is to be stored. The total measurement time shall be a minimum of 16s. Set the apparatus to the calibration mode. Introduce a portion of the first calibration sample into a tared sample tube up to the optimum height and transfer the sample mass from the balance to the NMR apparatus.

Enter the value of either the moisture content (as a percentage by mass) or the oil content (as a percentage by mass) into the spectrometer. Introduce the sample tube containing the first calibration sample into the measuring head and determine the moisture or oil content. Repeat the steps for the two (or more) calibration samples. Calculate automatically or manually the calibration parameters of the calibration curve. The correlation coefficient shall normally be greater than 0.95. By following this procedure the calibration curves for moisture and oil contents can be prepared. These curves are used to obtain either moisture or oil content of unknown samples.

Note

1. Calibration curves need to be prepared once every year or when the instrument is disturbed.
2. This method applies to seeds/ samples with a moisture content of less than 10%.

For seeds with higher moisture contents, drying is necessary before the oil content can be determined by pulsed NMR.

Specific Gravity/ Density of Fats and Oils

Principle

Specific gravity of any substance is it's weight of a unit volume, in comparisan to that of water. The weight of 1ml of water is taken as 1g.

Apparatus

1. Specific gravity bottle or pycnometer
2. Analytical balance

Procedure

The oil or fat sample may be preserved under an atmosphere of nitrogen in the cold. The sample should be allowed to reach the room tempertature and mixed well before being subjected to the various analytical tests.

The specific gravity of a liquid is the weight of a given volume of liquid at the specified temperature, compared with the weight of an equal volume of water at the same temperature, all weights being taken in air.

The specific gravity bottle (or pycnometer) is weighed empty (W_B)

The bottle is then filled completely with the oil/ fat sample and weighed (W_1).

The oil is removed and after cleaning, the bottle is filled completely with distilled water and weighed (W_w).

The temperature of the liquid is noted.

Calculation

$$\text{Specific gravity} = \frac{W_1 - W_B}{W_w - W_B}$$

The temperature at which the specific gravity is determined should be specified.

Colour of The Oil

Principle

The method determines the colour of oils by comparison with Lovibond glasses of known colour characteristics. The colour is expressed as the sum total of the yellow and red slides used to match the colour of the oil in a cell of the specified size in the Lovibond Tintometer.

Apparatus

1. Lovibond Tintometer
2. Glass Cells

Procedure

Filter the oil through a filter paper (or centrifuge the oil) to remove any impurities. Clean the glass cell with carbon tetrachloride and dry. Fill it with the oil and place the cell in position in the tintometer. Match the colour with sliding red, yellow and blue colours.

Report the colour of the oil in terms of Lovibond units.

Colour reading = Y+5 R or Y+10 R

Where, Y+5 R is the mode of expressing the colour of light coloured oils; and

Y+10 R is for the dark-coloured oils

Y= sum total of yellow slides used.

R= sum total of red slides used

Determination of Free Fatty Acids (FFA)/ Acid Value

Principle

Fats in grains are readily broken down by lipases into free fatty acids and glycerol during storage, particularly when the temperature and moisture contents are high. Fat hydrolysis gets accelerated due to mold growth in seeds and oil. The breakdown of polar lipids is more rapid. The acid value of a fat is the number of mg of KOH required to neutralise the free acid in 1g of the substance.

Reagents

1. 0.1 N KOH or N/10 aqueous NaOH solution accurately standardized with N/10 potassium hydrogen phthalate using phenolphalein indicator. Light pink colour is the end point. Dilute to N/40 solution.
2. Solvent: Mix 1litre ethanol and 1 litre diethyl ether in equal proportions. Neutralize shortly before use with N/10 NaOH solution using phenolphthalein as an indicator.
3. 1% phenolphthalein in alcohol

Procedure

About 10g of the oil or fat is weighed, accurately into a 250 ml conical flask to which is added 50 ml of a mixture of equal volumes of alcohol and ether previously neutralised, after the addition of 1 ml of phenolphthalien solution. If necessary, the contents may be warmed in a water bath until the substance is completely dissolved. The solution is titrated with N/10 KOH with constant shaking until pink colour persists for 15 sec. The titre value in ml (V) is noted.

Calculation

$$\text{Acid Value} = \frac{56.1\ VN}{W}$$

Where,

V = Volume in ml of Standard Potassium hydroxide or Sodium Hydroxide used

N = Normality of Sodium hydroxide/ Potassium hydroxide solution; and

W = Weight in g of the oil sample.

$$\text{Free fatty acid as Oleic acid \%} = \frac{28.2VN}{W}$$

$$\text{Free fatty acids as Lauric acid \%} = \frac{20.0\ VN}{W}$$

$$\text{Free fatty acids as Ricinoleic acid \%} = \frac{29.8\ VN}{W}$$

$$\text{Free fatty acids as Palmitic acid \%} = \frac{25.6VN}{W}$$

1

Principle

Cupric acetate reacts with the free fatty acids dissolved in benzene solution to give the blue coloured copper salts of fatty acids.

Reagents

1. *Cupric acetate-pyridine reagent*: 5% (w/v) aqueous solution of cupric acetate is prepared. The pH of the solution is adjusted to 6.0-6.2 with pyridine before making the final volume.
2. Chloroform
3. Benzene

Apparatus

1. Screw-capped culture tubes
2. Water bath
3. Centrifuge
4. Cyclomixer
5. Colorimeter or spectrophotometer

Procedure

1-3 mg of lipid sample in chloroform is taken in the screw-capped culture tube (30ml capacity). The lipid sample is evaporated to dryness. Five ml benzene is added and swirled to dissolve the lipid sample. One ml of cupric acetate-pyridine reagent is added to each tube. The contents are shaken in a cyclomixer for 2 minutes and then centrifuged for 5 min to get two layers. The upper layer is separated with the help of a dropper and the absorbance is read at 715 nm. A standard curve is prepared by taking palmitic acid in the range of 2-14 micromoles. The concentration of fatty acids is read from the graph and further calculations are made based the value.

Refractive Index

Principle

The refractive index (n) of a substance is the ratio of the velocity of light in vacuum to its velocity in the substance. It may also be defined as the ratio of the sine of angle of incidence to the sine of the angle of refraction. Refractive index generally varies with temperature and wave length of light. Refractive index of the oil sample is measured with the help of a suitable refractometer. It increases with the increase in unsaturation and also chain length of fatty acids in oils. It is useful in understanding the structure and purity of oils/fats.

Apparatus

1. *Refractometer*: Abbe refractometer (Refractive indices vary between 1.4220 to 1.4859 and can be read from the instrument to the fourth decimal place accurately. The temperature of the refractometer should be controlled to within ± 0.1^{0}C and for this purpose it should be provided with a thermostatically controlled water-bath and to circulate water through the instrument).

Calibration of the Instrument

The instrument is calibrated with a glass prism of known refractive index (an optical contact with the prism being made by a drop of Bromonaphthalene) or using distilled water which has a refractive index of 1.3330 at 20.0^{0}C.

Procedure

Clean the refractometer with alcohol and ether. A drop of oil or fat (in case of a solid fat the temperature should be suitably adjusted by circulating hot water) is placed on the prism. The prism is closed by the ground glass-half of the instrument. The dispersion screw is adjusted so that no colour line appears between the dark and illuminated halves. The dark line is adjusted exactly on the cross wires and the refractive index is read on the scale.

Usually commercial instruments are constructed for use with white light but are calibrated to give the rafractive index in terms of sodium light of wavelength, 589.3 nm at a temperature of 20°C unless otherwise specified.

Saponification Value

Principle

The saponification value is the number of mg of potassium hydroxide required to saponify 1g of oil/fat. The saponification value is an index of mean molecular weight of the fatty acids of the glycerides. Lower saponification value indicates higher molecular weight of fatty acids and vice-versa. The oil sample is saponified by refluxing with a known excess of ethanolic KOH. The alkali required for saponification is determined by titration of the excess potassium hydroxide with standard hydrochloric acid.

Apparatus

1. 250ml capacity conical flask with ground glass joints.
2. Air condenser (65 cm to one meter length).
3. Reflux condenser, to fit the flask.
4. Hot water bath or thermostatic electric hot plate

Reagents

1. KOH solution (0.5 N): 28 g KOH pellets are dissolved in 20 ml of water to which is added sufficient alcohol (95%) to make upto 1litre. The solution is allowed to stand overnight and the clear supernatant is used for the estimation. Store in brown bottle
2. 0.5 N HCl
3. 1% Phenolphthalien solution in 95% alcohol

Procedure

About 2g of the oil/fat is weighed accurately in a 250ml round bottom flask. Twenty five ml of the alcohoic KOH solution is added, a reflux air condenser is attached and the flask is kept on a boiling water bath for 1 hr. The contents of the flask are frequently mixed. When the saponification is complete, oily matter disappers and the solution becomes clear. Clarity may be achieved within one hour of boiling. While the solution is still hot, 1 ml of phenolphthalein solution is added and the excess alkali titrated against 0.5 N HCl (B). The experiment is repeated without the oil or fat to obtain the blank value (A). Since 1 ml of 0.5 N HCl is equivlent to 0.02805g of KOH, the following equation is used to calculate the saponification value.

Calculation

$$\text{Saponification value(S.V.).} = \frac{56.1 \times N \times (A-B)}{W.}$$

Where N = Normality of HCl, A = Titre value with blank,

B = Titre value with sample, W = Weight of oil

If the normality of HCl is 'N' then the above equation is multiplied by a factor of N/0.5.

Unsaponifiable Matter

Prinicple

Unsaponifiable matter is the fraction of substances in oils and fats which is not saponified by caustic alkali, but is soluble in ordinary fat solvents. It includes sterols, higher aliphatic alcohols, pigments, vitamins and hydrocarbons.

Apparatus

1. Flat bottom flask or conical flask with a ground glass joint, 250 ml capacity
2. Air condenser to fit the flask
3. Separating funnel, 500ml capacity.

Reagents

1. *Alcoholic Potassium hydroxide solution*: Dissolve 7 to 8 g of potassium hydroxide in an equal quantity of distilled water and add sufficient aldehyde free ethyl alcohol and make up to 100ml.
2. *Ethyl alcohol*: Ninety-five per cent.
3. *Phenolphthalein indicator solution*: Dissolve one gram of phenolphathalein in 100ml of ethyl alcohol.

4. *Petroleum ether (60-80^{0}C):* Analytical reagent (A.R.) grade.
5. *Aqueous alcohol*: 10 per cent of ethyl alcohol in water
6. Standard sodium hydroxide solution: Approximately 0.02N.
7. *Acetone*: Analytical reagent grade.
8. Anhydrous sodium sulphate.

Procedure

Weigh accurately 2.0 to 2.5 g of the oil/fat sample into a 250ml conical flask. Add 25ml of alcoholic potassium hydroxide solution. Boil the contents under reflux air condenser for one hour or until the saponification is complete (complete saponification gives a homogeneous and transparent medium). Care should be taken to avoid loss of ethyl alcohol during the saponification. Wash the condeser with about 10ml of ethyl alcohol.

Transfer the saponified mixture while still warm to a separting funnel, wash the flask first with some ethyl alcohol and then with cold water, using a total of 50ml of water to rinse the flask. Cool to 20 to 25^{0}C. Add to the flasks 50 ml of petroleum ether, shake vigorously, and allow the layers to separate.

Transfer the lower soap layer into another separating funnel and repeat the ether extraction three more times using 50 ml portions of petroleum ether. Wash the combined ether extracts three times with 25ml portions of aqueous alcohol followed by washing with 25ml portions of distilled water to ensure that ether extract is free from alkali and dry it with anhydrous sodium sulphate. Evaporate the ether extract on a water bath. Dry to constant weight, preferably in a vacuum oven at 70^{0}C and not over 200mm of mercury. Cool in a desiccator and weigh. After weighing, dissolve the residue in 50ml of warm ethanol which has been neutralized to a phenolphthalein end point. Titrate with 0.02N NaOH.

Calculation

Weight in g of the free fatty acid in the extract as Oleic acid = 0.282 VN

Where, V = volume in ml of standard sodium hydroxide solution
N = normality of standard sodium hydroxide solution

$$\text{Unsaponifiable matter (per cent by weight)} = \frac{100\,(A-B)}{W}$$

Where, A = weight in g of the residue
B = weight in g of the free fatty acids in the extract ,
W = weight in g of the sample

Iodine Value

Principle

The iodine value (I.V.)of a substance is the weight of iodine absorbed by 100 parts by weight of substance. This value indicates the degree of unsaturation of the oil. The stability or nutritional values of oil can be easily guessed from I.V. in the absence of fatty acid profile information.

Reagents

1. Iodine monochloride solution: 5 ml iodine monochloride is dissolved in 500 ml glacial acetic acid

   If iodine monochloride is not readily available, the solution can be prepared by following any of the procdures given below:

a) Iodine trichloride, 8g, is dissolved in about 200ml of glacial acetic acid with heating in a water bath, under moisture free condition (stoppered with a tube containing fused calcium chloride). Into another flask is take 9g of iodine. Dissolve in 300 ml of carbon tetrachloride. The solutions are mixed and sufficient glacial acetic acid is added to make the volume to 1 litre.

b) Wij's solution : 13.0g iodine (resublimed) is dissolved in 1 l of acetic acid and pass dry Cl_2 to make the titre value with N/10 hypo double the original one.

2. Sodium thiosulphate (0.1N) : Dissolve approximately 25 gNa_2 S_2O_3 solution and accurately determine the normality using potassium dichromate solution.
3. Potassium iodide, 10% aqueous solution add a pinch of Na_2 CO_3.
4. Starch (2%) indicator: Make a paste of 2g soluble starch with water and pour in 100 ml of boiling water. Keep on boiling for 2-3 min, cool and filter.

Procedure

1. Weigh oil in a glass stoppered flask (about 2.5 g)
2. Add carbon tetrachloride 15 ml.
3. Add 20ml Wij's solution. Keep the mixture for 30 min to one hr in the dark.

4. Later, take out the mixture from dark.
5. Add 15 ml 10% KI and add 100 ml distilled water.
6. Titrate with N/10 $Na_2S_2O_3$. When light yellow colour develops, add starch indicator and again titrate till white colour appears at the bottom. Note down the titration value and calculate. Thorough shaking is essential during titration.
7. Conduct a blank titration without the oil.

Calculation

$$I.V. = \frac{12.69 \times N \times (B-S)}{W}$$

Where, N = Normality of thiosulphate,
B = Blank titre value,
S = Sample titre value,
W = Weight of oil in g.

Iodine Value by Dam's Reagent

Reagents

1. Dam's reagent (0.05N pyridine dibromide solution): Add a solution of 2.0 g (2.06 ml) of pyridine in 5 ml of glacial acetic acid to a solution of 2.5 g (1.85ml) concentrated sulphuric acid in 5ml glacial acetic acid. Cool and add 2.0 g (0.63 ml) bromine. Dilute the solution to 500 ml with glacial acetic acid.
2. KI solution 10%: 10g KI in 100 ml distilled water.
3. Starch Indicator (1%): 1 g soluble starch in 100 ml of 13% KCl, boil and cool.
4. Standard $Na_2S_2O_3$: 0.02N hypo.

Procedure

To a known (5.0 ml) aliquot of lipid solution in $CHCl_3$ containing a known amount (2-5 mg) of lipid, add 5.0 ml of the pyridine dibromide solution in a 50 ml glass stoppered Erlenmeyer flask. Mix and leave at room temperature in the dark for 15 min. Add 0.5 ml KI solution, 0.5ml H_2O and a few drops of starch indicator and titrate the liberated iodine with 0.02N hypo. A blank consisting of 5 ml of $CHCl_3$ alone is run simultaneously.

Calculation

$$I.V. = (A-B)/W \times \frac{1.27}{5}$$

A = blank titration

B = sample titration

W = weight of lipid in mg.

$$\text{No. of double bonds per molecule} = \frac{I.V.}{127 \times 2} \times \text{molecular weight of the oil.}$$

Peroxide Value (P.V)

(Paquot,1979)

Definition

The peroxide value is the quantity of those substances in the sample, expressed in terms of milliequivalents of active oxygen per kg, which oxidize potassium iodide under the operating conditions described.

Apparatus

1. 3-ml glass scoop.
2. Flasks, with ground necks and stoppers, of about 250 ml capacity, dried before hand and filled with a pure, dry inert gas (nitrogen or, preferably, carbon dioxide).
3. 25-or 50-ml burette, graduated in 0.1 ml

Reagents

1. Chloroform, analytical reagent quality, free from oxygen by bubbling a current of pure, dry inert gas through it.
2. Glacial acetic acid, analytical reagent quality, free from oxygen
3. Potassium iodide, saturated aqueous solution, recently prepared, free from iodine and iodates.

4. Sodium thiosulphate, 0.01 or 0.002 N accurately standardized aqueous solution, standardized just before use.
5. Starch solution, 10g/1 aqueous dispersion, recently prepared from natural soluble starch.

Sample

Take care that the sample is taken and stored away from the light, kept cold and contained in completely filled glass containers, hermetically sealed with ground-glass or cork stoppers.

Procedure

The test has to be carried out in diffuse daylight or in artificial light. Weigh in a glass scoop or in a flask to the nearest 0.001 g, a mass of the sample in accordance with the following table, according to the expected peroxide value:

Expected peroxide value (m.e.q)	**Mass of test portion (g)**
0 to 12	5.0 to 2.0
12 to 20	2.0 to 1.2
20 to 30	1.2 to 0.8
30 to 50	0.8 to 0.5
50 to 90	0.5 to 0.3

Unstopper a flask and introduce the glass scoop containing the test portion. Add 10 ml of chloroform. Dissolve the test portion rapidly by stirring. Add 15 ml of acetic acid, then 1ml of potassium iodide solution, insert the stopper quickly, shake for 1 minute, and leave for exactly 5 minutes away from the light at a temperature from 15 to 25°C.

Add about 75 ml of distilled water. Titrate the liberated iodine with the sodium thiosulphate solution (0.002 N solution for expected values less than 12, and 0.01 N solution for expected values above 12) shaking vigorously, using starch solution as indicator. Have a duplicate analysis and a blank determination.

Calculation

The peroxide value (P.V.), expressed in milliequivalents of active oxygen per kg, is given by the formula:

$$P.V. = \frac{V \times T \times 1000}{m}$$

Where, V is the number of ml of the standardized sodium thiosulphate solution used for the test, corrected to take into account blank test.

T is the exact normality of the sodium thiosulphate solution.

m is the mass, in g, of the sample.

Take the mean of the two determinations.

Fatty Acid Methyl Esters (FAME)

(BF_3 Method)

Fatty acids in biological material can easily be identified and qualified by gas-liquid chromatography of their methyl esters prepared from total lipids, or other lipid fractions. The methyl esters can be extracted into hexane, and quantitative measurements of each fatty acid can be made with a gas liquid chromatograph (G.L.C.) equipped with the appropriate columns.

Scope

The method is applicable to common fats, oils and fatty acids. Unsaponifiables are not removed and, if present in large amounts, they may interfere with subsequent analyses. The procedure will result in partial or complete destruction of epoxy, hydroperoxy, cyclopropenyl, cyclopropyl and possibly hydroxyl groups, and is not suitable for the preparation of methyl esters of fatty acids containing these groups.

Apparatus

1. 50 ml. and 125 ml. Flat bottom boiling flasks or Erlenmeyer flasks with ST 19/ 38 or 24/40 outer necks.
2. Water cooled condensers, Liebig or West design, 20 or 30 cm Jacket, with ST 19/38 or 24/40 inner joint.
3. 250 ml separatory funnels.
4. 200 ml boiling flask for solvent removal.
5. Boiling chips – free of fat.

Reagents

1. BF_3:Methanol reagent (125g/l of methanol), commercial or may be prepared using the gas and methanol.

2. Sodium hydroxide, 0.5 N in methanol.
3. Sodium chloride, saturated solution in water.
4. Petroleum ether, redistilled, b.p. 30-60^{0}C.
5. Heptane, gas chromatographically pure.
6. Sodium sulphate, anhydrous, A.R.
7. Methyl red indicator, 0.1% in 60% ethanol.
8. Nitrogen gas, high purity.

Procedure

Accurate weighing is not required. Sample size need be known only to determine the size of flask and amounts of reagents as per the following table:

Sample (mg)	Flask (ml)	Na OH 0.5N (ml)	BF_3: Methanol (ml)
100-250	50	4	5
250-500	50	6	7
500-750	125	8	9
750-1000	125	10	12

1. For fatty acids

Introduce the fatty acids into the 50 or 125 ml. reaction flask. Add the specified amount of BF_3-methanol reagent, attach a condenser and boil for two minutes. Add 2 to 5 ml of heptane through the condenser and boil for one minute longer. Remove from heat, remove condenser and add enough saturated salt solution to float the heptane solution of the methyl esters into the neck of the flask. Transfer about 1 ml of the heptane solution into a test tube and add a small amount of anhydrous sodium sulphate. The dry heptane solution may then be injected directly into a gas chromatograph.

2. For fats and oils

Introduce the fat into the 50 or 125 ml reaction flask. Add the specified amount of 0.5 N methanolic sodium hydroxide and add a boiling chip. Attach a condenser and heat the mixture on a steam bath until the fat globules go into solution. This step takes 5-10 minutes. Add the specified amount of BF_3:methanol reagent through the condenser and proceed as described above for the fatty acids.

Notes

1. There is danger of losing some of the more volatile esters if the solvent romoval step is prolonged or if too vigorous a stream of nitrogen is used. The method may be extended to fatty acids with 8-carbon atoms if the solvent is not completely removed.

2. The methyl esters should be analyzed as soon as possible. They may be kept in an atmosphere of nitrogen in a screw cap vial at 2^0C for 24 hours. For longer storage they should be sealed in a glass ampoule under vacuum and placed in a freezer.

Other methods of FAME preparation

FAME can be prepared using many different procedures. The following is a list of some of them. The analyst can select any one depending on the ease and convenience.

1. *Simple method*: Into a tet tube, take one drop of oil, add two drops of CCl_4 and one ml of 1% sodium methoxide (0.5 g Na dissolved in 100 ml methanol). Stopper the test tube and heat on a water bath for 3min or shake the contents vigorously for 5min by hand or in a wrist action shaker. Allow to separate the layers and use the upper layer for analysis of fatty acids in a GLC.

2. Reflux (50 mg fat) with 5 ml of concentrated H_2SO_4: toluene: methanol (1:10:20) for one hr. Add 5ml H_2O and 5ml hexane. Separate upper hexane layer, dry over 3 to 5g anhydrous Na_2SO_4. Resulting dry solution is ready for injection into a gas chromatograph.

3. *FFA alone*: Diazomethane is added in slight excess to the fatty acid solution in methanol (5ml) and allowed to react for no more than 5 min. Reagent excess is destroyed by adding a few drops of 30% formic acid in methanol. The resulting solution can be directly used for GLC.

4. The oil (20 mg) in 2ml hexane is taken in a stoppered test tube. Add 0.5M Sodium methoxide in methanol (0.1ml) and shake for 5min at room temperature, then add 5µl acetic acid and 1g powdered anhydrous $CaCl_2$. Allow to stand the mixture for 1hr. Centrifuge at 2000-3000 rpm for 2-3 min. An aliquot of the supernatant can be used for GLC.

5. *KOH Method*: 100 mg oil +10ml hexane + 0.5ml methanolic KOH (2N). Mix and invert. An aliquot of the supernatant can be used for GLC (IUPAC, 1979).

6. *Castor methyl esters*: Oil (50mg) in methanol (5ml), add 0.5M Sodium methoxide in methanol (1ml). Shake for 5min while heating, cool and add glacial acetic acid (0.1 ml). Later add 0.5g Na_2SO_4 (anhydrous). Use the clear methanolic solution for injection into G.L.C.
7. Take 1 g finely powdered oilseed and extract the lipids with 20ml of $CHCl_3$: isopropanol (2:1 v/v). Separate the extract and evaporate the solvent completely and dissolve the lipids in 5 ml 5% HCl in methanol (super dry) and 5 ml dry benzene. Reflux at 80-100^{0}C for 2 hr. Cool and add 2 volumes of distilled water and neutralize with Na_2CO_3. Extract with 3 ml of petroleum ether (40-60^{0}C) thrice. Pool the extracts and dry with anhydrous Na_2SO_4. If necessary, it can be neutrlized with solid Na_2CO_3. Finally concentrate the supernatant to 1ml. Use this solution for injection to the gas chromatograph.
8. Transfer 0.5 to 1 mg of the purified total lipids to glass-stoppered vials, and evaporate the solvent under a stream of nitrogen at 30 to 37^{0}C. Add 4 ml of 5% (wt/vol) sulfuric acid in methanol to the dried lipid, and heat for 2hr at 70^{0}C. After cooling, add 4 ml of n-hexane and shake the tube for 10min. Remove and save the hexane phase. Re-extract the methanol phase with 4 ml of hexane. The pooled hexane phases are used for gas chromatographic analysis. The hexane solution needs to be concentrated, if necessary.

Fatty Acid Composition by Gas Chromatography

Principle

The separation of compounds in a gas chromatograph depends on the difference in partition coefficients between the liquid (stationary) and gaseous (mobile) phases of the constituents in a mixture. The constituents with a high affinity for the stationary phase will tend to remain in it for a long time. These will travel slowly through the column. Those with little or no affinity for the stationary phase will travel faster and get eluted quickly. Methyl esters of fatty acids are separated based on their affinities and relative mobilities between the stationary and mobile phases and determined quantitatively (Metcalfe, *et al.*, 1966, Young, *et al.*, 1984).

Scope

The method is applicable to methyl esters of fatty acids having 8 to 24 carbon atoms. The method permits quantitative separation of mixtures containing saturated and unsaturated methyl esters.

Apparatus

1. The gas chromatographic instrument should have as a minimum the following characteristics.

a. Column oven, capable of constant temperature operation to ± 0.0⁰C., at a temperature between 170 and 210⁰C.

b. Sample inlet port, independently heated to about 50⁰C higher temperature than the column oven.

c. Columns, 5 to 10 feet long 1/8 inch to ¼ inch in outside diameter, made of stainless steel, glass, copper, or aluminum tubing. Columns of 1/8 inch O.D. are packed with 8 to 12% polyester liquid phase coated on 80 to 100 mesh acid washed chromosorb W or the equivalent. Diethylene glycol succinate (DEGS) is the recommended liquid phase although for specific separations other liquid phases are employed. It is recommended that the analysis be made on two columns, each with different separating properties (polyesters of differing polarity) to check for the presence of coincident peaks.

d. Detectors, thermal conductivity (T.C.) or flame ionization (F.I.D), if separately thermostated, should be maintained above 50⁰C hotter than column temperature.

e. Recorder, 0 to 1.0, 2.5 or 5.0 mv range 1 second full scale deflection with a chart speed of ½ to 1 inch per minute. Attenuator switch to change recorder range as required. The 0-1.0 mv recorder is used when a thermal conductivity detector is employed; any one of the three may be used with the flame ionization detector.

f. Gases: For T.C.: Helium, minimum purity 99.95 mol %

For F.I.D.: Carrier gas: Helium, nitrogen, or argon, purity 99.95 mol %, Hydrogen, pruity 99.95 mol % and Air, dry, (G.C grade).

g. Syringe for injecting sample 1 to 5 ìl capacity with good reproducibility

Procedure

1. With carrier gas flowing through the apparatus, adjust to operating temperature and record a base line to check for stability of the instrument. With F.I.D., it is recommended that the aparatus be operated at medium sensitivity rather than near its maximum sensitivity. A new column should be conditioned by holding it at about 10^{0}C above operating temperature (170-200°C) with helium flowing for at least 24 hours. The column should be disconnected from the detector during the conditioning period.
2. Proper gas flow rate and temperature will permit elution of methyl linolenate, and shorter chain length esters in less than 30 minutes. When esters of fatty acids of greater chain length are present, the gas flow and/or temperature should be increased so that the retention time of the last component is reduced. In addition, the largest sample size consistent with attenuation requiremens should be used in order to better detect and quantitate these slow moving esters. A constant gas flow should be maintained throughout the analysis. When using a T.C. detector, the gas flow should be measured at the exit with a soap bubble flow meter or other suitable device. It is more difficult to measure the flow in an instrument equipped with a flame ionization detector. It is better to follow the manufacturer's instructions.
3. Measure the methyl ester sample in a syringe. When using a T.C. detector, the usual sample size is between 0.5 and 4 µ1. For F.I.D., about 0.1 to 0.01 µ1 of actual esters are measured but this is usually in a suitable solvent so that 0.5 to 4µ1 of solution is injected. Pierce the septum of the sample inlet port and quickly discharge the sample. Withdraw the needle and note on the recorder chart the peak due to air when using T.C. or solvent when using F.I.D., which marks the sample introduction reference point. Sample size must be adjusted so that the major peak is not attenuated more than eight times.
4. Watch the recorder pen to see that the peaks do not go off scale. Change the setting of the attenuator as necessary to keep the peaks on the chart paper. Mark attenuator setting on the chart.
5. After all the peaks have been traced and the pen has returned to base line, remove the chart for calculation

Calculations

1. Identify the peaks by relative position on the chart. The esters appear on the chromatogram in order of increasing number of carbon atoms and

of increasing unsaturation for the same number of carbon atoms. The fatty acid methyl esters elute in the order – palmitic (16:0), stearic (18:0), oleic (18:1). linoleic (18:2), arachidic (20:0). eicosenoic (20:1), behenic (22:0), and lignoceric (24:0), in a polar column. The C20 saturated (arachidic) ester usually appears before C18:3 (linolenic) ester but may be coincident or appear after C18:3 on some columns (Coincident peaks are usually revealed by repeating the analysis at different column temperatures or employing a second column with a different polyester).

With constant operating conditions, the retention times (or chart distances) from the air (solvent) peak to various sample component peaks can be used for identification of the peaks. However, relative retentions are more reproducible. Relative retentions are determined by dividing the observed retention time for each peak by the retention time obseved for the peak of methyl palmitate (or any other selected standard peak). Compare the observed retention times or relative retention times with those calculated from known mixtures run periodically on the same column under the same conditions.

2. Determine the area of each peak. If the instrument is equipped with an electro-mechanical or electronic integrator, the area is best measured by following the manufacturer's instructions. When an integrator is employed, the base line must remain constant or the integrator must be equipped with a base line corrector. Otherwise the area is obtained by drawing lines tangent to the sides of the peak and intersecting the base line.

   Determine the area of the resulting triangle by multiplying the height by one-half the base. For an attenuated peak the outer sides of the peak must be full chart span and the tangents drawn intersecting the base line should utilize the upper two-thirds to obtain the peak width. Determine the area by multiplying the height by one-half the base. The height must be corrected for attenuation including the correction required if the base line is not the recorder zero. Divide the area of each component by its calibration factor. The percentage of each component is calculated from the ratio of each area to the sum of the areas under all of the component peaks and reported as percent by weight.

   Area = 1.064 X h X w ½ *where* 'h' is the peak height and 'w ½' the width of the peak at half the maximum height (Littlewood, 1970).

3. Calibration factors are determined relative to methyl palmitate to correct for nonlinearity of instrument response and for molecular weight

differences. Such factors are determined by analyzing known mixtures preferably having composition similar to that of the unknown sample. Divide the area of each peak by the true weight percentage of that component, then by dividing each value by the value for methyl palmitate, the calibration factors are obtained.

4. Instrument and column performance are monitored by noting the separation of the oleate and stearate ester peaks which is expressed as peak resolution.

$$\text{Peak Resolution} = \frac{2Y}{S+O}$$

Where:

Y is the distance between the peak maxima for stearate & oleate esters

S is the base width of the stearate peak, O is the base width of the oleate peak

These values should be determined on a sample containing approximately equal quantities of oleate and stearate esters using a sample size such that peak heights are 24-50% of the chart width. If the peak resolution is equal to or greater than 1.0, then the column and instrument are in satisfactory condition. All columns when used will show a gradual loss in peak resolution. When the value becomes less than 1.0, a new column should be installed.

Precision

1. Two single determinations of major components (>5%) performed in one laboratory shall not differ by more than 1.0 percentage unit.
2. Two single determinations performed in different laboratories shall not differ by more than 3.0 percentage units.

Lipid Profile Studies

Reagents

1. Chloroform,
2. Isopropanol,
3. Acetone,

4. Heptane,
5. H_2SO_4,
6. Cyclohexane,
7. $AgNO_3$.

Equipment

1. Balance,
2. Thin layer applicator set,
3. Hot air oven,
4. Hot air blower
5. TLC solvent tank,
6. Chromatography sprayer,
7. Densitometer.

Extraction of lipids: Take 1.0g seed in a mortar, powder finely and add 20 ml of chloroform: isopropanol (2:1). Mix thoroughly in a homogenizer and filter/ centrifuge and decant Re-extract twice with 10 ml of solvent each time and pool all the filtrates/ supernatants. Concentrate the supernatant/ filtrate under vacuum to 1ml. (A)

TLC : Spot 10 µl on a silica gel G TLC plate and develop using acetone: heptane (2:88) solvent system. Spray with 20% aqueous H_2SO_4. Charr it in an oven at high temperature (300^0 C) and measure the concentration of different spots densitometrically.

Argentation TLC (for mono, di, & tri unsaturated lipids): Evaporate the solvent from extract (A) and re-dissolve it in 5 ml of cyclohexane. Spot 5 µl of the solution on a silica gel G $AgNO_3$ plate (5% $AgNO_3$ to silicagel G) along with marker solution of a standard lipid. Develop the plate in chloroform: methanol (98:2) in dark. Spray with 20% aqueous H_2SO_4 and charr in an oven at 300^0C. Measure the concentration of different spots densitometrically.

Argentation TLC (plate preparation) : Take 2g $AgNO_3$ and dissolve in 81 ml distilled water and add to 40 g silica gel G. Shake and prepare a slurry. Prepare the 0.25 mm TLC plate using a spreader. Use immediately and do not expose to light.

Determination of Lipase Activity

Principle

The method consists of incubating the enzyme solution with an olive oil emulsion and titrating the free fatty acid produced with sodium hydroxide.

Reagents

1. Gum acacia(5 %): 5 g gum acacia is suspended in 100 ml of warm water containing 0.2 g sodium benzoate. It is mixed and allowed to stand over-night.
2. Olive oil emulsion: A mixture of equal parts olive oil and 5 % of gum acacia, (reagent 1) homogenised and kept in refrigerator.
3. Buffer solution: Anhydrous sodium phosphate (Na_2 HPO_4)- 4.7 g is dissolved in 25 ml distilled water and final volume is made upto 100 ml with water.
4. 1 per cent phenolphthalein in 95 per cent alcohol.
5. 0.05 N Sodium hydroxide in distilled water.

Equipment

1. Centrifuge.
2. Homogenizer.
3. Hypodermic syringe.
4. Incubator.

Procedure

Preparation of the homogenate: The plant samples are powdered and then homogenised in 0.1 M potassium phosphate buffer, pH 7.5. The buffer is thoroughly pre-chilled to avoid enzyme destruction and the homogenization chamber during homogenization is kept immersed in crushed ice. The homogenization is carried out for 6 minutes, using three regimes of 2 min homogenization and 2 min rest to prevent rise in temperature. The slurry is filtered through four layers of muslin cloth and is centrifuged in a refrigerated centrifuge @ 20,000 g for 30 minutes at 4^0C. The filtrate supernatant is removed with a hypdermic syringe for the determination of the enzyme activity.

Assay of lipase activity: Disstilled water (3ml) and 1 ml of enzyme solution are added to two test tubes. One of these test tubes is placed in bioling water for 5 min and is then cooled to obtain the blank. Buffer solution (0.5 ml) and 2 ml of olive oil emulsion are added to both the test tubes. The contents are shaken well and incubated at 37^{0}C for 24 hours. 3 ml of 95% alcohol and 2 drops of phenolphthalein solution are then added and the contents are mixed. Each tube is then titrated with 0.05 N NaOH using phenolphthalein as an indicator until the appearance of a permanent pink colour (pH 10).

Expression of the Results

The enzyme activity is expressed on the basis of ml of 0.05 N NaOH used to titrate the FFA per g of seed material.

Phospho Lipids

Principle

These compounds contain a P atom and are constituents of membranes. The quantity of phosphorus serves as an indirect measure of the phospholipid content.

Reagents

1. *2.5% Ammonium molybdate:* Dissolve 25 g of reagent grade ammonium molybdate in 200 ml H_2O, Transfer to 1 litre volumetric flask containing 300 ml of 10N H_2SO_4 and make up.
2. *Fiske- Subbarao reagent*: To 195 ml of 15% sodium bisulphite solution add 0.5g of 1, 2, 4 amino naphthosulfonic acid. Add 5 ml of 20% sodium sulfite and mix well. (filter and store in amber bottle, stable for 4 weeks).
3. *Standard P solution* : Dissolve 0.351 g monopotassium phosphate in 10 ml of 10N H_2SO_4 and make upto 1litre in volumetric flask with distilled water D.W. (0.4mg P/5ml).

Equipment

1. Laboratory oven.
2. Spectophotometer with quarz cuvettes.
3. Waterbath.

Procedure

Evaporate lipid to dryness, add 0.5 ml of 10 N H_2SO_4. Samples are placed in an oven at 150 –160^{0}C for 3 hr. Two drops of fuming HNO_3 are added.

Keep in oven for 90 min to complete combustion. Remove and cool. Ammonium molybdate solution, 4.6 ml, + 0.2ml of Fiske- Subbarao reagent are added. Tubes covered with marbles, heated on a water bath for 7 min. The O.D. at 660 nm recorded.

Calculations

Phospholipid = O.D. value X 25

Note

The colour is proportional to the concentration of P up to 1.5μ moles in the reaction mixture.

Estimation of Crude Proteins

Principle

The biological material (plant products or tissues) is boiled (digested) with concentrated sulfuric acid. The nitrogenous compounds are converted into ammonium sulphate. The ammonium sulphate is then hydrolysed with strong alkali to ammonia which is distilled and trapped into boric acid. It is then titrated against standard acid (Mckenzie and Wallace, 1954).

$$(NH_4)_2SO_4 + 2NaOH \text{ ----------------- } Na_2SO_4 + 2H_2O + 2NH_3$$

$$3NH_3 + H_3BO_3 \text{ ----------------- } (NH_4)_3\, BO_3$$

$$(NH_4)_3\, BO_3 + 3\, HCl \text{ ----------------- } 3NH_4\,Cl + H_3\, BO_3$$

Apparatus

1. Kjeldahl flasks
2. Heaters/Burners
3. Micro Kheldahl distillation apparatus

Reagents

1. Concentrated sulphuric acid
2. Potassium sulphate (A.R.)
3. Copper sulphate (A.R.)
4. Sodium Hydroxide (40%): Dissolve 400 g of NaOH in distilled water and make up to 1 litre.

5. *Mixed indicator*: Dissolve 0.5 g of bromocrsol green and 0.1 g of methyl red in 100 ml of 95% alcohol and adjust the pH to 4.5 with dilute NaOH or HCl.
6. *Boric acid (4%):* Dissolve 40 g of boric acid (A.R.) in istilled water and make volume to 1 litre.
7. *0.01 N HCl:* Standardise against 0.01 N NaOH which has been standaridised against 0.01 N oxalic acid.

Procedure

Weigh accurately 0.5 to 1.5 g of the sample and transfer to a 300 ml Kjeldahl flask. Add 25 ml concentrated H_2SO_4, 10g potassium sulphate and a few crystals of copper sulphate (0.5 to 1.0 g) as catalyst. Potassium sulphate is used to raise boiling temperature of H_2SO_4. Digest till the sample is clear and no carbonisation remains. Cool, add a few ml of distilled water and transfer the sample to a 250 ml. volumetric flask. Rinse many times and transfer the washings to the volumetric flask. Make upto volume when contents have cooled to room temperature. Take 1-5 ml. aliquots into micro-kjeldajahl distillation apparatus and carry out the steam distillation.

Dip the tip of the condenser in 5ml boric acid solution with a drop of mixed indicator, contained in a 50 ml graduated conical flask. Before dipping the tip of the condenser in boric acid, have a blank distillation to clear the apparatus of any contamination with ammonia. Place 2-5 ml aliquots of digested sample in the distillation chamber giving a washing with 1-2 ml, distilled water. Before lowering the sample aliquot in the distillation chamber, dip the tip of the condenser in boric acid. Add sufficient concentrated NaOH to the digest in the chamber to more than neutralize the amount of acid present (determine by trial, generally 3-4ml 40% NaOH is sufficient).

Start steam distillation and steam distill the sample until 20 ml distilate has been received in the receiver. Lower the receiver and wash the tip of the condenser with distilled water. Before adding another aliquot of the digest into the distillation chamber, disconnect the flow of the steam into the chamber by pressing rubber pipe leading from boiling water flask to the distillation chamber. The residue of the sample will be removed automatically by siphoning in a few moments. A few rinsings should be given by the addition of distilled water into the chamber, starting distillation and siphoning off the chamber contents.

Titrate the contents of the receiver flask with 0.01 N HCI to light pink end point. Also test the accuracy by using standard ammonium sulphate solution.

Calculations

1 ml 0.01 N HCI = 0.140 mg N.

$$\% N = \frac{0.14 X T X 100}{1000 X W X 100\text{-}M}$$

Where,

T = Titre value, W= weight of air dry sample, and M= moisture per cent in the sample.

Protein and Total N by Micro-Kjeldahl Method

Principle

The nitrogen in the sample is converted to ammonia. In the presence of acid it becomes an ammonium salt, for example ammonium sulphate with H_2SO_4. The ammonia is steam distilled and measured by titration with acid to estimate the nitrogen content.

Reagents and Solutions

1. *Salicylic – sulphuric acid mixture:* 5 g of pure salicylic acid is dissolved in 100 ml of concentrated H_2SO_4.
2. *Catalyst*: 1g of $CuSO_4$ $5H_2O$, 8g of K_2SO_4 and 1g of SeO_2 are ground to a fine powder separately in a mortar and then mixed together.
3. *Indicator (Conway)*: A mixture of 6 ml methyl red (0.16% in 95% ethanol) 12 ml bromo-cresol green (0.04% in water) and 6 ml 95% ethanol (turns faint pink at the end point, pH 4.9).
4. *HCl (N/28):* 31.8 ml of concentrated HCl is diluted to 10 litre with water and standardized with standard Na_2CO_3 solution.
5. *Boric acid (2%)* : 20 g boric acid (A.R) is dissolved in water and made up to 1 liter.
6. Standard Na_2CO_3 solution

Apparatus

1. Kjeldahl flasks
2. Heaters/Burners
3. Micro Kheldahl distillation apparatus

Procedure

Determination of protein - N_2 0.2 to 0.5g of plant sample is taken in a 100 ml Pyrex beaker. About 20 ml of 0.5% acetic acid solution is added. The beaker is heated to boil on a hot plate to extract non-colloidal soluble nitrogenous components and to precipitate proteins. The solution is filtered through Whatman no. 44 filter paper. The residue is washed several times with hot 0.5% acetic acid solution until the filtrate is colourless. Filter with the residue is dried in an oven and then folded and placed in a 50 ml micro Kjeldahl flask and proceed as follows:

Digestion

6ml of salicylic acid-H_2SO_4 mixture is added to the filter with residue in 50 ml Kjeldahl flask. Approximately 0.5 g of sodium thiosulphate (powder) is added after allowing the salicylic-sulfuric acid mixture to soak the plant sample for half and hour. The flask is heated in a fume hood until fumes appear. The flame is removed and the flask is allowed to cool for 5 minutes. A small amount of catalyst is added and the digestion continued. Blank using only reagents without sample powder is also run simultaneously with every set of ten determinations. After cooling the digest, 2 to 3 ml of distilled water is streamed along the sides of the flask and the contents are mixed. The flask is closed with a rubber stopper and allowed to cool on a wooden stand before distillation. The total contents can be taken for distillation in a Kjeldahl distillation apparatus or made upto a suitable volume and aliquots taken for distillation in a micro Kjeldahl distillation assembly.

Total N_2: Weigh about 0.4 g of the seed / plant sample into a Kjeldahl flask. Add 100 ml concentrated H_2SO_4. Add 1ml 10% Cu SO_4 and 1g K_2 SO_4. The flask is heated in a fume hood until complete digestion takes place. The flame is removed and the flask is allowed to cool for 5 minutes.

After cooling the digest, 2 to 3 ml of distilled water is streamed along the sides of the flask and the contents are mixed. The flask is closed with a rubber stopper and allowed to cool on a wooden stand before distillation. The total contents can be taken for distillation in a Kjeldahl distillation apparatus or made upto a suitable volume and aliquots taken for distillation in a micro Kjeldahl distillation assembly.

Distillation

10 ml of above digested sample is put in a distillation chamber. Add 15-20 ml of 30 % NaOH. In a 100 ml of conical flask take 5ml of 2% boric acid and add a few drops of Conway indicator. Start distillation and collect 20-25 ml of distillate into boric acid in the conical flask. Titrate the collected distillate with N/28 HCl.

Calculation

1ml of N/28 HCl = 0.5 mg N

$$\% \text{ Nitrogen in the sample} = \frac{5 \times T \times 100}{(A\,D\,M) \times (100-M)}$$

Where T = Titre value (of N/28 HCl) – blank.

A D M = Air dry material of plant sample taken.

M = % moisture in the plant sample

Factor for Protein Content

The factor 6.25 can be used for a general purpose. This is true for proteins with 16 % nitrogen. Oilseeds generally contain globulins, whose nitrogen content is around 18%. Hence the factor 6.25 is not valid for oilseed protein calculation. The factors that need to be used for different oilseeds are as follows: groundnut= 5.46, sunflower and safflower= 5.3, soyabean= 5.71 and rapeseed= 5.67 (Mosse and Pernollet, 1983).

Protein Content

(Lowry *et al.*, 1951)

Principle

The most widely used colorimetric method for estimation of proteins is that of Lowry *et al*, (1951), using the Folin reaction. The blue color is due to the reaction of protein with copper ion in alkaline solution (the biuret reaction) and the reduction of the phosphomolybdate- phosphotungstic acid in the Folin reagent by the aromatic amino acids in the treated protein.

Equipment

1. Centrifuge
2. Spectrophotometer with quartz cuvettes

Reagent

1. Reagent A: 2% Na_2CO_3 (sodium carbonate) in 0.1 N NaOH
2. Reagent B: 0.5% $CuSO_4$ $5H_2O$ in 1% sodium citrate.
3. Reagent C: 1 ml of reagent B mixed with 50 ml of reagent A

4. *Reagent D*: Folin Ciocalteu reagent: Into a 1500 ml Florence flask introduce 100g of sodium tungstate(Na_2WO_4 $2H_2O$), 25 g of sodium molybdate (Na_2MoO_4 2 H_2O), 700 ml of water, 50 ml of 85 percent phosphoric acid, and 100 ml of concentrated hydrochloric acid, and reflux gently for 10 hrs. Add 150 g of lithium sulphate, 50 ml of water and few drops of bromine. Boil the mixture for 15 min without condenser, to remove excess bromide. Cool, dilute to 1 litre, and filter. The reagent should not have greenish tinge. Protect from dust. Dilute a portion with an equal volume of water before use.
2. *Standard solution:* Bovine serum albumin is used for preparing a standard curve. Weigh 100 mg bovine serum albumin, dissolve in 1N NaOH and make upto 100 ml in a volumetric flask. Take 10 ml and dilute to 100 ml in 1N NaOH.

Procedure

Weigh 12.5 g of finely powdered oilseed / cake sample. Add 250 ml of 0.2% NaOH. Shake for one hr. Centrifuge 50 ml of the suspension. Take 5 ml and dilute to 100 ml with distilled water. To 0.5 ml of the diluted protein solution, 5 ml of reagent C is added. The solution is allowed to stand for 10 min at room temperature and 0.5 ml of reagent D is pipetted rapidly into it with thorough mixing. The absorbance is read at 520 nm after 30 min. The protein content of the homogenate is then calculated from the absorbance using the standard curve.

Standard curve

Take 0.1, 0.2, 0.4, 0.6, 0.8, 1.0, 1.5 and 2.0 ml of diluted bovine serum albumin solution. Add 5 ml of reagent C. The solution is allowed to stand for 10 min at room temperature and 0.5 ml of reagent D is pipetted rapidly into the mixture with thorough mixing. The absorbance is read at 520 nm after 30 min. The standard curve is prepared by plottting the concentration on X axis and absorbance on the Y axis.

Calculation

$$\% \text{ Protein} = \frac{\text{ppm} \times 2.5}{W\,(100-M)}$$

Where,

ppm = value read from standard graph, W= weight of sample, M= moisture %

Note

1. When proteins are insoluble in dilute alkali, it is necessary to heat the suspension at 90°C in 1 N NaOH for 10 min to obtain complete solubilization. The standard should be treated in the same way because this treatment reduces the intensity of the color. When this procedure is applied to dissolve or extract proteins, a reagent containing 20 g of Na_2CO_3 in 1 liter of water (no alkali) is substituted for reagent A. The protein analysis procedure is otherwise identical.
2. The same quantities of various proteins give different readings with this reagent because the colour depends in part on the aromatic residues in the protein molecule. The standared curve should be prepared with samples of known concentration of the protein of interest for reliable results.

Protein by Biuret Reaction

Principle

Peptide bonds of proteins form blue-colored chelates with copper ions in alkaline solutions. This reaction is much less sensitive than the Lowry procedure, but has the advantage that it is less time-consuming and can be used in the presence of amonium salts and other materials that interfere with the Lowry procedure. There is also less variability between proteins in the amount of colour formed per unit weight of protein.

The procedure (Gornall, *et al.*, 1949) involves the use of tartrate to form a copper complex that is soluble in NaOH. The protein displaces the copper ion from the complex to form a copper protein complex with a different colour and absorption intensity.

Reagents

1. *Biuret reagent*: Dissove 1.5 g of cupric sulfate ($CuSO_4 5H_2O$) and 6 g of sodium potassium tartrate, $4H_2O$ in 500 ml of water. Add with stirring, 300 ml of a 10% (wt/vol) solution of sodium hydroxide in water. Adjust the volume of the mixture to 1 litter with distilled water. Discard if a reddish or black precipitate appears.

Equipment

1. Ultraviolet spectrophotometer. Calibrate with an absorbance standard.
2. Cuvettes: Quartz or silica,

Procedure

Weigh 12.5 g of finely powdered oilseed / cake sample. Add 250 ml of 0.2% NaOH. Shake for one hr. Centrifuge 50 ml of the suspension. Take 5 ml and dilute to 100 ml with distilled water. This dilute protein extract can be used for further analysis. Add 4.0 ml of the above biuret reagent to a neutralized sample (1.0ml) containing 1 to 10 mg of protein. Mix thoroughly, and incubate at room temperature for 30 min. Time and temperature for all samples, standards, and blanks should be the same. Measure the absorbance at 500 nm. A blank is prepared using 1.0 ml of water or aqueous solution used to dissolve proteins and 4.0 ml of biuret reagent. A standard curve can be prepared using any completely soluble protein such as crystalline bovine serum albumin. The response is not linear with protein concentration because of competition between the tartrate and protein for copper ions. From the concentration read from the standard curve, calculate the % preteins in the sample.

Protein by Ultraviolet Absorption

Principle

The absorption maximum of proteins is 280nm and is due to the presence of the aromatic amino acids, tyrosine and tryptophan which are present in all proteins. The absorption of ultraviolet light is a suitable means of estimating proteins in solution if the protein solution does not contain more than 20% by weight of other ultraviolet light absorbance compounds, such as nucleic acids or phenols. The solution also should not be turbid.

Reagent

1. Reagent A: 2% Na_2CO_3 (sodium carbonate) in 0.1 N NaOH

Equipment

1. Ultraviolet spectrophotometer. Calibrate with an absorbance standard.
2. Cuvettes: Quartz or silica, which do not prevent transmittance of ultraviolet light.
3. Centrifuge

Procedure

Weigh 12.5 g of finely powdered oilseed/cake sample. Add 250ml of 0.2% NaOH or Reagent A. Shake for one hr. Centrifuge 50ml of the suspension.

Take 5 ml and dilute to 100 ml with distilled water. This dilute protein extract can be used for further analysis. Absorbance measurements are made at 280 nm and 260 nm with the protein dissolved in a suitable buffer. A solvent blank containing the buffer is used to zero the instrument at each wavelength. If the ratio of absorbance at 280 nm to absorbance at 260 nm is not greater than 1.70, except with solutions known to contain pure protein, the following equation can be used to calculate the concentration of protein.

Calculation

Protein concentration (mg/ml) = $1.45\ A_{280} - 0.74A_{260}$

Free Amino Acids

Principle

Protein free amino acids are made to react with ninhydrin and the colour complex is estimated colorimetrically (Yemm and Cooking, 1954)

Reagents

1. Leucine standard: Dissolve 100 mg of leucine in distilled water and make upto 100 ml in a volumetric flask. Dilute this stock solution 1:10 to get 100 ppm solution.
2. Citrate buffer (pH5.0): Weigh 21g citric acid and dissolve in 1N NaOH and make upto 500 ml with distilled water. This solution could be stored in the cold with a thymol crystal added to prevent bacterial growth.
3. Ninhydrin reagent: Dissolve 4.0g of ninhydrin in 100 ml of 2-methoxy ethanol
4. $SnCl_2$: Dissolve 800mg of hydrated stannous chloride in 500 ml of citrate buffer
5. Diluent : Isopropanol : Distilled water (1:1)

Apparatus

1. Spectrophotometer with quartz cuvettes

Standard curve

Take 9 test tubes and add 0.0, 0.1, 0.2, 0.3, 0.4, 0.5, 0.6, 0.8, and 1.0ml of 100 ppm leucine solution. Add 1.0, 0.9, 0.8, 0.7, 0.6, 0.5, 0.4, 0.2, and 0.0 ml of citrate buffer to the above tubes. Add 0.5 ml of ninhydrin reagent (equal proportions of reagents 3 and 4 above mixed just before use). Shake well, heat on a boiling water bath for 20 min. Make up to 5ml with diluent. Cool and measure the absorbence at 570 nm in a spectrophotometer. Plot the readings against concentrations in ppm.

Extraction of Amino Acids

Weigh a suitable quantity of the oilseed sample (0.1g kernel or defatted meal or 0.5 g of leaf powder). Macerate the sample in 10ml of 75-80% aqueous ethanol. Filter and re-extract the residue twice more with 5ml portions of aqueous ethanol and pool all the filtrates. Make up to a suitable volume (20 or 25ml).

Amino Acid Estimation

Take 0.5 ml of the extract into a test tube. Add 1 ml citrate buffer and 0.5 ml of ninhydrin reagent (equal portions of reagents 3 and 4 above, mixed just before use). Shake well and heat to boiling on a water bath for 20 min. Make up to 5ml with diluent. Cool and measure the absorbence at 570 nm in a spectrophotometer. Read the concentration from the standard curve prepared above. Make the calculations based on the quantity of sample taken and dilution of the extracts.

Protein Digestibility (*in vitro*)

Principle

Pepsin enzyme is added to protein sample to simulate the digestion. After a specific digestion period, the enzyme activity is arrested and the decrease in protein content is used to measure the digestibility.

Reagents

1. Pepsin: Dissolve 40mg pepsin in 0.1N HCl and make upto 100ml.
2. 10%Trichloroacetic acid (TCA): Dissolve10g trichloroacetic acid in 100ml water.
3. Trichloroacetic acid (5%) : Dissolve 5g trichloroacetic acid in 100ml water
4. Borate buffer: a)Boric acid (0.2M) : Dissolve 12.4 g boric acid in water and make upto 1 litre. b)Borax solution (0.05M): Dissolve 19.05g borax in water and make upto 1 litre. Borate buffer is prepared by adding

140ml boric acid (0.2M), 50ml D.W. and a few drops of borax (0.05M) to adjust the pH to 6.8.

5. Borate buffer (0.1M, pH 6.8) containing 0.025 M calcium chloride is prepared by dissolving 3.6755 g calcium chloride in 1 litre of the buffer.
6. Pancreatin: Dissolve 50mg pancreatin in 10ml of 0.1 M borate buffer (pH6.8) containing calcium chloride.

Apparatus

1. Centrifuge.
2. Water bath.
3. Incubator.

Procedure

Two hundred mg of dried plant/seed sample is taken and incubated with 5ml of pepsin at 37^{0}C for 16 hrs in a water bath shaker. A few drops of toluene is added to check the growth of microbes. After neutralisation with 0.2N sodium hydroxide and addition of 2ml of panceratin, the digestion mixture is incubated for an additional 24 hrs at 37^{0}C in a waterbath shaker. Enzyme blanks are prepared by incubation under the described conditions with the protein sample omitted.

At the end of the incubation period the enzymes are inactivated by addition of 7ml of 10% TCA. The contents are centrifuged at 12,000 rpm for 20 min. The supernatants are pooled up and the volume made upto 25ml with 5% TCA. Five ml aliquot is taken and dried at 80-90^{0}C in a hot air oven and the nitrogen content is determined. The protein content is obtained by multiplying nitrogen content with 6.25.

Calculations

$$\text{Protein digestibility} = \frac{\text{Digested protein}}{\text{Total protein}} \times 100$$

Total Carbohydrates (Phenol: Sulphuric Acid Method)

The total soluble sugars are estimated by the method of Dubois *et al.*, 1956.

Reagents and Equipment

1. Phenol reagent: 5 g of reagent grade phenol in 100 ml of distilled water.
2. Reagent-grade concentrated sulfuric acid.
3. Thick-walled Pyrex boiling tubes: 6 inches to 1 inch diameter.
4. Spectrophotometer or colorimeter with quartz cuvettes.
5. *Glucose standard* :Dissolve 100mg of glucose in 100 ml of 0.15% (wt/vol) benzoic acid which is added as a preservative. Store at 5^0C. This solution is stable for several months. Dilute 1:10 in distilled water just before use to give a solution containing 100 µg/ml. Prepare standards, 10 to 100 µg/ml from the diluted solution.
6. *Mannose and galactose standard* : 20 mg mannose and 10 mg galactose are taken in a 100ml volumetric flask and made upto volume (300ppm solution). 10 ml of this solution is diluted to 100ml to obtain 30 ppm (30 µg/ ml)solution. The latter standard solution is more representative and preferable.
7. Aautoclave.
8. Centrifuge.

Procedure

Weigh 0.5 g plant (seed/defatted cake) sample in a 100 ml flask, add 40 ml of distilled water, autoclave at 15 lb pressure for 4 hr. Cool and filter the supernatant. Make up the filtrate to 50 ml in a volumetric flask. For test take 0.1 ml of extract in a test tube, add 2.9 ml distilled water, add 0.5 ml of saturated aqueous phenol and shake thoroughly. Add 5.0 ml of concentrated sulphuric acid carefully. Amber golden colour develops. Keep it for 30 min and measure the absorbance at 490 nm in a colorimeter. For blank take 3 ml of distilled water in a test tube, add 0.5 ml saturated aqueous phenol, shake thoroughly and add 5.0 ml of sulphuric acid, keep for 30 min. Measure the intensity of colour at 490 nm in a colorimeter. Extrapolate the concentration from the standard curve and calculate the percent sugars in the sample.

Standard Curve

Take 7 test tubes and add 0.0, 0.5, 1.0, 1.5, 2.0, 2.5 and 3.0 ml of 30 ppm standard, add 3.0, 2.5, 2.0, 1.5, 1.0, 0.5, and 0.0 ml distilled water. Shake well and add 0.5 ml of saturated aqueous phenol. Shake and add 5.0 ml concentrated sulphuric acid, shake, plug and keep for 30 min. Measure the intensity of colour at 490 nm. Plot the readings against concentrations in ppm.

Calculation

$$\mu g\ sugar = \frac{\text{Absorbance of sugar sample} \times \mu g \text{ standard.}}{\text{Absorbance of standard}}$$

Extraction of Sugars

Sugars can also be extracted from plant material with the following procedure. Take 0.5 g of the defatted oilseed sample in a mortar. Add 10 ml of 60:25:15 methanol : chloroform: water. Grind and centrifuge. Separate the supernatant. Repeat extraction of the residue with 5ml each of the above solvent mixture two more times. Centrifuge and pool the supenatants. Make up the supernatants to 25 or 50 ml.

Caution

Phenol causes severe skin and eye burns. Protect by wearing goggles, gloves, etc. when using this reagent. Do not pipette by mouth.

Total Water Soluble Sugars (T.S.S.)
(Anthrone Reagent Method)

Principle

The carbohydrates are hydrolysed by concentrated H_2SO_4 to form furfural and other degradation products. The furfural condenses with anthrone to form blue coloured complex which is subjected to quantification (Fong, *et al.*, 1953, A.O.A.C. 1965). This method is also based on Molische test.

Reagents

1. Saturated lead acetate (neutral)
2. Sodium oxalate
3. Anthrone reagent: 0.2 or 0.1 percent (w/v) anthrone in 70 percent H_2SO_4 (A.R.). The reagent is freshly prepared each day and allowed to stand for 30-40 min before use.
4. Glucose solution (10-100 μg/ml) for standard curve.

Apparatus

1. Water bath
2. Spectrophotometer with cuvettes

Extraction

One to two grams of the dried sample material is taken into thick walled pyrex test tubes. About 25 ml of distilled water is added in each tube. The contents are heated in a boiling water bath (for 15 minutes) and the supernatant, after decantation for 2 minutes, is filltered through whatman No.1 filter paper. The material which settles in the tubes is heated again with distilled water (about 20 ml) for 10 minutes and the supernatant is filtered. This process of extraction is repeated ten times after which no sugar components can be detected in the residue. All the filtrates are collected in a standard volumetric flask (250 ml). The filtrate is then clarified.

Clarifications

Before making the volume of the aforesaid volumetric flasks upto mark, enough saturated neutral lead acetate solution is added to produce flocculent precipitates. After thorough shaking, the flasks are allowed to stand for 15 minutes. The supernatant is tested with a few drops of lead acetate solution. When no further precipitate is formed the contents are diluted to 250 ml mark with distilled water. Mix thoroughly and filter through a dry filter paper into a dried conical flask. The excess of lead is precipitated with sodium oxalate and the clear solution obtained is used for analysis.

Procedure

Ten ml of anthrone reagent is pipetted into each of the thick walled pyrex test-tubes (150 x 250 mm) and diluted in ice cold water. The solution under test (exactly 1 ml) is layered on the acidic reagent, cooled for a further 3-5 minutes and then thoroughly mixed while still immersed in ice cold water. The tubes containing reaction mixture is fitted with small capillary tubes and heated in a vigorously boiling water bath for 7.5 min (in case of 0.2 percent anthrone) and 10 minutes (in case of 0.1 percent anthrone) and then immediately cooled in ice cold water. The percent transmission optical density (O.D.) of the solution is read in a spectrophotometer at 625 nm wave length against a blank. A standard curve is obtained by plotting the spectrophotometric readings of standard sugar solutions (10-100 μg/ml). The concentration of the test solution is calculated from the standard curve.

Calculations

$$\text{T.S.S. (\%)} = \frac{\text{O.D. for test solution x dilution factor}}{\text{O.D for standard x 100 x mg of glucose in the standard}}$$

Total Reducing Sugars

Principle

Reducing sugars are oxidized by alkaline potassium ferricyanide. This reduced potassium ferrocyanide is precipitated by $ZnSO_4$ as Zinc salts. Excess of potassium ferricyanide oxidises KI to I_2 which is titrated with sodium thiosulphate solution in the presence of starch indicator.

Reagents

1. Potassium ferricyanide: Take 8.25 g of potassium ferricyanide and 10.6 g of anhydrous sodium carbonate and make the solution to one litre in water. This solution is kept in a dark bottle and is not used until 2 or 3 days after preparation.
2. Glacial Acetic Acid (5%).
3. Iodine solution (KI+ ZnSO4 + NaCl). Dissolve 12.5 g KI, 25.0 g $ZnSO_4$ and 125 g NaCI in 500 ml of water. This solution is filtered through a double filter paper. Keep in dark.
4. 0.01 N sodium thiosulphate.
5. 1 percent starch as indicator.

Apparatus

1. Water bath

Procedure

5 ml of the test sugar solutions are mixed with 5 ml. of potassium ferricyanide solution in each test tube. The test tubes are closed with rubber corks (fixed with capillary tubes bent at right angle). Heat on a boiling water bath for 15 minutes and then cool immediately in ice cold water. To each tube 5 ml of iodine solution is added followed by 3 ml of 5 per cent glacial acetic acid. The liberated iodine is immediately titrated against 0.01 N sodium thiosulphate using 1 per cent starch as an indicator near the end point. Blank experiment is also run alongside.

Calculations

mg reducing sugars = b (X + 0.05)

b= a factor for invert sugar which is 0.338

X = volume of 0.01 N sodium thiosulphate

(Difference of the two titre values i.e. blank minus titer value of test sample)

Sugars (Alkaline Ferricyacide – Colorimetric Method)

Principle

Sugars above pH 10.5 reduce ferricyanide to ferrocyanide which react with ferrous ions to produce Prussian blue (Whistler, and Wolfrom, 1962).

Reagents

1. Alkaline cyanide: 0.53% w/v Na_2CO_3 containing 0.065% KCN.
2. Potassium ferricyacide : 0.05% w/v in water
3. Ferric ammonium sulphate: 1.5 g dissolved in 1 l of 0.05 NH_2SO_4

Apparatus

1. Spectrophotometer with quartz cuvettes
2. Water bath

Procedure

Mix 1-3ml of sugar solution containing 1 to 9µg reducing sugars. Add 1 ml alkaline cyanide + 1 ml potassium ferricyanide. Heat the mixture for 15 min on a water bath. A blue colour develops. Measure at 700 nm in a spectrophotometer. A standard curve is obtained by plotting the spectrophotometric readings of standard sugar solutions (10-100 µg/ml) against their absorbence. The concentration of the test solution is read from the standard curve and further calculations are made based on the sample taken and extract dilutions.

Determination of Total and Reducing Sugars

(Somogyi, 1940)

Reagents

1. *Solution A:* (Phosphate - copper reagent): 14g of anhydrous disodium phosphate (Na_2HPO_4) and 20 g of Rochelle salt are dissolved in 350 ml of distilled water. 50 ml of 1N NaOH solution and 40 ml of 10% Cu SO_4 solution are slowly added to it while stirring. Finally 90 g of anhydrous sodium sulphate and 12.5 ml of 1N potassium iodate (3.567 g KIO_3/100 ml) solution are added. The entire solution is then made up to 500 ml and allowed to stand for two days during which time impurities separate out. The clear top part of the solution is decanted and the remainder filtered through a good grade filter paper. If kept in a stoppered Pyrex bottle and protected from strong light, the solution will remain stable for months.
2. *Solution B:* 2.5% KI solution: Alkalinised by a knife tip of Na_2CO_3. Prepare solution afresh each week and store in a dark coloured bottle.
3. *Solution C:* Approximately 2N H_2SO_4 made by diluting concentrated H_2SO_4.
4. *Solution D:* 0.005 N Hypo (sodium thiosulphate) solution. This is prepared by diluting 50 ml of 0.1 N hypo solution to a litre, incorporating 2ml of 10% NaOH solution in it for protection against atmospheric carbon dioxide and for increasing the stability of the solution.

Apparatus

1. Hot plate
2. Water bath

Extraction

2.5g of powdered sample is placed in a 250 ml Pyrex beaker. Add 0.5 g of $CaCO_3$ and 100ml of distilled water and boil gently for 15 minutes on a hot plate. The beaker is cooled to room temperature and the contents are quantitatively transferred to a 250 ml volumetric flask and 5 ml of saturated neutral lead acetate solution are added and the contents of the flask are thoroughly mixed and allowed to stand for 15 minutes. Volume is made up to the mark with D.W. The contents are shaken well. 30 to 50 ml portion is decanted off. Sufficient anhydrous disodium phosphate (Na_2HPO_4) is added to the decanted solution to precipitate all the lead. The contents are shaken well once again and filtered through dry Whatman no. 42 fluted filter paper. A few drops of the filtrate are tested with disodium phosphate to make sure that all the lead is precipitated. Filtrate is re-filtered if necessary using a fresh dry filter paper. The filtrate thus obtained is called "clarified extract" and is stored in stoppered flask in a refrigerator and can be used for the determination of total and reducing sugars.

Determination of Reducing Sugars

10 ml of clarified extract is placed in 50 ml volumetric flask, diluted to volume and mixed. 5 ml of this solution is used for determination of reducing sugars using Somogyi's phosphate copper reagent on the same day as per the Iodimotric procedure given below.

Determination of Total Sugars

10ml of clarified extract is placed in 50 ml volumetric flask. 5ml of (1:4v/v) HCl are added. The contents are mixed well and allowed to stand for 24 hours at room temperature to invert the sugars. The solution is neutralized with normal NaOH solution using a piece of red litmus paper as indicator and keeping the flask in ice-cold water bath while NaOH solution is added slowly. Solution is made upto volume and mixed. 5 ml of this solution are used for the determination of total sugars using Somogyi's phosphate-copper reagent as per the Iodimetric procedure given below.

Iodimetric procedure

5 ml of solution A (phosphate-copper reagent) and 5ml of sugar solution are mixed in a 25x200 mm Pyrex test-tube, covered with a cotton plug, and heated by immersion in a vigorously boiling water bath for exactly fifteen minutes. They are firmly held in metal racks during the heating to avoid undue agitation. At the end of the heating period the tubes are removed to a container with running tap water and cooled for three minutes to about 30^{0}C. After cooling, 3 ml of solution B are added by running it from a pipette down the wall of the test tube, without stirring or agitation. Following this, about 1.5 ml of solution C is rapidly dropped (rather than permitted to flow into the test tube) with simultaneous agitation, so that the entire contents of the tube are mixed and acidified at once. A graduated pipette with a cracked off tip serves the purpose. The cotton plug is replaced and the solution is allowed to stand for five minutes. It is then titrated with solution D using starch solution (1 ml of 1%) as indicator at the end stages.

A pair of blanks (5ml of water with 5 ml of phosphate-copper reagent) is run with each set of ten determinations.

Calculations

$$\% \text{ Reducing sugars (as glucose)} = \frac{338 \times (B-T)}{A.D.M. \times (100-M)}$$

Where B = Titre value for blank

T = titre value for the sample.

A.D.M = Air dry material of sample, M= % moisture in it

The above calculation is based on the assumption that 1 mg of pure glucose solution gives a (B-T) value of 7.4 ml of 0.005 N thiosulphate.

N.B

1. With these reagents it is possible to determine reducing sugars equivalent to 2.0 to 3.0 mg of glucose in 5 ml of sugar solution.
2. The reagent is quite suitable when the (B-T) value falls between 15 and 20 ml of 0.005 N hypo.
3. For 2-5% reducing sugars- add 5 ml of N KIO_3 per 500 ml for 1-2 % or less than 1% - Add 2.5 ml of N KIO_3 per 500 ml 6-32% – Add 12.5 ml of N KIO_3 per 500 ml.

Determination of Crude Fibre

Principle

All soluble/digestible material are removed in a fat free sample using strong acids and alkalies. Insoluble fibre with minerals remain which is weighed. Minerals are separately determined by burning and deducted to obtain the crude fibre.

Reagents

1. Sulphuric acid (0.255 N or 1.25% w/v) : To 198. 58 ml of water, 1.42ml of concentrated H_2SO_4 is added to obtain 200ml of 0.255 N H_2SO_4
2. NaOH (0.313 N or 1.25. w/v) : Dissolve 2.504 g of NaOH in 200 ml of D.W
3. Ethyl alcohol 95%.
4. Petroleum ether/hexane (65^{0}C).

Apparatus

1. Hot plate
2. Muffle furnace

Procedure

Two grams of moisture and fat free sample is weighed into a 500 ml beaker. To this 200ml of boiling 0.255N H_2SO_4 is added and the mixture is boiled for 30 min keeping the volume constant by the addition of water at frequent intervals. At the end of this period, the mixture is filtered through a muslin cloth and the residue is washed with hot water till free from acid. The material is then transferred to the same beaker and 200 ml of boiling 0.313 N NaOH is added. After boiling for 30min (keeping the volume constant as before) the mixture is filtered through a muslin cloth. The residue is washed with hot water till free from alkali followed by washing with some alcohol and ether. The residue is then transferred to a crucible, dried overnight at 80-100^{0}C and weighed (M_1). The crucible is heated in a muffle furnace at 600^{0}C for 2 to 3 hrs, cooled and weighed again (M_2). The difference in the weights (M_1-M_2) represents the weight of crude fiber.

Calculation

$$\text{Crude fiber (on dry basis) \% by mass} = \frac{100\,(M_1 - M_2)}{M}$$

Where M_1= mass in g of gooch crucible contents before ashing.

M_2= mass in g of the gooch crucible contents after ashing.

M = mass in g of the dry fat free sample taken for the test.

Estimation of Cellulose

Principle

The cellulose content of the sample can be determined after it's hydrolysis to simple sugars and their estimation carried out by anthrone reagent method.

Reagents

1. Acetic/nitric reagent: Mix 150 ml of 80% acetic acid and 15ml of concentrated nitric acid

2. Anthrone reagent: Dissolve 200mg anthrone in 100 ml concentrated H_2SO_4. It is prepared freshly and cooled for 2 hrs before use.
3. H_2SO_4 (67%).

Apparatus

1. Water bath
2. Vortex mixer
3. Spectrophotometer with quartz cavettes.

Procedure

To 500mg of the sample, 3ml of acetic/nitric reagent is added and mixed in a vortex mixer. Then the test tubes are placced in a boiling water bath for 30 min, cooled and then the contents are centrifuged for 15-20min. The supernatant is discarded and the residue is washed with distilled water. To this add 10ml of 67% H_2SO_4 and allowed to stand for 1 hr. One ml of this solution is diluted to 100 ml. To 1ml of this diluted solution, 10 ml of anthrone reagent is added and mixed well. The test tubes are then heated in a boiling water bath for 10min, cooled and the colour developed is read at 630nm. A blank is also set up with anthrone reagent and distilled water.

Standard Curve

To 100mg of cellulose, 10ml of 67% H_2SO_4 is added and allowed to stand for 1hr. For standard curve, a series of volumes of 0.4 to 2ml corresponding to 40-200 μg of cellulose are taken, and 10ml of anthrone reagent is added and mixed well. It is then heated for 10min in a boiling water bath, cooled and the intensity of the colour developed is read at 630nm. A standard graph is prepared by plotting concentration versus O.D. of standard cellulose.

Calculations

The content is read from the standard graph and further calculations are made to obtain the percent cellulose in the sample based on weight taken and dilution made, if any.

Starch – Method-I

Principle

Starch is hydrolysed to simple sugars, which are estimated by the anthrone method.

Reagents

1. Anthrone : Dissolve 100mg of anthrone in 100ml of ice cold 95% sulphuric acid.
2. Ethanol (80%).
3. Perchloric acid (52%).
4. Glucose standard(Stock solution): Dissolve 100mg of glucose in 100ml of water.
5. Working standard: Dilute 10 ml of the stock solution to 100ml with water.

Apparatus

1. Centrifuge
2. Water bath
3. Spectrophotometer with quartz cuvettes.

Procedure

Hundred mg of the sample is homogenized in hot 80% ethanol to remove sugars. This is then centrifuged and the residue is retained. The residue is washed repeatedly with hot 80% ethanol till the washings give no colour with anthrone reagent. The residue is then dried over a water bath. To the dried residue is added 5ml of water and 6.5ml of 52% perchoric acid. It is then extracted at 0^0C for 20 min, centrifuged and the supernatant is saved. The extraction is repeated using fresh perchloric acid. Again it is centrifuged and the supernatants are pooled and made upto 100ml. Pipette 0.1 or 0.2ml of the supernatant and make upto 1ml with water. Standard curve is prepared by taking 0.2,0.4,0.6,0.8, and 1ml of the working standard and volume is made upto 1ml in each test tube with water. Four ml of anthrone reagent is added to each tube followed by heating for 8 min in a boiling water bath. It is then cooled rapidly and the intensity of green to dark green colour that developes is read at 630 nm.

Calculation

The glucose content in the sample is calculated from the standard graph. The value is multiplied by a factor 0.9 to arrive at the starch content.

Starch - Method-II

Principle

Starch can also be estimated after it's enzymatic hydrolysis by the phenol -sulphuric acid method (Southgate, 1976 and Dubois *et al.*, 1956). Starch comprises of amylose and amylopectin. On hydrolysis by glucosidase enzyme, starch is converted into maltose and finally to glucose. The sugar thus obtained is analyzed quantitatively using phenol - sulphuric acid reagent.

Equipments

1. Balance,
2. Autoclave,
3. Water bath,
4. Spectrophotometer,
5. pH meter
6. Water bath shaker.

Reagents

1. Ethyl alcohol
2. 2M sodium acetate buffer, pH 4.8: Dissolve 164 g sodium acetate in water and add 120 ml glacial acetic acid, adjusting the pH to 4.8 and make up to 11 litre.
3. Enzyme- Amyloglucosidase (Sigma)
4. Nitrogen free paper.

Procedure

1. Weigh 75mg defatted flour into a 50 ml conical flask.
2. Add a few drops of ethyl alcohol to disperse the flour, if necessary
3. Add 10 ml of distilled water
4. Cover the flask with nitrogen free paper.
5. Autoclave the contents for 90 minutes at 19 1b (126^0 C) to gelatinizing the starch.
6. Cool and add 1 ml of 2M sodium acetate buffer.
7. And 25mg amyloglucosidase and approximately 15 ml distilled water

8. Incubate the contents after covering with parafilm at 55^{0}C for 2h in a water bathshaker (for hydrolysis)
9. After incubation, make up the volume to 250ml.
10. Pipet 10 ml aliquot and dilute to 100 ml.
11. Pipet 1 ml of aliquot of the diluted extract and add 1 ml 5% phenol and 5ml 96% sulphuric acid, and mix well.
12. Cool and read the absorbance at 490 nm against the blank.
13. Also develop color with standard glucose (10, 20, 30, 40, 50 μg/ml and read the absorbence.
14. Develop the color with water, phenol and sulphuric acid (for blank).
15. Measure absorbances against a blank in a spectrophotometer at 490 nm.

Calculations

$$\text{Distribution factor} = \frac{\text{Concentration of standard. } (\mu g)}{\text{Absorbence of standared}} \times \frac{1 \text{ (conversion for g)}}{10,00,000}$$

$$\times \frac{250 \text{ ml (sample dilution)}}{0.1 \text{ ml (taken for color development)}} \times \frac{100 \text{ (per cent)}}{0.075 \text{ g (sample weight)}} \times 0.9$$

Distribution factor x Absorbance of sample = % of starch.

0.9 is the conversion factor for sugars.

Antinutritional Factors

Phenols by *Folin Ciocalteau* Method

Principle

The estimation of phenols is based on the reaction with an oxidizing agent phosphomolybdate forming a blue complex. The intensity of the complex is measured colorimetrically.

Reagents

1. Chloroform: methanol : water : (25:60:15) or Alcohol
2. 20% Na_2CO_3 in water

3. Diluent : Isopropanol : water (1:1)
4. F.C. (*Folin – Ciocalteau*) Reagent (F.C, : water: 1:1, dilute ready made reagent with distilled water). If not, prepare the solution as follows: Dissolve 100 g of sodium tungstate and 25g of sodium molybdate in 700ml distilled water. Add 50ml of 85% orthophosphoric acid and 100 ml concentrated hydrochloric acid and 150g lithium sulphate dissolved in 50 ml water and 4-5 drops of liquid bromine. Boil the mixture without any condenser for about 15 minutes to remove excess bromine. Cool and dilute to one litre with water and filter. The reagent should be golden yellow in colour (store the solution in a brown bottle, preferably in a refrigerator). Just before use dilute 1 volume with 2 volumes of water. If the reagent acquires a green colour, it is not fit for use. However, it can be regenerated by boiling with a few drops of bromine. Boil out excess bromine.
5. Standard: Catechol or phenol or caffeic acid (10 ppm solution, 10 mg/1000ml).

Apparatus

1. Water bath
2. Spectrophotometer with quartz cuvettes
3. Centrifuge.

Standard Curve

Take 0.0 0.1, 0.2, 0.3, 0.4, 0.5, 0.6, 0.7 and 0.8 ml of standard catechol (10 ppm solution) to a 25 ml volumetric flask and add 0.5 ml of F.C. reagent (1:1) add 2.0 ml of Na_2SO_4 solution, shake thoroughly and boil over a hot water bath for one minute. Take out and cool under running tap water. Make up to volume with diluent and measure the absorbance at 650 nm. Plot a graph with absorbance versus concentration.

Extraction of Phenols

Weigh 0.5g finely powdered plant sample in a centrifuge tube and add 5.0 ml of reagent No.1 and shake for 10 min. Centrifuge and take the supernatant in a 10 ml volumetric flask. Wash the residue twice with the solvent mixture (2ml each) and pool the supernatants and make upto 10 ml.

Estimation of Phenols

Take 0.1 ml of the above extract, add 0.5 ml F.C. reagent (1:1) and 2 ml of 20% Na_2CO_3 in 25 ml volumetric flask. Shake thoroughly and keep the sample in boiling water bath for 1 min. Immediately take out and cool under running tap water. Make up the volume to 25 ml by adding the diluent, and read the absorbance at 650 nm in a spectrophotometer.

Calculations

% Phenols = ppm/ W x 100

ppm = Concentration in ppm from the graph,

W = Weight of sample.

Chlorogenic Acid (HPLC Method)

(Srinivas *et al.*, 2007)

Principle

This method is useful in separating phenolic acids like chlorogenic acid through HPLC. Identification and quantification also can be carried out.

Reagents and Apparatus

1. HPLC equipment with C-18 column, a UV-Visible spectrophotometric detector and a data recorder.
2. Methanol 90% aqueous.
3. Potassium hydroxide 5% aqueous.
4. Sulphuric acid 5% aqueous.
5. Diethyl ether.
6. Mobile phase: Methanol 100% (HPLC grade).

Isolation of Phenols

The phenols can be isolated as per the method of Stedmann, (1957), Stedmann and Swain (1962). 20g of defatted oil seed powder is extracted with 50 ml portions of 90% methanol thrice. The pooled extract is filtered. The methanol solution is concentrated to 50 ml and re-extracted with diethyl ether (50-100 ml). The phenols are extracted with 5% KOH thrice (25, 20 and 20 ml portions). The pooled aqueous solution is then acidified using 5% H_2SO_4.

The acidic layer is saturated using sodium bicarbonate. The aqueous solution is extracted thrice with diethyl ether(10, 5 and 5ml portions). Finally the combined ether layers are concentrated to 1 ml.

Separation and Characterization of Phenolic Acids (HPLC)

20μl of the concentrated phenolic extract is injected into the pump and separated on a C-18 column (2.5 X 30 cm) of the HPLC instrument. The flow rate is 0.5ml per minute under isocratic conditions with the mobile phase being 100% methanol. Identification of phenolic acids is based on comparison of RT values of standard phenols-cinnamic acid, ferulic acid, syringic acid, coumaric acid, chlorogenic and delphic acids.

Estimation of Oxalic Acid

Principle

Oxalic acid is extracted and estimated by titration with $KMnO_4$ solution.

Reagents

1. Hydrochloric acid (2 N): To 82.91 ml distilled water 17.09 ml concentrated HCl is added.
2. Phospho tungstate reagent : Dissolve 24 g sodium tungstate in water. To this 40ml of phosphoric acid is added and the solution made upto 1 litre with water.
3. Methyl red (0.1%).
4. $KMnO_4$ (0.01N) : Dissolve 0.0316 g $KMnO_4$ in distilled water and make upto 100ml.

Procedure

Extraction of the sample: To 5 g of the well ground sample, 100ml of 2N HCl is added and the mixture shaken well for about 22hrs in a mechanical shaker. It is then centrfuged and filtered. The filtrate is transferred into a beaker and boiled for 15 min. It is then cooled to room temperature. Some distilled water is added followed by 2N HCl to make upto 100ml.

To 20ml of th filtrate, 2 or 3 drops of methyl red is added and neutralised with ammonia. Five ml of calcium chloride buffer is added and stirred well. The mixture is allowed to stand overnight and later it is filtered through

Whatman No. 40 filter paper and washed thoroughly to free the chlorides with distilled water. The precipitate along with the filter paper is transferred to a beaker and some distilled water is added followed by the addition of 5ml of 2N sulphuric acid. The mixture is heated to 80°C over a burner and titrated against N/100 potassium permagnate solution.

Calculations

1ml of 0.01 N $KMnO_4$ = 0.45 mg oxalic acid.

mg oxalic acid/100g sample =

$$\frac{\text{Titre value x Normalityof } KMnO_4 \text{ x dilution factor}}{\text{Weight of the sample}}$$

Estimation of Phytic Acid
(Wheeler and Ferrae, 1971)

Principle

The phytic acid and phytic acid chelates react with ferric chloride and form ferric phytate. The available ferric ion after reaction is determined by developing blood red colour with KSCN.

Reagents

1. $FeCl_3$ solution: Containing 2 mg ferric iron per ml in 3% TCA
2. 5% TCA: Dissolve 5g TCA in 100ml water
3. 1.5M KSCN solution.

Apparatus

1. Centrifuge
2. Incubator
3. Spectrophotometer with quartz cuvettes.

Procedure

250mg of defatted sample is taken in a 15ml centrifuge tube. 7.5ml of 5% TCA (trichloroacetic acid) is added while shaking. The mixture is incubated for 10 min at 60°C. The supernatant is collected after centrifuging at 5000 rpm and kept for 10min. The supernatant is then transferred into 25ml standard flask and made up to the volume by using distilled water. 20ml is pipetted from the flask, to which 5ml of ferric chloride solution is added and heated in a water bath at 95°C for 45 min. The reaction mixture is cooled and filtered using Whatman 42-filter paper. 2ml of KSCN is added and the colour developed is measured at 485nm. Calculations are made based on values obtained with standard phytic acid.

Estimation of Polyphenols

Principle

Polyphenolics are estimated by Folin Denis method (AOAC, 1984).

Reagents

1. Folin - Denis reagent: To 750ml water, add 100g sodium tungstate ($Na_2WO_4.2H_2O$), 20g phosphomolybdic acid and 50 ml phosphoric acid. Reflux for 2 hr, cool and make upto 1 litre with water.
2. Saturated sodium carbonate solution: Dissolve 45 g of Na_2CO_3 in 100ml distilled water at 70-80^{0}C and cool overnight. The supersaturated solution is then filtered through glass wool.
3. Tannic acid standard solution: Dissolve 100 mg tannic acid in 1 litre water. Fresh solution is prepared for each determination.
4. Methanol - HCl: Mix 10 ml concentrated HCl with methanol and make up to 1 litre with methanol.

Apparatus

1. Vortex mixer/cyclomixer
2. Water bath
3. Spectrophotometer with quartz cuvettes.

Procedure

Weigh 200 mg defatted sample into a 250 ml round bottom flask. To this 100 ml of methanol - HCl is added and the contents are refluxed for 2 hrs and allowed to cool. The extract is filtered through Whatman No. 40 filter paper into a 100 ml volumetric flask and volume made up with methanol-HCl, after a few washings. The extract is taken for estimation of polyphenols.

Standard Curve

Pipette 0.2, 0.4, 0.6, 0.8, and 1 ml of standard tannic acid solution into 20ml test tubes. Volume is made upto 5ml with water. Add 0.5 ml of Folin-Denis reagent and vortex it. To all the tubes 1 ml of saturated sodium carbonate solution is added. Make up the final volume to 10 ml and mix well using cyclomixer. Allow to stand for 30 min and read the absorbence at 760 nm.

Sample

One ml of sample extract is taken in triplicate in 20ml test tubes. Volume is made upto 5ml with water. Add 0.5 ml of Folin- Denis reagent and vortex it. To all the tubes 1 ml of saturated sodium carbonate solution is added. Make up the final volume to 10 ml and mix well using cyclomixer. Allow to stand for 30 min and read the absorbence at 760 nm. Using the standard curve intercept the concentration and calculate the polyphenols in the experimental material.

Estimation of Tannins

Principle

Tannins are estimated by Vanillin Hydrochloric acid method (Price *et al.*, 1978). The colour complex formed absorbs at 500nm.

Reagents

1. Hydrochloric acid:methanol(8%):Mix 8ml concentrated HCl with 92ml methanol
2. Vanillin(1%): Dissolve 1g vanillin in methanol and make upto 100ml.
3. Vanillin - HCl reagent: Mix equal volumes of solutions 1 and 2 before use.
4. HCl in methanol (4%): Mix 4 ml concentrated HCl with 96 ml methanol.
5. HCl in methanol (1%): Mix 1ml concentrated HCl with 99ml methanol.
6. Standard solutions: A stock solution containing 1mg/ml catechin is prepared in methanol (0.1ml of stock is equivalent to 100 μg catechin).

Apparatus

1. Centrifuge
2. Spectrophotometer with quartz cuvettes.

Tannin extraction

Five hundred mg defatted sample is weighed and transferred to a centrifuge tube. To this 10ml acidic- methanol (1%) is added and shaken for 20 min on a shaker. The tubes are centrifuged for 10 min and the supernatants transferred to a 25ml volumetric flask.

To the residue in the centrifuge tubes 5ml acidic methanol (1%) is added and shaken for 20min. The tubes centrifuged for 10min and the supernatants are transferred to the first extraction. The volume is made upto 25ml and mixed well.

Standard Curve

The standard solution is taken at levels of 0.1,0.2, 0.3, 0.4,0.5 0.6, 0.7, and 0.8 ml in different test tubes. The volume is made upto 1ml in all the tubes. Five ml of vanillin – HCl reagent is added and the intensity of the colour developed is read in the spectrophotometer at 500nm against a sample blank. Sample blank is prepared by the addition of 5 ml of 4% HCl in methanol to 1ml of the sample solution. The absorbence is plotted against different concentrations of standard.

Procedure

One ml extract is taken in triplicate into three 20ml test tubes. To this 5ml of freshly prepared Vanillin – HCl reagent is added slowly and the intensity of the colour developed is read in the spectrophotometer at 500 nm. The concentration of the sample is intercepted from the standard graph.

Calculations

Catechin equivalents (CE%) =

$$\frac{(\text{mg catechin/ml}) \times \text{volume made up(25 ml)} \times 100}{\text{Volume of extract taken (1 ml)} \times \text{weight of the sample (500mg)}}$$

Aflatoxins by Thin Layer Chromatography (TLC)

(Pons *et al.*,1966)

Principle

The aflatoxins are extracted into chloroform and separated by TLC. Aflatoxins fluoresce under U.V. light. The spots after extraction are estimated spectrphotometrcally. The developed spots can be identified based on Rf values of known aflatoxin standards. Their quantitation can also be made based on visual intensities of the spots.

Reagents

1. Acetone,
2. Chloroform,
3. Methanol,
4. Lead acetate 10% solution,
5. Anhydrous sodium sulphate,
6. Silica gel G,
7. Aflatoxin standards.

Apparatus

1. Balance,
2. Thin layer applicator set,
3. Wrist action shaker,
4. Water bath,
5. Hot air oven,
6. Hot air blower,
7. TLC solvent tank,
8. Chromatography sprayer,
9. Forced draught air oven,
10. U. V. Cabinet with 365 nm U.V. light,
11. Spectrophotometer (U.V-Visible).

Procedure

1. Take 50 g of groundnut seeds and grind to a fine powder.
2. Place the 50g sample in a stopped conical flask, add 250 ml of 70% aqueous acetone, and seal the stopper with cellotape.
3. Shake the flask for 30 min at 300 rpm.
4. Filter the extract through No. 4 Whatman paper.
5. Transfer 150 ml of the filtrate to a 250 ml beaker.

6. Add 30 ml of 10% lead acetate solution and boil on a steam bath to reduce the volume to 150 ml and then cool to room temperature.
7. Filter the extract through N0.4 Whatman paper.
8. Transfer the filtrate to a 250 ml seperating funnel.
9. Extract the filtrate twice with chloroform (each time with 50ml of chloroform).
10. Drain the chloroform layers through a layer of 15 –20 g of anhydrous sodium sulphate in a funnel lined with a filter paper.
11. Wash the sodium sulphate with 20ml chloroform.
12. Combine the chloroform extracts in a 150 ml beaker.
13. Evaporate the chloroform extracts to near dryness on steam a bath.
14. Dissolve the residue in a suitable aliquot of chloroform.
15. Detect aflatoxins by TLC.

Preparation of TLC Plates

1. Weigh out silica gel G (30 g) into a stoppered flask, add 60 ml distilled water and shake for 1 min.
2. Coat the plates with a layer of slurry 0.25 mm thick and leave the silica gel plates until gelled (15 min.)
3. Heat the plates for one hour in a forced draught oven at 110^{0}C
4. Cool the plates in dust –free atmosphere for 30 min and store them in a plate cabinet.

Development of TLC Plates

Apply 20 µl extract of samples to the chromatoplates using a micropipette. Spot the extract samples in a line 2 cm from the bottom of the plate and at least 2 cm inside from either side. Spotting should be done as quickly as possible and in subdued light. Also spot a qualitative standared of aflatoxin.

Develop the chromatoplate in a solvent system of chloroform: acetone (1:9) in a developing tank. Develop the plates until the solvent front has run 10-12 cm from the base line. Remove the plates from the developing tank and allow them to dry in air in subdued light. Examine plates in a dark room at a distance of 30 cm from a ultraviolet lamp (peak emission 365nm and observe the presence or absence of blue or green fluorescent spots corresponding to aflatoxin B and G. Compare the fluorescent spots from the sample extracts with those from the standards which have similar Rf values.

Spectrophotometry (Nabney and Nesbitt, 1965)

Determintion of aflatoxins, particularly aflatoxin B1, is based on the intensity of the ultraviolet light at 363 nm. After separation by thin-layer chromatography, scrape the silica gel 'G' containing the aflatoxin B_1 spot from the plate and extract with cold methanol for 3 minutes. Filter off the methanol into a 50 ml beaker and wash the silica gel 'G' three times with methanol, making the combined filtrates up to 5 ml. Record the ultraviolet absorption specturm of the methanolic solution in a 2 cm cell and calculate the amount of aflatoxin B_1 present in the sample. The optical density at 363 nm, minus that at 420 nm, is used for calculations.

The aflatoxin B1 concentration is given by $\dfrac{D \times M \times 10^6 \ \mu g/kg}{C \times 200 \times 2}$

Where D is the corrected optical density at 363 nm, M is the molecular weight (312) of aflatoxin B1, and C is the molar extinction coefficient (22,000).

Visual Method of Estimation of Aflatoxins

The spectrophotometric method described above is useful in the case of fairly high concentrations of aflatoxins. However, in most of the samples, and in particular when dealing with natural contamination, levels may be low. For such samples, the visual method of quantitation is preferred.

Spot successively 5 µl, 10µl, and 20 µl aliquots of the sample extract on the TLC plates. On the same plate, spot 5 µl, 10µl, 15 µl, and 20 µl, of aflatoxin standard containing 0.02 µg, 0.05 µg, and 0.1 µg etc. of aflatoxin B_1 respectively. Develop the plate in a solvent system of chloroform:methanol (1:9) in a developing tank. Develop the plates until the solvent front has run 10-12 cm from the base line. Remove the plates from the developing tank and allow them to dry in air in subdued light. Examine the plates in the dark room under long wave UV light and observe the presence or absence fo aflatoxins. Compare the fluorescent intensities of the spots of sample extracts with those of the standards. If the spot of smallest aliquot (5µl) of the sample is too intense to match the standard spot, the sample extract can be diluted. The process of dilution and spotting is repeated over and again if the spot of smallest portion of the sample extract is again too intense to match the standard.

Calculation

$$\text{Aflatoxins in } \mu g/kg = \frac{S \times Y \times V}{X \times W}$$

S = µl aflatoxin standard which matches the unknown (spot from the sample extract)

Y = concentration of aflatoxin standard, µg/ml

V = µl of final dilution of sample extract

X = µl of sample extract spotted giving fluorescent intensity equivalent to S (standard aflatoxin)

W = weight of sample (g).

Aflatoxins by Elisa

(Trucksess, *et al.*, 1990)

Apparatus and Reagents

1. High-speed blender; micropipet and tips; filter paper; test tube (12 x 75 mm, borosilicate); timer; syringe (20 ml) etc.
2. Fluorodensitometer-366 nm excitation, 420 nm emission.
3. TLC plates-Silica Gel 60.
4. Fluorimeter- with 360 nm excitation filter, 450 nm emission filter.
5. Affinity column-Aflatest.
6. LC System- with injector, pump, UV detector, and C18 column.
7. ELISA test kit-Immuno Dot Screen Cup consisting in part of the following: antibody-coated cup; aflatoxin-enzyme conjugate; phosphate-buffered solution (pH 7.2); tetramethylbenzidine aqueous solution (0.4 g/L); hydrogen peroxide solution (0.02% in 0.13 aqueous citric acid).

TLC Methods

Aflatoxin contaminated samples (corn, peanuts, and peanut butter) are extracted with chloroform and water. All final extracts are analyzed and quantitated by a TLC-fluorodensitometric technique.

Monoclonal Antibody Column Method

Extraction of peanuts and peanut butter: Weigh 50g test portion into a blender. Add 5g NaCl and 125 ml methanol-water (6: 4). Blend 1min at high speed. Add 125 ml water, and blend at low speed for 2s. Filter and collect 5ml (equivalent to 1g test sample). Proceed with column chromatography.

Column chromatography: Couple monoclonal antibody column to 30ml syringe barrel with connector. Add test extract to syringe barrel, and use syringe plunger to push extract through column. Flow rate is about 6ml/ min. Remove syringe, take plunger out of barrel, and reconnect syringe to column. Add 20ml water to barrel and apply pressure to push water through. Remove as much water from packing as possible. Place small test tube under column. Add 1ml methanol directly to column and use syringe to push it through column (1-2 drops/s). Use fluorimeter cuvette as receiver to collect eluate and continue with quantitation by either bromination solution fluorometry or LC (eluate is equivalent to 1g test sample).

Bromination solution flouorometry: Set instrument excitation at 360 nm and emission at 450 nm. Insert blank and adjust zero knob to display 0. Insert standard solution (34 mg quinine sulfate/ ml 0.1N H_2SO_4. This solution gives same fluorescence intensity as 20ng B_1 /2 ml) and adjust span knob to display 20. Add 1ml 0.001 % bromine solution to eluate (total volume, 2 ml). Mix, and place test tube in instrument. Wait exactly 1 min and record reading (ng total aflatoxin/g test sample).

Quantitation by LC: Add 0.2 ml water to eluate, mix, and evaporate under a stream of nitrogen to 100-200 µl. Apply entire test extract to C18 column. Mobile phase is methanol-water (45 : 55) and flow rate is 1ml/min. Set UV detector at 365 nm and 0.01 AUFS, Approximate retention times for aflatoxin B1, B2, and G1 are 16, 12, and 10 min, respectively.

ELISA Screening Method

Extraction of corn, raw peanuts, cottonseed, and poultry feed. Weigh 50g test portion in a blender jar. Add 100 ml methanol-water (8+2). Blend 3 min at high speed, filter, and collect 5ml. Measure 200 µl filtrate. Add 400 µl dilution buffer, mix, and save for immunoassay.

Add 150 µl aflatoxin extract to ELISA cup, wait for 1min, and apply additional 150 µl test solution to ELISA cup, wait for 1min and apply additional 150 µl test solution to cup. Wait 1min and add 100 µl test aflatoxin conjugate solution. Wait for 1min and add 1.5 ml wash solution. Ad 1.0ml substrate solution. Wait 1min and observe color development.

Determination

Observe center of cup for blue color or no color development exactly 1min after adding substrate. Test samples containing $\geq$ 20 ng aflatoxisn/g show no color development; test samples containing <20 ng aflatoxing/g develop blue or gray color in center of the cup.

Aflatoxins by HPLC Method

(Park *et al.*,1990)

Principle

Aflatoxins are extracted, purified and derivatized with trifluoroacetic acid (aflatoxins B1 and G1 to B2a and G2a, respectively). They are separated by reverse-phase liquid chromatography, and detected by fluorescence. Method can measure 0.1 ng of aflatoxins B1, B2, G1, and G2. Detection limit is 0.3 ng/g for each aflatoxin.

Apparatus

1. Liquid chromatograph - with fluorescence detector, excitation filters (360 nm) and glass emission filters (440 nm), and column.- 15 cm x 4.6 mm id, Supelcosil, 5 µm, or equivalent.
2. Adjustable pipets.- 10-1000 µl and 100-1000 µl with disposable tips (Eppendorf).
3. Filter tube.- Glass, 15 cm id, with coarse frit bed support (glass wool not useful).

Reagents

1. Solvents.- Distilled-in-glass: methanol, hexane, methylene chloride, benzene, acetone, acetonitrile. Anhydrous ethyl ether stored in metallic container. Glass-bottled ether may form peroxides soon after opening which degrade aflatoxins.
2. LC elution solvent.- H_2O:CH_3CN:methanol (700 :170 : 170). Adjust ratio of water to obtain baseline resolution of aflatoxins B2a and G2a.
3. Silica gel for column chromatography: Activate silica gel by drying 4 h at 100°C. Cool to room temperature. Weigh desired quantity (100g) in to glass-stopper container. Add 1 ml H_2O in small increments; agitate silica gel between additions. Shake or tumble mechanically 4-6h. if

mechanical tumbler or shaker is not available, shake manually 5 min/h over 8 h. Let stand 16 h. Perform suitability test as in D.

4. Trifluoroacetic acid (TFA):Transfer 1 or 2ml TFA to a vial with Teflon-lined cap. Keep in freezer when not in use. Discard if discoloration appears.

5. Sodium sulphate.-Anhydrous, coarse. Heat 2-3 h at 600 °C to remove organic impurities.

6. Aflatoxin standard solution. 1) Aflatoxin stock solution.- 10 μg/ml. Prepare individual stock solution in benzene:CH3CN (98:2) 2) Working standard solutions- Use Eppendorf pipette to transfer appropriate quantity of stock solutions to each of the vials to obtain the amounts of afltoxins in each vial.

Vial	**B1 and G1 ng**	**B2 and G2 ng**
1	250	125
2	500	250
3	1000	500
4	2000	1000

Evaporate solutions to dryness under gentle stream of nitrogen (drying may be facilitated by warming to 40 °C). Using Eppendorf pipet, add 200 μl hexane and 50 μl TFA to each vial, cap, and vortex-mix for 30 s. Let solutions stand 5min; then add 10.0ml H_2O:CH_3CN (9:1) and vortex-mix 30s. Let layers separate for 5-10 min, or centrifuge at 1000 rpm for 30s. Final concentration of aflatoxins is as follows:

Vial	**B1 and G1** μg/10.05ml	**B2 and G2** μg/10.05 ml
1	0.25	0.125
2	0.5	0.25
3	1.0	0.50
4	2.0	1.00

Silica Gel Suitability Test

Prepare 2.0g silica gel column as in G.Transfer appropriate quantity of aflatoxin stock solutions to 50 ml beaker to contain 500ng each of aflatoxin B1 and G1 and 250 ng each of aflatoxins B2 and G2. Evaporate mixture to dryness under gentle stream of nitrogen. Add 2-3ml CH_2Cl_2 and swirl for 10s.

Quantitatively transfer to silca gel column with 2 about 1 ml portions of CH_2Cl_2:acetone (9:1). Derivatize as in H and perform LC analysis. Compare chromatogram with that of derivatized standard. Peak heights (areas) should be nearly identical to those for working standard, i.e., B1, G1= 0.25 μg/10.05 ml. Recovery should be $\geq$ 90%.

Extraction

Transfer 50 g corn or 50g peanut butter to 1l blender jar, add 200 ml methanol followed by 50 ml 0.1N HCl, and blend 3 min at high speed. Filter through 24 cm Whatman No.1 paper. Filtrate may not be completely clear. Collect 50 ml filtrate.

Partition

Transfer 50 ml filtrate to 250ml separatory funnel. Add 50ml 10% NaCl solution, swirl, add 50 ml hexane, and shake gently for 30s.Let phases separate and drain lower aqueous layer into another 250 ml separatory funnel. Discard hexane layer. Add 25 ml CH_2Cl_2 and shake moderately 30s. If emulsion occurs, break up with clean Pasteur pipette. Let phases separate and drain lower CH_2Cl_2 layer through 4 cm coarse granular, anhydrous Na_2SO_4 in glass filter tube. Collect eluate in 250 ml beaker. Repeat partition with 2 additional 25ml portions of CH_2Cl_2 and vigorously shake and drain as above. Collect eluate in the 250 ml beaker. Evaporate eluate on steam bath under gentle stream of nitrogen to 2-3 ml.

Column Chromatography

Slurry 2.0g silica gel with 10 ml ether:hexane (3:1) in 30 ml beaker, pour slurry into cleanup column, and wash beaker with additional 5ml ether-hexane solvent to effect transfer. Keep stopcock closed and let silica gel settle without tamping. Wash sides of column with 2-3 ml ether:hexane (3:1), using wash bottle. After gel settles, open stopcock, and, while column drains, add 1 cm anhydrous granular Na_2SO_4. Transfer extract from partition, F, to column. Wash tip of beaker with 0.5 ml CH_2Cl_2 using wash bottle, and collect wash in column. Wash beaker with 2 ml CH_2Cl_2 and add wash to column. Do not use more than 5-6 ml CH_2Cl_2 to transfer extract to column.

With stopcock fully open, add 25ml benzene:acetic acid (9:1), and then add 30ml ether:hexane (3:1) to column, draining each wash to top of Na_2SO_4. Discard washes. Elute aflatoxin with 100ml CH_2Cl_2:acetone (90:10) and collect eluate in 250 ml beaker. Evaporate solvent on steam bath under gentle stream of nitrogen to about 6ml. Quantitatively transfer to a vial, using 2-3 ml CH_2Cl_2 as wash.

Evaporate eluate almost to dryness on steam bath in an aluminum block under gentle stream of nitrogen. Evaporate remaining 200 µl just to dryness under gentle stream of nitrogen by holding vial in palm of hand and slowly rotaing vial.

Derivatization

Add 200µl hexane to column extract, G, Then, add 50µl TFA (using Eppendorf pipette), cap vial, and vortex-mix vigorously for 30 s exactly). This procedure must be followed closely to ensure consistent reaction yields. Let mixture stand 5 min.Using Eppendorf pipette, add 1.950 ml $H_2O:CH_3CN$ (9:1). Vortex-mix vigorously for 30s (exactly), and let layers separate for 10 min or centrifuge at 1000 rpm for 30 s. Concentration is 10 µg/2 ml aqueous CH3CN.

LC Determination

Using instrument parameters inject 25µl of derivatized standard solutions. Prepare standard curve to check linearity of responses. Inject 25 µl TFA-treated sample solution (lower aqueous phase). If sample peaks are outside linear range dilute aliquot of TFA-treated sample solution to suitable volume with H_2O-CH_3CN (9+1), re-mix on vortex mixer, and inject another 25 ìl portion. Calculate individual aflatoxin concentrations as follows. Use responses of standard containing 500 ng B1 and G1, and 250ng B2 and G2 for calculations.

$$\text{Aflatoxins, ng/g} = (P/P') \times C \times (2/10) \times 1000 \times D$$

Where P and P′ = peak areas (integrator counts) or heights for sample and standard, respectively, per 25 µl injection; C = concentration of individual aflatoxins in standard solution (0.5 or 0.25 µg/10.5 ml); and D =dilution factor if 2 ml solution for injection is diluted.

Crude Ricin

Principle

Ricin being an albuminous protein is extracted into water and fractionated. Crude ricin is precipitated using ammonium sulphate and weighed gravimetrically.

Reagents

1. Petroleum ether(40-60^{0}C),
2. Diethyl ether,
3. HCl,
4. NaCl
5. Tris- HCl buffer,
6. Ammonium hydroxide 2% solution,
7. Ammonium sulphate.

Apparatus

1. Buchner funnel,
2. Vacuum pump,
3. Balance,
4. Dialysis membrane.

Isolation, Purification and Quantification

Ricin from the castor seeds can be isolated as per the procedure of Funatsu, *et al.,* (1971). The method, slightly modified by the author, is given below. The outer covering of the seeds (hulls) are removed, the kernel is powdered in a Knifetec sample mill. It is completely defatted using ice-cold petroleum ether (40-60°C) and diethyl either. The defatted powder is suspended in double distilled water and adjusted to pH 4.0 using HCl. The suspension is agitated for 7-8hrs. The solution is filtered under vaccum. The yellow supernatant liquid is collected and the residue is decanted. The proteins in the supernatant are precipitated by adding solid NaCl till its saturation. The precipitate is dissolved and extensively dialysed for 72hrs in double distilled water and in Tris-HCl buffer, pH 7.5 at 4°C. The insoluble particles and other impurities separated in dialysis are decanted. The clear protein solution is precipitated using 2% ammonium hydroxide. The resulting precipitate is decanted. The clear protein solution is finally precipitated with saturated ammonium sulphate salt solution. Crude ricin is thus separated and weighed for purposes of quantification. The ricin can be dissolved in either distilled water or in pH 4 aqueous HCl, for any further studies.

Ricin: RID Assay
(Pinkerton *et al.*,1999)

Principle

A RID (radial immunodiffusion) assay is used to quantify levels of specific antigens in a mixture of several antigens (Garvey *et al.,* 1977). A specific antibody developed against ricin is incorporated into a thin layer of agar. Seed extracts containing the antigens for ricin and RCA are placed into wells cut into the agar layer. The antigens diffuse evenly outward and bind with specific antibodies to form an antigen-antibody precipitation ring around the well. The square of the diameter of the precipitation ring is plotted on a standard curve to estimate antigen concentration (Kemeny and Chantler, 1988). The RID technique described by Stansfied (1981) and Mancini *et al.,* (1965) is modified slightly and presented below.

Reagents and Apparatus

1. Anti ricinus communis lectin,
2. Agarose,
3. PEG,
4. R.I.D. Plates,
5. Incubator.

Procedure

One hundred sixty seven microliters of rabbit anti-*Ricinus communis* lectin diluted with 34 mg/ ml of distilled water. This solution is placed in 9.5ml of a 1.2% (v/v) agarose/3% (v/v) PEG 6000 solution which has cooled to 50°C. Antibody-agar solution is added to RID plates and allowed to solidify before the 2-mm wells are cut. Ricin solutions of 3.5 and 2.9mg /l are used to develop six standard concentrations. Ten microliters of these standards are added to the wells. The plates are allowed to incubate at 26°C for 24h, and then placed at 4°C for another 24h. The precipitate ring around each well is measured on two different axes and the square of the average diameter is used to estimate the ricin + RCA concentration.

Ricin: HPLC Method

Principle

A solution containing proteins are chromatographed on a non-polar column. A more polar liquid that serves as the mobile phase is passed over the matrix and solute molecules (proteins) get eluted in proportion to their solubility (Lottspeich and Henschen, 1981).

Reagents and Apparatus

1. HPLC equipment with C-18 column, a UV-Visible spectrophotometric detector (SPD) and a data recorder.
2. Mobile phase: Tris-HCl buffer, pH 7.5
3. Methanol 100% (HPLC grade).
4. Standard proteins like BSA and trypsin inhibitor
5. Nylon membrane.

HPLC Separation

The crude and pure ricin fractions are evaluated to check their purity using reverse phase HPLC (Shimadzu LC-10AT, Japan) under isocratic conditions on C-18 column (4.6 x 25cm. Supelco products, USA). Before applying the ricin samples, the column is calibrated using standard proteins like BSA, Trypsin inhibitor etc.The samples are dissolved in 100% mobile phase (Tris-HCl buffer, pH 7.5) and then filtered using 0.45μm nylon membrane (Millipore, USA). 20μl of the protein sample is injected into the column at a flow rate of 0.5ml per minute and monitored at 280nm using SPD. The chromatogam shows a major ricin peak, wthin 5 min of elution. Both identification and quantification can be carried out using this HPLC separation. BSA can be used for quantification.

TLC of Sugars (Raffinose, Stachyose and Verbacose)

Principle

There are two phases in TLC, i.e., stationary phase (e.g., silica gel) and mobile phase (solvent mixtures). Mixture of sugars are separated depending upon their Rf value which is the ratio of component front to solvent front. The individual components can be qualitatively and quantitatively determined.

Equipment

1. Balance,
2. Thin layer applicator set,
3. Hot air oven,
4. Hot air blower
5. TLC solvent tank,
6. Chromatography sprayer.

Reagents

1. Mixture of chloroform, acetic acid, water (for sugars), (3:3.5:0.5, v/v) ratio.
2. Silica gel with 13% calcium sulphate-silica gel G-60.
3. Diphenylamine spray reagent: Dissolve 1g diphenylamine in 100 ml acetone. Add 1ml of aniline, 8.5ml orthophasphoric acid and 1.5 ml of water. Mix well before use.

Procedure

1. Arrange 5 clean TLC plates (20 x 20cm) on the TLC applicator (There should be no gaps in between plates).
2. Weigh 40g of silica gel (or silica gel-G-60) into a conical flask.
3. Add 80-85 ml distilled water and close the flask with a rubber stopper.
4. Shake vigorously for 5 to 7 min. The slurry is ready for spreading when particles swell and become more viscous.
5. Pour the complete slurry into the spreader at once and move the spreader forward with uniform force (slow spreading causes uneven distribution).
6. Leave the coated plates undisturbed for 1h at room temperatue.
7. Activate the plates by keeping them in an oven at 100°C for 1h.
8. Remove the plates and store in desiccator when not in use. Leave an inch gap at the bottom of the coated plate and spot the required volume (e.g., 5mg equivalent sugars in aqueous or alcohol medium) of the extract. Also spot standard sugars like glucose raffinose, stachyose, verbacose etc.)
9. During spotting use hot air blower for drying spots.

10. Saturate TLC chamber 2-3 hr or overnight with suitable solvent system (e.g., chloroform: acetic acid: water – 3.0:3.5:0.5, v/v).
11. Keep the spotted plates in the chamber for developing.
12. Remove the plates when the solvent front is one inch below the top of the plate.
13. Dry the plates with the help of a hot air blower.
14. Spray the dried plates with diphenylamine reagent.
15. Dry the plates in an oven until coloured spots develop.

Calculations

1. Calculate the Rf value using a scale by measuring the distance covered or eluted by the spot and the solvent front. (Rf= spot front/solvent front). Identify the unknown spots using Rf values of known sugars.
2. Measure the color of the spot on a densitometer using appropriate filter. Compare with densitometer readings of a known standard sugar developed on TLC plates as above.

Glucosinolates-Quick Test for the Determination of Glucosinolates

(Institute of Agronomy and Plant Breeding University Gottingen, Germany)

Principle

This is a quick test method for determination of glucosinolates.

Equipment

1. Polypropylene tubes,
2. Water bath,
3. Centrifuge,
4. Microtitre tray,
5. Vibrator,
6. Hot air oven,
7. Multi-channel photometer.

Chemicals

1. Methanol,
2. Diazobenzene sulphonic acid,
3. O- Phosphoric acid,
4. Linetta standard.

Procedure

1. Take 200mg finely powdered seeds into poly propylene test tubes
2. Add 300µl 60% methanol
3. Keep for 5min at 80°C (water bath)
4. Add 2ml water, mix in a VORTEX mixer for 15min at 80°C
5. Mix and centrifuge
6. Take 50 µl of the above extract in a well of a microtitre tray
7. Add 200 µl reagent solution: 20mmol = 1.535g 4-diazobenzene sulfonic acid. Dissolve in 250ml water + 250ml 85% O-phosphoric acid. Solution is slightly brown. Stable at 4°C for several weeks.
8. Mix in a vibrator.
9. Coupling reaction: Keep for 30 min at 70 °C
10. Photometry against a blank (300µl reagent) with a multi-channel photometer at 510 nm
11. Conversion to the concentration unit "µmol/g seed": Concentration (µmol/g seed) = (A_{510}/0.2g seeds –0.4) X 546.
12. The control value "A_{510}/0.2g seed" for the standard "Linetta" should be between 1.15 and 1.26 (O =1.20).

Gluco Test Method for Glucosinolates

(Thies, 1985)

Apparatus

1. Reflectometer
2. Glucotest paper
3. Myrosinase (100 units enzyme activity)
4. Charcoal filter papers

5. Filter supports
6. Hamilton syringes.

Chemicals

1. Chlorohexidine: 10% in methanol.

Procedure

1. Weigh 5g of finely powdered seed into a 250 ml glass beaker
2. Add boiling water over the sample filling up to the 200ml mark.
3. Stirr the suspension for 1 min to speed up the extraction of the glucosinolates (GSLs).
4. Cool 1ml of the extract to less than 60°C in a polystyrol tube of 3.5 ml voume.
5. Pour the cooled extract into a tube containing 100 units of myrosinase. Close with thumb and mix by 4 quick turns. After a reaction period of 5min, add 5 ml of a protein precipitating agent (10% chlorohexidin diacetate in methanol) and shake the tube again thoroughly by hand.
6. Draw the contents of the tube into a 1 ml syringe and press the same tightly onto the filter support armed with activated charcoal paper.
7. Pull out the piston from the syringe: the syringe must thereby be kept horizontal to facilitate the passage of the extract into the upper part of the filter unit.
8. Place filter unit with syringe onto a 3.5 ml polystyrol tube and collect at least 5 drops of a clear filtrate. Discard opaque or coloured liquid.
9. Tranfer 17 μl of the filtrate onto a clean slide forming 1cm long strip.
10. Calibrate the reflectometer with a 17 mm long piece of GLUCOTEST paper and place the same immediately afterwards onto the liquid strip on the slide. Cover with cardboard box to exclude light effects. After 1 minute repeat.
11. Callibration of the reflectometer with a second dry test-paper strip (necessary because the commercial reflectometer is programmed with 1minute reaction time only).
12. Insert the treated test paper with a forceps into the reflectometer after a total reaction time of 2 min (in the dark) and note the result of the measurement in digital form.

13. Repeat steps 10 to 13 and calculate the mean x of both measurements.
14. Use the formula : $C = (X - 90)\ 0.5$.

Where C= the content of GSLs in the sample in μ mole per g air dry seeds.

The value 90 is obtained as a mean when extracts are measured which are prepared as above but without myrosinase addition.

Glucosinolates-Desulphation and HPLC

(Buchner,1987)

Principle

Crude glucosinolate extracts are prepared without defatting of the meal and addition of internal standards before extraction. Glucosinolates in the crude extracts are enzymatically desulphated on-column. The resulting solution of desulphoglucosinolates is separated and measured using reversed-phase high-performance liquid chromatography (HPLC) with a solvent gradient. Calculations are based on HPLC peak areas, use of relative response factors and internal standard of known purity.

Apparatus

1. Pasteur pipettes, glass wool placed in the tips as ion exchange columns.
2. Water bath or heating plate.
3. Small laboratory centrifuge.
4. Millipore dip-in-filter type Immersible Cx-10.
5. Low temperature oven.
6. UV spectorphotometer.
7. HPLC chromatograph: HPLC equipment with C-18 column, a UV-Visible spectrophotometric detector(SPD) and a data recorder.

Reagents

1. 15 m mol/l glucotropaeolin, 15 mmol/l sinigrin and 3 mmol/l sulfanilic acid as internal standards.
2. 20 mM/l tris buffer pH 7-8, 20 mM/l sodium carbonate buffer pH 6.5-7, 20mM/l sodium acetate buffer pH 4.
3. 20 mM/l potassium phosphate buffer pH 4.

4. Anion exchangers: DEAE-Sephadex A-25, Servacel Ecteola 23, Serdolit AW-2, AW-4, AW-14, Duolite A 30 B. Swollen in 2 M acetic acid, filled in the pasteur pipettes, rinsed with the 10 fold volume 6 M imidazole formiate, rinsed again with the 20 fold volume of water.
5. Sulfatase type H-1.
6. DEAE-Sepharose C1-6B formiate form .
7. 0.2 mol/1 sodium acetate.
8. Ultra pure water and acetonitrile for HPLC.

Purification of the Sulfatase

1. 200mg sulfatase is dissolved in 20 ml water.
2. The solution is passed through 20 ml Sepharose column (diameter 26 mm).
3. Rinse the same with 40ml water.
4. Elute with 60 ml of 0.2 mol/l sodium acetate.
5. Concentrate the eluate to about 4 x 100µl using four dip-in-filter.
6. Dilute to 4 x 2 ml with water
7. Concentrate to 4 x 100µl once more.
8. Rinse the filters, combine and dilute the filtrates to 20 ml sulfatase stem solution.
9. Store in a deep freeze.
10. Regeneration of column: rinse with 10 ml 6 M/l imidazole formiate and 50 ml water subsequently (one column filling can be used up to 10 times).
11. 50 µl of sulfatase stem soulution + 2 ml 0.15 m mol/1 sinigrin (stem solution 1:10 diluted) should result at least in an absorption decrease at 227 nm of 0.0338 absorption untis/minute.

Sample Preparation and HPLC Analysis

1. Weigh 0.2 g of finely powdered seed.
2. Add 1.5 l hot 70% methanol/water mixture and immediately afterwards 200 µl of internal standard glucotropaeolin (1 part standard to 2 parts final volume diluted rapeseed extract). Shake.
3. Extract 10 minutes at 80°C
4. Centrifuge 3 minutes at 3000 rpm.

5. Decant supernatant and repeat step (2) to (5) for the precipitate adding more internal standard two times.
6. Combine supernatants and fill up to 5ml
7. Put 500 to 1000 µl of the extract to 20 mg sephadex columns.
8. Rinse two times with 1ml 20 mmol/l sodium acetate buffer pH 4.0
9. Add 75 µl sulfataste solution (1 part sulfatase stem solution + ½ part water)
10. Cap the columns.
11. Desulfation takes about 6 hrs at 39°C
12. Elute with 1.5 ml water. Shake the samples
13. If necessary filter the samples.
14. Inject 30-80µl into HPLC. Programme: 0-16.5 min 0-16.5% acetonitrile, 16.5-18 min 16.5-0% acetonitrile (rest: water). Flow speed: 2.0 ml/min.
15. Detection at 229 nm.
16. Calculate the amounts of glucosinolates using the area of the internal standard and the relative response factors.

Analysis of Sesame Lignans

(Hemalatha and Ghafoorunnissa, 2004)

Principle

The lignans in the unsaponifiables of sesame oil are extracted and separated on a C-18 column of HPLC equipment. They are qualitatively and quantitatively estimated with the help of standards.

Apparatus

1. HPLC equipment with C-18 column, a UV-Visible spectrophotometric detector and a data recorder.

Reagents

1. Mobile phase: Methanol : water, 75:25 (HPLC grade).
2. Sesamol – authentic sample, available commercially.
3. Sesamin and sesamolin, extracted and prepared as described below.

Sesamin and Sesamol Extraction

Sesame oil (500g) in acetone (1:8 v/v) and cooled over night at -70°C. The glyceride crystals formed are separated by filtration. The mixture kept at -20°C. Removal of acetone from the filtrate gives a yellow oil, which on mixing with iso-octane and left at 4°C for 4-5 days sesamin crystallises, which is separated through centrifugation. The filtrate after evaporation of iso-octane is saponified with ethanolic KOH. The mixture is diluted and extracted with ether. The filtrate after removal of ether is dissolved in chloroform and petroleum ether (80-100°C). Upon cooling at 5°C, sesamolin crystallises. Sesamin and sesamolin are recrystallised in ethanol. These are used as standards in HPLC separation.

Procedure

The oil at 1:100 dilution in hexane is directly injected using a reversed phase HPLC system into a C-18 column (4.6 mm x 250 mm) through an injector. The peaks are recorded using the UV-Visible spectrophotometric detector. The mobile phase is methanol : water, 75:25 (HPLC grade). flow rate 0.8 ml/min. Peaks can be identified and quantified by comparison with authentic lignans (sesamol, sesamin and sesamolin). Sesamol can be used as an internal standard for calculating the percentage recovery of the lignans.

Vitamin A (HPLC Method)

Principle

Vitamin A is extracted with a suitable organic solvent after precipitation of the protein with ethanol. An aliquot of the organic phase which is evaporated under nitrogen is injected into a revesed phase HPLC column, followed by an eluting solvent of suitable polarity. A reversed phase column has a non-polar stationary phase with a polar mobile phase. Retinal is detected at 326 nm using a sensitive UV detector. An internal standard (Retinyl acetate) is used to account for processing losses.

Reagents

1. Retinyl acetate and Retinol standards.
2. Absolute alcohol,
3. Hexane,
4. Methanol,
5. Water (all solvents HPLC grade)

Apparatus

1. Vortex mixer
2. Centrifuge
3. HPLC equipment with C-18 column, a UV-Visible spectrophotometric detector and a data recorder.

Chromatographic Conditions

Column: C_{18} reverse phase, 5 mm particle size, 25cm colum (or 10µm, 15 cm column) Mobile phase: Methanol-water (95:5 methanol :water), Flow rate : 1ml/min.

Detector : UV detector 326 nm, sensitivity 0.01 AUFS

Chart speed: 1 cm/min for recording

Temperature: Ambient

Sample Preparation

Take 100 ml of oil sample and add 100 ml internal standard of retinyl acetate (1µ g/ml), mix the contents for 1 min and then add 0.5 ml of n-hexane. Vortex for 1 min and centrifuge at 3000 rpm for 5-min. Remove the hexane layer into another tube. Re-extract the sample with another 0.5 ml of n-hexane and centrifuge. Remove the hexane layer and pool with the previous aliquot in the tube and evaporate the total contents to dryness under nitrogen. To the dry tube, add 100 µl of HPLC grade methanol and mix thoroughly. Inject 20 µl into HPLC instrument.

Preparation of Standard Retinol

A few crystals or small quantity of retinol (Sigma) is dissolved in absolute alcohol and the absorbence is determined by spectrophotometric method using the extinction coefficient of retinol. (E 1% - 1800) in alcohol (An O.D. of 0.18 at 324 nm is equivalent to 1µg/ml).

Standard Retinyl Acetate

In the same way the activity in retinyl acetate standards is calibrated by the extinction value (E1% = 1550). An O.D. of 0.155 at 326 nm is equivalent to 1µg/ml.

Procedure

By use of an internal standard losses due to incomplete extraction, inaccurate aliquots or oxidation are automatically corrected, provided that the internal standard has physical and chemical properties sufficiently similar to retinol, is suitably separated from retinol on HPLC, does not coincide with other 326 nm absorbing material in the sample and not converted to retinol under the assay conditions.

A precisely known amount of internal standard added to the aliquot of sample is analyzed. By determining the relative extraction efficiency and detector response of retinol and the internal standard, a standard curve is fashioned in which the ratio of peak heights (or areas) are plotted against the retinol concentration in the sample.

A standard curve is prepared by adding varying amounts of retinol to a fixed amount (ie 50mg) of internal standard on a final volume of 50 μl of eluting solvent. The solvent is injected into HPLC under conditions mentioned above, the peak height is measured and peak height ratio is calculated.

The peak height ratio is then plotted against retinol concentration.

Calculation

From the concentration of retinol in the standard and its relationship with Area-Under the Curve (AUC) of the peak, the concentration in the unknown is obtained from its corresponding AUC

$$\text{Concentration in } \mu g/100\text{ ml} = \frac{\text{AUC of sample x conc. of std x 5000}}{\text{AUC of standard X 1000}}$$

Vitamin A

(Spectrophotometric Method)

Principle

Vitamin A in the oil sample is extracted into chloroform and estimated colorimetrically using thiobarbituric acid reagent.

Reagents

1. Solvent: Acetone:water (9:1 v/v) or 90% ethanol.
2. Thiourea reagent: Dissolve 4.0 g of thiourea in 100 ml of glacial acetic acid and filter the solution through glass wool or centifuge the solution.

3. Thio barbituric acid reagent: Dissolve 600 ml of 2-thiobarbituric acid in 100 ml of absolute ethanol, filter and store the solution at 4^{0}C.

4. Vitamin A aldehyde standard: Dissolve 10 mg all-trans retinal in 100 ml absolute ethanol (stock solution, store in dark at 4^{0}C). Before analysis dilute 2 ml of stock solution to 25 ml with 90% ethanol to give a working solution of 8 µg ratinal/ml.

Apparatus

1. Spectrophotometer with quartz cuvettes.

Procedure

Place the sample of chloroform solution of lipids (containing 1-10 µg retinal) in a 15 ml glass stoppered tube and remove solvent in a stream of N_2 below 25^{0}C. Immediatley dissolve residue in 3 ml of 90% ethanol. Add 1.5ml each of thiourea and thiobarbituric acid reagents, mix and leave at room temperature in the dark for 30 min. Centifuge the mixture to make it clear and measure absorbance of the red solution at 530 nm in a 1cm stoppered cell against a reagent blank. Standardize the procedure using 1-10µg standard all trans- retinal (molar extinction coefficient = 58,000 at 530 nm).

Calculation

$$\mu g \text{ vitamin A aldehyde} = \frac{\text{Absorbence of sample x } \mu g \text{ standard}}{\text{Absorbence of standard}}$$

Beta Carotene

Principle

Carotenoids are separated on a column of calcium hydroxide and estimated colorimetrically.

Reagents and Apparatus

1. Petroleum ether(40-60^0 C)
2. Acetone
3. Calcium hydroxide
4. Anhydrous sodium hydroxide

5. Rotary flash evaporator
6. Glass columns(1.5X30 cm)
7. Spectrophotometer
8. Vacuum pump.

Method

About 20g of fresh oil is allowed to stand overnight in a mixture of 100ml of petroleum ether:acetone (1:1). The extract is filtered and the residue on the filter washed twice to remove the entire yellow colour. The pooled filtrate is shaken with 50ml portions of water twice and then discarded. The organic layer is then dried over anhydrous sodium sulphate and conentrated under reduced pressure to a final voume of 5–6ml.

A column (1.5x30cm) is packed with alumina containing 3% anhydrous sodium sulphate. An aliquot of the sample is put on the column, eluted with 3% acetone in petroleum ether. The eluate is concentrated and made upto 5 ml. The OD of the sample is determined spectophotometrically at 460nm. The concentration is read from a graph prepared with standard beta carotene.

Calculation

1 OD=4µg/ml of beta carotene in a 1 cm cell.

Thiamine (Vitamin B1)

Principle

The thiamine is oxidized to thiochrome which fluoresces under U.V. light. Under standard conditions and in the absence of any other fluorescing substance the fluorescence is proportional to the thiamine concentration.

Reagent

1. Acetate buffer(pH 4-4.2): 0.1N acetic acid (66.05 ml glacial acetic acid in one litre water), 0.1M sodium acetate (136g sodium acetate $3H_2O$ in 1litre), Mix 70 ml of 0.1N acetic acid and 30ml of 0.1M sodium acetate and make up to 500ml
2. Enzyme suspension: 150mg taka diastase and 75mg papain in 5ml acetate buffer
3. Standard thiamine : 25mg thiamine dissolved in 250 ml 0.01 N HCl

4. Working standard: 1ml of stock standard thiamine diluted to 100ml
5. 3% H_2O_2
6. 1N NaOH
7. 2% Potassium ferricyanide
8. Oxidising agent: 2 parts of 40% NaOH + 1 part of 2% potassium ferricyanide
9. Isobutyl alcohol.

Apparatus

1. Water bath.

Procedure

Take 5-10 g sample, add 100ml acetate buffer (pH 4-4.2) and 5ml enzyme suspension. Keep the contents overnight at 37°C. Heat on a waterbath at 80°C for 10 min to inactivate the enzymes. Filter and take 10ml of filtrate into three separating funnels (SF)

Blank (SF)	Experimental (SF)	Recovery (SF)
10ml filtrate+ 1ml of D.W.+ 1ml 1N NaOH	10ml filtrate+ 2ml distilled H_2O	10ml filtrate+ 1ml standard B1

Shake the contents of all the SFs. Then add 2.5 ml of oxidising agent to each of separating funnels and again shake. Add 3-4 drops of 3% H_2O_2 to the three funnels. This discolourises the solution. Keep for for 2 min. Add a mixture of 7.5 ml H_2O and 7.5ml of isobutyl alcohol to each separating funnel. Shake vigourously. Throw away the aqueous layer and take the alcohol layer into a test tube. Read fluorescence in a fluorimeter or a spectrofluorimeter at 336 nm primary wave length and 445 nm secondary wave length. Note the readings separately for the sample (S), blank (B) and recovery sample (R) and calculate the thiamine content.

$$\text{Thiamine, mg / 100g sample} = \frac{\text{S–B} \;\; \text{X Dilution x 100}}{\text{R– S} \;\; \text{Weight of sample}}$$

Tocopherols (HPLC Method)

Principle

Tocopherols are extracted and analysed as per the method of Indyke, (1990).

Apparatus

1. HPLC equipment with C-18 column, a UV-Visible spectrophotometric detector and a data recorder
2. Centrifuge.

Reagents

1. Ethanol
2. Potassium hydroxide 0.5 N
3. Hexane : diethyl ether(3:1)
4. Mobile phase: Methanol (HPLC grade).

Extraction

0.1g of the oil sample is accurately weighed into a test tube. 10ml of ethanol is added to the sample and the mixture is agitated to avoid agglomeration. 0.5N KOH (2ml) is added quickly and the tubes are stoppered and incubated for 10 min at 70°C to complete the saponification process. Later the product is cooled to room temperature and extracted with hexane/ diethyl ether (3:1) and kept for shaking for 5-10min. 30ml of water is added into the mixture and inverted 15-20 times and centrifuged at 1000 rpm (180g) for 10 minutes. An aliquot (10ml) of the upper layer is evaporatd to dryness under vacuum at 45°C. The residue is dissolved in ethanol (1.0ml), filtered through 0.45mm nylon membrane (Filtech Pharmalab, USA).

HPLC Separation of Tocopherols and Their Quantification

The sample is separated using HPLC on a C-18 column, 4.6x25cm. 2μl of the sample is injected into the column. An isocratic mobile phase of methanol (100%) is used at a flow rate of 1.0ml/min at ambient temperatures after filtration and degassing. Detection is done at 210nm using a spectrophotometric detector. Identification and quantification is based on comparison of RT values of standard tocopherols.

Total Tocopherols- Colorimetric Method

(Wong *et al.*, 1988)

Principle

By using a preliminary heat-bleach at 250°C the Emmere-Engel method has been adapted for determination of total tocopherols (including tocotrienols) in crude as well as refined palm oil, olein and stearin. Tocopherols and tocotrienols reduce ferric ions to ferrous, which reacts to form a coloured complex with dipyridyl.

Reagents

1. Toluene
2. Ethanol
3. Bipyridine: 0.07% 2-2′ bipyridne in 95% of ethanol
4. $FeCl_3$: 0.2% $FeCl_3$ 6 H_2O in 95% ethanol

Apparatus

1. Spectrophotometer with quartz cuvettes.

Method

1. 200 mg of oil sample weighed accurately into a 10 ml volumetric flask.
2. 5ml of toluene added by pipette and the oil taken into solution.
3. 3.5ml of (2,2′-bipyridine- 0.07% w/v in 95% aqueous ethanol) and 0.5ml of $FeCl_3$ 6 H_2O (0.2% w/v in 95% aqueous ethanol) are added in that order.
4. Make up the solution to 10ml with 95% aqueous ethanol.
5. Let stand for one minute and read absorbence at 520 nm against a blank (prepared as above without the oil). Protect solutions from strong light during colour development.
6. Calibrate the method by preparing standards containing 0-240 μg of pure α-tocopherols in 10 ml of toluene and analyse as above.

	Toluene ml	**Bipyridine**	**$FeCl_3$ $6H_2O$**	**Toluene**	**Read at**
Sample 200mg	5ml	3.5	0.5	1	520 nm.
Blank -	5ml	3.5	0.5	1	

Calculations

Total tocopherols (ppm) = A-B/ MxW.

Where, A= Absorbence of sample in 10mm cell.

B= Absorbence of blank in 10 mm cell.

M= Gradient of absorbance, it is approximately 8.0×10^{-3} absorbence per μg α- tocopherol.

W= weight of sample in g.

Gross Calorific Value (GCV)

Definition

The gross calorific value (GCV) of a substance is the heat released on combustion of that substance in an oxygen-filled bomb calorimeter at a constant volume. The conditions of combustion are such that the products of combustion are ash, gaseous CO_2, liquid water, SO_2 and N_2. GCV is usually reported in units of KJ/Kg.

Principle

In order that the GCV of a sample be found, a known quantity is burned under strictly controlled conditions. The sample is burned in oxygen to ensure complete conversion to its combustion products. The heat released by this combustion is determined on the basis of the equation.

$$\text{Heat released} = \frac{\text{Mass of the apparatus x specific heat capacity}}{\text{of the apparatus x temperature rise.}}$$

Since the mass of the apparatus is constant from test to test (same calorimeter and bomb and same water), it need not be known quantitatively. Instead, the effective heat capacity is used in this equation. The effective heat capacity of the apparatus is found by burning a sample of thermo-chemical standard benzoic acid in the apparatus and measuring the temperature rise. The effective heat capacity is then the heat released by the acid divided by the temperature rise and this value is used in every subsequent test until the apparatus is changed in some way. So the GCV of a sample can be found by burning a known mass in an apparatus with a known effective heat capacity, and measuring the temperature rise.

Apparatus

The sample is burned in a crucible, which is contained in a vessel called a bomb. During a GCV determination, the assembled bomb rests inside a metal container called a calorimeter. This calorimeter is filled with a quantity of water sufficient to cover the bomb, and this quantity of water is the same for each determination. The calorimeter vessel is completely enclosed by a water jacket, but separated from it by an air gap of approximately 10 mm. This water jacket can be of 3 types.

1. *Adiabatic* : As the name implies, this type of jacket does not allow any heat to flow from the calorimeter to the jacket. It does this by means of a control system which keeps th jacket at the same temperature as the calorimeter. Two thermistors detect any temperature differences between the calorimeter and the jacket, and a heater warms the jacket as neccesary.
2. *Isothermal* : This is a jacket which is thermostatically maintained at constant temperature. Heat flows from th calorimetor to the jacket and has to be compensated for in the calculation of the GCV.
3. *Static* : This type of jacket also remains at constant temperature and does so by being large enough to absorb heat produced on firing the bomb without significant temperature rise. The effect of heat flow is compensated for by the same method.

Procedure

A general procedure for a GCV determination is outlined below:

1. Prepare a sample.
2. Weigh the crucible .
3. Introduce the sample into the crucible
4. Weigh the filled crucible
5. Assemble the bomb, and fill with O_2 to 30 bar.
6. Fill calorimeter with pre-determined quantiy of water
7. Place bomb in a calorimeter, and calorimeter inside jacket
8. Assemble stirrers, thermometers etc.
9. Allow time for calorimeter and jacket to reach the same temperature.
10. Record firing temperaure (caution: type of jacket and temperature varies)

11. Fire the bomb.
12. After the required interval record the final temperature.
13. Dis-assemble the apparatus.
14. Carry out analysis of bomb contents, if required.

Calculations

$$GCV = \frac{ÄT\,(Ca) - e_1 - e_2 - e_3 - e_4}{M}$$

ÄT = Corrected temperature rise.

Ca = Eeffective heat capacity of apparatus

e_1 = correction for heat of combustion of wire

e_2 = correction for heat of combustion thread

e_3 = correction for heat of formation of H_2SO_4

e_4 = correction for heat of formation of HNO_3

M = mass of sample.

Calorific Value

Calorific value of an organic sample is obtained by multiplying the sum of carbohydrates and proteins by four and adding to it the figure obtained by multiplying the hexane extractives(fat/oil) by nine.

Calorific value =4 (Carbohydrate + Protein) + 9 (Hexane extractives)

Carbon and Hydrogen Content

Definition

The carbon content a sample is the amount of carbon contained in it. The hydrogen content is the total amount of hydrogen in a sample minus that due to the moisture content of the sample.

Principle

A sample of known mass is burnt at 1350^{0}C in a stream of oxygen such that complete combustion occurs. All the carbon is converted into CO_2 and all H_2 to H_2O. The CO_2 is absorbed by soda asbestos and H_2O by Mg $(ClO_4)_2$.

Chemicals

1. $Al_2 O_3$
2. O_2 cylinder
3. Mg $(ClO_4)_2$
4. Soda asbestos pellets.

Apparatus

1. *Combustion system:* The sample is burnt inside an aluminous percelain combustion tube, the centre section of which is situated inside a furnace. This furnace should be capable of maintaining a temp. of 1350^{0}C. The sample is contained in an unglazed porelain boat and covered with a layer of $Al_2 O_3$ to prevent violent oxidation. A silica rod is used to push the boat into the combustion tube.
2. *Oxygen supply system:* The sample is burned in a rapid stream (300 ml/min) of O_2. The O_2 is passed through a drying tube before introduction (glass tube filled with magnesium perchlorate). CO_2 in O_2, if any, needs to be removed.
3. *Absorption train :* The H_2O vapour and CO_2 released on the combustion of the sample are collected by assorbing them in three packed columns. *Tube 1:* magnesium perchlorate - absorbs all H_2O. From mass gain the H_2 content is calculated. The C-content is calculated from mass gain of other tubes.

   Tube2 : filled with soda asbestos pellets – CO_2 absorption

   Tube 3: also with soda asbestos + short section of magnesium perchlorate. This is to absorb H_2O produced during absorption of CO_2. Glasswool is used to prevent dust from the adsorbent materials being carried between tubes. The tubes are joined to combustion tube via a heat resistant stopper, and to each other by short lengths of rubber hose.
4. *Air purge:* After absorption the tubes are purged with air supplied through cylinder or drawn through a vaccum pump . Air must be passed through magnesium perchlorate and soda asbestos. Air has to be vented via the flow mater.

Procedure

1. To prepare the apparatus for use connect up the apparatus without combustion boat and raise the temperature of furnace to 1350^{0}C

2. Pass O_2 through the apparatus at 300 ml / min for 10 min
3. Disconnect the absorption train and purge with air for 2-3 min. disconnect the train and seal from atmosphere. Allow to cool in a dust-free atmosphere for 15-20 min.
4. When cool, and immediately before use, remove rubber tubing and wipe each absorption tube to remove dust and grease. Weigh each unit accurately.
5. Take an oven-dried combustion boat, allow to cool and weigh.
6. Add about 0.5 g sample to boat and weigh accurately.
7. Cover the sample with 0.5-1.0g Al_2O_3. Reweigh and place in desiccator.
8. Reconnect the weighed absorption train.
9. Insert the combustion boat containing the sample to a point 240 mm from centre of furnace. Connect the O_2 supply and pass O_2 at 300 ml / min.
10. During the test follow the schedule for boat position given below using silica rod to push the boat along and withdrawing the rod after each movement.

Time elapsed (Min)	Movement forward (mm)	Distance from centre of furnace (mm)
0	Insert boat	240
1	40	200
2	20	180
3	20	160
4	20	140
5	20	120
6	20	100
7	20	80
8	20	60
9	20	40
10	40	40
14	End of test	0

11. After the end of the 14th min shut off O_2.
12. Disconnect absorption train, purge with air for 2-3 min, seal and leave to cool as before but for one hr. When cool, wipe and weigh as before.

Remove the boat and discard. It may be useful to keep a nickel- chrome wire hook for this purpose.

If the moisture content of $Al_2 O_3$ has not been determined within a month before test, determine the same.

Calculation of Results

M_1= mass of sample

M_2 = increase of mass in tube 1.- less due to Al_2O_3

M_3= total increase in mass of tubes 2 &3

C_{ad}= % C content of sample wet basis

H_{ad}= % hydrogen content of sample wet basis

M_{ad}= % moisture content of sample wet basis

$$\text{Then } C_{ad} = \frac{27.29 \times M_3}{M_1} \text{ and}$$

$$H_{ad} = \frac{11.19\ M_2 - 0.1119\ M_{ad}}{M_1}$$

Mineral Analysis
Ash Content

Principle

The sample is burnt (wet or dry) to remove completely the organic compounds. Only the mineral components remain which are estimated gravimtrically.

Apparatus

1. Burner,
2. Muffle furnace,
3. Kjeldahl flask, 100 ml,
4. Glass beads,
5. Polythene bottles.

Chemicals

1. Concentrated HCl
2. Concentrated H_2SO_4
3. Concentrated HNO_3
4. Concentrated $HClO_4$

Procedure

Two g dried sample is taken in a weighed silica crucible. The crucible is placed on a burner and heated till the material is completely charred. Then this is placed in the muffle furnace and heated to 550^0-700^0C for 4 hrs. The crucible is transferred to a dessicator to cool it to room temperature and the weight of the crucible is later recorded.

Calculations

$$\text{Per cent ash} = \frac{\text{weight of crucible with ash- weight of empty crucible} \times 100}{\text{weight the sample} \times 100\text{-M}}$$

M= moisture percent in the sample.

a. Ash solution

The ash is moistened with a small amount of glass distilled water (0.5 1.0ml) and 5 ml of distilled hydrochloric acid is added to it. The mixture is evaporated to dryness on a boiling water bath. Another 5ml of hydrochloric acid is added again and the solution evaporated to dryness as before. Four ml of hydrochloric acid and a few ml of water are then added and the solution warmed over a boiling water bath and filtered into a 100 ml volumetric flask using whatman No 40 filter paper. After cooling, the volume is made upto 100ml and suitable aliquots are used for the estimation of minerals like phosphorus, iron and calcium.

b. Wet digestion

Organic matter can be destroyed and oxidised by boiling with sulphuric, perchloric and nitric acids.

1. Weigh about 3g of dry ground organic material in a 100ml Kjeldahl flask.
2. Add 25 ml 3:2:1 (conc. nitric acid-conc. perchloric acid-conc. sulphuric acid) tri-acid mixture and shake well ensuring that no dry lumps (solids) remain. Add a clean glass bead (acid washed).

3. Leave aside for 3-4 hr or preferably overnight in a fume cupboard. (watch for foaming during the first hour especially with samples high in fat. If foaming occurs and threatens to overflow, immerse the bulb of th flask in cool water till froth subsides).
4. Heat for about 30 min cautiously until the initial vigorous reaction has subsided (dense yellow fumes will be evolved).
5. Heat more storngly for 4 hr until most of the nitrous fumes are removed.
6. Continue heating until white fumes of perchloric acid are evolved. If charring occurs or flask contents tend to become dry, remove from heat, cool and add 5ml of nitric acid and continue heating.
7. Carry a blank with each set of 10 samples.
8. When cool, transfer quantitatively with 3-4 washings of deionised water the contents of the digestion flask to 15 ml graduated test tube. Make upto the 10 ml mark with deionised water and mix thoroughly. Allow the heated solution to cool and make up to the 10 ml mark again if necessary.
9. The tubes are centrifuged for 30 min.
10. Transfer the ash solution to acid-washed polythene bottles and store in a cool place prior to elemental analysis by atomic absorption spectrophotometry.

Note

1. Use only glass distilled or deionised water.
2. Nitric acid diluted 1:100 in deionised water is used as glassware washing liquid.
3. All operations should be carried out in a fume cupboard with proper ventilation.

Atomic Absorption Spectrophotometry

Principle

Representative sample in a suitable liquid form is sprayed into the flame of an atomic absorption spectrophotometer and the absorption or emission of the mineral to be analysed is measured at a specific wave length.

Reagents

1. Standard stock solution (1000 μg/ml): Dissolve 1g of the mineral salt in a minimum amount of redistilled HCl (10ml). Dilute to 1 l with deionised water.
2. Working standard solutions: Dilute aliquots of the stock solutions with deionised water to make at least 4 standard solutions within the range of determinations.

Equipment

1. Atomic absorption spectrphotometer.

Procedure

1. Depending upon the mineral to be determined standards and blank solutions may be aspirated into the flame directly or after suitable dilutions to attain working range of the instrument.
2. The operating conditions recommended by the manufacturer should be followed.
3. Read at least 3-4 ranges of standard solutions before and after sample readings. Flush burner with deionised water between samples and check for zero setting.
4. Prepare calibration curve from the readings of standards.
5. Determine the concentration of samples from the standard graph.

Calculation

$$\text{ppm mineral} = \frac{(\mu g\ \text{mineral/ml}) \times \text{dilution factor}}{\text{ml aliquots} \times \text{g sample}}$$

Operation Parameters

Mineral	**Wavelength nm**	**Flame**	**Optimum working range, μg/ml**
Ca (dilute with 0.1% lanthanum			
Chloride solution)	422.7	Rich air - C_2H_2 (Oxidising)	0.5-2.5
Cr	357.9	Rich air -C_2H_2 (reducing)	2.5-10.0
Cu	324.7	Air - C_2H_2 (oxidising)	2.0-8.0
Fe	248.3	Rich air - C_2H_2 (oxidising)	2.0-8.0
Mg	285.2	Rich air - C_2H_2 oxidising	0.1-0.6
Mn	279.5	Air - C_2H_2 (oxidising)	1.0-4.0
Zn	213.9	Air - C_2H_2 (oxidising)	0.4-1.6

Calcium
(Titrimetric method)

Principle

Calcium is precipitated as oxalate and titrated with standard potassium permaganate.

Reagents

1. 4% ammonium oxalate solution
2. Dilute ammonia solution (2ml of liquor ammonia + 98 ml water)
3. 1 N H_2SO_4
4. 0.01 N Potassium permanganate solution
5. 0.01 N Oxalic acid: Sodium oxalate is dried in an oven at 100-105°C for 12hr. Exactly 0.67g is dissolved in redistilled water, 5ml concentrated H_2SO_4 is added and the solution made upto 1 litre after it has cooled down.
6. Standardisation of potassium permanganate solution: 25ml of 0.01 N oxalic acid is transferred to an Erlenmeyer flask. One ml of concentrated H_2SO_4 is added, warmed to about 70°C and titrated against $KMnO_4$ soultion till the faint pink colour persists.

Procedure

Two ml of the ash solution is taken into a 15ml centrifuge tube. Add 2ml of distilled water and 1ml of 4% ammonium oxalate solution. Mix thoroughly and leave overnight. Again the contents are mixed and centrifuged for 5min at 1500 rpm. The supernatant liquid is poured off and the centrifuge tube drained by inverting the tube for 5 min on a rack (care should be taken not to disturb the precipitate). The mouth of the centrifuge tube is wiped with a piece of filter paper. The precipitate is stirred and the sides of the tubes are washed with 3ml of dilute ammonia. It is centrifugeed again and drained as before. The precipitate is washed once more with dilute ammonia to ensure the complete removal of ammonium oxalate. The precipitate is dissolved in 2ml of 1N H_2SO_4. The tube is heated by placing it in a boiling water bath for 1 min and titrated against 0.01 N $KMnO_4$ solution to a definite pink colour persisting for at least 1 min.

Calculation

1ml of 0.01 N $KMnO_4$ is equivalent to 0.2004 mg of calcium.

$$\text{mg of calcium/100 ml} = (X\text{-}b) \times \frac{100}{2} \quad \text{where,}$$

X= number of ml of 0.01 N $KMnO_4$ required to titrate the sample,

b= number of ml of 0.01 N $KMnO_4$ required to titrate 2ml of 1N H_2SO_4 (blank).

If the normality of $KMnO_4$ is N, the value obtained in the above formula should be multiplied by the factor, N/ 0.01

Note: Potassium permanganate solution needs to be frequently standadised.

Calcium
(Colorimetric Method)

Principle

Calcium forms a coloured complex with the O-cresolphthalein dye, which is made more specific in the presence of 8-quinolinol.

Reagents

1. Standard $CaCl_2$: 10µg Ca++/ ml prepared by dissolving required amount of anhydrous $CaCO_3$ in dilute HCl and making upto a definite volume with water.
2. Ammonia/ammonium chloride buffer, pH 10.5: Dissolve 0.24 g NH_4Cl in 100 ml of 5% NH_4OH solution (V/V).
3. O-Cresolphthalein solution, 0.1%: 100mg of O- cresolphthalein dissolved in 28ml of ammonia/ammonium chloride buffer and diluted to 100ml with water .
4. 8-Quinolinol: 1% solution in absolute ethanol.

Equipment

1. Spectrophotometer with quartz cuvettes

Procedure

Different aliquots of the $CaCl_2$ solution with Ca++ content varying from 2-10µg are taken in graduated stoppered tubes and volume in each case is made upto 5.5ml with water. Subsequently, 1ml of 8-quinolinol solution and 2.5ml ammonia/ammonium chloride buffer are added and the volume adjusted to 9.0 ml. The contents are finally treated with 1ml of O-cresolphthalein solution. After addition of each reagent mentioned above, the contents are thoroughly mixed. Water blank in the place of $CaCl_2$ solution is simultaneously carried throughout the procedure. Standards are read against blank in a spectrophotometer at 565 nm. Readings should be recorded after 5min (stable upto 3hr).

Concentration of sample in µg is read from the standard graph using the absorbency reading. Further calculations are made based on the weight of the sample and related dilutions etc.

Phosphorus
(Fiske and Subba Row Method)

Principle

Organic matter is destroyed by digestion with H_2SO_4 (or fuming nitric acid). The phosphate containing solutions are treated with molybdic acid to produce phosphomolybdic acid. This is reduced by the addition of 1,2, 4-aminonaphtholsulphonic acid reagent to obtain a blue colour, the intensity of which is proportional to the amount of phosphate present.

Reagents

1. 10 and 5N H_2SO_4
2. 30% H_2O_2 or fuming nitric acid
3. Aminonaphtholsulphonic acid reagent: 195 ml of 15% sodium bisulphite solution is taken in a stoppered cylinder and 0.5 g of 1,2,4-aminonapholsulphonic acid is added to it followed by 5ml of 20% sodium sulphite. It is stoppered and shaken until the powder is dissolved. If the solution is not complete, more of sodium sulphite, 1ml at a time is added with shaking. The solution is then transferred to a brown glass bottle and stored in the cold. This solution is stable for four weeks. Prepare afresh later.

4. Standard phosphate: 351 mg of pure dry monopotassium phosphate (KH_2PO_4) is dissolved in water and transferred quantitatively to a 1litre volumetric flask. Ten ml of 10 N H_2SO_4 is added and the solution diluted to the mark with water and mixed. This solution contains 0.4 mg of phosphorus in 5ml. It is stable indefinitely.
5. 2.5% Ammonium molybdate: 2.5g of reagent grade ammonium molybdate is dissolved in water, transferred to 100ml volumetric flask, filled to the mark and mixed.

Equipment

1. Spectrophotometer with quartz cuvettes.

Procedure

Two ml of ash extract containing 2-20 µg of phosphorus is transferred to a test tube or micro-Kjeldahl flask and 2.5 ml of 5N H_2SO_4 is added; also add a quartz chip to minimise bumping. The contents of the flask are digested on a micro burner. After evaporation is complete and the mixture turns brown or black with no further change, the tube is removed, cooled slightly and 1 drop of 30% H_2O_2 (if it gives too high a blank then 1 or 2 drops of fuming nitric acid) added to fall directly into the digestion mixture. The heating is then continued until the contents of the tube become colourless. When the mixture is clear, the tube is cooled and a few ml water is added and the solution is heated to boiling momentarily. The contents are then transferred to a 25ml volumetric flask quantitatively with washings. To the flask is added 2.5 ml of 2.5% ammonium molybdate solution, followed by 1ml of aminonaphtholsulphonic acid reagent. The solution is made upto 25ml with water and mixed well. It is allowed to stand for 5min, then the intensity of the colour is read in a colorimeter or a spectrophotometer. For photometric measurements, the most accurate procedure is to run a digested blank and series of digested standard solutions along with each series of unknowns. The photometer is set to zero OD with blank and colour of standard and unknown solutions are read at 660nm. The standard curve is plotted and the unknown read off from the curve.

Calculation

mg of total acid soluble phosphorus per 100ml solution =

$$\frac{\text{mg P in unknown} \times 100}{\text{ml of solution taken}}$$

Iron
(Wong's method)

Principle

Iron is determined colorimetrically using the fact that ferric iron gives a blood red colour with potassium thiocyanate.

Reagents

1. 30% H_2SO_4
2. 7% Potassium persulphate soultion: 7 g potassium persulphate is dissolved in glass distilled water and the solution made upto 100 ml.
3. 40% Potassium thiocyanate solution: 40g KCNS is dissolved in 90 ml glass distilled water, 4ml acetone added and the volume made upto 100 ml.
4. Standard iron solution: 702.2 mg ferrous ammonium sulphate is dissolved in 100 ml glass distilled water. Add 5ml of 1:1 HCl, make upto 1 litre and mix thoroughly (0.1 mg Fe/ml). The standard solution is prepared fresh once in 6 months.
5. Working standard: (10µg Fe/ml) is prepared by diluting the above solution 10 times.

Equipment

1. Spectrophotometer with quartz cuvettes.

Procedure

To an aliquot (6.5 ml or less) of the mineral solution enough water is added (if necessary) and make upto a volume of 6.5 ml followed by 1.0ml of 30% H_2SO_4, 1.0ml potassium persulphate solution and 1.5 ml 40% KCNS soultion. The red colour that develops is measured within 20 min at 540 nm. A standard (10-100µg) is run similarly and a standard graph is prepared. Calculations are made based on the concentration read from the graph.

Potassium
(Flame Photometric Method)

Principle

The most commonly used method for K determination is by flame photometry. It is based on the principle that atoms of some specific elements take energy from flame and get excited to the higher orbit. Such atoms release energy of a wavelength which is specific for that element and is proportional to the concentration of atoms of that element.

Apparatus

1. Flame photometer.

Reagents

1. Standard stock solution: Weigh 1.9069 g of analytical grade KCl. Dissolve in distilled water and make upto 1litre. This solution contains 1000 ppm K. Prepare 100 ppm K solution by diluting the 1000 ppm K solution 10 times (10ml in 100ml final volume). Final standard solutions of 0, 5 and 10 ppm are prepared from 100 ppm K.

Standard curve

To prepare the standard curve, the instrument is set at highest concentrationof 5 ppm using a standard filter. The manufacturers specify the linear range of K (normally 5 ppm) and suitable factor is calculated for finding out K in plant samples.

Procedure

The plant samples for K estimation can be digested by triacid method (as described previously). The digest is diluted to a suitable concentration range so that the final concentration lies between 0 to 5 ppm. The samples are then read in a flame photometer at 548 nm wave length or using suitable filter for K.

$$K(\%) = R \times \frac{5}{100} \times \frac{100}{\text{sample}} \times \frac{100}{1000000}$$

R= readiing of flame photometer,

where 5 ppm K=100 R and if further dilution is made, then appropriate modification is required in the calculation.

References

A.O.A.C., 1965. Association of official Analytical Chemists. 1965. Official Methods of Analysis 10th Ed. Washington D.C.

Bligh, E.G. and Dyer, W.J. 1959. Canadian Journal of Biochemistry and Physiology. 37: 911-917.

Buchner, 1987. Comparison of Procedures for Optimum Extraction, Purification and Analysis of Desulfo Indolyl Glucosinolates. Glucosinolates in Rapeseeds: Analytical Aspects. Martinus Nijhoff Publishers. Dordrecht. Pp 77-79.

Colborne, A. J. and Laidman, D.L. 1975. Phytochemistry. 14: 2639.

Dubois, M.K., Giles, J.K., Hamilton, P.A., Rebers and Smith, F. 1956. A Colorimetric Method for the Determination of Sugars and Related Substances. Analytical Chemistry. 28:350-356.

Folch, J., Less, M. and Stanley, G.H.S. 1957.Journal of Biological Chemistry. 226: 497.

Fong, J., Scheffer, P.L. and Kirk, P.K. 1953. The Ultramicro Determination of Glycogen in Liver. A Comparison of Anthrone and Reducing Sugar Methods. Archives of Biochemistry and Biophysics. 45: 319.

Funatsu, M., Funatsu, G., Ishiguro, M., Nanno, S. and Hara, K. 1971. Structure and Toxic Function of Ricin D. Proceedings of Japan Academy of Science. 47:715.

Garvey, J.S., Cremer, N.E. and Sussdorf, D.H. 1977. Radial Immuno-Diffusion. Methods in Immunology. W.A. Benjamin, Inc., Boston.

Gornall, A.G., Bardawill, C.S. and David, M.M. 1949. Journal of Biological Chemistry. 177:751-756.

Hauser, G. and Eichberg, J.1973. *Biochim. Biophysical Acta*. 326: 201.

Haverkate, F. and Van Deenen, L.L. M. 1965. *Biochim. Biophysical Acta*, 106: 78.

Helmy, F.M. and Hack, M.H. 1966. Lipids. 1:279.

Hemalatha, S. and Ghafoorunissa. 2004. Lignans and Tocopherols in Indian Sesame Cultivars. Journal of the American Oil Chemists' Society. 81(5): pp. 470.

Indyke, H.E. 1990. Simultaneous Liquid Chromatographic Determination of Cholesterol, Phytosterols and Tocopherols. Analyst. 91: 115.

ISO 8292: 1991, 1992. Section 2.22 Determination of Solid Fat Content (Pulsed NMR Method). Methods of Analysis of Fats and Fatty Oils – Part 2: Other Methods, BSI Standards: London. pp 1-4.

I.U.P.A., C. 1987. Standard Methods for the Analysis of Oils, Fats and Derivatives. 2.150 (ex 2.323) Solid Content Determination in Fats by NMR (Low Resolution Nuclear Magnetic Resonance), 7th Ed: C. Paquot and A. Hautfenne. Blackwell Scientific Publication: Oxford. pp 59-70.

Jambunathan, R., Madhusudhana Raju, S. and Subhadra P. Barde. 1985. Analysis of Oil Content of Groundnuts by Nuclear Magnetic Resonance Spectrometry. Journal of the Science of Food and Agriculture. 36:162-166.

Kemeny, D.M. and S. Chantler, 1988. An introdction to ELISA. ELISA and Other Solid Phase Immunoassays. John Wiley and Sons, Ltd., London.

Littlewood, A.B. 1970. Gas Chromatography-Principles, Technique and Applications. Academic Press. pp. 289.

Lottspeich, F. and Henschen, A. 1981. HPLC in Proteins and Peptide Chemistry, Ed. Hupe. Walter Degrothy.

Lough, A.K., Felinski, L. and Garton, G.A. 1962. Journal of Lipid Research. 3: 478.

Lowry. O.H., Resebrough, N.J., Farr, A.L. and Randall, R.J. 1951. Protein Measurement with Folin-Phenol Reagent. Journal of Biological Chemistry. 193:265.

Lucas, C.C. and Ridout, J.H. 1970. Progress in Chemistry or Fats. 10:1.

Mancini, G., Carbonara, A.O. and Heremans, J.F. 1965. Immunochemical Quantification of Antigens by Single Radial Immunodiffusion. Immunochemistry 2: 235-241.

McKenzie, H.A. and Wallace, H.S. 1954. The Kjeldahl's Determination of Nitrogen. Australian Journal of Chemistry. 17:55.

Metcalfe, L.D., Schmitz, A. and Pelka, J.R. 1966. Rapid Preparation of Fatty Acid Esters from Lipids for Gas Chromatographic Analysis. Analtycal Chemistry. 38:514-415.

Morrison, W.R., Tan, S.L. and Hargin, K.D. 1980. Journal of the Science of Food and Agriculture. 31: 329.

Mosse, J. and Pernollet, J.C. 1983. Storage Proteins of Legume Seeds. In. Chemistry and Biochemistry of Legumes. Ed. S.K. Arora. Edward Arnold, London. pp 111-193.

Nabney, J. and Nesbitt, B.R. 1965. A spectrophotometric method for determining the aflatoxins. Analyst. 90:155-160.

Official Methods of Analysis of the Association of Official Analytical Chemists. 1984. 14th Edition. Associationof Official Analytical Chemists, INC 1111, North Nineteenth Street, Suite 210, Arlington, VA 22209, U.S.A. p.187-188.

Paquot, 1979. Standard Methods for the Analysis of Oils, Fats and Derivatives. Pergamon Press, Oxford. U.K. pp.1-170.

Park, D.L., Nesheim, S., Truckess, M.W., Stack, M.E. and Newell, R.F. 1990. Liquid Chromatographic Method for Determination of Aflatoxins B1, B2, G1, and G2, in Corn and Peanut Products: Collaborative Study. Journal of the Association of Official Analytical Chemistry. 73(3): 425-428.

Pinkerton S.D., Rolfe, R., Auld, D.L., Ghetie, V. and Kaytervacg, B.F. 1999. Selection of Castor for Divergent Concentrations of Ricin and Ricinus Communis Agglutinin. Crop Science. 39:353-357.

Pons, W.A., Cuculu, A.F., Lee L.S., Franz, A.O. and Goldblatt, L.A. 1966. Determination of Aflatoxins in Agricultural Products: Use of Aqueous Acetone for Extraction. Journal of Association of Official Analytical Chemists. 49:554-562.

Price, M.L., Van Scoyoc, S. and Butler, L.G. 1978. A critical Evaluation of the Vanillin Reaction as an Assay for Tannins in Sorghum Grain. Journal of Agricultural and Food Chemistry. 26.

Somogyi, M.I. 1940. A New Reagent for the Determination of Sugars. Journal of Biological Chemistry. 160:61-8.

Srinivas, V.S. Chakravartula and Nagaraj Guttarla, 2007. Identification and characterization of phenolic compounds in castor seed. Natural Products Research. 21(12):1073-1077.

Stansfield, W.D. 1981. Serology and Immunology. Macmillan Publushing Co., NewYork.

Stedmann, R.L., and Swain, A.P. 1962. Tabacco Hexane Extractives. Tobacco Science. 6: 1.

Stedmann, R.L. 1957. Polyphenolic Constituents of Tobacco. Tobacco Science. 1: 155.

Southgate, D.A.T.1976. On Determination of Food Carbohydrates. Applied Science Publishers Ltd. pp 52-55.

Thies, W. 1985. *Glucosinolatgehalt in Handelspartien von 00-Raps – Schnelle Bestimmung durch Taschen-Reflektometer. Raps* 3: 112-114.

Trucksess, M.W., Young, K., Donahue, K.F., Morries D.K. and Lewis, E. 1990. Comparison of Two Immuno Chemical Methods with Thin-Layer Chromatographic Methods for Determination of Aflatoxins. Journal of the Associaltion of Official Analytical Chemistry 73(3): 425-428.

Wheeler, E.L. and Ferrae, R.E. 1971. A Method for Phytic Acid Determination in Wheat Fractions. Cereal Chemistry. 48: 312- 315

Whistler R.L. and Wolfrom. M.L. 1962. Methods in Carbohydrate Chemistry. 1. Academic Press, London.

Wong, M.L., Timms R.E. and Goh, E.M. 1988. Colorimetric Determination of Total Tocopherols in Palm Oil, Olein and Stearin. Journal of the American Oil Chemists' Society. 65:(2).

Yemm, E.W. and Cooking, E.C. 1954. The Determination of Amino Acids with Ninhydrin. Analyst. 80: 209-13

Young, C.T., Mazingo R.W., Sanders, T.H. and Taber, R.A. 1984. Quality Methods. American Peanut Research and Education Society, 376 Ag. Hall, Oklahoma State University, Stillwater, OK 74078, U.S.A.

Appendix

Conversion Factors

To obtain	Multiply	By
Gram	Pounds	453.5924
Grams (cm / sec)	Centipoises	0.01
Grams/cu cm	Lb/cu ft	0.016018
Grams /cu cm	Lb/cu in	27.680
Grams/cu cm	Lb/gal	0.119826
Inches	Centimetres	0.3937
Inches of Hg @ 32^{0}F	Atmospheres	29.9213
Inches of Hg @ 32^{0}F	Lb/sq in	2.0360
Inches/deg F	Cm/deg C	0.21872
Kg	Pounds	0.45359
Kg/cumeter	Lb/cu ft	16.018
Kg/(hr) (meter)	Centiposises	3.60
Kg/meter	Lb/ft	1.488
Kg/sq cm	Lb/sq in	0.073
Kg/sq meter	Lb/sq ft	4.8824
Liters	Cu ft	28.316
Liters	Gal	4.546

contd...

contd...

Liters/kg	Cu ft/lb	62.42621
Liters/min	Cu ft/sec	1698.963
Liters/sec	Cu ft/min	0.47193
Liters/sec	Gal/min/cu cm	0.063088
Meters	Feet	0.3048
Meters/sec	Ft/(sec)	0.3048
Pounds (avoir)	Kg	2.2046
Ponds/cu ft	Grams/cu cm	62.428
Pounds/cu ft	Pounds/gal	7.48
Pounds/ cu in.	Gram/cu cm	0.036127
Pounds/(hr) (ft)	Centioses	2.42
Pounds/inches	Grams/cm	0.0056
Pounds/(sec) (ft)	Centipoises	0.000672
Pounds/Gal	Pounds/cu in	231
Sq centimetres	Sq. ft	929.0
Sq centimeters	Sq.in	6.4516
Sq ft.	Sq. meters	10.764
Sq in.	Sq. centimeters	0.155
Sq meters	St ft	0.0929
Centimeters	Inches	2.540
Cm of Hg @ 0 deg C	Atmospheres	76.0
Cm/(sec) (sec)	Gravity	980.665
Centipoises	Centistokes	Density
Centistokes	Centipoises	1/density
Cu cm	Cu ft	28,317
Cu cm	Cu in.	16.387
Cu cm	Gal (USA, liq.)	3785.43
Cu cm	Liters	000.03
Cu cm	Quails (USA, liq.)	946.358
Cu cm/sec	Cu ft/min	472.0
Cu ft	Cu meters	35.314
Cu ft	Gal (USA, liq.)	0.13368
Cu ft	Liters	0.03532
Cu ft/ min	Cu meters/sec	2118.9
Cu ft/min	Gal (USA) liq./sec	8.0192
Cu ft/sce	Liters/min	0.0022280

contd...

contd...

Cu ft/sec	Gal/min	0.0005886
Cu in	Cu centimeters	0.061023
Cu in	Liters	61.03
Cumeters	Liters	0.001000028
Cu meters/hr	Gal/min	0.22712
Cu meters/kg	Cu ft/ lb	0.062428
Cu meters min	Cu ft/min	0.02832
Cu meters/sec	Gal/min	0.000063088
Feet	Meters	3.281
Ft/min	Cm/sec	1.9685
Ft/sec	Meters/sec	3.2808
Ft/(sec) (sec)	Meters (sec) (sec)	3.2808

Summary of SI Units and Convenient Metric Equivalents

Dimension	SI Base or Derived unit	Convenient SI or metric unit	SI Equivalent of Metric units
Time	Second (s)	Minute (min) hour(h), day (d) year (a)	
Length (depth)	meter (m)		
Area	m^2	hectare (ha)	$10^4 m^2$
Population, stocking Density	m^{-2}	ha^{-1}	$10^{-4} m^{-2}$
Mass	Kilogram (kg)	Tonne (t)	10^3 kg
Amount of substance (number of entities)	mole (mole)		
Mass per unit area	kgm^{-2}	kg ha^{-1}, t ha^{-1} $10^{-2}kg\ m^{-2}$	$10^{-1}\ gm^{-2}$
Mass per unit area and time	kg m^{-2} s^{-1}	g m^{-2} d^{-1}	
Volume	m^3	liter (L)	$10^{-3} m^{-3}$
Density, concentration	kg m^{-3} mol m^{-3}	mol L^{-1}	$10^{-3} m^3$

contd...

contd...

Force	Newton (N)		
Pressure	pascal (pa		
Temperature	Kelvin (K), Celsius (°C)		
Energy	Joule (J)		
Energy / unit Area	J m^{-2}	MJ ha^{-1}	$10^2 Jm^{-2}$
Power	watt (W)	MJ d^{-1}	
Power per unit area	W m^{-2}	MJ m^{-2} d^{-1}	
Photon flux density	mol m^{-2} s^{-1}	mol m^{-2} d^{-1}.	

Prefixes for Multiples and Sub-Multiples of Units Relative Size

Prefix	Abrevation	Relative size of unit with prefix	Example
Nano	n	10^{-9}	ns nanaosecond
Micro	m	10^{-6}	μg microgram
Milli	m	10^{-3}	mL millilitre
Centi	c	10^{-2}	cm^3 cubic centimetere
Deci	d	10^{-1}	dm decimetre
Deca	da	10	dam decametre
Hecta	h	10^2	hm^2 square hectometre
Kilo	k	10^3	kw kilowatt
Mega	M	10^6	Mpa megapascal
Giga	G	10^9	GJ gigajoule
Tera	T	10^{12}	Tg tetragram

Pressure Conversion Chart

To obtain Units x factor	ATM	kg/cm^2	Units of Pressure psi	mm.Hg.	bars	in.H_2O
1 atm = 760mm Hg	1	1.033	14.70	760	1.013	407.1
1 kg/cm	0.9678	1	14.22	735.6	0.9807	394.1
1 psi	0.06805	0.07031	1	51.71	0.6895	27.70
1 torr =1/760 atm = 1mm Hg.	1.316×10^{-3}	1.360×10^{-3}	19.34×10^{-3}	1	1.333×10^{-3}	0.5357
1 bar=10^6 dyne/cm^2	0.9869	1.020	14.50	750.1	1	401.8
1 inch of H_2O at 15°C	2.458×10^{-3}	2.538×10^{-3}	0.0361	1.867	2.489×10^{-3}	1

* 1F° = 9/5°C + 32 * °K = 0°C + 273.16

Concentrations-Conversion Table

Proportion	Potency	%	g/kg mg/g g/mg	ppm mg/kg pg/g ng/mg	ppb pg/kg ng/g pg/mg	ng/kg pg/g
1:100	1×10^{-2}	1	10	10000		
1:1000	1×10^{-3}	0.1	1	1000		
1:10000	1×10^{-4}	0.01	0.1	100		
1:100000	1×10^{-5}	0.001	0.01	10		
1:1000000	1×10^{-6}	0.0001	0.001	1	1000	
1:10000000	1×10^{-7}	0.00001	0.0001	0.1	100	
1:100000000	1×10^{-8}	0.000001	0.00001	0.01	10	
1:1000000000	1×10^{-9}	0.0000001	0.000001	0.001	1	1000
1:10000000000	1×10^{-10}			0.1	100	
1:100000000000	1×10^{-11}			0.01	10	
1:1000000000000	1×10^{-12}			0.001	1	

US, British and Metric Equivalents

Liters	x	0.2642	=	US gallons
US gallons	x	3.785	=	Liters
Imperial gallons	x	1.201	=	US gallons
Imperiol gallons	x	4.546	=	Liters
US gallons	x	0.8327	=	Imperial gallons
Cubic meters	x	35.31	=	Cubic feet
Cubic feet	x	0.0283	=	Cubic meters
Cubic meters	x	264.2	=	US gallons
US gallons	x	0.0039	=	Cubic meters
Metres	x	3.281	=	Feet
Feet	x	0.3048	=	Meters
Metres	x	39.37	=	Inches
Inches	x	0.0254	=	Meters
Centimeters	x	0.3937	=	Inches
Inches	x	2.540	=	Centimeters
Millimeters	x	0.0394	=	Inches
Inches	x	25.4	=	Millimeters
Kilograms	x	2.2046	=	Lbs.

contd...

contd...

Lbs.	x	0.4536	=	Kilograms
Tons (long)	x	1016.05	=	Kilograms
Tons (long)	x	2240	=	Lbs.
Tones (metr.)	x	1000	=	Kilograms
Tones (metr.)	x	2204.6	=	Lbs.
Tons (short)	x	907.185	=	Kilograms
Tons (short)	x	2000	=	Lbs.
Grams	x	15.43265	=	Grains
Grams	x	0.0647989	=	Grams
Grams	x	0.0352740	=	Ounces (US)
Ounces (US)	x	28.349527	=	Grams

Energy Dimensions-Conversion Factors

Given Dimension Unit	Required dimension with conversion factor (1)					
	J	kWh	MeV	mkp	$kcal_{15}^{o}$	erg
1 J	1	$2.77778.10^{-7}$	$6.242.10^{12}$	0.1019716	$2.38920.10^{-4}$	10^{7}
1 kWh	3600000	1	$2.247.10^{19}$	367097.8	860.11	$3.6.10^{13}$
1 MeV	$1.602.10^{-13}$	$4.45.10^{-20}$	1	$1.634.10^{-14}$	$3.827.10^{-17}$	$1.602.10^{-6}$
1 mkp	9.80665	$2.72407.10^{-6}$	$6124\text{-}10^{13}$	1	$2.34301.10^{-3}$	$9.806665.10^{7}$
1 $kcal_{15}^{o}$	4185.5	$1.16264.10^{-3}$	2.613.1016	426.80	1	$4.1855.10^{10}$
1 erg	10^{-7}	$2.77778.10^{-14}$	$6.242.10^{5}$	$0.1019716.0^{-7}$	$2.38920.10^{-11}$	1

(1) Examples: $1\ J = 2.38920.10^{-4}$ kcal, $1\ MeV = 1.602.10^{-13}$ J

Dietary Requirements

Estimated energy and fat requirements/day

Occupation	Work level	Energy requirement (Kcal)	Projected fat intake (g)	
			40%	20%
Male clerk (65kg)	Light Activity	2580	115	58
Subsistence farmer (58 kg)	Moderate activity	2780	124	62
Male heavy worker 65kg	Heavy activity	3490	155	78
Retired male (60kg)	Light activity	1960	87	43
House wife (55 kg)	Light activity	1990	88	44
Female (Dev. Country)	Light activity	2235	99	50

FAO/WHO Recommended Daily Intake of Energy

	Adult man (65 kg) Per day	Per kg/day
Energy (Kcal)	3000	46
(MJ)	12.55	0.19
Protein (Egg) (g)	37	0.57
(Animal source/ mixed)	46	0.71
Cereal diet (g)	62	0.95
Iron (mg)		5
Calcium (mg)		400-500
Vitamin A (ìg)		750
Vitamin D (ìg)		2.5
Thiamine (mg)		1.2
Reboflavin (mg)		1.8
Niacin (mg)		19.8
Folic acid (ìg)		200
Vitamin B12 (ìg)		2.2
Ascorbic acid (mg)		30.0

Vitamin Requirement

Vitamin	Qnt	Children 0.5/4-6 yrs.		Males 11-14;23-50		Females (Pregnant)
A	µg	420	500	1000	1000	1000
D	µg	10	10	10	5	10
E	mg	3	6	8	10	10
K	µg	12	20-40	50-100	70-140	70-140
Thiamine	mg	0.2	0.9	1.4	1.4	1.5
Riboflavine	mg	0.4	1.0	1.6	1.6	1.6
Niacin	mg	6	11	18	18	18
Vit. B6	mg	0.3	1.3	1.8	2.2	2.6
Pantothenic	mg	2	3-4	4-7	4-7	4-7
Biotin	ìg	35	85	100-200	100-200	100-200
Folic acid	µg	30	200	400	400	800
B12	µg	0.5	2.5	3.0	3.0	4.0
C	mg	35	45	60	60	80

Mineral Element Requirement

Element	Infacts	Children	Males		
	0.5-1.0 yr	4-6	15-18	25-30	51+
Sodium mg	500	900	1800	2200	2200
Calcium mg	540	800	1200	800	800
Potassium mg	1100	1550	3250	3900	3900
Phosphorous mg	360	800	1200	800	800
Magnesium mg	70	200	400	350	350
Copper mg	0.85	1.7	2.5	2.5	2.5
Iron mg	15	10	18	10	10
Zinc mg	5	10	15	15	15
Manganese mg	0.85	1.75	3.75	3.75	3.75
Iodine μg	50	90	150	150	150
Selenium mg	0.02-0	0.03-0.12	0.05-0.2	0.05-0.2	0.05-0.2
Molybdenum mg	0.04-0.08	0.06-0.15	0.15-0.5	0.15-0.5	0.15-0.5

Role of Essential Amino Acids and Essential Fatty Acids In Human Nutrition

Component	Dietary sources	Major body functions	Deficiency
Aromatic Aminoacids Phenylalanine, Tyrosine Basic : Lysine hestidine	From protein Good sources: Legumes, dairy products, meat, fish	Precursors of structural protein, enzymes, antibodies, hormones and metabolically active compounds.	Deficient protein intake leads to development of kwashiorkor and coupled with low energy
Branched chain (A.A.) Isoleucine Leucine, valine	Adequate sources: rice, corn , wheat Poor sources: cassava, sweet potato	Specific functions: a) tyrosine-precursor of epinephrine & thyroxin	intake to *marasmus* Excess protein intake aggravates chronic disease status
Sulfur A.A. Methionine/cystine Other A.A. Tryptophan Threonine		b) Arginine-precursor of polyamines c) methionine-methyl group metabolism d) Tryptophan-precursor of serotinine	
Essential fatty acids 20:4, 18:3, 18:2	Vegetable fats, wheat germ, etc.	Involved in cell membrane structure and function Precursors of prostra - glandins (regulation of gastric function, release of hormones, smooth muscle activity).	poor growth and skin lesions

Role of Vitamins in Human Nutrition

Vitamin	Dietary source	Major body function	Deficiency
B1 (thiamine)	Pork, organ meat, whole grains, legumes	Co-enzyme-removal of CO_2	Beri beri
B2 (riboflanvin)	wide occurence	FAD-FMN energy metabolism	Reddened lips, cracks at corner of mouth lesions of eye
Niacin	Liver, lean meats, grains, legumes	Co-enzyme constituent NAD & NADP	Pellagra,
Vitamin B6 (pyridoxine)	Meats, vegetables, whole grain cereals	Co-enzyme(pyridoxal phosphate) involved in amino acid meabolism.	Irritability, convulsions muscular twitching dermatitis near eyes, kidney stones
Pantothenic acid	Wide occurrence	Co-enzyme A and in energy metabolism	Fatigue, sleep disturbances impaired co-ordination
Folic acid	Legumes, greens whole grains products	Co-enzyme, transfer of 'C' units in nucleic & amino acid metabolism	Anemia, gastrointestinal disturbance, diarrhea, red tongue
Vitamin B12	Muscle meats, eggs dairy products	Co-enzyme, transfer of single 'C' atoms in nucleic acids & A.A. met)	Pernicious anemia, neurological disorders
Biotin	Legumes, vegetables Meats	Co-enzyme-fat synthesis Amino acid met.& glycogen formation	Fatigue, depression, nausea, dermatitis, muscular pains
Choline	All foods of phospholipids, egg yolk, liver, grains, legumes	Constituents of phospholipids precursor of the neuro transmitter acetyl choline	not reported in humans
Vitamin 'C' (ascorbic acid)	Citrus fruits, tomatoes, pepper and greens and dentine) important in	Maintains inter-cellular matrix of cartilage(bone hemorrhages. collagen synthesis	Scurvy,degeneration of skin, teeth blood vessels, epithelial
Vitamin A (Retinol)	Provit A (carotene) wide distribution in greens Retinol in milk, butter cheese, margarine)	Constituent of rhodopsin maintainance of epithelial cells, muco polysaccharide synthesis	Xerophthalmia, night blindness, permanent blindness
Vitamin D	Cod-liver oil, eggs, diary products etc.	Promotes growth and mineralization of bones increases absorption of Ca	Rickets-bone deformity, osteomalacia
Vitamin E (tocopherol)	Seeds, greens (leafy). margarine	Antioxidant, prevents cell membrane damage	Possibly anemia
Vitamin K	Green leafy vegetables cereals, fruits & meats	Imp. in blood clotting (formation of prothrombin)	Severe bleeding, internal hemorrhage

Role of Minerals in Human Nutrition

Mineral	Dietary sources	Major body function	Deficiency excess
Ca	Milk, cheese greens Legumes	Bone & tooth formation Blood clotting, nerve Transmission	Stunted growth, rickets, osteoporosis, convulsions
P	Milk, cheese, meat, poultry, grains	Bone & tooth formation Acid base balance	Weakness, demineralization, loss of Ca
S	S-Amino acids (methionine cyst)	Constituent of active tissue and cartilage	Related to intake and deficiency sulfur A.A.
K	Meats, milk, many fruits	Acid base body water balance and nerve function	Muscular weakness, paralysis
Cl	Common salt	Formation of gastric juice and acid- base balance	Muscle cramps mental apathy, reduced appetite
Na	- do -	Acid base and body water balance, nerve function	- do -
Mg	Whole grains, greens	Activates enzymes involved in protein synthesis	Growth failure behavioral disturbances, weakness, spasms
Fe	Eggs, beans, meats, legumes, whole grains, greens	Constituent of Haemoglobin energy metabolism	Anemia & reduced resistance to infection
F	Drinking wate,r tea, seafood	Maintains bone structure	Higher: frequent tooth decay
Zn	Widely disturbed	Constituent of energy involved in digestion	Growth failure, small sex glands, skin lesions
Cu	Meats, water	Eenzymes, Fe metabolism	Anemia
Se	Seed foods, grains	Parallel to vitamin E	- do - (rare)
Mn	Wide	Enzymes- fat synthesis	– –
I	Marine food & vegetables	Thyroid hormones	Goiter (enlarged thyroid)
Co	Organ & muscle meats, milk	Vit B12	– -
Water	Solid food, liquids, water	Transport of nutrients, Temp. & metabolic regulation	Thirst, dehydration

Common Biological Fatty Acids

Carbons	Common name	Systematic	Name Symbol	Structure
Saturated fatty acids				
12	Lauric	Dodecanoic	12:0	$CH_3(CH_2)_{10}COOH$
14	Myristic	Tetradecanoic	14:0	$CH_3(CH_2)_{12}COOH$
16	Palmitic	Hexadecanoic	16:0	$CH_3(CH_2)_{14}COOH$
18	Stearic	Octa decanoic	18:0	$CH_3(CH_2)_{16}COOH$
20	Arachidic	Eicosa noic	20:0	$CH_3(CH_2)_{18}COOH$
22	Behenic	Docosanoic	22:0	$CH_3(CH_2)_{20}COOH$
24	Lognoceric	Tetracosanoic	24:0	$CH_3(CH_2)_{22}COOH$
Unsaturated fatty acids (CIS)				
16	Palmitoleic	9-hexadecenoic	16:1	$CH_3(CH_2)_5$ CH=CH(CH_2)COOH
18	Oleic	9- octadecenoic	18:1	$CH_3(CH_2)_7$ CH=CH$(CH_2)_7$COOH
18	Linoleic	9-12, octadecenoic	18:2	$CH_3(CH_2)_4$ (CH =CH$CH_2)_2$-$(CH_2)_6$COOH
18	Linolenic (α)	9-12,15octadecenoic	18:3	CH_3CH_2(CH=CH$CH_2)_3(CH_2)^6$COOH
18	γ Linolenic	6,9,12,octadecenoic	18:3	$CH_3(CH_2)_4$(CH=CH$CH_2)_3(CH_2)_3$COOH
20	Arachidonic	5,8,11,14 eicosa tetraenoic	20:4	$CH_3(CH_2)_4$(CH=CH$CH_2)_4(CH_2)_2$COOH
24	Nervonic	15 tetracosenoic	24:1	$CH_3(CH_2)_7$CH=CH$(CH_2)_{13}$COOH

Some Unusual Fatty Acids

A. Cylic fatty acid

CH_2

CH_3-$(CH_2)_7$ C = C-$(CH_2)_7$ COOH — Plants of Malvales family

Sterculic

B. Epoxy fatty acid

O

$CH_3(CH_2)_4$-CH - CH-CH_2CH=CH$(CH_2)_7$COOH — Epoxy derivative of oleic acid

Vernolic

C. Trans fatty acid

CH_3-$(CH_2)_{11}$CH=CH CH_2-COOH — Characteristic of phospehidyl glycerol in chloroplasts

Trans-hexadecenoic

D. Cis-fatty acids — Precursors for detergent and nylone manufacture

$CH_3(CH_2)_{10}$CH=CH$(CH_2)_4$COOH

Petroselinic

$CH_3(CH_2)_7$ CH=CH$(CH_2)_{11}$COOH — Trierucin - lubricant

Erucic

D. Hydroxy fatty acid

CH_3-$(CH_2)_5$-CH-CH_2CH=CH$(CH_2)_7$ COOH — Triricinolein - lubricant

OH Ricinoleic

Amino Acids

1. L-Glycine $H_2N\text{-}CH_2\text{-}COOH$
2. L-Alanine $CH_3\text{-}CH(NH_2)\text{-}COOH$
3. L-Valine $CH_3\text{-}CH(CH_3)\text{-}CH(NH_2)\text{-}COOH$
4. L-Leucine $CH_3\text{-}CH(CH_3)\text{-}CH_2(NH_2)\text{-}CH\text{-}COOH$
5. L-Isolencine $CH_3\text{-}CH_2\text{-}CH(CH_3)\text{-}CH(NH_2)\text{-}COOH$
6. L-Serine $HO\text{-}CH_2\text{-}CH\ (NH_2)\text{-}COOH$
7. L-Threonine $CH_3\text{-}CH(OH) - CH(NH_2)\text{-}COOH$
8. L-Cysteine $HS - CH_2 - CH(NH_2) - COOH$
9. L-Cystine $HOOC\text{-}CH(NH_2)\text{-}CH_2 - S - S - CH_2 - CH(NH_2)\text{-}COOH$
10. L-Methionine $CH_3 - S - CH_2 - CH_2 - CH(NH_2) - COOH$
11. L-Glutamic acid $HOOC\text{-}CH_2\text{-}CH_2 - CH(NH_2) - COOH$
12. L-Aspartic acid $HOOC\text{-}CH_2 - CH(NH_2) - COOH$
13. L-Lysine $H_2N\text{-}CH_2\text{-}CH_2\text{-}CH_2\text{-}CH_2\text{-}CH(NH_2) - COOH$
14. L-Hydrox lysine $H_2N\text{-}CH_2\text{-}CH_2(OH)\ CH_2\text{-}CH_2\text{-}CH(NH_2)\text{-}COOH$
15. L-Arginine $H_2N\text{-}C(=NH)\text{-}NH\text{-}CH_2\text{-}CH_2\text{-}CH_2\text{-}CH(NH_2) - COOH$
16. L-Histidine

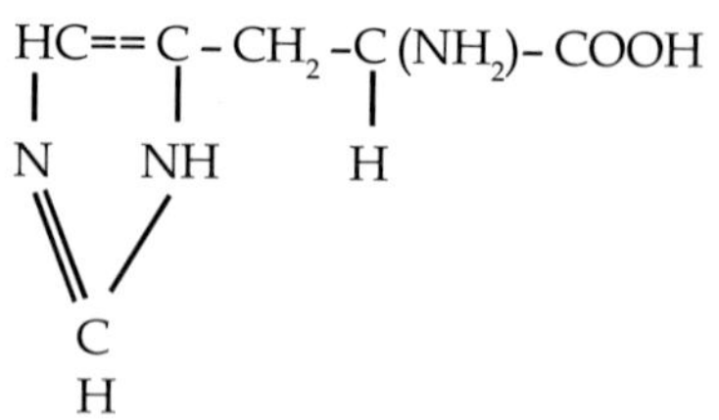

17. L-Phenylalanine $C_6H_5 - CH_2 - CH(NH_2)\text{-}COOH$
18. L-Tyrosine $HO\text{-}C_6H_4\text{-}CH_2 - CH(NH_2) - COOH$
19. L-Tryptophan

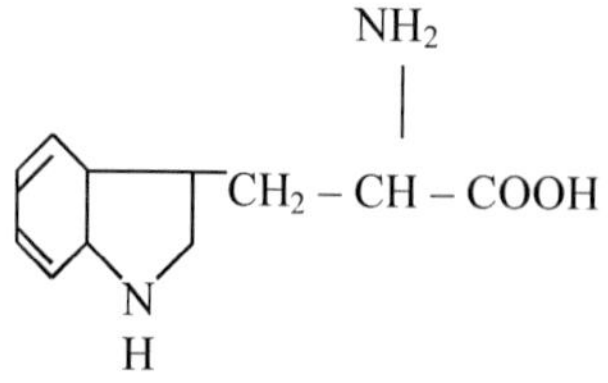

20. L-Proline

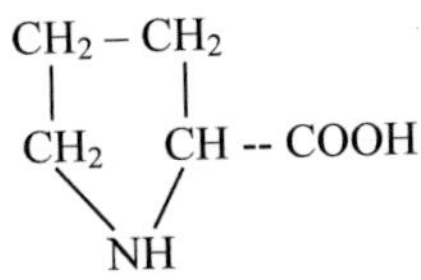

21. L- Hydroxy proline

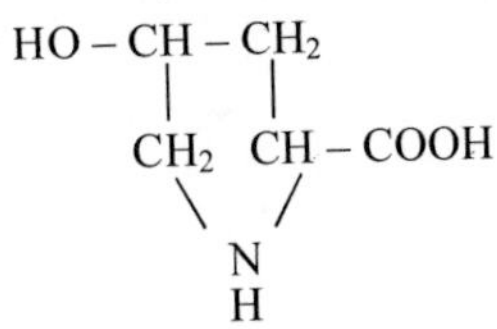

22. L- Aspargine (HOOC)-CH(NH_2)- CH_2 -CO- NH_2

23. L- Glutamine : HOOC – CH(NH_2)-CH_2 -CH_2 –CO- NH_2

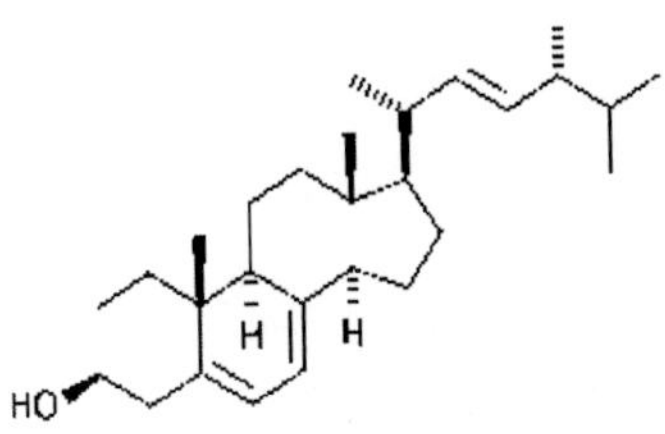

Ergosterol

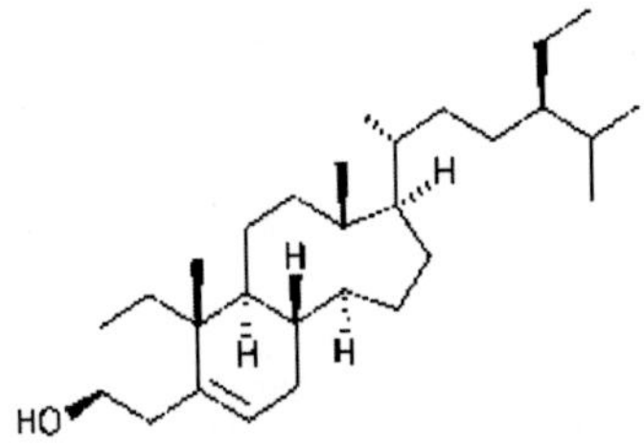

Beta Sitosterol

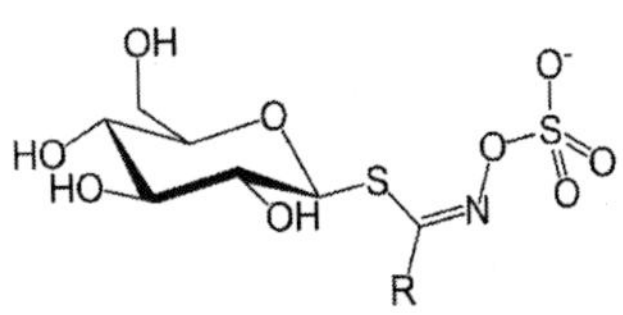

Glucosinolate

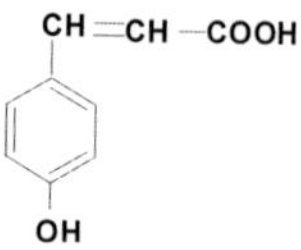

p-coumaric acid

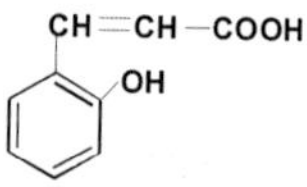

o-coumaric acid

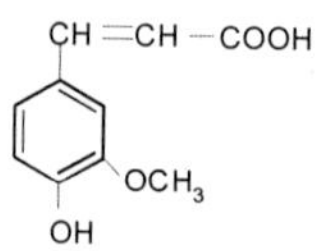

ferulic acid

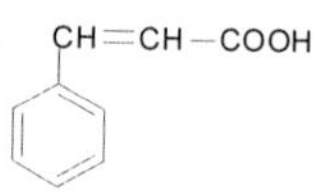

cinnamic acid

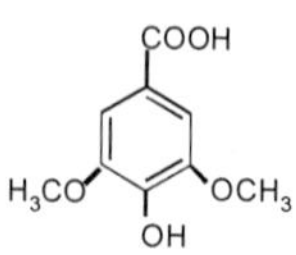

Syringic Acid

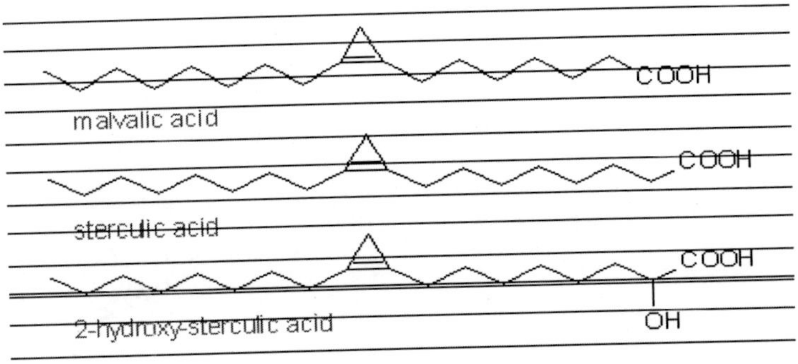

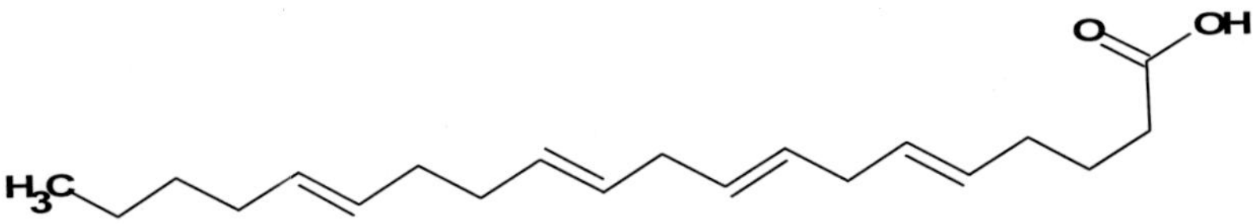

C 20:4-Arachidonic acid

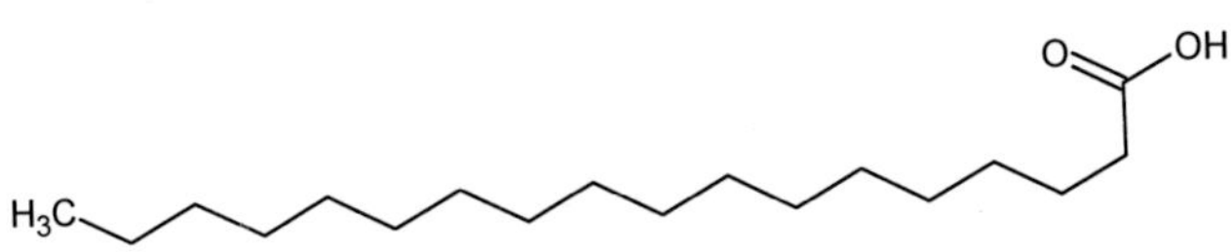

C 18:0

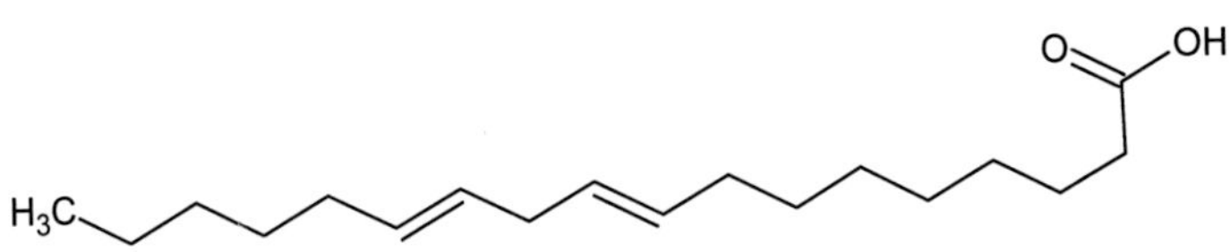

C 18:2

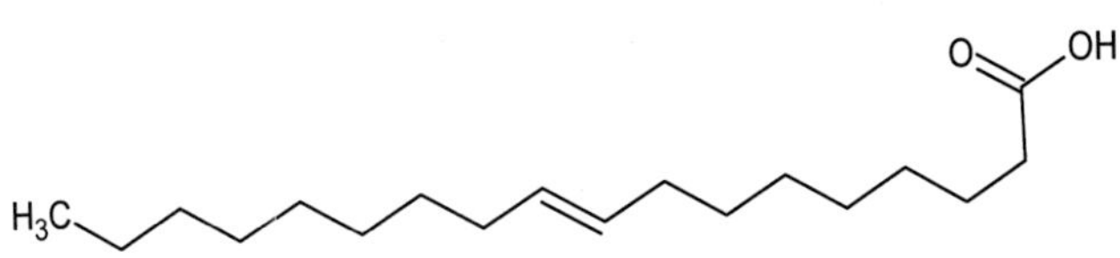

C 18:1

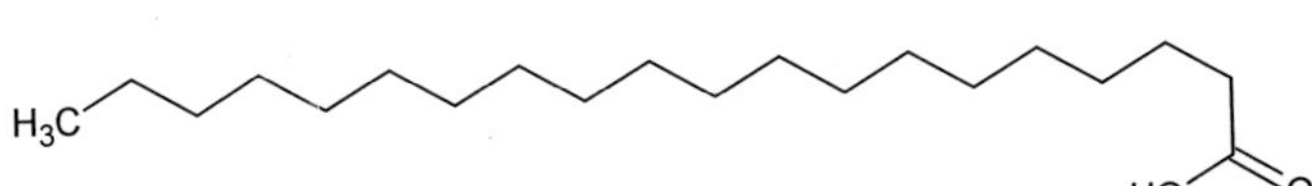

C 20:0

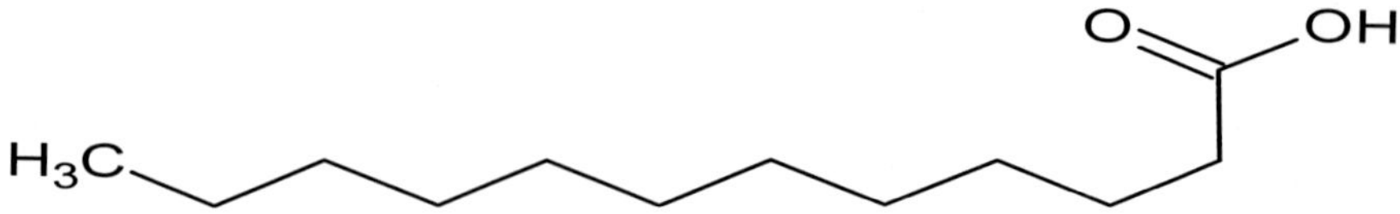

C 12:0

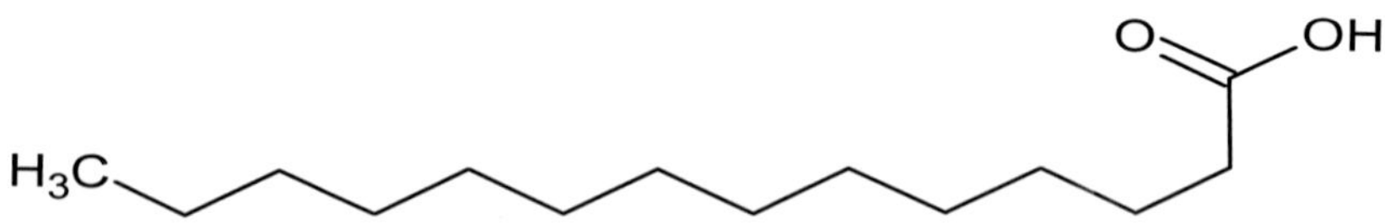

C 14:0

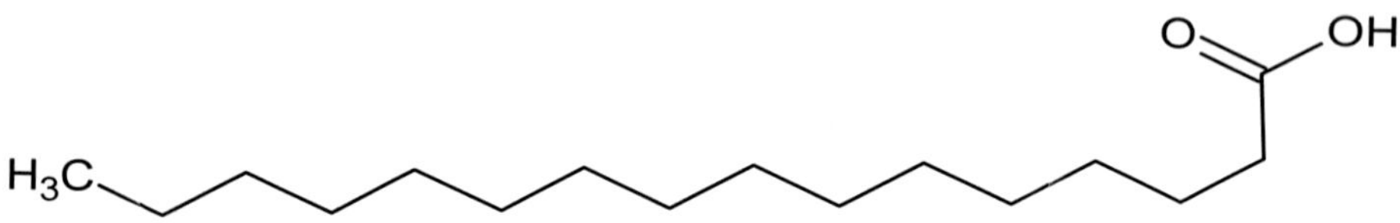

C 16:0

Index

A

Absorption 90, 115, 145, 172, 179, 209, 326, 350, 351, 357, 367, 385, 402, 420, 421, 422, 424, 425, 506, 507, 533, 548, 558, 571, 572, 574

Aflatoxins 306, 374, 417, 428, 439, 440, 441, 442, 443, 444, 445, 447, 448, 449, 450, 451, 452, 461, 466, 530, 532, 533, 534, 536, 537, 539, 574

Alkaloids 122, 253, 428

Allergens 25, 122, 124, 135, 137, 150, 162, 376, 426, 434, 436

Amino acid composition 3, 19, 37, 38, 51, 67, 80, 90, 104, 108, 123, 133, 146, 152, 162, 164, 194, 236, 241, 243, 251, 259, 260, 269, 390, 393, 432

Amino acids 4, 14, 17, 18, 19, 37, 40, 50, 55, 67, 68, 71, 75, 79, 80, 90, 91, 95, 101, 108, 109, 120, 121, 133, 162, 179, 194, 201, 220, 221, 243, 260, 269, 274, 275, 289, 327, 363, 371, 372, 374, 375, 377, 379, 392, 398, 399, 420, 423, 453, 504, 507, 508

Analysis 20, 46, 65, 67, 69, 82, 86, 87, 90, 96, 105, 106, 149, 150, 153, 155, 159, 161, 218, 257, 290, 299, 347, 454, 458, 463, 465, 466, 469, 474, 488, 490, 491, 492, 493, 494, 505, 510, 511, 517, 541, 553, 558, 565, 568, 570

Animal fats 3, 98, 356, 357

Antinutrients 4, 5, 14, 17, 25, 40, 43, 99, 231, 260, 268, 314, 388, 391, 398, 419, 420, 456, 466

Antinutritional factors 99, 373, 420, 421, 523

Arachis hypogaea 7, 12, 143, 157, 164, 347, 348, 418

Ash content 20, 30, 44, 57, 201, 258, 564

Aspergillus flavus 25, 441, 459

Atomic absorption 565, 566

Avenasterol 37, 49, 240

Azadirachta indica 214

B

Bakery products 82, 85, 207, 379, 380, 394, 395, 398

Beta carotene 177, 418, 553, 554
Bleaching 174, 187, 206, 210, 228, 233, 234, 235, 238, 273, 338, 342, 343, 344, 366
Brassica species 33, 38, 142, 162, 455, 461
Butter 21, 22, 27, 65, 75, 83, 98, 174, 175, 238, 239, 246, 249, 250, 251, 255, 256, 257, 258, 270, 273, 276, 349, 356, 361, 367, 370, 379, 380, 381, 394, 395, 397, 404, 418

C

Cake 13, 16, 17, 18, 23, 24, 27, 30, 37, 38, 39, 41, 44, 47, 49, 50, 51, 58, 59, 67, 69, 75, 78, 79, 80, 81, 84, 85, 90, 92, 101, 107, 108, 109, 110, 113, 119, 120, 121, 122, 123, 124, 125, 131, 133, 135, 136, 137, 141, 144, 149, 154, 157, 159, 160, 161, 162, 163, 168, 170, 177, 178, 184, 196, 201, 207, 214, 215, 216, 218, 219, 220, 221, 222, 230, 231, 235, 236, 237, 240, 241, 246, 247, 251, 252, 258, 259, 261, 262, 266, 267, 268, 269, 271, 277, 284, 296, 302, 311, 315, 316, 322, 323, 324, 325, 327, 328, 330, 335, 336, 423, 431, 436, 437, 438, 443, 444, 446, 457, 464, 504, 506, 507, 511
Campesterol 36, 37, 49, 106, 130, 183, 193, 229, 277
Candy 386
Canola oil 31
Carbohydrates 3, 4, 9, 16, 19, 23, 24, 27, 38, 39, 52, 61, 67, 68, 71, 73, 75, 78, 79, 81, 87, 91, 92, 93, 96, 108, 112, 120, 121, 125, 126, 140, 161, 165, 172, 178, 179, 210, 214, 219, 222, 240, 241, 242, 259, 262, 268, 271, 288, 351, 366, 371, 372, 378, 381, 382, 386, 388, 390, 430, 436, 465, 510, 512, 560
Carthamin 54, 55, 56, 59, 148, 159
Carthamus tinctorius 43, 136, 148, 149, 417
Castor 2, 4, 110, 111, 112, 113, 114, 115, 116, 117, 118, 119, 120, 121, 122, 124, 135, 136, 137, 141, 143, 144, 145, 148, 150, 151, 153, 154, 155, 156, 157, 158, 159, 160, 162, 165, 199, 225, 226, 260, 261, 294, 306, 309, 311, 312, 318, 321, 373, 375, 376, 378, 404, 405, 406, 409, 411, 413, 414, 416, 421, 426, 427, 429, 430, 431, 432, 433, 434, 436, 437, 438, 457, 458, 459, 464, 474, 491, 540
Caustic refining 341
Chemical composition 152
Chemical composition of 9, 45, 74, 112, 125, 135, 136, 138, 140, 144, 147, 149, 155, 159, 161, 164, 181, 190, 200, 219, 288, 294, 393, 464
Chikki 85, 384, 393, 394
Coconut 2, 3, 116, 128, 171, 173, 174, 202, 203, 204, 205, 206, 207, 208, 209, 210, 228, 240, 274, 291, 294, 300, 337, 338, 342, 349, 351, 359, 361, 370, 377, 389, 390, 391, 402, 409, 413, 414, 415, 416, 417, 418
Coir 176, 210
Colour 9, 11, 17, 31, 43, 45, 54, 55, 61, 62, 66, 69, 72, 74, 75, 83, 88, 98, 103, 104, 108, 112, 114, 125, 131, 132, 154, 155, 169, 171, 174, 181, 186, 189, 190, 191, 192, 195, 197, 198, 203, 204, 211, 214, 216, 225, 226, 232, 233, 234, 237, 238, 246, 248, 250, 255, 260, 261, 267, 271, 275, 276, 278, 286, 302, 303, 305, 309, 315, 330, 337, 338, 341, 344, 346, 352, 360, 361, 370, 377, 380, 387, 390, 391, 396, 397, 446, 448, 477, 478, 481, 485, 498, 499, 505, 506, 507, 511, 515, 520, 521, 524, 527, 529, 530, 554, 558, 568, 570, 571, 572
Composition 3, 4, 5, 9, 10, 11, 12, 13, 14, 16, 19, 20, 22, 29, 30, 33, 34, 35, 36, 37, 38, 39, 44, 45, 46, 48, 49, 50, 51, 52, 56, 57, 60, 61, 63, 64, 66, 67, 68, 72, 73, 74, 75, 76, 77, 78, 79, 80, 81, 87, 88, 89, 103, 104, 105, 107, 108, 109, 112, 120, 123, 124, 125, 127, 129, 133, 135, 136, 137, 138, 139, 140, 141, 142, 143, 144, 145,

146, 147, 148, 149, 150, 151, 152, 153, 154, 155, 157, 158, 159, 160, 161, 162, 163, 164, 168, 172, 173, 175, 179, 180, 181, 182, 189, 190, 191, 192, 193, 194, 196, 198, 199, 200, 201, 206, 212, 214, 215, 216, 219, 225, 226, 227, 230, 231, 232, 233, 236, 237, 239, 241, 242, 243, 246, 248, 250, 251, 252, 255, 256, 258, 259, 260, 262, 264, 267, 268, 269, 270, 272, 274, 275, 276, 277, 278, 279, 280, 281, 282, 284, 285, 287, 288, 290, 291, 292, 294, 295, 296, 297, 298, 299, 300, 304, 336, 347, 349, 357, 359, 367, 382, 390, 392, 393, 402, 412, 413, 414, 416, 418, 423, 432, 436, 464, 465, 466, 491, 495

Confectionery products 372, 379

Cooking 7, 11, 15, 30, 43, 59, 65, 75, 94, 95, 97, 98, 110, 124, 174, 179, 186, 189, 191, 194, 197, 198, 199, 202, 203, 207, 213, 238, 239, 256, 312, 313, 314, 315, 318, 322, 327, 346, 348, 351, 352, 356, 359, 360, 361, 362, 363, 366, 371, 375, 378, 384, 386, 388, 389, 404, 421, 424, 425, 426, 427, 428, 429, 436, 456, 508

Copra 2, 203, 204, 205, 209, 307, 389, 390

Copra meal 390

Corn 3, 26, 65, 117, 168, 189, 190, 191, 192, 193, 194, 195, 247, 290, 292, 297, 299, 300, 338, 343, 361, 396, 405, 441, 446, 449, 534, 535, 538

Cotton 54, 55, 101, 117, 125, 132, 134, 135, 137, 155, 186, 195, 196, 197, 199, 200, 201, 202, 220, 228, 247, 251, 282, 283, 313, 314, 349, 358, 373, 375, 378, 402, 405, 411, 417, 422, 423, 426, 427, 449, 471, 517

Crambe 3, 168, 266, 267, 268, 269, 290, 291, 293, 296, 297, 298, 404, 461

Cream 22, 97, 203, 214, 223, 246, 271, 273, 288, 389, 413

Cyclopropenoid fatty acids 406

D

Defatted germ flour 194, 195, 292

Degumming 191

Dehulling 45, 79, 82, 96, 157, 160, 200, 300, 310, 311, 312, 315, 373, 375, 378, 384, 385, 386, 388, 392, 417

Dehusking 373, 423, 425

Deodourization 191, 255, 344, 346

Desiccated coconut 203, 390

Detoxification 26, 123, 144, 157, 163, 201, 233, 247, 294, 300, 375, 420, 421, 428, 436, 437, 438, 444, 445, 449, 457, 459, 460, 464

Dewaxing 65, 343

Dhupa 292, 300

Digestibility 17, 23, 27, 35, 38, 42, 80, 91, 231, 250, 261, 325, 370, 375, 377, 378, 380, 383, 390, 420, 429, 437, 509, 510

Digestion 170, 179, 334, 350, 420, 427, 502, 503, 509, 510, 565, 570, 571

Drying 8, 15, 31, 47, 52, 58, 59, 60, 88, 99, 102, 112, 116, 117, 126, 127, 131, 132, 181, 211, 228, 239, 245, 250, 252, 254, 255, 262, 274, 275, 284, 302, 303, 305, 306, 307, 313, 324, 325, 334, 335, 336, 342, 376, 384, 386, 387, 388, 389, 390, 404, 405, 406, 409, 410, 411, 428, 445, 469, 473, 475, 476, 536, 537, 544, 561

E

Elisa 431, 462, 534, 535, 574

Erucic acid 30, 31, 32, 33, 34, 35, 36, 37, 41, 136, 142, 147, 162, 163, 267, 268, 298, 356, 374, 404, 426, 474

Estimation 83, 158, 164, 481, 499, 504, 508, 519, 523, 524, 526, 527, 528, 529, 533, 565, 573

Expeller 11, 16, 49, 78, 79, 109, 110, 113, 147, 149, 178, 195, 200, 204, 215, 219, 226, 232, 237, 238, 240, 245, 246, 261, 263, 271, 286, 312, 313, 315, 317, 318,

320, 321, 322, 323, 326, 327, 336, 337, 371, 373, 389, 390, 428, 446

Extraction 1, 4, 11, 16, 17, 21, 23, 29, 41, 47, 49, 55, 66, 67, 69, 70, 75, 78, 81, 88, 89, 95, 100, 103, 114, 135, 148, 154, 168, 169, 170, 171, 177, 181, 190, 195, 196, 200, 203, 204, 207, 211, 214, 215, 218, 219, 232, 235, 236, 241, 242, 245, 247, 250, 255, 256, 257, 258, 259, 260, 261, 267, 271, 277, 281, 284, 285, 286, 290, 298, 302, 305, 306, 307, 310, 311, 312, 313, 314, 315, 316, 317, 318, 322, 323, 324, 325, 326, 328, 329, 330, 331, 334, 335, 336, 337, 339, 347, 348, 371, 373, 374, 375, 380, 386, 387, 390, 391, 393, 397, 415, 418, 427, 428, 430, 436, 437, 452, 457, 459, 464, 466, 467, 469, 470, 471, 472, 483, 496, 508, 511, 512, 516, 521, 524, 526, 529, 530, 535, 538, 546, 547, 550, 552, 573

F

Fabricated foods 371

Fats 3, 4, 98, 108, 110, 138, 141, 144, 145, 146, 151, 153, 154, 156, 161, 162, 175, 178, 186, 194, 197, 199, 208, 257, 264, 273, 274, 291, 292, 293, 294, 295, 299, 300, 301, 337, 343, 345, 346, 347, 348, 349, 350, 351, 352, 353, 356, 357, 358, 359, 361, 362, 363, 364, 366, 367, 368, 374, 399, 400, 401, 402, 403, 404, 405, 406, 409, 412, 413, 418, 426, 427

Fatty acid composition 142, 151, 160

Fatty acid methyl esters 246, 400, 401, 407, 408, 410, 488, 494

Fatty acids 4, 11, 12, 13, 14, 15, 24, 32, 33, 34, 35, 36, 46, 47, 48, 56, 57, 62, 63, 64, 65, 71, 76, 77, 78, 84, 85, 88, 89, 92, 98, 100, 102, 103, 104, 105, 106, 113, 114, 115, 116, 127, 128, 131, 136, 145, 156, 157, 164, 169, 171, 172, 173, 174, 175, 176, 178, 180, 182, 183, 184, 186, 188, 189, 193, 194, 196, 197, 198, 199, 206, 207, 208, 209, 212, 213, 215, 221, 227, 233, 234, 236, 238, 239, 240, 243, 245, 246, 250, 252, 256, 262, 264, 265, 272, 275, 280, 282, 287, 293, 305, 330, 332, 336, 337, 338, 341, 342, 343, 345, 349, 350, 351, 356, 357, 359, 360, 361, 362, 363, 364, 365, 366, 368, 369, 392, 397, 400, 401, 402, 404, 406, 407, 408, 409, 410, 411, 412, 414, 415, 426, 427, 429, 466, 478, 479, 480, 481, 484, 488, 489, 490, 492

Fatty alcohols 175, 180, 189, 209, 287, 400, 401, 408, 410

Fatty amines 209, 400, 408, 409, 410

Fermented products 93, 382, 386

FFA 11, 47, 62, 73, 127, 170, 172, 180, 181, 185, 198, 215, 235, 238, 245, 255, 261, 264, 273, 276, 304, 305, 306, 309, 332, 337, 338, 339, 341, 342, 343, 344, 346, 362, 478, 490, 498

Flaking 312, 313, 314, 322, 386, 388

Flatus factors 420, 423

Flour 22, 24, 42, 51, 69, 85, 90, 91, 93, 95, 100, 110, 136, 162, 178, 194, 195, 203, 209, 210, 220, 274, 292, 297, 300, 372, 381, 383, 384, 385, 388, 390, 393, 394, 395, 396, 397, 398, 399, 413, 414, 416, 425, 522

Food uses 21, 43, 92, 99, 100, 149, 174, 175, 194, 299, 376, 399, 418

Free amino acids 179, 507

Free tatty acid 88, 169, 172, 184, 191, 197, 211, 255, 280, 332, 336, 360, 362, 479, 483, 497

Frying medium 352, 359, 360

Frying operation 359, 360

G

Garcinia indica 249, 292, 300

Gas chromatography 574

Germination 59, 161, 165, 180, 190, 269, 305, 310, 348, 421, 422, 424, 425, 433

Ghani 40, 47, 80, 141, 237, 312, 315, 316, 317

Ghee 75, 109, 174, 238, 239, 256, 349, 352, 356, 361, 362, 394, 398, 454

Glucosinolates 40, 41, 268, 298, 373, 374, 378, 391, 422, 451, 452, 453, 454, 455, 456, 457, 458, 460, 461, 462, 463, 464, 466, 544, 545, 546, 547, 549, 573

Glucotest 546

Glycerol 401, 410

Goitrogens 25, 374

Gossypol 197, 200, 201, 202, 314, 338, 342, 373, 375, 378, 420, 422, 423

Groundnut 7, 8, 9, 10, 11, 12, 13, 14, 15, 16, 17, 18, 19, 20, 21, 22, 23, 24, 25, 26, 27, 28, 39, 40, 51, 69, 81, 85, 86, 101, 109, 114, 126, 128, 137, 141, 143, 145, 147, 148, 150, 152, 153, 154, 155, 156, 157, 160, 161, 163, 164, 165, 181, 183, 185, 186, 201, 211, 246, 262, 282, 283, 284, 302, 303, 306, 308, 309, 311, 317, 318, 330, 347, 348, 349, 351, 358, 359, 361, 371, 372, 374, 375, 377, 378, 379, 380, 381, 382, 383, 384, 388, 390, 394, 409, 412, 414, 417, 418, 422, 427, 439, 442, 443, 444, 446, 459, 461, 503, 531

Guizotia abyssinica 101, 142, 143, 150, 155, 158, 159, 161,

H

Helianthus annuus 144, 153, 158

Hemagglutinins 99

HPLC 20, 68, 82, 83, 92, 122, 151, 163, 218, 227, 454, 525, 526, 536, 542, 547, 548, 549, 550, 551, 552, 556, 574

HPLC method 525, 536, 542, 550, 556

Hull 29, 30, 38, 42, 44, 45, 47, 51, 60, 61, 66, 68, 72, 74, 75, 78, 82, 87, 103, 104, 112, 143, 159, 162, 164, 375, 395, 398

Husk 75, 167, 178, 179, 203, 210, 215, 389, 392

Hydraulic oils 410, 411

Hydrogen 121, 132, 234, 350, 352, 354, 357, 363, 364, 365, 425, 437, 450, 472, 474, 478, 493, 534, 561, 563

Hydrogenated fats 178, 363, 364

Hydrogenation 89, 97, 98, 116, 128, 148, 164, 175, 186, 198, 199, 223, 234, 240, 273, 286, 288, 300, 352, 356, 362, 363, 364, 365, 366, 367, 368, 404, 408, 426, 427, 429

I

Immunotoxin 435, 464

Industrial uses 36, 99, 116, 129, 401

Iron 15, 21, 24, 39, 40, 53, 65, 80, 85, 121, 126, 131, 133, 156, 178, 179, 219, 227, 278, 311, 315, 316, 317, 330, 373, 377, 381, 392, 396, 423, 425, 473, 527, 565, 571, 572

J

Jatropha 3, 168, 224, 225, 226, 227, 229, 230, 290, 292, 295, 300, 408

Jojoba 3, 168, 285, 286, 287, 288, 289, 290, 291, 292, 294, 295, 296, 297, 298, 299

K

Karanj 3, 168, 231, 232, 233, 235, 236, 237, 292, 408

Kokum 3, 168, 239, 249, 250, 251, 292, 300

Kusum 3, 168, 263, 264, 265, 266, 300

L

Lecithin 37, 66, 88, 98, 99, 100, 137, 368, 370

Lignans 83, 84, 130, 150, 212, 358, 396, 549, 550, 574

Linseed 2, 4, 44, 47, 86, 109, 124, 125, 126, 127, 128, 129, 130, 131, 132, 133, 135,

138, 139, 143, 146, 151, 157, 158, 262, 263, 284, 309, 312, 313, 318, 359, 372, 373, 378, 404, 409, 410

Lipase activity 497, 498

Lipid profile 496

Lipids 13, 16, 33, 56, 70, 77, 78, 83, 88, 100, 102, 105, 130, 136, 137, 141, 151, 154, 169, 180, 181, 183, 184, 193, 272, 281, 290, 293, 295, 296, 298, 309, 347, 350, 353, 356, 401, 415, 417, 418, 448, 466, 467, 468, 469, 470, 471, 472, 478, 488, 491, 496, 497, 498, 553, 574

Lubricants 3, 4, 15, 36, 116, 117, 131, 402, 404, 405, 407, 408, 409, 410, 411

M

Macaroni products 97

Mahua 3, 168, 235, 237, 238, 239, 240, 241, 242, 243, 250, 295, 300

Maize 190, 191, 192, 271, 277, 279, 291, 293, 300, 379, 383, 384, 386, 388, 414, 441, 442, 444, 449

Mango 3, 168, 248, 249, 270, 271, 273, 274, 290, 291, 292, 295, 296, 297, 299, 404

Mango kernel 3, 168, 270, 271, 273, 274, 404

Margarine 15, 98, 186, 193, 194, 199, 207, 239, 245, 248, 363, 367, 368, 369, 370

Meal 16, 19, 26, 37, 38, 39, 40, 41, 42, 43, 45, 49, 50, 51, 52, 53, 66, 67, 68, 69, 75, 79, 80, 81, 89, 90, 91, 92, 100, 108, 113, 120, 121, 124, 133, 136, 138, 139, 141, 142, 144, 147, 148, 149, 150, 156, 157, 159, 161, 162, 164, 178, 181, 190, 194, 195, 200, 201, 202, 216, 219, 231, 236, 240, 241, 242, 246, 251, 258, 259, 260, 263, 267, 268, 269, 271, 274, 284, 285, 288, 289, 290, 291, 292, 296, 298, 299, 311, 315, 322, 323, 324, 325, 328, 330, 337, 342, 348, 371, 373, 374, 375, 376, 377, 379, 380, 389, 390, 391, 393, 395, 396, 397, 398, 399, 412, 413, 422, 423, 433, 436, 437, 438, 442, 449, 450, 451, 453, 455, 456, 457, 459, 460, 461, 463, 508, 547

Meat products 380, 387

Melia 3, 168, 251, 252

Micella refining 341, 347

Milk 195, 201, 203, 207, 219, 263, 356, 367, 368, 369, 371, 374, 379, 380, 381, 385, 388, 389, 390, 392, 394, 412, 413, 415, 418, 426, 440, 441, 443, 445, 447, 449, 451, 453

Milk products 369

Milling of rice 180

Minerals 3, 4, 16, 19, 20, 23, 27, 39, 40, 55, 57, 68, 73, 80, 82, 85, 87, 91, 92, 99, 119, 120, 121, 125, 126, 133, 155, 178, 179, 194, 205, 210, 219, 226, 231, 237, 240, 262, 266, 268, 270, 278, 281, 371, 372, 378, 381, 388, 392, 397, 424, 425, 436, 447, 465, 466, 518, 565

Miso 93, 97, 372

Moringa 3, 168, 274, 275, 276, 277, 278, 294, 296, 297, 299

Mufa 89, 129, 174, 276, 282, 350, 351, 352, 354, 356, 358, 359, 361, 363

Mycotoxins 158, 374, 428, 438, 439, 440, 442, 443, 445, 451, 457, 458, 459, 463

N

Natto 97, 372

Neem 3, 121, 168, 214, 215, 216, 217, 218, 219, 220, 221, 222, 223, 224, 231, 235, 250, 252, 296, 299, 300, 375

Neutralization 143, 294, 338, 343, 428, 459

Niger 1, 2, 4, 101, 102, 103, 106, 107, 108, 109, 110, 136, 140, 141, 142, 143, 144, 150, 152, 153, 154, 156, 158, 159, 160, 161, 165, 218, 304, 309, 312, 318, 349, 351, 361, 372, 378, 398, 399, 415, 417

NMR method 114, 472, 574

Non-food uses 399

Nutritional quality 13, 14, 33, 34, 46, 51, 61, 77, 97, 104, 108, 129, 150, 154, 160, 282, 300, 320, 349, 363, 372, 373, 379, 391, 413, 415, 418, 448, 449

O

Occurrence 49, 112, 154, 220, 433, 442, 444

Ochratoxin 439, 440, 441, 442, 443

Oil 1, 2, 3, 4, 5, 7, 9, 10, 11, 12, 13, 14, 15, 16, 20, 21, 22, 23, 24, 26, 28, 29, 30, 31, 32, 33, 34, 35, 36, 37, 39, 40, 41, 43, 44, 45, 46, 47, 48, 49, 50, 51, 56, 57, 58, 59, 60, 61, 62, 63, 64, 65, 66, 67, 68, 69, 70, 71, 72, 73, 74, 75, 76, 77, 78, 79, 82, 83, 84, 86, 87, 88, 89, 91, 92, 93, 96, 97, 98, 99, 100, 101, 102, 103, 104, 105, 106, 107, 110, 111, 112, 113, 114, 115, 116, 117, 118, 119, 120, 124, 125, 126, 127, 128, 129, 130, 131, 132, 133, 134, 135, 136, 137, 138, 139, 140, 141, 142, 143, 144, 145, 146, 147, 148, 149, 150, 151, 152, 153, 154, 155, 156, 157, 158, 159, 160, 161, 162, 163, 164, 165, 167, 168, 169, 170, 171, 172, 173, 174, 175, 176, 177, 178, 179, 180, 181, 182, 183, 184, 185, 186, 187, 188, 189, 190, 191, 192, 193, 194, 195, 196, 197, 198, 199, 200, 202, 203, 204, 205, 206, 207, 208, 209, 210, 211, 212, 213, 214, 215, 216, 217, 218, 219, 221, 222, 223, 224, 225, 226, 227, 228, 229, 230, 231, 232, 233, 234, 235, 236, 237, 238, 239, 240, 241, 243, 245, 246, 247, 248, 249, 250, 251, 252, 254, 255, 256, 257, 258, 259, 260, 261, 262, 263, 264, 265, 266, 267, 268, 270, 273, 274, 275, 276, 277, 278, 279, 280, 281, 282, 283, 284, 285, 286, 287, 288, 289, 290, 291, 292, 293, 294, 295, 296, 297, 298, 299, 300, 314, 321, 351, 352, 356, 357, 358, 359, 360, 361, 362, 363, 364, 365, 366, 367, 368, 369, 370, 371, 372, 373, 374, 375, 376, 377, 378, 379, 385, 386, 388, 389, 390, 392, 393, 395, 398, 399, 401, 402, 404, 405, 406, 409, 410, 411, , 412, 413, 414, 415, 416, 417, 418, 422, 423, 426, 427, 428, 433, 436, 437, 442, 446, 447, 452, 455, 458, 459, 460, 461, 462, 464, 465, 466, 469, 471, 472, 474, 475, 476, 477, 478, 479, 480, 481, 482, 483, 484, 485, 486, 490, 491, 497, 498, 525, 549, 550, 551, 553, 554, 556, 557, 558, 560, 574

Oil characteristics 115, 191, 346

Oil extraction 4, 29, 47, 66, 67, 69, 70, 75, 78, 168, 170, 178, 190, 196, 200, 203, 204, 207, 214, 219, 235, 247, 302, 306, 307, 311, 312, 313, 314, 315, 316, 318, 330, 331, 334, 336, 339, 348, 371, 373, 380, 390, 452, 459

Oil palm 168, 169, 176, 177, 294, 295, 331, 332

Oil refining 100, 177, 186, 194, 337, 341, 344

Oil storage 340, 346

Oilseed meal 37, 66, 68, 162, 377

Oilseed products 1, 349

Oleochemicals 69, 153, 162, 400, 404, 408, 410

Oligosaccharides 25, 95, 466

Olive 2, 26, 34, 35, 117, 169, 210, 211, 212, 213, 214, 247, 275, 296, 331, 338, 361, 378, 402

Oncom 382, 383, 413

Oxalic acid 25, 78, 79, 82, 392, 424, 500, 526, 527, 567, 568

P

Palm kernel cake 177, 178

Palm kernel meal 178

Palm kernel oil 168, 169, 171, 172, 175, 178, 293, 331, 332, 334, 361, 402

Palm oil 157, 168, 169, 170, 171, 172, 173, 174, 175, 176, 177, 178, 279, 291, 293, 297, 331, 332, 333, 334, 335

Peanut 18, 21, 22, 25, 65, 81, 117, 135, 136, 138, 139, 140, 141, 144, 145, 148, 149,

150, 154, 155, 156, 160, 161, 165, 379, 380, 381, 383, 384, 394, 397, 402, 405, 411, 412, 413, 414, 415, 416, 417, 418, 534, 535, 538

Peanut butter 21, 65, 394, 397, 418, 535, 538

Peanut milk 412

Peroxide value 88, 197, 283, 344, 486, 487, 488

Phospholipids 180, 193, 272, 339, 357

Phosphorus 14, 20, 24, 39, 50, 53, 74, 79, 82, 113, 133, 156, 160, 178, 179, 181, 201, 219, 241, 242, 271, 278, 285, 286, 288, 330, 339, 365, 372, 381, 392, 393, 498, 565, 570, 571

Phytic acid 43, 82, 83, 155, 375, 383, 413, 420, 425, 527

Plastics 116, 119, 209, 399, 402, 404, 405, 409, 410

Polyphenols 25, 99, 121, 157, 212, 213, 373, 528, 529

Post harvest technology 5, 301, 302, 348

Processing 4, 5, 47, 51, 73, 88, 91, 97, 98, 100, 133, 136, 137, 139, 142, 146, 147, 148, 160, 163, 169, 170, 175, 177, 186, 188, 190, 191, 194, 195, 200, 202, 204, 207, 210, 211, 213, 218, 245, 271, 274, 292, 296, 298, 299, 300, 301, 302, 307, 311, 312, 313, 317, 318, 324, 327, 328, 332, 333, 334, 338, 339, 340, 345, 346, 347, 348, 371, 373, 376, 378, 387, 408, 410, 412, 413, 417, 418, 419, 421, 422, 428, 439, 446, 447, 452, 458, 461, 550

Production 1, 2, 3, 14, 26, 28, 41, 69, 71, 72, 78, 86, 89, 93, 96, 100, 101, 106, 108, 111, 117, 124, 128, 135, 136, 139, 145, 146, 147, 148, 150, 152, 153, 154, 155, 161, 164, 165, 167, 168, 174, 175, 176, 177, 186, 190, 192, 193, 195, 196, 200, 201, 203, 204, 207, 209, 210, 221, 224, 231, 248, 270, 279, 281, 284, 287, 288, 289, 292, 295, 299, 301, 302, 313, 331, 332, 339, 340, 346, 347, 351, 354, 361, 362, 363, 371, 374, 377, 380, 384, 385, 389, 392, 394, 395, 399, 401, 402, 406, 409, 413, 414, 415, 416, 418, 425, 439, 442, 444, 445, 449, 450, 474

Properties 9, 19, 23, 29, 38, 51, 52, 66, 81, 85, 90, 93, 95, 99, 114, 118, 119, 121, 123, 128, 131, 132, 133, 137, 141, 146, 155, 157, 162, 171, 174, 176, 182, 183, 186, 187, 188, 189, 191, 195, 204, 208, 211, 213, 214, 216, 217, 220, 222, 223, 224, 226, 228, 229, 230, 233, 240, 241, 247, 253, 257, 258, 260, 262, 273, 275, 279, 282, 284, 285, 286, 287, 288, 292, 294, 296, 337, 347, 372, 393, 398, 399, 403, 404, 406, 407, 408, 410, 412, 413, 414, 417, 429, 432, 434, 445, 446, 448, 460, 462, 464, 492, 552

Prostaglandins 352, 353, 354

Protein 3, 4, 7, 9, 10, 13, 14, 15, 16, 17, 18, 19, 22, 23, 25, 28, 30, 32, 37, 38, 41, 42, 43, 44, 45, 46, 47, 49, 51, 55, 60, 61, 64, 65, 67, 68, 69, 70, 71, 73, 74, 75, 78, 79, 80, 81, 85, 86, 87, 89, 90, 91, 92, 93, 95, 96, 97, 99, 100, 101, 102, 103, 104, 105, 106, 107, 108, 109, 113, 121, 122, 123, 124, 125, 126, 133, 137, 138, 139, 140, 141, 142, 144, 145, 146, 149, 150, 151, 153, 155, 156, 157, 158, 159, 160, 161, 162, 163, 164, 165, 172, 173, 177, 178, 179, 190, 194, 195, 196, 200, 201, 202, 203, 205, 210, 214, 219, 220, 222, 226, 229, 231, 232, 236, 237, 240, 243, 245, 246, 251, 252, 255, 258, 259, 260, 261, 262, 266, 267, 268, 269, 270, 271, 274, 275, 277, 278, 280, 281, 285, 288, 295, 296, 297, 299, 300, 303, 305, 311, 312, 313, 314, 315, 322, 323, 325, 327, 328, 336, 342, 346, 348, 349, 350, 366, 370, 371, 372, 373, 374, 375, 376, 377, 378, 379, 380, 381, 382, 383, 384, 385, 386, 387, 388, 389, 390, 391, 392, 393, 395, 396, 397, 398, 399, 412, 413, 414, 415, 416, 417, 418, 420, 422, 423, 426, 428, 429, 430, 431, 432, 433, 434, 435, 436, 437, 464, 465, 466, 502, 503, 504, 505, 506,

507, 509, 510, 539, 540, 542, 546, 550, 561, 574

Protein concentrate 38, 75, 92, 95, 391, 392, 393, 396, 398, 417

Protein isolate 17, 18, 22, 23, 51, 69, 79, 81, 92, 95, 100, 121, 137, 138, 155, 157, 159, 160, 163, 195, 299, 380, 381, 387, 391, 393, 396, 397, 416, 418, 428, 450

Protein quality 17, 67, 69, 91, 108, 133, 142, 145, 155, 190, 195, 375, 380, 383, 412, 414, 416, 448, 456

PUFA 89, 91, 97, 129, 192, 215, 239, 276, 282, 283, 350, 351, 352, 354, 355, 358, 359, 361, 363, 397

Pulsed NMR 473, 476, 574

Purification 468

Q

Quantification 468, 512, 525, 540, 541, 542, 556, 557, 574

R

Rancidity 84, 89, 128, 206, 207, 240, 286, 352, 360, 362, 363, 379

Rapeseed 1, 2, 4, 28, 29, 30, 31, 32, 33, 34, 35, 36, 37, 38, 39, 40, 41, 42, 43, 117, 125, 129, 133, 135, 136, 137, 138, 139, 141, 142, 146, 147, 149, 161, 162, 268, 269, 311, 312, 322, 338, 345, 348, 503

Reducing sugars 9, 68, 205, 228, 248, 288, 513, 514, 515, 516, 517, 518

Refining of oil 337

Refractive index 11, 31, 47, 62, 73, 76, 88, 103, 104, 114, 115, 127, 171, 186, 191, 192, 205, 211, 215, 226, 232, 238, 245, 249, 252, 256, 262, 267, 272, 276, 280, 286, 480, 481

Removal 13, 21, 52, 69, 70, 124, 149, 157, 196, 206, 210, 222, 234, 241, 242, 254, 265, 298, 311, 312, 323, 324, 331, 337, 338, 339, 340, 344, 347, 365, 381, 390, 397, 398, 390, 430, 446, 447, 448, 456

Resveratrol 24, 358

Rice bran 178, 179, 180, 181, 182, 183, 184, 185, 186, 187, 188, 189, 239, 290, 294, 298, 300, 323, 343, 349, 351, 358, 359, 361, 402

Rice bran oil 180, 181, 182, 183, 184, 185, 186, 187, 290, 294, 300, 358

Ricin 120, 121, 122, 123, 124, 162, 226, 260, 429, 430, 431, 432, 433, 434, 435, 436, 437, 438, 457, 458, 459, 460, 461, 462, 463, 464, 466, 539, 540, 541, 542, 574

Ricinine 120, 121, 122, 124, 378, 434, 436, 459, 464

Ricinus communis 110, 122, 151, 158, 159, 160, 430, 431, 458, 460, 462, 463, 464, 541

Rid assay 431, 541

Roasted coconut 205

Roasting 21, 25, 70, 71, 79, 82, 85, 94, 231, 270, 374, 378, 379, 394, 395, 396, 412, 418, 428, 447

Rubber 3, 99, 111, 168, 175, 204, 207, 224, 228, 261, 262, 287, 290, 294, 296, 298, 399, 410, 411, 427, 501, 503, 514, 543, 562

Rubber additive 411

S

Safflower 1, 2, 4, 35, 40, 43, 44, 45, 46, 47, 48, 49, 50, 51, 52, 53, 54, 55, 56, 57, 58, 59, 65, 104, 106, 129, 135, 136, 137, 138, 140, 141, 142, 144, 145, 146, 147, 148, 149, 150, 151, 152, 153, 154, 155, 156, 157, 158, 159, 160, 161, 162, 163, 164, 165, 281, 282, 304, 306, 309, 311, 312, 318, 338, 349, 351, 352, 359, 361, 378, 397, 398, 402, 413, 415, 416, 417, 503

Saffron 43, 59

Sal 3, 123, 168, 239, 240, 254, 255, 256, 257, 258, 259, 260, 261, 273, 291, 294, 298, 427, 437, 459

Saponification value 11, 31, 47, 62, 88, 102, 103, 115, 126, 127, 171, 192, 203, 206,

208, 211, 212, 215, 226, 252, 256, 258, 262, 271, 276, 280, 286, 481, 482
SC-CO_2 195, 329, 330, 331, 459
Seed dryers 307
Sesame 1, 2, 3, 4, 64, 71, 72, 73, 74, 75, 76, 77, 78, 79, 80, 81, 82, 83, 84, 85, 86, 91, 110, 114, 116, 126, 137, 139, 141, 142, 143, 144, 145, 146, 147, 149, 151, 152, 153, 154, 155, 156, 157, 159, 160, 161, 163, 164, 165, 236, 302, 303, 309, 312, 313, 318, 349, 351, 358, 359, 361, 371, 372, 375, 377, 378, 392, 393, 394, 395, 396, 397, 399, 402, 404, 412, 413, 414, 415, 416, 417, 418, 424, 549, 550, 574
Sesamin 83, 84, 142, 147, 358, 414, 416, 550
Sesamol 83, 84, 142, 146, 358, 550
Sesamolin 83, 84, 142, 147, 358, 550
Sesamum 71, 72, 74, 84, 139, 140, 144, 146, 147, 151, 152, 157, 158, 159, 160, 164, 165
SFA 77, 129, 354
SHEA 239, 273, 404
Shell 7, 9, 27, 69, 311, 336, 389, 395
Shorea robusta 254, 273
Shortenings 15, 88, 98, 99, 175, 186, 200, 346, 366, 367, 368, 370
Significance 34, 65, 140, 149, 160, 412, 418, 424, 438, 439, 461
Simarouba 3, 168, 243, 244, 245, 246, 247, 248, 291, 295, 296, 298, 299, 300
Sitosterol 36, 37, 49, 106, 174, 183, 193, 221, 229, 240, 247, 257, 277
Soaps 3, 4, 110, 117, 118, 119, 124, 131, 208, 217, 223, 228, 234, 239, 240, 248, 252, 256, 265, 344, 399, 402, 403, 404, 405, 409, 410
Soluble sugars 10, 28, 46, 53, 73, 102, 103, 109, 113, 270, 510, 512
Solvent extraction 11, 16, 47, 49, 70, 75, 88, 89, 100, 103, 154, 200, 215, 236, 237, 245, 255, 256, 258, 261, 267, 271, 285, 290, 298, 305, 310, 313, 315, 317, 318, 322, 323, 324, 325, 328, 331, 337, 371, 373, 374, 375, 387, 390, 427, 428, 436, 457
Soxhlet extraction 245, 469, 471
Soybean 35, 39, 65, 67, 71, 81, 82, 85, 86, 87, 88, 89, 90, 91, 92, 93, 94, 95, 96, 97, 98, 99, 100, 101, 117, 129, 140, 145, 148, 149, 164, 168, 178, 187, 199, 201, 281, 295, 302, 303, 304, 306, 309, 310, 311, 312, 315, 320, 323, 325, 327, 328, 330, 339, 341, 346, 347, 348, 349, 351, 352, 358, 359, 361, 368, 369, 370, 372, 373, 374, 375, 377, 378, 384, 385, 386, 387, 388, 391, 393, 402, 405, 407, 410, 411, 420, 421, 424, 425, 426, 427, 428, 442
Specific gravity 31, 62, 76, 88, 102, 114, 132, 191, 192, 205, 206, 211, 214, 215, 224, 226, 230, 232, 253, 264, 267, 272, 286, 335, 336, 476, 477
Splitting 207, 234, 265, 386, 388, 400, 402
Starch 3, 9, 19, 91, 95, 108, 179, 190, 192, 193, 194, 195, 226, 232, 254, 258, 259, 260, 270, 271, 274, 288, 381, 390, 408
Sterols 36, 37, 49, 106, 125, 163, 183, 184, 186, 187, 188, 193, 206, 229, 238, 240, 256, 277, 283, 337, 345, 357, 420, 466, 482
Stigmasterol 37, 49, 106, 173, 183, 193, 229, 277
Storage 8, 15, 24, 26, 47, 48, 73, 75, 83, 104, 112, 113, 127, 128, 141, 151, 154, 163, 165, 302, 303, 305, 306, 307, 308, 309, 310, 323, 332, 335, 336, 338, 340, 346, 347, 419, 433, 439, 444, 463, 465, 468, 475, 478, 490, 574
Sugars 3, 4, 9, 10, 19, 28, 38, 39, 40, 44, 46, 53, 61, 68, 73, 81, 82, 91, 102, 103, 106, 107, 108, 109, 113, 120, 126, 137, 160, 179, 201, 204, 205, 228, 237, 241, 242, 243, 248, 266, 270, 278, 288, 309, 327, 347, 378, 379, 382, 386, 423, 424, 464, 510, 511, 512, 513, 514, 515, 516, 517, 518, 519, 520, 521, 523, 542, 543, 544, 573

Sunflower 1, 2, 4, 35, 59, 60, 61, 62, 63, 64, 65, 66, 67, 68, 69, 70, 71, 104, 117, 129, 136, 138, 139, 140, 141, 142, 143, 144, 146, 147, 150, 151, 153, 154, 156, 157, 158, 161, 163, 164, 165, 201, 281, 282, 304, 309, 311, 312, 338, 343, 351, 352, 359, 361, 368, 369, 375, 377, 378, 395, 396, 397, 402, 405, 413, 414, 415, 416, 417, 418, 442, 474, 503

T

Tannins 123, 140, 143, 253, 254, 255, 259, 260, 266, 270, 271, 274, 374, 423, 437, 459, 529

Tempeh 93, 97

Thiamine 19, 24, 40, 53, 57, 68, 74, 80, 91, 125, 126, 242, 381, 554, 555, 556

Thin layer chromatography 105, 114, 256, 464, 530

Thumba 279

Tobacco seed 3, 168, 280, 281, 282, 283, 284, 293, 299, 300

Tocopherols 15, 37, 39, 49, 58, 66, 75, 82, 83, 84, 91, 92, 104, 129, 151, 153, 158, 169, 173, 174, 177, 179, 180, 182, 183, 184, 185, 186, 188, 206, 212, 215, 240, 277, 317, 330, 337, 342, 345, 346, 357, 358, 396, 418, 556, 557, 558, 574

Tocotrienols 169, 174, 177, 178, 182, 183, 184, 185, 186, 188, 342, 357, 358, 413, 418, 557

Toxic adulterants 427

Toxic compounds 430, 449

Toxicology 145, 414, 442, 462

Traditional products 383

Trichothecines 442, 443

Trypsin inhibitors 25, 373, 386, 387, 388, 420

U

Unsaponifiable matter 75, 171, 173, 182, 183, 192, 212, 233, 249, 250, 262, 271, 286, 294, 337, 357, 482, 484

Utilization 4, 16, 40, 56, 58, 69, 84, 85, 95, 100, 110, 120, 127, 139, 141, 142, 146, 147, 149, 150, 152, 154, 157, 158, 161, 164, 165, 167, 176, 247, 250, 269, 271, 274, 278, 281, 284, 289, 291, 292, 295, 296, 297, 299, 303, 347, 348, 355, 375, 376, 378, 383, 391, 392, 397, 398, 399, 409, 412, 413, 415, 416, 417

V

Vanaspati 199, 239, 246, 256, 352, 356, 357, 361, 362, 367

Vitamins 3, 4, 15, 19, 20, 23, 39, 55, 57, 68, 71, 82, 85, 91, 92, 98, 125, 126, 179, 194, 242, 278, 338, 349, 350, 368, 369, 371, 378, 380, 381, 388, 392, 397, 409, 420, 465, 466, 482

W

Winterization 89, 98, 191, 338, 342, 343